NONLINEAR MICROWAVE CIRCUITS

IEEE Press
445 Hoes Lane, P. O. Box 1331
Piscataway, NJ 08855-1331

Technical Reviewers

Dr. Barry R. Allen, *TRW, Inc.*

Dr. Les Besser, *Besser Associates*

Prof. Antti V. Räisänen, *Helsinki University of Technology*

NONLINEAR MICROWAVE CIRCUITS

Stephen A. Maas
Nonlinear Technologies, Inc.

The Institute of Electrical and Electronics Engineers, Inc., New York

This book may be purchased at a discount from the publisher when ordered in bulk quantities. For more information contact:

IEEE Press Marketing
Attn: Special Sales
445 Hoes Lane, P. O. Box 1331
Piscataway, NJ 08855-1331
Fax: (908) 981-9334

For more information about IEEE Press products, visit the IEEE Home Page: http://www.ieee.org/

This is the IEEE reprinting of a book previously published by Artech House Inc., under the title *Nonlinear Microwave Circuits*.

Printed in the United States of America

10 9 8 7 6 5 4 3 2 1

ISBN 0-7803-3403-5

IEEE Order Number: PP5385

Library of Congress Cataloging-in-Publication Data

Maas, Stephen A.
Nonlinear microwave circuits / Stephen A. Maas
p. cm.
Includes bibliographical references and index.
ISBN 0-7803-3403-5
1. Microwave circuits. I. Title.
TK7876.M285 1996
621.381'32—dc20 96-9355
CIP

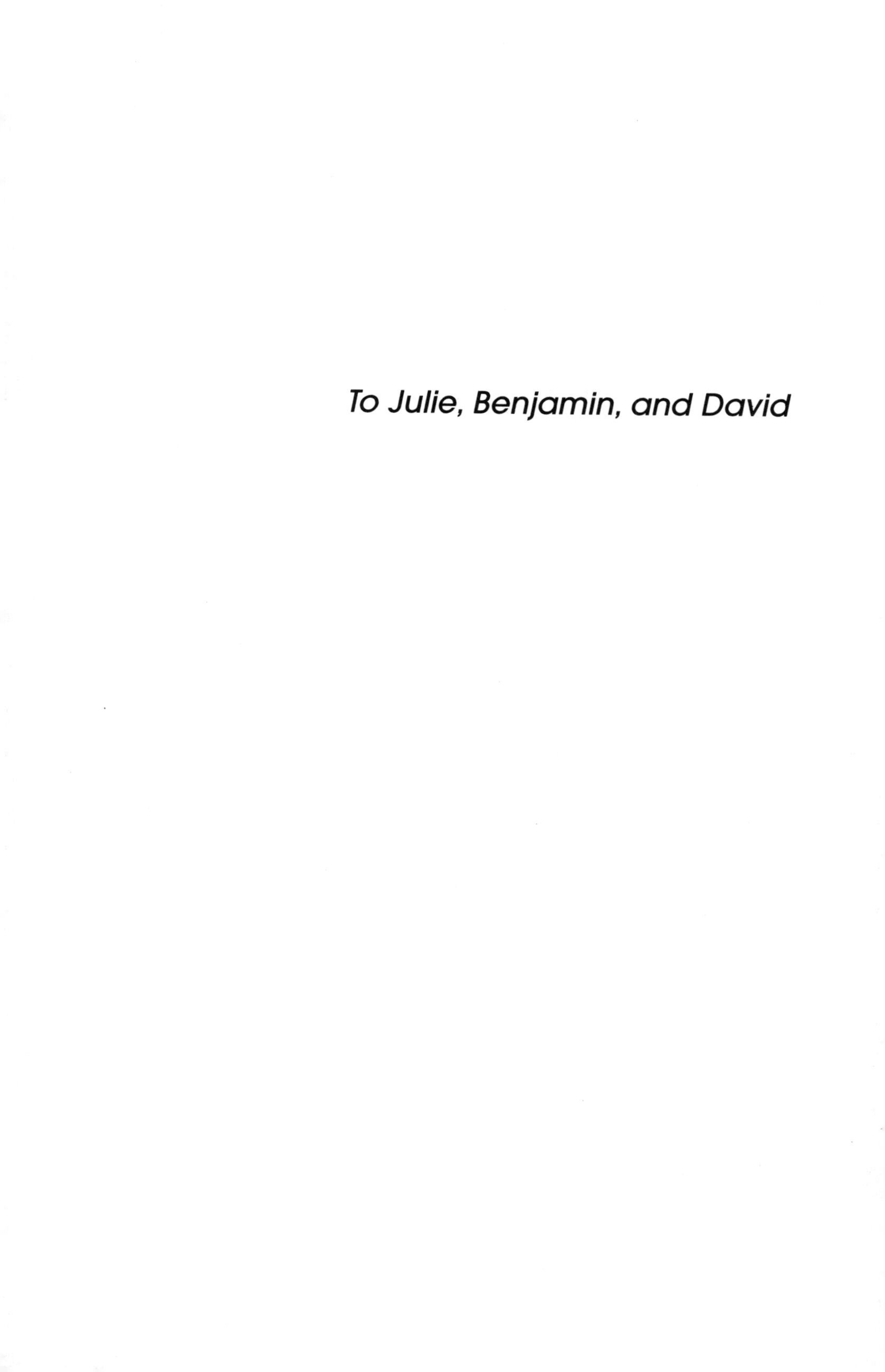

To Julie, Benjamin, and David

Contents

PREFACE

The engineer who designs nonlinear microwave circuits is faced with a quandary. He or she knows that such components as mixers, frequency multipliers, and power FET circuits, often critical to the performance of the system in which they are used, are subject to many kinds of instability and spurious effects that are likely not to appear before the system is delivered to a customer. However, such nonlinear components may seem quantitatively and intuitively difficult to understand and to design; the engineer who designs a nonlinear circuit (which the engineer may or may not have intended to be nonlinear) will get very little help from conventional sources because most textbook network theory is based on an assumption of linearity. After searching in vain for reliable design information, the engineer usually makes use of what little data he or she can find and "tweaks" the circuit empirically (sometimes successfully, sometimes not, for no clear reason) to make it work. The frustration, anger, and insecurity engendered by this situation is responsible for many sleepless nights, ruined marriages, and hateful children.

There is another dimension to this problem: the components that are most troublesome to design and to force into operation, especially diode mixers and frequency multipliers, never seem to be completely supplanted by more modern components. As better devices (e.g., GaAs MESFETs) replace diodes at lower frequencies, diode components simply are pushed to higher frequencies, where the problems involved in designing them are even more exasperating. Consequently, nonlinear circuits—even the nastiest ones—remain in use despite all advances in microwave technology; they are designed by trial and error; and they often do not work very well.

Clearly, practical engineers need a method or set of methods specifically for designing nonlinear circuits. These methods should be accurate, having only numerical or other obvious (and mild) limitations, and should not depend ultimately upon approximations or guesswork. In short, we need nonlinear circuit design to be moved from the realm of empiricism to the realm of science. Techniques for designing nonlinear circuits do, in fact, exist, but they are not well known throughout the microwave industry. As a result, most engineers find their first nonlinear circuit design project to be one of life's major challenges, a rite of passage into microwave adulthood, when it should be just another project.

The primary purpose of this book is to demystify the two techniques that are most useful for analyzing nonlinear microwave circuits, harmonic-balance analysis and the Volterra series. The former technique is best applied to strongly nonlinear circuits having large-signal single-tone (and, in some cases, multitone) excitation; the latter is useful for the converse, weakly nonlinear circuits having small-signal multitone excitation. One of my goals is to change the perception of nonlinear effects in microwave circuits from something that nobody understands or can possibly understand, yet everybody accepts in the same way that medieval peasants accepted the bubonic plague, to one where nonlinearity is simply another consideration in the design process, along with the usual linear considerations. Imparting a general knowledge of these two techniques to a large number of design engineers will do much to achieve that goal.

The book is organized into two parts. The first part, Chapters 1 through 5, is concerned with nonlinear circuit theory and solid-state device models. The second part, Chapters 6 through 12, shows how that theory can be applied both to prosaic circuits (e.g., diode mixers and FET amplifiers) and to less common circuits (e.g., FET mixers and frequency multipliers) that should be used in systems more widely than they are at present. A common theme in all of the chapters in the second part is that we can use approximate techniques to generate an initial design of a component and then fine-tune the approximate design by using the more sophisticated numerical methods described in the earlier chapters. This is the logical process in a field where true synthesis procedures do not exist (and probably never will exist) and analysis procedures are admittedly a little complicated and time-consuming.

I am deeply indebted to many people for their help and encouragement in this lengthy writing project. The first, of course, is my family, who have the wisdom to recognize that my need to pursue wild ideas and long projects is probably not a symptom of mental or emotional instability and, even if it is, are willing to tolerate it. I must also express my gratitude to the management of the Aerospace Corporation, especially Herb Wintroub and Mike Daugherty, for their encouragement, support, and tolerance, and to my coworkers for the same. Especially, I would like to thank Mike Meyer, for his painstaking care in correcting my writing gaffes and patient lessons on things I should have learned a long time ago; without his help, the book would be very different, and certainly no better. I am also indebted to Dale Kind, Al Young, Jeff Crawford, and Malcolm McColl for reading and commenting on individual chapters and participating in helpful discussions.

CHAPTER 1

INTRODUCTION, FUNDAMENTAL CONCEPTS, AND DEFINITIONS

Before we can describe the unique properties of nonlinear microwave circuits and the analytical methods necessary to understand them quantitatively, the author and reader must be certain that they both are speaking the same language. This is no small problem because many of the terms and concepts inherent in nonlinear circuit theory are completely foreign to linear circuits, and many engineers harbor preconceived ideas about these circuits, ideas that often are not altogether correct. Accordingly, in order to establish a common basis for the following discussions, we begin by mixing some definitions into a heuristic introduction to microwave nonlinearity.

1.1 LINEARITY AND NONLINEARITY

All electronic circuits are nonlinear. This is a fundamental truth of electronic engineering. The linear assumption that underlies most modern circuit theory is in practice only an approximation. Some circuits, such as small-signal amplifiers, are only very weakly nonlinear, however, and are used in systems as if they were linear. In these circuits, nonlinearities are responsible for phenomena that degrade system performance and must be minimized. Other circuits, such as mixers and frequency multipliers, are exploited for their nonlinearities; these circuits would not be possible if nonlinearities did not exist. In these, it is often desirable to maximize nonlinearities and even to minimize the effects of annoying linear phenomena. The problem of analyzing and designing such circuits is usually considerably more involved than it is for linear circuits and is the subject of much special concern.

The statement that all circuits are nonlinear is not made lightly. The nonlinearities of solid-state devices are well known, but it is not generally recognized that even passive components such as resistors, capacitors, and inductors, which are expected to be linear under virtually all conditions, are nonlinear in the extremes of their operating ranges. When large voltages or currents are applied to resistors, for example, their resistances change as a result of thermal and other effects. The same is true of capacitors, especially those designed for hybrid circuit applications, made with semiconductor materials, and the nonlinearity of iron- or ferrite-core inductors and transformers is legendary. Even RF connectors have been found to generate intermodulation, evident at high power levels, caused by the nonlinear resistance of the contact between dissimilar metals in their construction. Thus, the linear circuit concept is an idealization, and a full understanding of electronic circuits, interference, and other aspects of electromagnetic compatibility requires an understanding of nonlinearities and their effects.

Linear circuits are defined as those for which the superposition principle holds. Specifically, if excitations x_1 and x_2 are applied separately to a circuit having responses y_1 and y_2, respectively, the response to the excitation $ax_1 + bx_2$ is $ay_1 + by_2$, where a and b are arbitrary constants (a and b may, in concept, be real or complex constants or even time varying). This criterion can be applied to either circuits or systems. One important implication of this definition is that the response of a linear component or system includes only those frequencies present in the excitation waveforms. Thus, linear time-invariant circuits do not generate new frequencies. As nonlinear circuits usually generate a remarkably large number of new frequency components, this criterion provides an important dividing line between linear and nonlinear circuits.

Nonlinear circuits are often characterized as either *strongly nonlinear* or *weakly nonlinear*. Although these terms have no generally accepted formal definitions, a good working definition is that a weakly nonlinear circuit can be described with adequate accuracy by a power series expansion of its nonlinear current-voltage (I/V), charge-voltage (Q/V), or flux-current (Φ/I) characteristic. This definition implies that the characteristic is continuous, has continuous derivatives, and for most purposes does not require more than a few terms in the power series. (The excitation level, which affects the number of power-series terms required, also must not be too high). Additionally, it is usually assumed that the nonlinearities and RF drive are weak enough that the dc operating point is not perturbed. Virtually all transistors and passive components satisfy this definition if the excitation voltages are well within the components' normal operating ranges. Examples of components that do not satisfy this definition are strongly driven transistors and Schottky-barrier diodes because of their exponential *I/V* characteristics; digital logic gates, which have input-output transfer characteristics that vary abruptly with input voltage; and step-recovery diodes, which have very strongly nonlinear capacitance-voltage characteristics under forward bias. If a circuit is weakly nonlinear, it can be analyzed via relatively straightforward techniques, such as a power series or Volterra series. Strongly nonlinear circuits are all those that do not fit the definition of weak nonlinearity; they must be analyzed via harmonic-balance or time-domain approaches. These circuits are not too difficult to handle if they include only single-frequency excitation or are composed solely of lumped components. The most difficult case to analyze is a strongly nonlinear circuit that includes mixed lumped components, arbitrary impedances, and multiple excitations.

Another concept frequently used in nonlinear circuit analysis is that of *quasilinearity*. A quasilinear circuit is one that can be treated for most purposes as a linear circuit, although it may include weak nonlinearities. The nonlinearities are weak enough that their effect on the linear part of the circuit's response is negligible; however, the circuit may generate distortion products that often are great enough to be of concern. A small-signal transistor amplifier is an example of a quasilinear circuit, as is a Class-A power amplifier that is not driven into saturation.

Two final concepts we will employ from time to time are those of *two-terminal nonlinearities* and *transfer nonlinearities*. A two-terminal nonlinearity is a simple nonlinear resistor, capacitor, or inductor; its value is a function of one independent variable, the voltage or current at its terminals, called a *control voltage* or *control current*. In a transfer nonlinearity, the control voltage or current is somewhere in the circuit other than at the element's terminals. It is possible for a circuit element to have more than one control, one of which is usually the terminal voltage or current. Thus, many nonlinear elements must be treated as combinations of transfer and two-terminal nonlinearities. An example of a transfer nonlinearity is the nonlinear controlled current source in the equivalent circuit of a *field-effect transistor* (FET), where the drain current is a function of the gate voltage. Real circuits and circuit elements often include both types of nonlinearities. An example of the latter is the complete FET equivalent circuit described in Section 2.4, including nonlinear capacitors with multiple control voltages, transconductance, and drain-source resistance.

The need to distinguish between the two types of nonlinearities can be illustrated by an example. Consider a nonlinear resistor (Figure 1.1(a)), and a nonlinear but otherwise ideal transconductance amplifier (Figure 1.1(b)). Both are excited by a voltage source having some internal impedance R_s. The amplifier's output current is a function of the excitation

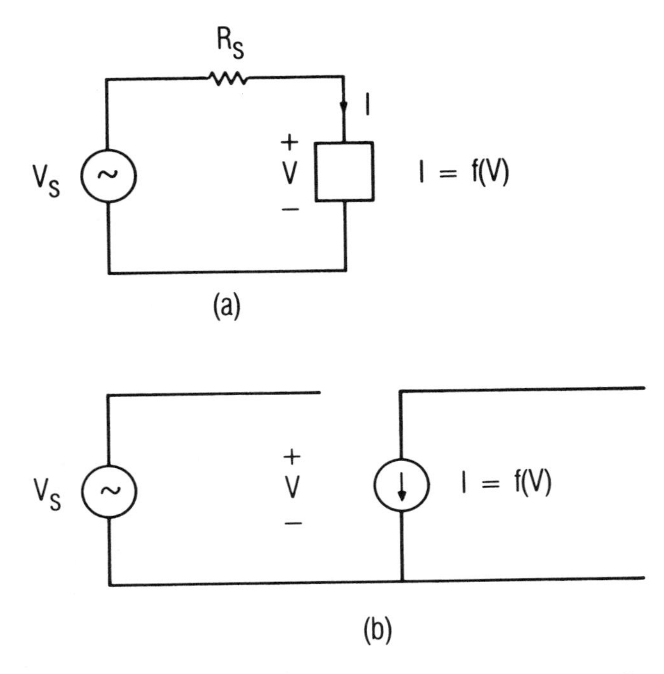

Figure 1.1 (a) Two-terminal nonlinearity; (b) transfer nonlinearity.

voltage and the nonlinear transfer function; it can be found by simply substituting the voltage waveform into the transfer function, regardless of the nature of the input or output circuits. In two-terminal nonlinearity, however, the excitation voltage generates current components in the resistor at few frequencies. These components circulate in the rest of the circuit, generating voltages at those new frequencies across R_s and therefore across the nonlinear resistor. These new voltage components generate new current components, and the process continues until current and voltage components at all possible frequencies are generated. The two-terminal nonlinearity is clearly more complicated analytically than the transfer nonlinearity.

1.2 FREQUENCY GENERATION

The traditional way of showing how new frequencies are generated in nonlinear circuits is to describe the component's I/V characteristic via a power series and to assume a multitone excitation voltage. We will repeat this analysis here, as it is a good intuitive introduction to nonlinear circuits.

However, our heuristic examination will illustrate some frequency-generating properties of nonlinear circuits that are sometimes ignored in the traditional approach and will introduce some analytical techniques that complement others we will introduce in later chapters.

Figure 1.2 shows a circuit with excitation V_s and a resulting current I. The circuit consists of a two-terminal nonlinearity, but because there is no source impedance, $V = V_s$, and the current can be found by simply substituting the source voltage waveform into the power series. Mathematically, the situation is the same as that of the transfer nonlinearity of Figure 1.1(b). The current is given by the expression:

$$I = aV + bV^2 + cV^3 \tag{1.2.1}$$

where a, b, and c are constant, real coefficients. We assume that V_s is a two-tone excitation of the form:

$$V_s = v_s(t) = V_1 \cos(\omega_1 t) + V_2 \cos(\omega_2 t) \tag{1.2.2}$$

Substituting (1.2.2) into (1.2.1) gives, for the first term:

$$i_a(t) = av_s(t) = aV_1 \cos(\omega_1 t) + aV_2 \cos(\omega_2 t) \tag{1.2.3}$$

After doing the same with the second term, the quadratic, and applying the well-known trigonometric formulas for squares and products of cosines, we obtain

$$\begin{aligned} i_b(t) &= bv_s^2(t) \\ &= \frac{1}{2} b \left\{ V_1^2 + V_2^2 + V_1^2 \cos(2\omega_1 t) + V_2^2 \cos(2\omega_2 t) \right. \\ &\quad \left. + 2V_1V_2 \left[\cos((\omega_1 + \omega_2)t) + \cos((\omega_1 - \omega_2)t)\right] \right\} \end{aligned} \tag{1.2.4}$$

and the third term, the cubic, gives

$$\begin{aligned} i_c(t) = cv_s^3(t) = \frac{1}{4} c\, \{ & V_1^3 \cos(3\omega_1 t) + V_2^3 \cos(3\omega_2 t) \\ & + 3V_1^2 V_2 [\cos((2\omega_1 + \omega_2)t) + \cos((2\omega_1 - \omega_2)t)] \\ & + 3V_1 V_2^2 [\cos((2\omega_2 + \omega_1)t) + \cos((2\omega_2 - \omega_1)t)] \\ & + 3(V_1^3 + 2V_1 V_2^2) \cos(\omega_1 t) \\ & + 3(V_2^3 + 2V_1^2 V_2) \cos(\omega_2 t)\} \end{aligned} \tag{1.2.5}$$

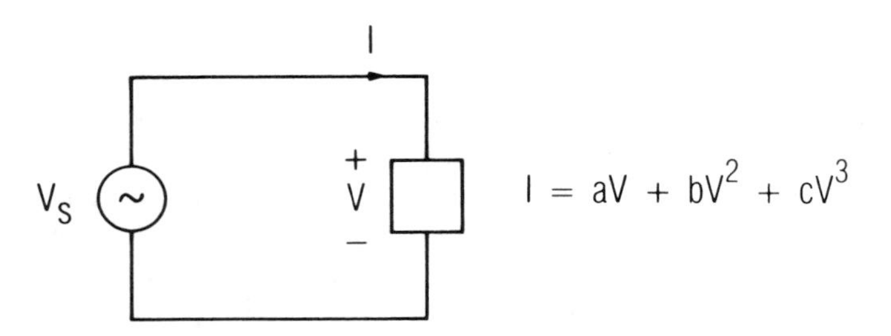

Figure 1.2 Two-terminal nonlinear resistor excited directly by a voltage source.

The total current in the nonlinear element is the sum of the current components in (1.2.3) through (1.2.5). This is sometimes called the *short-circuit current* in the element because there is no source impedance. It consists of a remarkable number of new frequency components, each successive term in (1.2.1) generating more new frequencies than the previous one; if a fourth- or fifth-degree nonlinearity were included, the number of new frequencies in the current would be even greater. However, in this case, there are only two frequency components of voltage, at ω_1 and ω_2, because the voltage source is in parallel with the nonlinearity. If there were a resistor between the voltage source and the nonlinearity, even more voltage components would be generated via the currents in that resistor, those new voltage components would generate new current components, and the number of frequency components would be very large indeed. In order to have a tractable analysis, it would be necessary to ignore all frequency components beyond some point; the number of components retained would depend upon the strength of the nonlinearity, the magnitude of the excitation voltage, and the desired accuracy of the result. The conceptual and analytical complexity of even apparently simple nonlinear circuits is the first lesson of this exercise.

A closer examination of the generated frequencies shows that all occur at a linear combination of the two excitation frequencies; in effect, at the frequencies:

$$\omega_{m,n} = m\omega_1 + n\omega_2 \tag{1.2.6}$$

where $m, n = \ldots -3, -2, -1, 0, 1, 2, 3, \ldots$, $\omega_{m,n}$ is called a *mixing frequency*, and the current component at that frequency is called a *mixing product*. The sum of the absolute values of m and n is called the *order* of the mixing product. For the $\omega_{m,n}$ to be distinct, ω_1 and ω_2 must be *noncommensurate;* a pair of frequencies is *commensurate* if their ratio is a rational number, and the entire set is commensurate if all possible pairs are commensurate. It will usually be assumed that the frequencies are noncommensurate when two or more arbitrary excitation frequencies exist.

An examination of (1.2.3) through (1.2.5) shows that a kth-degree term in the power series (1.2.1) produces new mixing frequencies of order k or below; those mixing frequencies are kth-order combinations of the frequencies of the voltage components at the element's terminals. This does not, however, mean that $|m| + |n| < k$ in every nonlinear circuit. In the earlier example, the terminal voltage components were the excitation voltages, so only two frequencies existed. However, if the circuit of Figure 1.2 included a resistor in series with the nonlinear element, the total terminal voltage would have included not only the excitation frequencies but higher-order mixing products. The nonlinear element then would have generated all possible kth-order combinations of those mixing products and the excitation frequencies. Thus, in general, a nonlinear element can generate mixing frequencies involving all possible harmonics of the excitation frequencies, even those where $|m| + |n|$ is greater than the highest power in the power series. It does this by generating kth-order mixing products between all the frequency components of its terminal voltage.

Another conclusion one may draw from (1.2.3) through (1.2.5) is that the odd-degree terms in the power series generate only odd-order mixing products and the even-degree terms generate even-order products. This property can be exploited by balanced structures (Chapter 5) that combine nonlinear elements in such a way that either the even- or odd-degree terms in their power series are eliminated, so only even- or odd-order mixing frequencies are generated. These circuits are very useful in rejecting unwanted even- or odd-order mixing frequencies.

The generation of apparently low-order mixing products from the high-degree terms in (1.2.1) is worth some examination; the terms at ω_1 and ω_2 in (1.2.5) exemplify this phenomenon. The existence of these terms implies that the fundamental current, for example, is not solely a function of the excitation voltage and the linear term in (1.2.1); it is dependent on all the odd-degree nonlinearities. Consequently, as V_s is increased, the cubic term becomes progressively more significant, and the fundamental-frequency current components either rise more rapidly or level off, depending on the sign of the coefficient c. A closer inspection of these terms shows that they can be considered to have arisen from the kth-degree term as kth-order mixing products; for example, the ω_1 terms in (1.2.5) arise as the third-order combinations:

$$\begin{aligned}\omega_1 &= \omega_1 + \omega_1 - \omega_1\\ &= \omega_1 + \omega_2 - \omega_2 \end{aligned} \tag{1.2.7}$$

The presence of the negative frequencies might be more convincing if the cosine functions were expressed in their exponential form, $\cos(\omega t) = [\exp(j\omega t) + \exp(-j\omega t)]/2$. Thus, when dealing with nonlinear circuits, we must always use a system of analysis that does not exclude the presence of negative frequencies; standard sinusoidal steady-state analysis using phasor concepts includes only positive-frequency terms and is therefore relegated to linear circuits only.

It is worthwhile to consider some specific examples in order to introduce one approach to nonlinear analysis and to gain further insights into the behavior of nonlinear circuits. Figure 1.3 shows a nonlinear circuit consisting of a resistive nonlinearity and a voltage source. The I/V nonlinearity includes only odd-degree terms:

$$I = f(V) = \frac{V}{2} + \frac{V^3}{7} + \frac{V^5}{15} \tag{1.2.8}$$

The 1-Ω resistor complicates things somewhat, but the current can still be found via power-series techniques. First, the series is reverted to find the voltage as a function of the current:

$$V = f^{-1}(I) = 2.00\,I - 2.286\,I^3 + 3.570\,I^5 + 3.184\,I^7 + \ldots \tag{1.2.9}$$

The formula for the series reversion can be found in Abramowitz (Reference 1.1, p. 16). The voltage across the resistor is $1 \cdot I$. Adding this to (1.2.9) (via Kirchhoff's voltage law), we obtain

$$V_s = 3.0\,I - 2.286\,I^3 + 3.570\,I^5 + 3.184\,I^7 + \ldots \tag{1.2.10}$$

Reverting (1.2.10) gives, for the current:

$$I = 0.3333\,V_s + 0.02822\,V_s^3 + 0.002271\,V_s^5 - 0.001375\,V_s^7 + \ldots \tag{1.2.11}$$

Equation (1.2.11) expresses I in terms of the known excitation, V_s. It includes only odd terms because all the circuit elements, the nonlinear and linear resistors, have only odd terms in their power series (we can think of the linear resistor as a special case of a nonlinear resistor, having a one-term power "series"). The series in (1.2.11) is infinite, but it has been truncated after the seventh-degree term; the series does, in fact, include all odd harmonics, thus all odd-order mixing products. To illustrate this point, we assume that $V_s = v_s(t) = 1 + 2\cos(\omega t)$; $v_s(t)$ and the resulting $i(t)$ waveform are shown in Figure 1.4, where the presence of harmonics in the current waveform is evident from its obvious nonsinusoidal shape. The actual harmonics could be found by substituting the expression $v_s(t) = 1 + 2\cos(\omega t)$ into (1.2.11) and by applying the same algebra as in (1.2.3) through (1.2.5). It is also evident at a glance that the dc component of the current is much greater than 0.364 A, the current that would be generated by the dc source alone if the ac source were zero. We must not forget that one of the low-order mixing frequencies generated by high-degree nonlinearities is a dc component; thus, the excitation of a nonlinear circuit may offset its dc operating point.

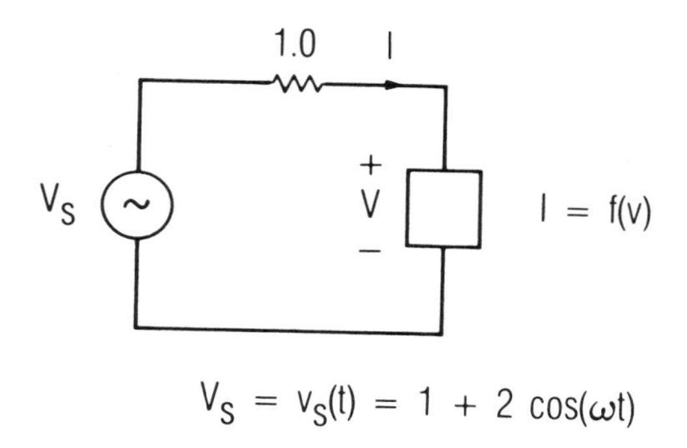

Figure 1.3 A nonlinear resistor, an excitation source, and a linear series resistor.

As a second example, consider again the circuit of Figure 1.3 with

$$f(V) = aV^2 \tag{1.2.12}$$

where a is a constant, as shown in Figure 1.5(a). Equation (1.2.12) describes an ideal square-law device. This is a strange situation at the outset, for two reasons: first, (1.2.12) cannot be reverted; second, it implies that, because the squared term generates only even-order mixing products and the excitation frequency is a first- (i.e., odd-) order mixing product, no excitation-frequency current is possible! It is possible that a true square-law device could be made; however, it would be active and unstable because its incremental resistance at some bias voltage V_0, $\mathrm{d}f(V)/\mathrm{d}V$, $V = V_0$, would be negative when $V_0 < 0$. Practical two-terminal "square-law" elements employ solid-state devices and have I/V characteristics like

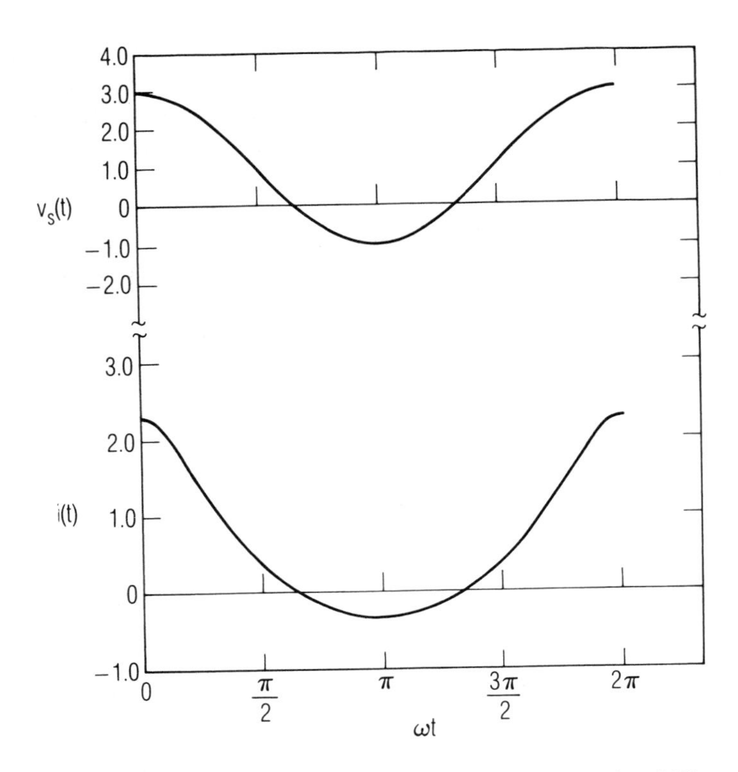

Figure 1.4 Voltage and current waveforms in the circuit of Figure 1.3.

that shown in Figure 1.5(b); the current follows a square law when $V > 0$ but is zero when $V < 0$. This characteristic still presents some analytical problems, because its *I/V* characteristic has undefined derivatives at $V = 0$. The device could, in concept, be operated in such a way that the voltage is always greater than zero by biasing it at a value V_0 great enough that no negative excitation peaks can drive the terminal voltage to zero. Its power series then becomes

$$f(v + V_0) = a(v + V_0)^2 = a(V_0^2 + 2V_0 v + v^2) \tag{1.2.13}$$

where a again is a constant and v is the voltage deviation from the bias point. Equation (1.2.13) includes the linear term $2V_0 v$. Thus, it is rarely possible, in practice, to obtain a true square-law device or, for that matter, a device having only even-degree terms in its power series; practical devices invariably have at least one odd-order term in their power series. This generalization applies to many devices that are often touted as square-law devices, such as FETs.

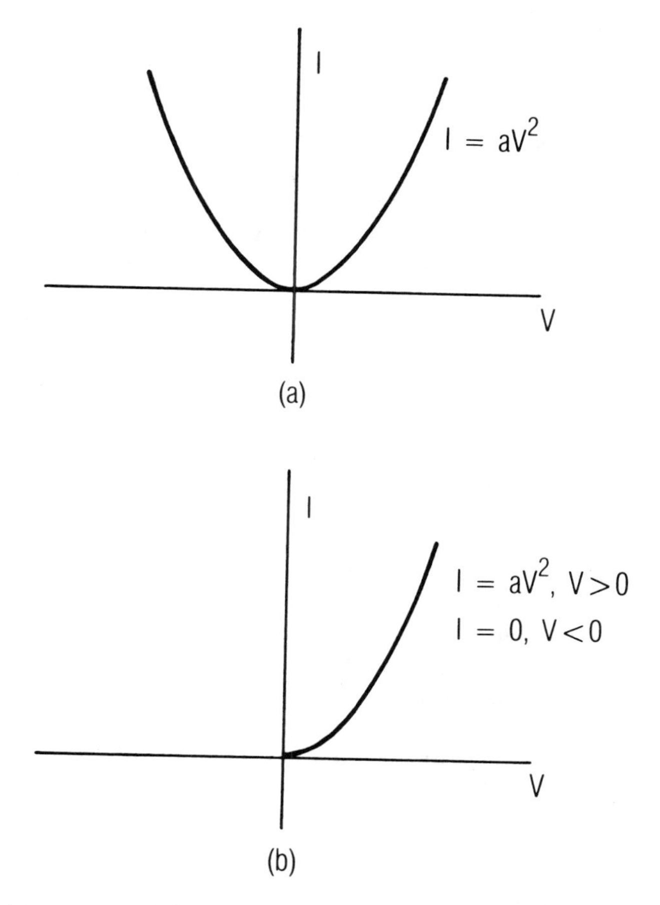

Figure 1.5 (a) *I/V* characteristic of the ideal square-law device; (b) *I/V* characteristic of a real "square-law" device.

Now that the pure square-law device has been ignominiously unmasked and shown to be a banal multiterm nonlinearity in disguise, it is interesting to see what happens to the circuit of Figure 1.3 when the nonlinearity includes even-degree terms plus one odd-degree term, the linear one. By choosing the coefficients carefully, one can define the characteristic over any arbitrary range without generating negative resistances. We assume that

$$I = f(V) = V + 2V^2 + 3V^4 \tag{1.2.14}$$

The reverted series, including the 1-Ω resistor, is

$$V_s = 2I - 2I^2 + 8I^3 - 43\,I^4 + 260\,I^5 + \ldots \quad (1.2.15)$$

which has all powers of I. Reverting this series again to obtain an expression for I in terms of V_s will clearly result in a series having all powers of V_s. Thus, even though the original series contained only one odd-degree term (the linear one), the current contains mixing frequencies of all orders, even and odd, including those orders greater than four, the degree of the original power series.

In summary, the I/V characteristic of a nonlinear circuit or circuit element can be characterized by a power series. The kth-degree term in the series generates kth-order mixing products of the frequencies in its control voltage or current. Some of these may coincide with lower-order frequencies. Mixing products may also coincide with higher-order frequencies; these are generated as kth-order mixing products between other mixing products. Thus, in general a nonlinear circuit having both even- and odd-degree nonlinearities in its power series generates all possible mixing frequencies, regardless of the maximum degree of its nonlinearities.

A special case of the nonlinear circuit having two-tone excitation occurs where one tone is relatively large and the other is vanishingly small. This situation is encountered in microwave mixers, where the large tone is the *local oscillator* (LO), and the small one is the RF excitation. Because the RF excitation is very small, its harmonics are negligibly small; and we can assume that only its fundamental component exists. The resulting frequencies are

$$\omega = \omega_{\mathrm{RF}} + n\omega_{\mathrm{LO}} \quad (1.2.16)$$

Equation (1.2.16) can also be expressed via our preferred notation:

$$\omega_n = \omega_0 + n\omega_{\mathrm{LO}} \quad (1.2.17)$$

where $n = \ldots -3, -2, -1, 0, 1, 2, 3, \ldots$, and $\omega_0 = |\omega_{\mathrm{RF}} - \omega_{\mathrm{LO}}|$ is the mixing frequency closest to dc; in a mixer, ω_0 is often the *intermediate frequency* (IF), the output frequency. In (1.2.16) and (1.2.17) the mixing frequencies are above and below each LO harmonic, separated by ω_0.

If the total small-signal voltage $v(t)$ is much smaller than the LO voltage $V_L(t)$, the circuit can be assumed to be quasilinear in the RF voltage. The total large-signal and small-signal current $I(t)$ in the nonlinearity of (1.2.1) is given by

$$I(t) = a[v(t) + V_L(t)] + b[v(t) + V_L(t)]^2 + c[v(t) + V_L(t)]^3 \quad (1.2.18)$$

Separating the small-signal part of (1.2.18) and assuming that $v^2(t) \ll v(t)$, we find the small-signal current $i(t)$ to be

$$i(t) \approx av(t) + 2bv(t)V_L(t) + 3cv(t)V_L^2(t) + \ldots \qquad (1.2.19)$$

This is a linear function of v, even though many of the current components in (1.2.19) are at frequencies other than the RF. Thus, a microwave mixer, which has an input at ω_{RF} and output at, for example, ω_0, is a quasilinear component in terms of its input-output characteristics under small-signal excitation.

1.3 NONLINEAR PHENOMENA

The examination of new frequencies generated in nonlinear circuits does not tell the whole story of nonlinear effects, especially the effects of nonlinearities on microwave systems. Many types of nonlinear phenomena have been defined; by means of the foregoing power series techniques, these can be related to the nonlinearities in individual components or circuit elements. Superficially, on the systems level, these are considered to be separate phenomena, but many are simply different manifestations of the same nonlinearities.

Harmonic Generation

One of the most obvious properties of a nonlinear system is its generation of harmonics of the excitation frequency or frequencies. These are evident as the terms in (1.2.3) through (1.2.5) at $m\omega_1$, $m\omega_2$. The mth harmonic of an excitation frequency is an mth-order mixing frequency. In many systems, for example, narrow-band receivers, the generation of harmonics is not a serious problem because the harmonics are far removed in frequency from the signals of interest and inevitably are rejected by filters. In others, such as transmitters, harmonics and other spurious outputs may interfere with other communications systems and must be reduced very carefully by filters or by other means.

Intermodulation

All the mixing frequencies in (1.2.3) through (1.2.5) that arise as linear combinations of two or more tones are often called *intermodulation* (IM) *products*. IM products generated in an amplifier or communications receiver often present a very serious problem, because they represent spurious signals that interfere with, and can be mistaken for, desired signals. IM products are generally much weaker than the signals that generate them; however, a situation often arises wherein two or more very strong signals, which may be outside the receiver's passband, generate an IM product that is within the receiver's passband and obscures a weak, desired signal. Even-order IM products usually occur at frequencies well above or below the signals that generate them and consequently are often of little concern. The IM products of greatest concern are usually the third-order ones that occur at $2\omega_1 - \omega_2$ and $2\omega_2 - \omega_1$, because they are the strongest of all odd-order products, are close to the signals that generate them, and often cannot be rejected by filters. Intermodulation is a major concern in microwave systems.

Saturation and Desensitization

The excitation-frequency current component in the nonlinear circuit examined in Section 1.2 was a function of power series terms other than the linear one; recall that (1.2.5) included components at ω_1 and ω_2, which varied as the cube of signal level. In order to describe saturation, we refer to (1.2.1) to (1.2.5). From (1.2.3) and (1.2.5), we find that the current component at ω_1, designated $i_1(t)$, is, (with $V_2 = 0$),

$$i_1(t) = (aV_1 + \frac{3}{4} cV_1^3) \cos(\omega_1 t) \tag{1.3.1}$$

If the coefficient c of the cubic term is negative, the response current saturates; that is, it does not increase at a rate proportional to the increase in excitation voltage. Saturation occurs in all circuits because the available output power is finite. If a circuit such as an amplifier is excited by a large and a small signal, and the large signal drives the circuit into saturation, gain is decreased for the weak signal as well. Saturation, therefore, causes a decrease in system sensitivity.

Cross Modulation

Cross modulation is the transfer of modulation from one signal to another in a nonlinear circuit. To understand cross modulation, imagine that the excitation of the circuit in Figure 1.1 is

$$V_s = v_s(t) = V_1 \cos(\omega_1 t) + [1 + m(t)]V_2 \cos(\omega_2 t) \tag{1.3.2}$$

where $m(t)$ is a modulating waveform; $m(t) > -1$. Equation (1.3.2) describes a combination of an unmodulated carrier and an amplitude-modulated signal. Substituting (1.3.2) into (1.2.1) gives an expression similar to (1.2.5) for the third-degree term, where the frequency component in $i_c(t)$ at ω_1 is

$$i_c'(t) = \frac{3}{2} cV_1V_2^2[1 + 2m(t) + m^2(t)] \cos(\omega_1 t) \tag{1.3.3}$$

which shows that a distorted version of the modulation of the ω_2 signal has been transferred to the ω_1 carrier. This transfer occurs simply because the two signals are simultaneously present in the same circuit, and its seriousness depends most strongly upon the magnitude of the coefficient c and the strength of the interfering signal ω_2. Cross modulation is often encountered on an automobile AM radio when one drives past the transmission antennas of a radio station; the modulation of that station momentarily appears to "come in on top of" every other received signal.

AM/PM Conversion

AM/PM conversion is a phenomenon wherein changes in the amplitude of a signal applied to a nonlinear circuit cause a phase shift. This form of distortion can have serious consequences if it occurs in a system in which the signal's phase is important; for example, phase-modulated communications systems. The response current at ω_1 in the nonlinear circuit element considered in Section 1.2 is, from (1.2.3) and (1.2.5):

$$i_1(t) = (aV_1 + \frac{3}{4} cV_1^3) \cos(\omega_1 t) \tag{1.3.4}$$

that is, $i_1(t)$ is the sum of first- and third-order current components at ω_1. Suppose, however, these components were not in phase. This possibility is not predicted by (1.2.1) through (1.2.5) because these equations describe a memoryless nonlinearity; it is, however, possible for a phase difference to exist in a circuit having capacitive nonlinearities. The response is then the vector sum of two phasors:

$$I_1(\omega_1) = aV_1 + \frac{3}{4} cV_1^3 \exp(j\theta) \tag{1.3.5}$$

where θ is the phase difference. Even if θ remains constant with amplitude, the phase of I_1 changes with variations in V_1. It is clear from comparing (1.3.5) to (1.3.1) that the AM/PM conversion will be most serious as the circuit is driven into saturation.

Spurious Responses

At the end of Section 1.2 it was shown that a mixer, with an RF input at ω_{RF} and an LO of ω_{LO}, has currents at the frequencies given by (1.2.16) or (1.2.17). It is easy to see that, if the RF is applied at any of those mixing frequencies, currents at all the rest are generated as well. Thus, the mixer has some response at a large number of frequencies, not just the one at which it is designed to work. In fact, if the applied signal is very strong, its harmonics are generated and the mixer has spurious responses at any frequency that satisfies the relation:

$$\omega_{IF} = m\omega_{RF} + n\omega_{LO} \tag{1.3.6}$$

where n and m are both positive and negative integers. Comparing (1.3.6) to (1.2.6) shows that spurious responses are a form of two-tone intermodulation wherein one of the tones is the LO. In microwave systems the concept of spurious responses is applied only to mixers.

1.4 APPROACHES TO ANALYSIS

One of the delights of the last few years has been the development of a theoretically sound approach to the analysis of nonlinear microwave circuits. Techniques used previously were questionable attempts to bend linear theory to nonlinear applications, were highly approximate, or were attempts at "black box" characterizations that did not include all the variables or parameters necessary to obtain meaningful results.

One straightforward way to characterize a large-signal circuit such as an amplifier is to graph on a Smith chart the contours of its load impedances that result in prescribed values of gain and output power. These approximately circular contours can then be used to select an output load impedance that represents the best trade-off of gain against output power. The contours are generated empirically by connecting various loads to the amplifier and by measuring the gain and output power at each value of load impedance. This process, called *load pulling,* has many limitations; the most serious practical one is the difficulty of measuring the load impedances at the device terminals. Load pulling has a major theoretical problem as well: The load impedance at harmonics of the excitation frequency can significantly affect circuit performance, but load pulling is concerned only with the load impedance at the fundamental frequency. Furthermore, load pulling is not useful for determining other important properties of nonlinear or quasilinear circuits, for example, harmonic levels or the effects of multitone excitation.

Another approach to the analysis of large-signal nonlinear circuits is to measure a set of two-port parameters, usually S parameters, at the large-signal excitation level. The standard small-signal equations for S-parameter design are then used to predict the performance characteristics of the circuit. This approach may have limited success if the circuit or device is not very strongly nonlinear and if it is not applied where it is obviously unsuited, for example, to frequency multipliers. Two-port parameters are fundamentally a linear concept, however, so the large-signal S-parameter approach represents a brutal attempt to force nonlinear circuits to obey linear circuit theory.

In order to see just one example of the problems that arise from bending linear concepts to fit nonlinear problems, consider the meaning of the output reflection coefficient, S_{22}, of a FET or bipolar transistor. For large-signal S-parameter analysis, S_{22} is measured by applying an incident wave to the output port at a power level comparable to that at which the device is used. Now imagine that the device is driven hard at its input and that the output reflection coefficient is again measured (ignore for a moment the obvious practical difficulties of making such a measurement). If the amplifier is significantly nonlinear, which in all likelihood it will be, we can hardly expect the reflection coefficient to be the same under these conditions or over the wide range of incident power levels the device is likely to encounter. However, the S-parameter concept is based on the assumption that it will be the same.

An intermediate approach, which is theoretically valid and is practical for low-frequency analog and digital design, is to use time-domain techniques. It is a straightforward matter to use conventional circuit theory to write time-domain differential equations that describe a nonlinear circuit. The resulting differential equations are nonlinear and can be solved numerically. Although time-domain techniques are most practical for analyzing circuits that include only lumped elements, they can be used with a limited variety of distributed elements such as ideal transmission lines; lossy or dispersive transmission lines cannot be modeled in the time domain. The two major limitations of time-domain analysis are its inability to handle frequency-domain quantities (e.g., impedances) and the difficulty of applying it to circuits having noncommensurate excitations. Indeed, one of the advantages of frequency-domain analysis is that a relatively complex circuit often can be reduced to one or more sets of impedances, called *embedding impedances,* at each important harmonic of the excitation frequency. When this reduction is possible, it significantly simplifies the task of analyzing a nonlinear circuit.

Many frequency-domain techniques for analyzing microwave circuits have become popular in recent years. The two most important are called *harmonic-balance analysis* and *Volterra-series* or *nonlinear transfer function* analysis. Harmonic-balance analysis is applicable primarily to strongly nonlinear circuits that are excited by a single large-signal excitation source; it can be applied to such circuits as transistor power amplifiers, mixer local oscillator circuits, and frequency multipliers using either diodes or transistors. Volterra-series analysis is applicable to the opposite problem: weakly driven, weakly nonlinear circuits having multiple small-signal excitations at noncommensurate frequencies. As such, it is most useful for evaluating intermodulation characteristics and other nonlinear phenomena in small-signal receiver circuits, such as amplifiers. With some modifications, the Volterra series can also be used to determine the IM properties of time-varying circuits such as mixers; similarly, harmonic-balance can be extended to certain situations involving noncommensurate signals. Demystifying the theory and practical use of these two techniques is the primary subject of this book.

All three methods—time-domain analysis, harmonic-balance analysis, and the Volterra series—require a circuit model consisting of lumped components and, for the latter two, impedance elements or multiports. Solid-state device models must consist of linear or nonlinear capacitors, inductors, resistors, and voltage or current sources (nonlinear inductors can be accommodated, although they are rarely encountered in solid-state devices or circuits). Underlying all the nonlinear models described in Chapter 2 is the *quasistatic assumption,* whereby all nonlinear elements are assumed to change instantaneously with changes in their control voltages. This assumption is also implicit in linear circuit theory; it requires, for example, that the charge on a capacitor is a function solely of the voltage at its terminals. If the capacitor is nonlinear, its incremental capacitance, as well as its charge, must change instantaneously with control voltage. A quasistatic circuit is not necessarily *memoryless;* a memoryless circuit is one in which no charge or flux storage elements (no capacitors or inductors) exist, so voltages and currents at any instant do not depend upon previous values of voltage or current. In a quasistatic circuit, the network voltages and currents may depend upon previous values of other voltages or currents, but the capacitances, inductances, resistances, and controlled sources do not depend directly upon their own histories.

The quasistatic assumption is critical to the entire business of both linear and nonlinear circuit analysis. It allows us, for example, to devise equivalent circuits for solid-state devices using only lumped linear and nonlinear elements, and makes many of the techniques of linear circuit theory applicable to at least the linear parts of nonlinear circuits. One of the nicest things about the quasistatic assumption is its range of validity.

Theoretical and experimental studies of silicon and gallium arsenide semiconductors and devices show that time-delay phenomena are usually on the order of picoseconds or are short compared to the inverse of the highest frequency at which we would attempt to use the device. Furthermore, the prohibition of time delays is not absolute; in some cases they can still be managed. For example, it is often possible to include in a quasistatic model the well known time delay between the gate voltage and drain current in a GaAs MESFET.

1.5 POWER AND GAIN DEFINITIONS

Although it is customary to speak loosely of gain and power in microwave circuits, these quantities can be defined in several different ways. The different definitions of gain are related to the concepts of available and dissipated power. These concepts are important in both linear and nonlinear circuits, although they are particularly important in nonlinear circuits where a waveform may have components at many frequencies that may or may not be harmonically related.

Available or *transferable power* is the maximum power that can be obtained from a source. The concept of available power is illustrated in Figure 1.6, in which a sinusoidal voltage source having a peak value V_s has an internal impedance of $R_1 + jX_1$ (unless we state otherwise, all frequency-domain voltages and currents in this book are phasor quantities; thus their magnitudes are equal to peak sinusoidal quantities, not rms).

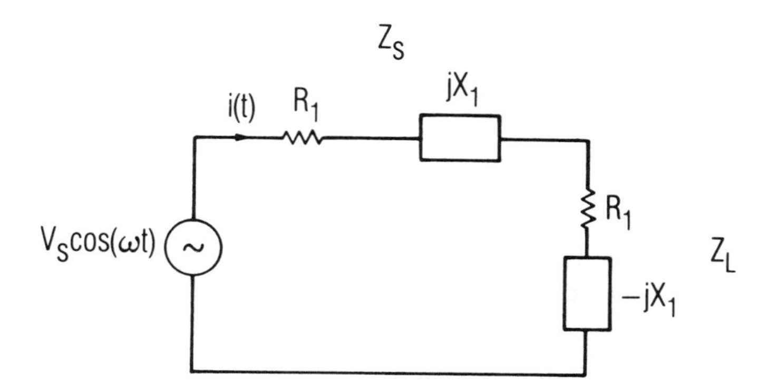

Figure 1.6 Circuit having a matched source and load, illustrating the concept of available power.

The maximum power is obtained from this source if the load impedance equals the conjugate of the source impedance, $Z_L = Z_s^* = R_1 - jX_1$. Under these conditions:

$$I = \frac{V_s}{2R_1} \tag{1.5.1}$$

where I is the peak value of the current, $i(t)$. The power dissipated in the load is

$$P_d = P_{av} = \frac{1}{2} I^2 R_1 = \frac{1}{2} I^2 \operatorname{Re}\{Z_s\} = \frac{V_s^2}{8R_1} \tag{1.5.2}$$

which is the maximum available from the source. *Dissipated*, or *transferred* power is the power dissipated in a load that may or may not be matched to the source. In Figure 1.7, the load is not conjugate-matched to the source, so the dissipated power is less than that given in (1.5.2). In this case,

$$I = \frac{V_s}{[(R_1 + R_2)^2 + (X_1 + X_2)^2]^{1/2}} \tag{1.5.3}$$

and the power dissipated in the load is

$$P_d = \frac{1}{2} I^2 R_2 = \frac{V_s^2 R_2}{2[(R_1 + R_2)^2 + (X_1 + X_2)^2]} \tag{1.5.4}$$

In a nonlinear circuit, the voltage source may contain many frequency components, and the source or load impedance may not be the same at each frequency. An example of this situation is the output circuit of a diode frequency multiplier. The multiplier generates many harmonics, only one of which is desired, so it has an output filter that allows only the desired harmonic to reach the output port. Thus, the impedance presented to the diode at the desired output frequency is the load impedance, but it is the out-of-band impedance of the filter at all other harmonics. The current in the loop is a function of frequency, as shown in Figure 1.8. Because the load and source are linear, each frequency component can be treated separately without concern for the others. Then, the available and transferred power are

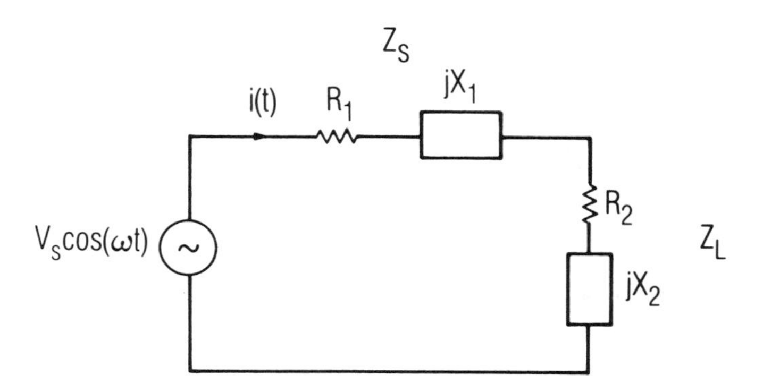

Figure 1.7 Circuit having an unmatched source and load.

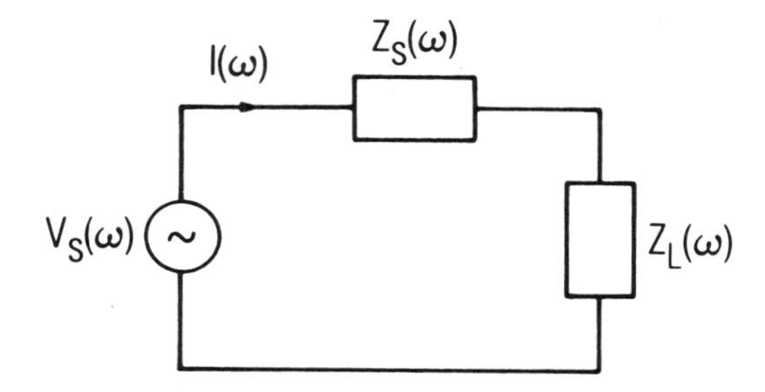

Figure 1.8 Unmatched circuit having a nonsinusoidal voltage-source excitation.

$$P_{av}(\omega) = \frac{|V_s(\omega)|^2}{8 \operatorname{Re}\{Z_s(\omega)\}} \tag{1.5.5}$$

$$P_d(\omega) = \frac{1}{2} |I(\omega)|^2 \operatorname{Re}\{Z_L(\omega)\} \tag{1.5.6}$$

An equivalent representation uses a current source and admittances as shown in Figure 1.9. Similarly, the available and dissipated powers are found to be

$$P_{av}(\omega) = \frac{|I_s(\omega)|^2}{8 \operatorname{Re}\{Y_s(\omega)\}} \tag{1.5.7}$$

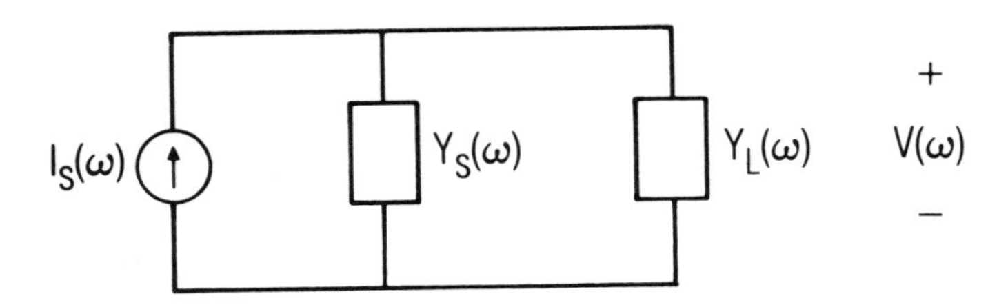

Figure 1.9 Unmatched circuit having a nonsinusoidal current-source excitation.

$$P_d(\omega) = \frac{1}{2} |V(\omega)|^2 \, \mathrm{Re}\{Y_L(\omega)\} \tag{1.5.8}$$

Figure 1.10 shows a model often used for the situation wherein a voltage (or current) source has many discrete frequency components. The load impedance at each frequency is represented by an impedance in series with a filter. The filters $F_1, F_2, \ldots, F_N$ are ideal series-resonant circuits; that is, they are short circuits at their resonant frequencies and open circuits at all other frequencies. Thus, the current component at only one frequency circulates in each branch. One of these branches is the output circuit; the rest may be arbitrary impedances that represent the combined effects of out-of-band filter or matching circuit terminations, package or other circuit parasitics, or in some cases resonances (called *idlers*) that are purposely introduced to optimize performance. These terminations at frequencies other than the output frequency may have a strong effect upon the circuit's performance, so the design of the output network may have to take into account these terminations as well as that at the output frequency.

The gain of a two-port network can be defined in terms of available and dissipated powers. The two most important ways of specifying gain are *transducer gain* and *maximum available gain*. When a microwave engineer speaks loosely of "gain," he or she usually means transducer gain, whether he or she knows it or not. To see why this is so, imagine a technician using a signal generator and power meter to measure the gain of an amplifier. First, the technician connects the power meter to the carefully matched output of the signal generator and notes the power. Because the signal source and the power meter are matched, this is the available power. The technician then connects the signal generator to the amplifier input and the power meter to its output and notes the output power. The output power is the power dissipated in the load, which is not necessarily conjugate-matched to the amplifier's output. The technician

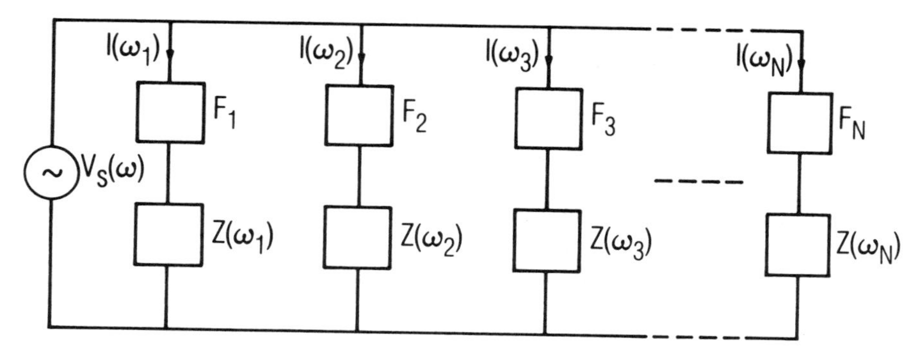

Figure 1.10 Model of a voltage source and load where the excitation has a number of discrete frequency components.

calls the ratio of these powers the *gain*, which in this case is the power delivered to the load divided by the power available from the source. This is precisely the definition of *transducer gain*.

Transducer gain is a very useful concept because, in microwave systems, the important thing to know is how much more or less power a circuit delivers to a standard load (e.g., a 50-Ω coaxial line), compared to the power that could have been obtained from the source alone. This is precisely what transducer gain tells us. Furthermore, transducer gain is almost always a defined quantity, because it requires only that the source and output powers be finite, and real sources always have finite available power. Thus, the concept is handy in nonlinear circuits where, as our discussion of large-signal *S* parameters illustrated earlier, it is often impossible to define input and output impedances or reflection coefficients.

Other gain definitions are often useless because they do not tell the engineer what he or she wants to know or occasionally result in meaningless or undefined quantities. One such concept is *power gain*, defined as power delivered to the load divided by power delivered to the two-port's input. We find that the power gain of a low-frequency MESFET amplifier, for example, is meaninglessly high: the FET's output power is modest but its input impedance is highly reactive, so the input power is close to zero. This result tells nothing about the way the amplifier works in a system. The concept of power gain can give even more bizarre results when applied to other circuits, such as a negative-resistance amplifier without a circulator. The input power of a negative-resistance device is difficult to define, but we could justifiably say that it is negative and equal to the output power. Thus, the power gain of a negative-resistance amplifier is always -1. Even with these strange results, however, the concept of power gain has some limited usefulness; one of these uses the design of linear amplifiers having prescribed values of transducer gain. This technique is described in Section 8.1.2.

Available gain is defined as the power available from the output divided by the power available from the source. It is intrinsically not a very useful concept (although it will costar with power gain in Section 8.1.2), but its maximum value, called the *maximum available gain*, which occurs when the input of the two-port is conjugately matched to the source, is very useful. The maximum available gain is, therefore, the highest possible value of the transducer gain, which occurs when both the input and output ports are conjugately matched. Maximum available gain is defined only if the two-port is unconditionally stable; that is, if the input and output impedances always have positive real parts when any passive load is connected to the opposite port.

1.6 STABILITY

The fundamental definition of a stable electrical network is that its response is bounded when its excitation is bounded. In the case of a linear two-port having a sinusoidal steady-state excitation, this definition leads to a stability criterion: if the input or output port has an impedance that is the negative of its terminating impedance, the two-port is unstable. If the termination is passive, instability (in fact, oscillation) occurs if the input or output impedance has a negative real part. If the input and output impedance have positive real parts when any passive load is connected to the opposite port, the network is unconditionally stable. If it is possible for the input or output impedance to have a negative real part when a passive termination is connected to the opposite port, the network is conditionally stable.

The situation is more complicated in the case of nonlinear circuits. Because the kinds of interactions that can occur in nonlinear circuits are more complex, such circuits often exhibit transient and steady-state phenomena other than sinusoidal oscillation that, although bounded, are loosely classed as instability. These include parasitic oscillations; spurious outputs that occur only under large-signal excitation; "snap" phenomena, in which the output level or bias conditions change abruptly as input level is varied; and the exacerbation of normal noise levels. Of course, plain, old-fashioned oscillation is also a possibility. Consequently, it is extremely difficult to devise a meaningful and practical stability criterion for nonlinear circuits.

Even without the academic advantage of a stability criterion, it is usually possible, with care, to design nonlinear or quasilinear circuits that are well behaved. For example, if a harmonic-balance analysis of a proposed circuit design converges without incident to a solution, one can be confident that it is stable, by all practical definitions of the term. (It is also stable in theory because harmonic-balance analysis is a process of perturbing the voltages across the nonlinear elements. If these perturbations do not cause larger perturbations, the circuit must be locally stable. The idea that a circuit is stable if such perturbations do not cause greater perturbations is equivalent to the concept of stability defined earlier.) The converse may not be true, however, because the failure of an iterative technique such as harmonic balance to converge may be caused by numerical problems, not by inherent instability. Other tricks and techniques for ensuring stability of individual types of circuits will be covered in their respective chapters.

REFERENCE

[1.1] M. Abramowitz and I.A. Stegun, *Handbook of Mathematical Functions*, Dover, New York, 1970.

CHAPTER 2

SOLID-STATE DEVICE MODELING FOR QUASISTATIC ANALYSIS

Inherent in the Volterra-series and harmonic-balance analyses of nonlinear circuits is the quasistatic assumption, the assumption that all the parameters of nonlinear elements (e.g., capacitance and transconductance) change instantaneously with a change in one or more control voltages or currents. As the dominant nonlinearities in a microwave circuit are inevitably those of solid-state devices, it is important to have quasistatic models of those devices. Models that consist of lumped linear and nonlinear elements are usually most convenient.

After defining the topology of the device model, we must attach values to the capacitances, resistances, or transconductances of the circuit elements. In linear circuits, defining these elements is not a major problem because each capacitor, resistor, or controlled source is characterized by only one or, at most, two constants. In a nonlinear circuit, the element values are functions of their control voltages or, less commonly, currents. These functions usually contain important constants that often cannot be determined with sufficient accuracy from the dimensions and material parameters of the device, and consequently those constants must be determined by measurement. Acceptable analytical expressions or numerical techniques for characterizing the circuit elements occasionally do not exist, so it is necessary to make repeated measurements and to generate a table of values that can be interpolated or to fit measured data to a polynomial expression. This chapter describes the circuit models and element characterizations of MESFETs, Schottky-barrier diodes and varactors, and step-recovery diodes; at present, these are among the most important solid-state devices used in nonlinear and quasilinear microwave circuits.

2.1 NONLINEAR DEVICE MODELS

Because they are based on a fundamental assumption of linearity, impedance concepts and multiport circuit theory cannot be used as the sole means of describing a nonlinear circuit. Accordingly, the most popular means for characterizing transistors—*S*, *Y*, or other multiport parameters—cannot be used to model nonlinear solid-state devices. Instead, the most successful (but by no means the only) method of characterizing such devices is to use a lumped circuit model that includes a mix of linear and nonlinear resistors, capacitors, and controlled sources (nonlinear inductors are possible as well, both theoretically and practically, but they are rarely encountered). The nonlinear elements are always assumed to be quasistatic; in the devices examined in this chapter, the quasistatic assumption is valid at frequencies up to at least 100 GHz. The nonlinear elements in FET and diode models are invariably voltage controlled, usually having one or at most two control voltages.

One advantage of the quasistatic approach is that small-signal models of devices can be converted easily to large-signal models. Small-signal devices are often described by lumped-element models as an alternative to multiport parameters; the advantage of using such models is that they require less characterization data than a large table of two-port parameters. Such models can be converted to large-signal models simply by including the voltage dependences of the circuit's elements and sometimes by making minor changes in the topologies of the models. It is not necessary to attach a voltage dependence to each element; usually many are very weakly nonlinear and often can be assumed to be linear.

Quasistatic modeling is usually not applicable to devices whose operation is dominated by time effects. These include transit-time devices such as IMPATTs and Gunn or transferred-electron devices. Such devices are so strongly nonlinear that they are rarely used in circuits that have amplitude-modulated or multiple CW excitations and are usually used as oscillators or amplifiers of CW or phase-modulated signals. Conversely, the models developed in this chapter are particularly useful in the nonlinear and quasilinear circuits commonly employed in communications and radar systems; such components include small-signal amplifiers, linear power amplifiers, and harmonic generators.

An obvious requirement of a good device model is that it be sufficiently accurate. Furthermore, it must maintain its accuracy over a wide frequency range because important mixing products and harmonics may span a wide range of frequencies. Solid-state devices, however, are not lumped circuits, so any such model is necessarily an approximation. Usually more complex circuit topologies are more accurate, but the natural desire to minimize complexity, and consequently computational difficulty, often dictates that the simplest adequate model be used. Concern for computational difficulty is crucial, because many nonlinear analyses require many, perhaps thousands, of evaluations of the circuit equations. Thus, the use of unnecessarily complex models may involve excessive computational cost.

Another requirement is that it must be possible to determine the model's parameters without undue difficulty. The nonlinear characterization of any solid-state device usually requires a number of measurements. If the number and difficulty of these measurements are excessive, the design cost of the resulting circuit is increased and accuracy may suffer, if only because of the greater chance of fatigue and error on the part of the poor soul who must make the measurements. A nonlinear design technique that requires laborious measurements is not likely to be widely accepted and will always tempt the designer to take shortcuts in the process. The result might be that a more complex technique, which is theoretically very accurate, may be less accurate in practice than a simpler one properly executed.

2.2 NONLINEAR LUMPED CIRCUIT ELEMENTS AND CONTROLLED SOURCES

The nonlinear device models we will consider consist of resistors, capacitors, and controlled sources. The circuit elements can be described by one of two kinds of characteristics: a large-signal global characteristic, or an incremental small-signal characteristic. The former describes the overall I/V or Q/V relationship and is used for modeling large-signal circuits; the latter describes the deviation of voltage and current or charge in the vicinity of a bias point, and is used for modeling small-signal, quasilinear circuits. In the large-signal case, the circuit element is effectively treated as a "black box" having the prescribed I/V or Q/V characteristic; in the small-signal case, it is a linear or nonlinear small-signal resistor, capacitor, or controlled source having a resistance, capacitance, or small-signal current that is a function of a dc (or occasionally time-varying) control voltage. In this section, we examine the relationship between the large-signal and small-signal characterizations, and in particular show how the small-signal characterization can be derived from the large-signal one.

Two concepts critical to the modeling of nonlinear solid-state devices are *voltage* and *current control,* and *incremental quantities.* A *voltage-controlled* element's value is dependent upon a voltage that may be applied to its terminals or may be elsewhere in the circuit. The element's value must be a single-valued function of the control voltage. For example, it is usually natural to express a diode junction capacitance as a function of junction voltage. The capacitance varies with the junction voltage and is a single-valued function of it. Conversely, a *current-controlled* element is one whose value is a single-valued function of a current.

Many elements can be treated as either current- or voltage-controlled. For example, the small-signal junction conductance of a Schottky-barrier diode can be expressed as an exponential function of voltage or as a linear function of current. Either way, the function is single-valued, and the choice of expressing the device as current- or voltage-controlled depends primarily on convenience. In contrast, the current in some types of diodes rises with junction voltage, then drops as voltage is further increased. Such nonlinearities must be treated as voltage-controlled, because the junction voltage is not a single-valued function of current.

An important question involves the precise definitions of the small-signal resistance and capacitance of nonlinear elements. For example, the global I/V characteristic of a linear resistor is given by Ohm's law, $V = RI$. But, suppose a current-controlled nonlinear resistor were used in an application where a small-signal ac current is applied and a dc control current I_0 exists. The ac component of its voltage should be given by $v(t) = r(I_0)\, i(t)$, where $v(t)$ and $i(t)$ are the small-signal voltage and current, respectively. Figure 2.1 illustrates this case, and it is clear that

$$v(t) = \left.\frac{dV}{dI}\right|_{I=I_0} i(t) \tag{2.2.1}$$

so

$$r(I_0) = \left.\frac{dV}{dI}\right|_{I=I_0} \tag{2.2.2}$$

Of course, (2.2.1) is exact only as the maximum value of $i(t)$ approaches zero. This definition of resistance, called the *incremental resistance*, is valid in small-signal quasilinear analysis. The same idea applies to controlled sources, such as those described by a linear transconductance; thus,

$$g_m(V_{g0}) = \left.\frac{dI_d}{dV_g}\right|_{V_g=V_{g0}} \tag{2.2.3}$$

in which we assumed the drain current I_d to be a function of the gate voltage V_g only. (If the drain current is assumed to depend on other voltages, a partial derivative must be used.)

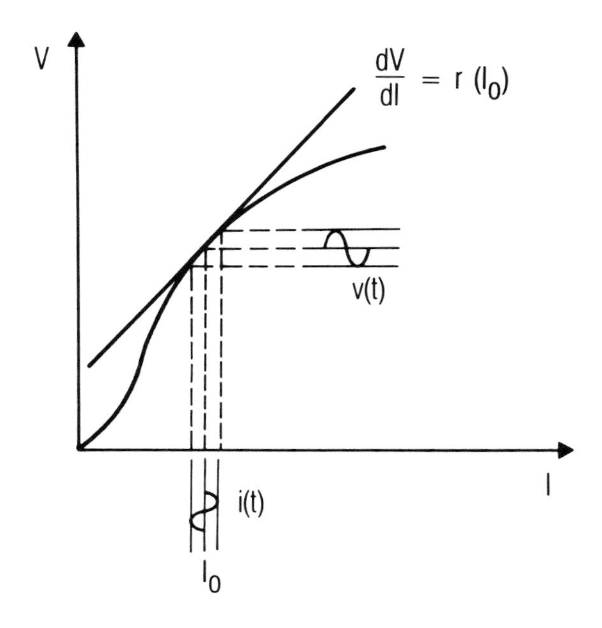

Figure 2.1 Incremental small-signal resistance of a nonlinear resistor.

2.2.1 The Substitution Theorem

Expressions (2.2.1) through (2.2.3) or the more complete ones that follow can describe either a single element or a controlled source. The difference between the two is only that in a nonlinear conductance, the control voltage is applied to the element's terminals; in a controlled source, the control voltage is somewhere else in the circuit. This point can be clarified via the *substitution theorem*, which defines an equivalence between a circuit element and a controlled source.

In Figure 2.2(a), a controlled current source is shown separately from the rest of the network in which it is embedded. Its large-signal current is GV, where V, the control voltage, is the voltage at its terminals. The current is clearly unchanged if a conductance of value G is substituted for the controlled source. The same is true if the I/V relationship of the source is a more complicated linear or nonlinear function of voltage: a conductance having the same I/V characteristic can be substituted, and the representations are equivalent. We now can state the substitution theorem precisely: A linear or nonlinear resistive circuit element having the characteristic $I = f(V)$ is equivalent to a controlled current source having the same characteristic, wherein V is the terminal voltage. Although this definition refers to large-signal V and I, the substitution theorem is equally

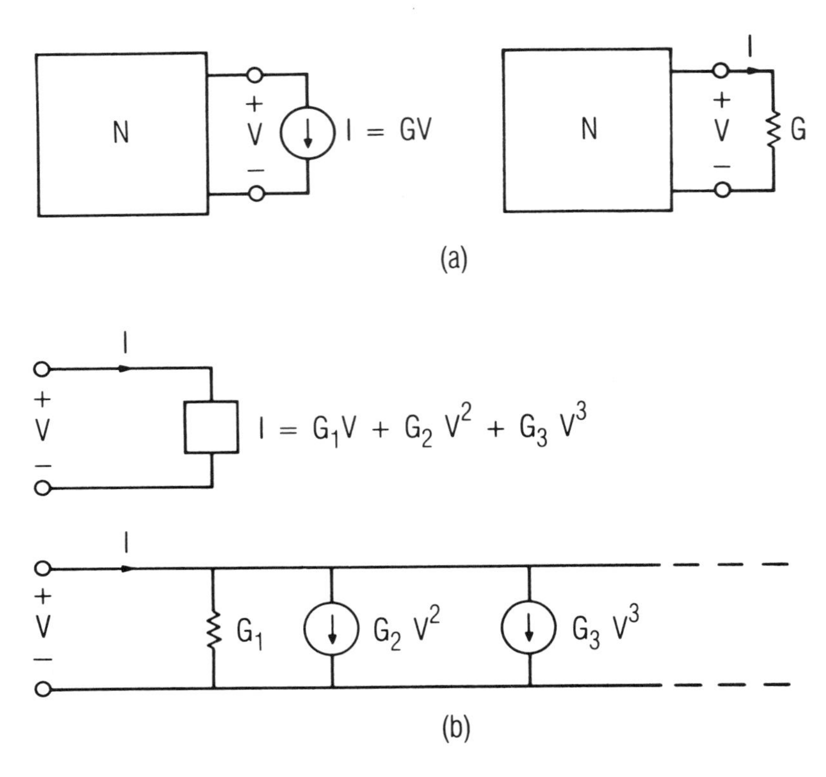

Figure 2.2 The substitution theorem: (a) source-conductance equivalence; (b) nonlinear element equivalence.

applicable to a small-signal incremental characteristic. Also, it is applicable, by analogy, to capacitive elements or current-controlled elements.

One important application of the substitution theorem is shown in Figure 2.2(b), where the I/V characteristic of a nonlinear conductance is described by the power series $I = f(V) = G_1V + G_2V^2 + G_3V^3 + \ldots$. The nonlinear element can be described by an equivalent circuit that includes a linear conductance G_1 and controlled current sources representing the higher-degree terms in the series. Of course, the linear component G_1V could also be represented by a current source if it were more convenient to do so. One should note that, although the term G_1V was modeled as a conductance in Figure 2.2(b), G_1 is not the incremental conductance; G_1 is equal to the incremental conductance only at the point $V = 0$.

2.2.2 Nonlinear Conductance or Resistance

An element that is described by an I/V characteristic $I = f(V)$, as shown in Figure 2.3, is a voltage-controlled conductance. We assume that it has a dc control voltage V_0, which in practice could be a bias voltage, and a small-signal ac voltage $v(t)$. We can expand the current in a Taylor series around V_0 to determine its ac part:

$$f(V_0 + v) = f(V_0) + \left.\frac{df(V)}{dV}\right|_{V=V_0} v + \frac{1}{2}\left.\frac{d^2f(V)}{dV^2}\right|_{V=V_0} v^2 + \frac{1}{6}\left.\frac{d^3f(V)}{dV^3}\right|_{V=V_0} v^3 + \ldots \tag{2.2.4}$$

where the indication of time dependence, "(t)," has been deleted from $v(t)$ for simplicity (the "(t)" will be deleted from all the small-signal voltage, current, and charge waveforms in all the equations in this section). We can assume that $v \ll V_0$ and that the nonlinearity is weak enough so that the series converges. Then, the small-signal current i is

$$i = f(V_0 + v) - f(V_0) = \left.\frac{df(V)}{dV}\right|_{V=V_0} v + \frac{1}{2}\left.\frac{d^2f(V)}{dV^2}\right|_{V=V_0} v^2 + \frac{1}{6}\left.\frac{d^3f(V)}{dV^3}\right|_{V=V_0} v^3 + \ldots \tag{2.2.5}$$

A subtlety of (2.2.5) is that the current i is the total small-signal current, including dc as well as ac components. Thus, i has dc components even though v has only ac components because the even-degree terms in (2.2.5) introduce them. Under the stated assumptions, the change in the dc operating point is small compared to $f(V_0)$, so the dc components in v^2, v^4, . . . , are usually negligible. In the quasilinear case, the terms of degree greater than one are assumed to be negligible, and

$$i = \left.\frac{df(V)}{dV}\right|_{V=V_0} v = g(V_0)v \tag{2.2.6}$$

and $g(V_0)$ is the *incremental conductance* at V_0. For the nonlinear case, (2.2.5) can be expressed as

$$i = g_1 v + g_2 v^2 + g_3 v^3 + \ldots \tag{2.2.7}$$

and, with the help of the substitution theorem, the nonlinear element can be modeled as shown in Figure 2.4. The linear term in (2.2.7) is the incremental conductance.

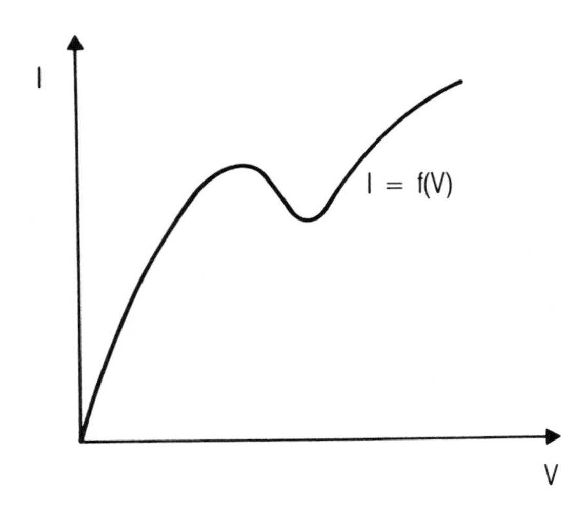

Figure 2.3 A voltage-controlled nonlinear conductance.

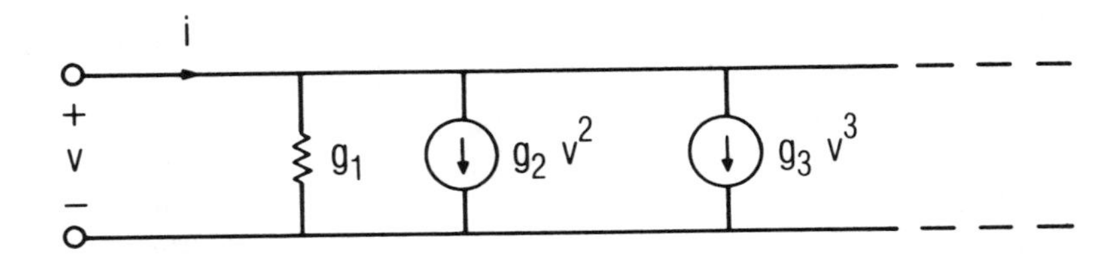

Figure 2.4 Small-signal nonlinear equivalent circuit for the nonlinear conductance.

Often a nonlinear circuit element is controlled by more than one current or voltage. An example of such a situation is the simplified FET equivalent circuit shown in Figure 2.5, in which the drain current I is a function of both the gate voltage V_1 and the drain voltage V_2; thus $I = f(V_1, V_2)$. In this case, V_2 is applied to the current source and V_1 is

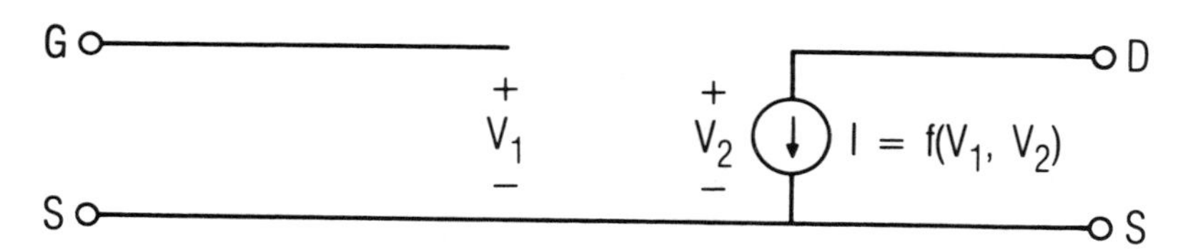

Figure 2.5 A multiply controlled nonlinear element.

a node voltage elsewhere in the circuit. In most practical nonlinear elements that have multiple control voltages, at least one of the voltages is the applied voltage, but the theory makes no such requirement. The function $f(V_1, V_2)$ can be expanded in a two-dimensional Taylor series and the dc current component subtracted, giving the rather sticky expression:

$$i = \frac{\partial f}{\partial V_1} v_1 + \frac{\partial f}{\partial V_2} v_2 + \frac{1}{2}\left(\frac{\partial^2 f}{\partial V_1^2} v_1^2 + 2\frac{\partial^2 f}{\partial V_1 \partial V_2} v_1 v_2 + \frac{\partial^2 f}{\partial V_2^2} v_2^2\right)$$
$$+ \frac{1}{6}\left(\frac{\partial^3 f}{\partial V_1^3} v_1^3 + 3\frac{\partial^3 f}{\partial V_1 \partial V_2^2} v_1 v_2^2 + 3\frac{\partial^3 f}{\partial V_1^2 \partial V_2} v_1^2 v_2 + \frac{\partial^3 f}{\partial V_2^3} v_2^3\right) + \cdots \tag{2.2.8}$$

In (2.2.8) the notation has been streamlined somewhat, and it is understood that the derivatives are evaluated at the bias points of V_1 and V_2; these are $V_{1,0}$ and $V_{2,0}$, respectively. In the small-signal quasilinear case, the high-degree terms are neglected and

$$i = \frac{\partial f}{\partial V_1} v_1 + \frac{\partial f}{\partial V_2} v_2 \tag{2.2.9}$$

or

$$i = g_m(V_{1,0}, V_{2,0}) v_1 + G_d(V_{1,0}, V_{2,0}) v_2 \tag{2.2.10}$$

where the partial derivatives have been replaced by their more conventional equivalent terms, the transconductance g_m and the drain conductance G_d.

The extension of (2.2.4) through (2.2.10) to resistances or current-controlled voltage sources is trivial: we need only interchange I and V, and i and v, in (2.2.4) through (2.2.10). The same expressions can be used for voltage-controlled voltage sources or current-controlled current sources by substituting the control voltage or current for V, the small-signal excitation for v, and the response for i.

A distressingly common error, which can be found in scandalously high places, is to assume that an expression for the small-signal current can be found by expanding the nonlinear conductance in a power series. Specifically, the approach is to find $g(V)$ from (2.2.6), and to say

$$i = g(V_0 + v)v \tag{2.2.11}$$

Expanding the conductance in (2.2.11) in a Taylor series gives

$$i = g(V_0) + \left.\frac{dg(V)}{dV}\right|_{V=V_0} v + \frac{1}{2}\left.\frac{d^2g(V)}{dV^2}\right|_{V=V_0} v^2 + \ldots \tag{2.2.12}$$

or, with $g(V_0)$ given by (2.2.6):

$$i = \left.\frac{df(V)}{dV}\right|_{V=V_0} v + \left.\frac{d^2f(V)}{dV^2}\right|_{V=V_0} v^2 + \frac{1}{2}\left.\frac{d^3f(V)}{dV^3}\right|_{V=V_0} v^3 + \ldots \tag{2.2.13}$$

This expression is clearly not the same as (2.2.5). There is no reason why (2.2.11) should be equivalent to (2.2.5); $g(V_0 + v)$ is just the linear conductance at a slightly different control voltage. This error is particularly insidious because the linear terms in (2.2.13) and (2.2.5) are fortuitously the same. The correct incremental I/V characteristic can be obtained from the $g(V)$ characteristic; the method is described in Section 2.2.4.

2.2.3 Nonlinear Capacitor

Just as any circuit element having a nonlinear I/V dependence can be treated as a conductance or a resistance, any element that stores charge in response to a voltage can be treated as a capacitor. A linear capacitor has the charge/voltage relation $Q = CV$. A nonlinear capacitor has the large-signal nonlinear dependence of charge on voltage:

$$Q_c = f_Q(V) \tag{2.2.14}$$

where Q_c is the capacitor charge. Initially we assume that the voltage $V = V_0$ is the sole dc control voltage and that it is applied to the capacitor's terminals. By expanding the charge function in a Taylor series, as with the conductance, and subtracting the dc component of the charge, we obtain the small-signal component of the charge:

$$
\begin{aligned}
q &= f_Q(V_0 + v) - f_Q(V_0) \\
&= \left.\frac{\mathrm{d}f_Q(V)}{\mathrm{d}V}\right|_{V=V_0} v + \frac{1}{2}\left.\frac{\mathrm{d}^2 f_Q(V)}{\mathrm{d}V^2}\right|_{V=V_0} v^2 \\
&\quad + \frac{1}{6}\left.\frac{\mathrm{d}^3 f_Q(V)}{\mathrm{d}V^3}\right|_{V=V_0} v^3 + \ldots
\end{aligned}
\tag{2.2.15}
$$

Again, for simplicity, the small-signal voltage $v(t)$ is written as v and $q(t)$ as q in (2.2.15). The small-signal current is the time derivative of the charge:

$$
\begin{aligned}
i &= \frac{\mathrm{d}q}{\mathrm{d}t} \\
&= \left.\frac{\mathrm{d}f_Q(V)}{\mathrm{d}V}\right|_{V=V_0} \frac{\mathrm{d}v}{\mathrm{d}t} + \left.\frac{\mathrm{d}^2 f_Q(V)}{\mathrm{d}V^2}\right|_{V=V_0} v \frac{\mathrm{d}v}{\mathrm{d}t} \\
&\quad + \frac{1}{2}\left.\frac{\mathrm{d}^3 f_Q(V)}{\mathrm{d}V^3}\right|_{V=V_0} v^2 \frac{\mathrm{d}v}{\mathrm{d}t} + \ldots
\end{aligned}
\tag{2.2.16}
$$

Equation (2.2.16) can be expressed as

$$
i = (C_1(V_0) + C_2(V_0)v + C_3(V_0)v^2 + \ldots)\frac{\mathrm{d}v}{\mathrm{d}t} \tag{2.2.17}
$$

which is the series form of the incremental capacitance. For the quasilinear case:

$$
i = C_1(V_0)\frac{\mathrm{d}v}{\mathrm{d}t} \tag{2.2.18}
$$

where

$$
C_1(V_0) = \left.\frac{\mathrm{d}f_Q(V)}{\mathrm{d}V}\right|_{V=V_0} \tag{2.2.19}
$$

Equation (2.2.19) is the standard definition of capacitance in the Schottky-barrier or *pn* junction of a solid-state device.

Capacitors, like conductances, can be controlled by more than one voltage. Capacitors having multiple control voltages are found in many solid-state devices, not the least of which is the GaAs MESFET. In this case, the large-signal Q/V characteristic is

$$Q_c = f_Q(V_1, V_2, V_3, \ldots) \tag{2.2.20}$$

As it is rarely necessary to consider more than two control voltages, $Q_c = f_Q(V_1, V_2)$ can be expanded in a two-dimensional Taylor series about the bias points $V_{1,0}$ and $V_{2,0}$ as before. After subtracting the dc charge components to get the small-signal charge, we have

$$\begin{aligned} q = {} & \frac{\partial f_Q}{\partial V_1} v_1 + \frac{\partial f_Q}{\partial V_2} v_2 + \frac{1}{2}\left(\frac{\partial^2 f_Q}{\partial V_1^2} v_1^2 + 2\frac{\partial^2 f_Q}{\partial V_1 \partial V_2} v_1 v_2 \right. \\ & \left. + \frac{\partial^2 f_Q}{\partial V_2^2} v_2^2\right) + \frac{1}{6}\left(\frac{\partial^3 f_Q}{\partial V_1^3} v_1^3 + 3\frac{\partial^3 f_Q}{\partial V_1 \partial V_2^2} v_1 v_2^2 \right. \\ & \left. + 3\frac{\partial^3 f_Q}{\partial V_1^2 \partial V_2} v_1^2 v_2 + \frac{\partial^3 f_Q}{\partial V_2^3} v_2^3\right) + \ldots \end{aligned} \tag{2.2.21}$$

where the partial derivatives are evaluated at the dc bias voltages $V_{1,0}$ and $V_{2,0}$. The current is obtained by taking the derivative with respect to time:

$$\begin{aligned} i = \frac{dq}{dt} = {} & \left(\frac{\partial f_Q}{\partial V_1} + \frac{\partial^2 f_Q}{\partial V_1^2} v_1 + \frac{\partial^2 f_Q}{\partial V_1 \partial V_2} v_2 + \frac{1}{2}\frac{\partial^3 f_Q}{\partial V_1^3} v_1^2 \right. \\ & \left. + \frac{1}{2}\frac{\partial^3 f_Q}{\partial V_1 \partial V_2^2} v_2^2 + \frac{\partial^3 f_Q}{\partial V_1^2 \partial V_2} v_1 v_2 + \ldots\right)\frac{dv_1}{dt} \\ & + \left(\frac{\partial f_Q}{\partial V_2} + \frac{\partial^2 f_Q}{\partial V_2^2} v_2 + \frac{\partial^2 f_Q}{\partial V_1 \partial V_2} v_1 + \frac{1}{2}\frac{\partial^3 f_Q}{\partial V_2^3} v_2^2 \right. \\ & \left. + \frac{1}{2}\frac{\partial^3 f_Q}{\partial V_1^2 \partial V_2} v_1^2 + \frac{\partial^3 f_Q}{\partial V_1 \partial V_2^2} v_1 v_2 + \ldots\right)\frac{dv_2}{dt} \end{aligned} \tag{2.2.22}$$

Fortunately, the dependence of Q on one voltage is often less strong than on the other, and, in those cases, (2.2.22) can be simplified considerably. However, before deleting terms wildly, one should be careful not to dispose of the baby with the bathwater: The terms in (2.2.22) generate different frequency components under sinusoidal steady-state conditions, so deleting certain terms, even if they are very small, may delete the intermodulation component of interest. One advantage to working in the frequency domain with capacitive nonlinearities is that much of the complexity evident in (2.2.21) and (2.2.22) is circumvented: In Sections 3.1 and 3.2, it will be shown that the process of taking the derivative in the time domain can be performed in the frequency domain merely by multiplying a matrix by a diagonal matrix.

The lowest-degree terms in (2.2.22) give the small-signal, quasilinear capacitance:

$$i = \frac{\partial f_Q}{\partial V_1} \left. \frac{\mathrm{d}v_1}{\mathrm{d}t} \right|_{\substack{V_1 = V_{1,0} \\ V_2 = V_{2,0}}} + \frac{\partial f_Q}{\partial V_2} \left. \frac{\mathrm{d}v_2}{\mathrm{d}t} \right|_{\substack{V_1 = V_{1,0} \\ V_2 = V_{2,0}}} = C_1(V_{1,0}, V_{2,0}) \frac{\mathrm{d}v_1}{\mathrm{d}t} + C_2(V_{1,0}, V_{2,0}) \frac{\mathrm{d}v_2}{\mathrm{d}t} \tag{2.2.23}$$

In both the capacitance and conductance cases, we assumed that the element was biased at some dc voltage and that a much smaller ac voltage was superimposed. This situation is common in many nonlinear microwave problems, for example, in calculating intermodulation distortion in small-signal amplifiers. However, in many circuits, a large ac signal may also exist, for example, the LO waveform in a mixer or a saturating signal in a small-signal amplifier. If the nonlinearity is strong or if the signal is very large (or both, as in a diode mixer) a large number of terms must be used in the series expansion to give adequate computational accuracy. Carrying expressions such as (2.2.22) to a high number of terms is difficult enough, but as we shall see in Chapter 4, the task of analyzing even a relatively simple nonlinear circuit by means of such a long series is nearly impossible.

It is possible, however, to circumvent these difficulties by expanding the Q/V or I/V characteristic in a Taylor series and using the large ac voltage (plus any dc bias voltage, of course) as the central "point." This expansion allows the small-signal voltage at any instant to be treated as a small deviation from the central value, so the minimum number of Taylor series terms can be used. The trade-off in this approach is that the Taylor series "coefficients" are time-varying and, thus, must be differentiated along with the small-signal voltage in such expressions as (2.2.15). This approach, which is examined further in Section 3.2.3, allows one to make an accurate and tractable analysis of such phenomena as intermodulation in mixers, which appears at first to be an extraordinarily difficult problem.

2.2.4 Relationship between I/V, Q/V and G/V, C/V Expansions

The series expansions developed in the previous section, describing the incremental conductances and capacitances, were derived from static I/V and Q/V characteristics. Sometimes, however, it is more convenient to begin with incremental C/V or G/V data, that is, the linear capacitance or conductance as a function of bias voltage. This situation arises often in the modeling of solid-state devices, in which C/V or G/V characteristics are often easier to measure. In this case, we must find the Taylor-series expansions of the I/V or Q/V characteristics from a series expansion of the C/V or G/V characteristic.

The Taylor-series expansion of the characteristic $I = f(V)$, from (2.2.4), is

$$\begin{aligned} f(V_0 + v) &= f(V_0) + \left.\frac{df(V)}{dV}\right|_{V=V_0} v + \frac{1}{2}\left.\frac{d^2f(V)}{dV^2}\right|_{V=V_0} v^2 \\ &\quad + \frac{1}{6}\left.\frac{d^3f(V)}{dV^3}\right|_{V=V_0} v^3 + \ldots \\ &= f(V_0) + g_1 v + g_2 v^2 + g_3 v^3 + \ldots \end{aligned} \tag{2.2.24}$$

and the expansion of $G(V)$ is

$$\begin{aligned} G(V_0 + v) &= G(V_0) + \left.\frac{dG(V)}{dV}\right|_{V=V_0} v \\ &\quad + \frac{1}{2}\left.\frac{d^2G(V)}{dV^2}\right|_{V=V_0} v^2 \\ &\quad + \frac{1}{6}\left.\frac{d^3G(V)}{dV^3}\right|_{V=V_0} v^3 + \ldots \\ &= \zeta_0 + \zeta_1 v + \zeta_2 v^2 + \zeta_3 v^3 + \ldots \end{aligned} \tag{2.2.25}$$

We note that

$$G(V) = \frac{df(V)}{dV} \tag{2.2.26}$$

and after substituting (2.2.26) into (2.2.25) and comparing the result to (2.2.24), we see immediately that

$$
\begin{aligned}
g_1 &= \zeta_0 \\
g_2 &= \zeta_1/2 \\
g_3 &= \zeta_2/3 \\
&\ \ \vdots \\
g_n &= \zeta_{n-1}/n
\end{aligned}
\tag{2.2.27}
$$

We can do the same with the Taylor-series expansion of the Q/V characteristic (2.2.15) and the expansion of the C/V characteristic. The Q/V expansion has the form:

$$Q(V_0 + v) = f_Q(V_0) + C_1 v + C_2 v^2 + C_3 v^3 + \ldots \tag{2.2.28}$$

and the C/V characteristic has the expansion:

$$C(V_0 + v) = \gamma_0 + \gamma_1 v + \gamma_2 v^2 + \gamma_3 v^3 + \ldots \tag{2.2.29}$$

Comparing their Taylor-series terms as in (2.2.24) through (2.2.26) gives the identical result:

$$
\begin{aligned}
C_1 &= \gamma_0 \\
C_2 &= \gamma_1/2 \\
C_3 &= \gamma_2/3 \\
&\ \ \vdots \\
C_n &= \gamma_{n-1}/n
\end{aligned}
\tag{2.2.30}
$$

2.3 SCHOTTKY-BARRIER AND JUNCTION DIODES

Virtually all microwave mixer diodes and many varactors use Schottky junctions instead of *pn* junctions or point contacts. Although they are often used to realize varactors, *pn* junction diodes are no longer used in microwave circuits as resistive diodes. A Schottky-barrier diode consists of a metal contact deposited on a semiconductor; such contacts can be made with far better uniformity than point contacts, and they do not have the recombination-time limitations of *pn* junctions. Inexpensive silicon Schottky-barrier diodes are capable of good performance as mixers at frequencies well into the millimeter-wave region. Gallium arsenide diodes, which are somewhat more expensive, can be used to realize mixers at terahertz (1000 GHz) frequencies. Gallium arsenide Schottky-barrier varactors, which generally have higher Q-factors than silicon varactors, are commonly used to realize efficient millimeter-wave frequency multipliers.

The Schottky-barrier diode is perhaps the simplest modern solid-state device in existence and the easiest to characterize accurately. The junction I/V and capacitance characteristics can both be expressed by simple closed-form equations that are highly accurate; there is little need to make a trade-off in the diode model between accuracy and simplicity. Furthermore, the diode model that is developed in this section is accurate to frequencies of at least a few hundred GHz and, with minor modifications, to even higher frequencies. As a result, circuit modeling of mixers and frequency multipliers, including noise, intermodulation, and conversion efficiency, has been highly successful and can now be considered a mature practice.

2.3.1 The Schottky-Barrier Diode Model

Figure 2.6 shows the general structure of a Schottky-barrier diode; most Schottky devices are similar. The diode is fabricated on a high-conductivity n-type (n^+) substrate; because the electron mobilities of all practical n dopants are much higher than those of p materials, n materials are used almost exclusively in microwave Schottky devices. A very pure, high-conductivity n^+ buffer layer is grown on top of the substrate to assure low series resistance and to prevent impurities in the substrate from diffusing into the epitaxial layer during processing. The buffer is usually a few microns thick, and the buffer and substrate are doped as heavily as possible, usually approximately 10^{18} atoms/cm^3 for GaAs, somewhat higher for silicon. An n epitaxial layer (sometimes called the *epilayer* or, simply, the *epi*) is grown on top of the buffer. In GaAs mixer diodes, the epilayer is doped to $1 \cdot 10^{17}$ to $2 \cdot 10^{17}$ cm^{-3} and is usually 1000–1500 Å thick.

The contact of the metal anode to the epitaxial layer forms the rectifying junction. Platinum and titanium are the most commonly used anode material for fabricating GaAs diodes; gold and aluminum have also been used, although these materials have poor reliability. A gold layer is usually plated onto the metal anode to prevent corrosion and to facilitate a bond wire, ribbon, or whisker connection. The anode metal rarely covers the entire top surface of the chip; the size and shape of the anode are selected to give the appropriate combination of junction capacitance and series resistance for the intended application. The circular anodes of microwave diodes vary in diameter from 1.5 microns for millimeter-wave diodes to 10–20 microns for microwave diodes. For practical reasons, in many diodes

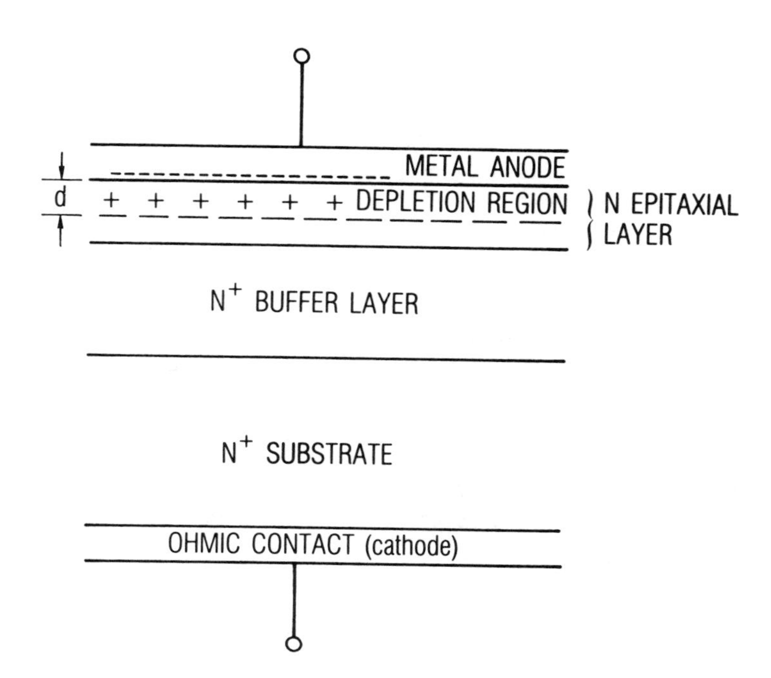

Figure 2.6 Schottky-barrier diode cross section.

a large number of anodes are defined on the top surface of a single chip and are isolated from each other by an oxide (SiO_2) layer. An ohmic contact must be made to the substrate; alloyed gold–germanium is commonly used on GaAs. The ohmic contact is usually formed on the bottom of the substrate, but it can be formed on the top of the diode (e.g., for beam-lead devices) if appropriate means are used to isolate the anode from the cathode and to minimize the parasitic capacitance that arises from their proximity.

The physics of conduction and capacitance in Schottky barriers will not be covered here; the interested reader should consult References 2.1 through 2.3. For present purposes it is enough to point out that the contact of the metal to the semiconductor allows some of the free electrons in the semiconductor to collect on the surface of the metal. A region of the semiconductor (imaginatively called the *depletion region*) thus is depleted of electrons and contains only positively charged donor ions. Because of these ions, an electric field, which opposes further movement of electrons, is set up between the anode and the semiconductor and a state of equilibrium is reached. Also because of this electric field, a potential difference, called the *diffusion potential* or *built-in voltage*, exists between the neutral semiconductor and the anode.

The width of the depletion region can be found from the doping density and material parameters of the semiconductor. The depletion width d of an ideal junction having uniform epitaxial doping is

$$d = \left(\frac{2\phi\epsilon_s}{qN_d}\right)^{1/2} \tag{2.3.1}$$

where the diffusion potential ϕ is the difference between the metal and semiconductor work functions; N_d is the doping density; ϵ_s is the electric permittivity of the semiconductor; and q is the electron charge, $1.6 \cdot 10^{-19}$ coulomb.

If a dc voltage is applied so that it reverse-biases the junction (i.e., the negative pole is connected to the anode), more negative charge collects on the anode and the depletion region widens. The width of the biased depletion region is

$$d = \left(\frac{2(\phi - V)\epsilon_s}{qN_d}\right)^{1/2} \tag{2.3.2}$$

where V is the applied voltage, which is defined as a negative quantity under reverse bias. As the reverse bias is increased, the depletion region becomes wider and more electrons move to the anode, leaving behind more positive charge in the form of ionized donor atoms. Similarly, if the diode is forward-biased, the depletion region narrows and less charge is stored. Thus, a negative voltage stores more negative charge on the anode, and a positive voltage reduces it. The junction, therefore, operates as a nonlinear capacitor.

As the forward bias is increased, the electric field in the junction becomes weaker and presents less of a barrier to electrons. Therefore, more electrons have sufficient thermal energy to cross the barrier, and forward conduction occurs. The current is proportional to the number of electrons having energy greater than the barrier energy; that number is an exponential function of barrier height. Thus, the I/V characteristic is an exponential function, one of the strongest nonlinear functions found in solid-state devices. Because conduction occurs almost entirely as the result of thermal emission of electrons (majority carriers) over a barrier, the Schottky-barrier diode is often called a *majority carrier device*. This mode of operation contrasts with that of a *pn* junction, wherein conduction is controlled by minority carriers and switching time is limited by minority carrier lifetimes.

In most conventional Schottky diodes, the epitaxial layer is never fully depleted of its charge in normal operation, even at the highest reverse voltages. Consequently, there is always some undepleted epitaxial material between the depletion region and the buffer layer, especially under forward bias, when the depletion region is narrow. Because this material has a relatively high resistivity, especially compared to the substrate, it represents a parasitic resistance in series with the diode junction. In mixers and frequency multipliers, it represents a potentially serious loss mechanism. Figure 2.7 shows the equivalent circuit of the Schottky-barrier diode. The diode consists of three elements, two of which are nonlinear, the junction capacitance and conductance. The third element, the parasitic series resistance, R_s, is also nonlinear, but because it varies only slightly under forward bias, it is usually successfully treated as a linear fixed-value resistance. The series resistance of a varactor diode, which is operated with reverse bias and rarely experiences forward conduction, varies somewhat more with junction voltage. However, making the assumption that the series resistance is nonlinear adds little accuracy to the analysis of most varactor frequency multipliers. Figure 2.7 therefore is adequate for modeling the intrinsic junction of a Schottky diode, as long as the $C(V)$ and $I(V)$ characteristics are specified appropriately. Package parasitics and additional parasitics, for example, beam-lead overlay capacitance, may also be included in the model if appropriate.

If the epitaxial region is uniformly doped, the junction charge function is

$$Q(V) = -2C_{j0}\phi(1 - V/\phi)^{1/2} \tag{2.3.3}$$

and the small-signal junction capacitance is

$$C(V) = \frac{dQ(V)}{dV} = \frac{C_{j0}}{\left(1 - \dfrac{V}{\phi}\right)^{1/2}} \tag{2.3.4}$$

where ϕ is the diffusion potential and C_{j0} is the zero-bias junction capacitance. V is the junction voltage shown in Figure 2.7, that is, excluding the voltage that is dropped across the series resistance. It is defined as positive if the junction is forward-biased. Equations (2.3.3) and (2.3.4) are strictly valid only if the epilayer doping is uniform. If the epilayer doping is not uniform, (2.3.4) often remains valid as long as the nonuniformity is not strong and the exponent in the denominator is changed. The general expression for diode junction capacitance is

$$C(V) = \frac{dQ(V)}{dV} = \frac{C_{j0}}{\left(1 - \frac{V}{\phi}\right)^{\gamma}} \tag{2.3.5}$$

where γ is chosen so that (2.3.5) has the best possible fit with measured junction capacitance; γ is usually close to 1/2. For example, a linearly graded junction, wherein the epitaxial-layer doping increases linearly with distance from the junction, has $\gamma = 1/3$. It is interesting to note that the reverse-biased junction has the same capacitance as a parallel-plate capacitor whose plate spacing equals the depletion width and whose dielectric constant equals that of the semiconductor.

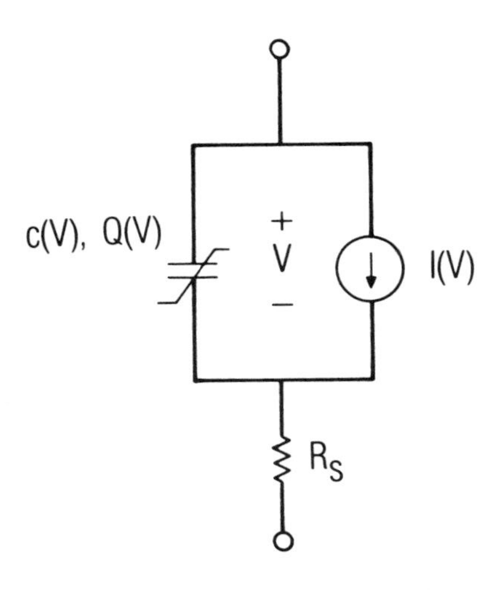

Figure 2.7 Equivalent circuit of the Schottky-barrier diode.

If the doping is very strongly nonuniform, (2.3.5) may not describe the capacitance adequately over a wide reverse-voltage range. In this case, (2.3.5) may be used piecewise, with different parameters for different voltage ranges, or an empirical expression, for example, a polynomial, may be used. Occasionally, a diode is purposely designed to have a strongly graded or otherwise modified doping profile, in order to either maximize or minimize its capacitive nonlinearity. One such diode is the *Mott* diode, which has a very thin, lightly doped epitaxial layer that is fully depleted at zero bias. The Mott diode exhibits a relatively weak capacitance variation with voltage; it is used to achieve good conversion loss and noise temperature in mixers at low LO levels, at low temperatures, or both. At the opposite extreme are varactor diodes, in which the capacitive nonlinearity is made as strong as possible, so they can be used as voltage-controlled tuning elements or as efficient frequency multipliers. One of the most extreme cases is that of the *hyperabrupt varactor*, in which the doping concentration actually decreases with distance from the junction. Hyperabrupt varactors can have $\gamma = 1.5$ or even $\gamma = 2.0$ over at least part of their reverse voltage ranges. These varactors usually have high series resistance, because the undepleted part of the epitaxial layer is very lightly doped, and are consequently unsuited for use in frequency multipliers. They are most useful in tuning applications, especially in voltage-controlled oscillators, where the strong, controlled nonlinearity can be used to achieve a wide and nearly linear frequency-voltage characteristic.

The *I/V* characteristic of a Schottky diode can be expressed by a simple relation, which is derived under the assumption that conduction occurs primarily via the thermionic emission of electrons over a barrier. Other mechanisms, such as tunneling, occur as well, but for most Schottky diodes of moderate doping densities, operated at close to room temperature, the thermionic-emission assumption is valid and agrees well with measurements. The *I/V* characteristic of a Schottky-barrier diode has the same general form as that of a *pn* junction diode:

$$I(V) = I_{sat}[\exp(qV/\eta KT) - 1] \tag{2.3.6}$$

Equation (2.3.6) is sometimes called the *law of the junction*. In (2.3.6), q is the electron charge; K is Boltzmann's constant, $1.37 \cdot 10^{-23}$J/K; and T is absolute temperature. The *ideality factor*, η, is used to account for unavoidable imperfections in the junction and for other secondary phenomena that thermionic emission theory cannot predict. Thus, η is always greater than 1.0 and, in a well made diode, η should be less than 1.20. I_{sat}, a proportionality constant, is called the *current parameter*, or, because (2.3.6) implies $I(V) = I_{sat}$ as $V \rightarrow -\infty$, the *reverse-saturation current*. An expression for I_{sat} is

$$I_{sat} = A^{**}T^2W_j \exp(q\phi_b/KT) \tag{2.3.7}$$

where A^{**} is the modified Richardson constant; W_j is the junction area; and ϕ_b is the barrier height in volts, a constant usually approximately 0.1 V greater than the diffusion potential. A^{**} is approximately 96 A cm^{-2} K^{-2} for silicon and 4.4 A cm^{-2} K^{-2} for GaAs. We should be careful about taking (2.3.7) too seriously; because of secondary effects such as charge generation and surface imperfections in the junction, I_{sat} can differ significantly from the value given by (2.3.7). Equation (2.3.7) can be used, however, to draw some general conclusions. For example, the value of the Richardson constant in GaAs is lower than in silicon, which implies that the knee of the *I/V* characteristic occurs at a higher voltage for GaAs diodes than for silicon diodes.

Figure 2.8(a) shows the *I/V* characteristic of a Schottky diode in Cartesian coordinates, and Figure 2.8(b) shows the same characteristic graphed on semilog axes. The semilog graph is a straight line having a slope of one decade of current per 58.3η mV of junction voltage change at low current levels. Imperfections in the diode design or fabrication can be identified readily by deviations from that straight line. For example, excessive tunneling current reduces the slope to nearly half the usual value, as does junction damage due to electrical overstress. The curve deviates from a straight line at the high current end because of the voltage dropped across the parasitic series resistance.

At high reverse bias, junction breakdown results from avalanching. Avalanche breakdown voltage increases as doping density is reduced, but series resistance also increases. Thus, there is a trade-off in diode design between low R_s and high reverse-breakdown voltage. GaAs diodes generally have greater reverse-breakdown voltages than silicon, partly because the higher electron mobility in GaAs allows lower series resistance to be achieved with lighter doping.

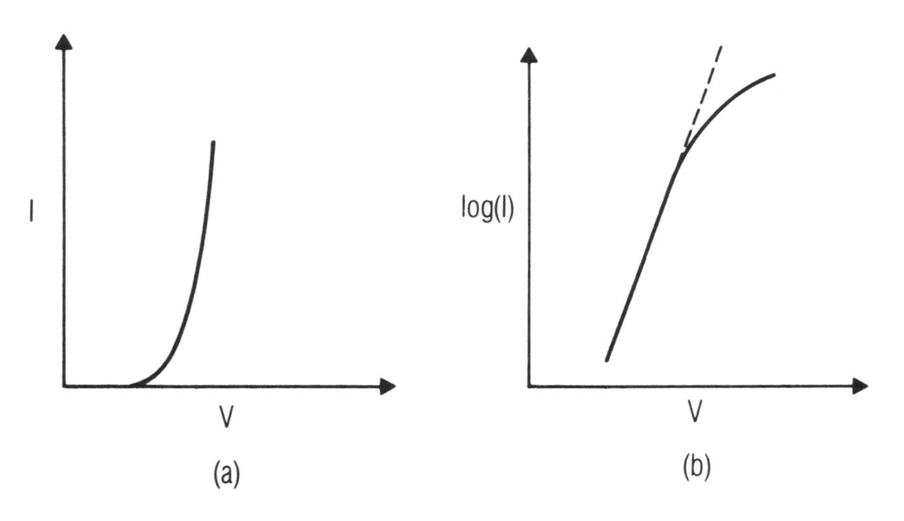

Figure 2.8 *I/V* characteristic of the Schottky-barrier diode: (a) in Cartesian coordinates; (b) in semilogarithmic coordinates.

2.3.2 Mixer Diodes

Because it almost completely lacks minority-carrier effects, the Shottky-barrier diode is a very fast-switching device. As such, it is ideal for use as a diode mixer, which is often idealized as a high-frequency switch. Mixer diodes are almost exclusively Schottky-barrier devices, and competing device structures of even a few years ago, such as point-contact diodes, now are considered obsolete. Very high-quality silicon Schottky diodes are available at low cost, and for applications requiring the best possible conversion loss and noise figure, GaAs diodes can be obtained at modest expense. Diode technology today is sufficiently mature to allow mixers at frequencies above 1000 GHz to be fabricated.

Figure 2.9 shows the cross section of a dot-matrix mixer diode. The vertical structure of the diode is identical to that shown in Figure 2.6, but the area of the junction is defined precisely by the anode. The anode is formed as a circular dot; many such dots, in some cases thousands, are arranged in a regular pattern over the top surface of the chip. The large number of dots facilitates the connection of a sharply pointed wire, or *whisker*, to one of the anodes, which are often so small that they cannot be seen individually.

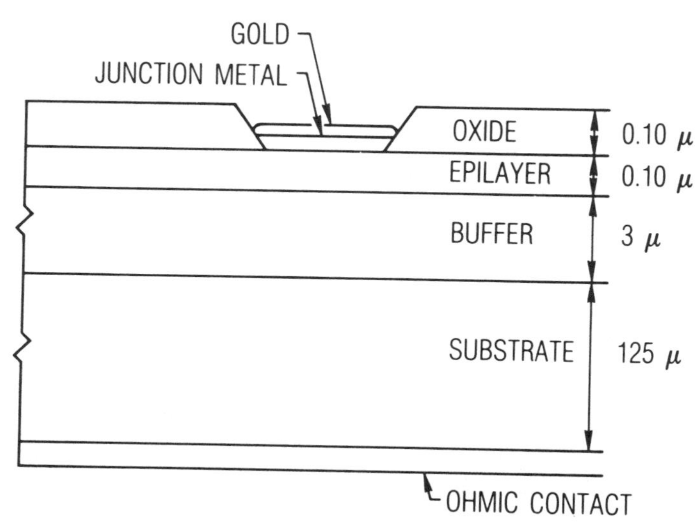

Figure 2.9 Cross section of a dot-matrix diode, with dimensions typical for mixer applications.

In operation, the mixer diode is used as a variable-resistance diode (thus, the infrequently used name *varistor*, a contraction of the term *variable resistor*) or as a switch, which in many respects is the same thing. The incremental small-signal conductance of the junction can be found by differentiating (2.3.6):

$$g(V) = \frac{dI(V)}{dV} = \frac{q}{\eta KT} I_{sat} \exp(qV/\eta KT) \tag{2.3.8}$$

or

$$g(V) = \frac{q}{\eta KT} I(V) \tag{2.3.9}$$

The quantity 1.0 subtracted in (2.3.6) has been neglected because it is much smaller than the exponential term. The result is that the small-signal junction conductance is equal to a constant times the large-signal junction current.

Virtually all high-frequency Schottky-barrier mixer diodes are uniformly doped, so the expression for junction capacitance in (2.3.5) is usually valid. However, it is often not valid to assume that the dc series resistance, which can be found from Figure 2.8(b), represents the series resistance at microwave or millimeter-wave frequencies. The high-frequency series resistance is significantly higher than the dc value because of skin effect: although the dc current is distributed throughout the substrate, the high-frequency current exists as a thin sheet at the surface of the chip and is nearly zero in the bulk substrate. The increased path length and reduced cross-sectional area of the high-frequency current "sheet" increase the series resistance of the diode.

The high-frequency series resistance depends on both the geometry and the material parameters of the chip; it can be estimated by assuming that the chip is cylindrical, and that its anode is in the center. Figure 2.10 shows the chip and the current path, which can be separated into three parts: the epitaxial region under the depletion region; the spreading current along the top of the substrate, under the epi; and finally the current in the chip's sidewall. The sum of these three components, designated R_{s1}, R_{s2}, and R_{s3}, respectively, is the series resistance R_s. The resistance components are

$$R_{s1} = \frac{4(t - d)}{\pi \sigma_e a^2} \tag{2.3.10}$$

$$R_{s2} = \frac{\ln(b/a)}{2\pi\delta\sigma_s} \tag{2.3.11}$$

$$R_{s3} = \frac{h}{\pi\delta b\sigma_s} \tag{2.3.12}$$

The variables in (2.3.10) through (2.3.12) correspond to those in Figure 2.10. The terms σ_e and σ_s are the conductivities of the epitaxial region and the substrate, respectively, and δ is the skin depth in the substrate. At dc, R_{s2} and R_{s3} are replaced by the bulk substrate resistance, which is usually negligible.

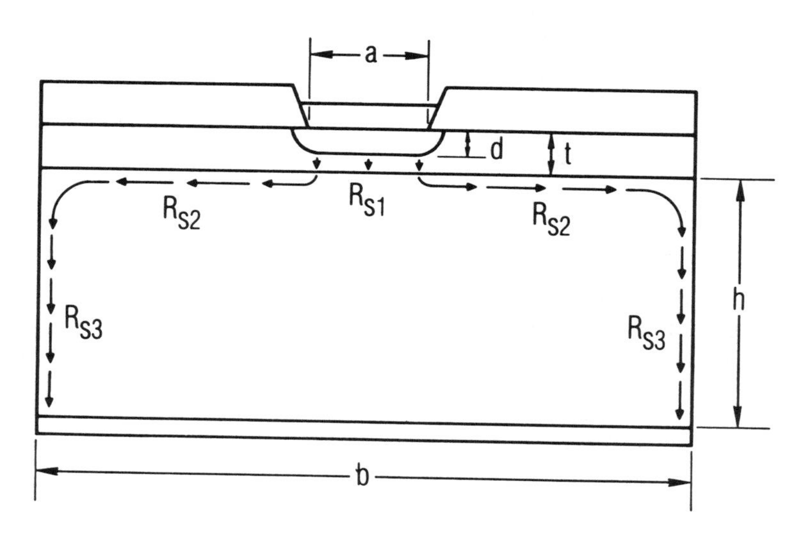

Figure 2.10 High-frequency components of the series resistance: the epitaxial layer resistance R_{s1}, spreading resistance R_{s2}, and sidewall resistance R_{s3}.

Another type of diode used for mixers is the *Mott* diode. The Mott diode is a variant of the conventional Schottky-diode structure and, as such, is qualitatively described by Figure 2.9. It differs in its epitaxial layer, which is very thin and lightly doped; as a result, the epilayer is fully depleted, even under relatively high forward bias, and the depletion region extends into the buffer layer. The junction capacitance then consists of two components that can be represented by two capacitors in series: the capacitance of the depleted epilayer, which is very small and fixed; and the larger capacitance, arising from the part of the depletion region that extends into the buffer layer. This latter component is voltage-variable, but because it is in series with a much smaller capacitor, the smaller, fixed capacitor dominates. The capacitance of the Mott diode, therefore, is only very weakly dependent on voltage. In the chapter on mixer design, this characteristic will be shown to minimize mixer LO power requirements; furthermore, experimental results indicate, although for unclear reasons, that cooled Mott diode mixers achieve very low noise temperatures.

It is probably best to measure rather than calculate the Mott diode's capacitance and to fit an empirical expression to the measured data. The capacitance of the Mott diode can also be modeled by splitting the junction capacitance $C(V)$ in Figure 2.7 into two series components, a fixed one and a variable one, where

$$C_1(V) = C_{1,0} \tag{2.3.13}$$

$$C_2(V_2) = \frac{C_{j0}}{\left(1 - \dfrac{V_2}{\phi}\right)^{1/2}} \tag{2.3.14}$$

and V_2 is the voltage drop across the variable part of the capacitance. The resulting model is shown in Figure 2.11. $I(V)$ is described by (2.3.6), but because of the Mott diode's lower doping density, its current parameter I_{sat} is lower than that of a conventional Schottky having the same diameter.

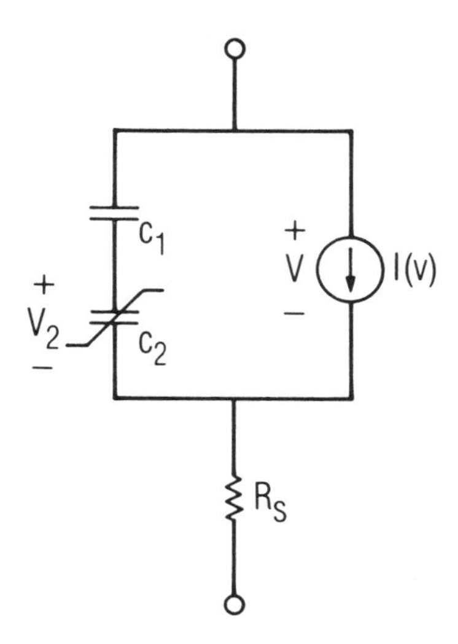

Figure 2.11 Mott diode equivalent circuit. C_1 is the epilayer capacitance and C_2 is the substrate depletion capacitance.

A figure of merit for the mixer diode, one that is applicable to both Mott and conventional Schottky diodes, is the *cutoff frequency*, f_c. The cutoff frequency is traditionally calculated from dc quantities (thus, the common misnomer *dc cutoff frequency*), without regard to skin-effect enhancement of the series resistance. The cutoff frequency is defined as

$$f_c = \frac{1}{2\pi R_s C_{j0}} \tag{2.3.15}$$

Equation (2.3.15) is strictly valid only as a figure of merit; however, diode cutoff frequency is often used as a mixer design parameter. The cutoff frequency given by (2.3.15) is valid for design purposes if the junction capacitance is calculated at the operating bias voltage and the series resistance includes skin effect.

2.3.3 Schottky-Barrier Varactors

Frequency-multiplier varactors are often realized as p^+ structures on n substrates in both GaAs and silicon. Because the diodes' p^+ regions are difficult to fabricate uniformly in the small anode sizes necessary for very high-frequency operation, these diodes are limited to the lower microwave and millimeter-wave regions. Furthermore, at high frequencies, the importance of minimizing series resistance and maximizing capacitance swing becomes progressively greater, and the Schottky-barrier varactor is generally superior in these respects. Frequency multipliers having input frequencies above approximately 50 GHz usually employ Schottky-barrier varactors; often, such multipliers are used to generate output power at frequencies of several hundred GHz. High-performance Schottky-barrier varactors for such applications are invariably realized in GaAs.

The structure of the Schottky-barrier varactor is qualitatively the same as that of the mixer diode shown in Figure 2.9. In order to achieve both good efficiency and high output power, varactors require higher breakdown voltages than mixer diodes; accordingly, the doping density in a varactor's epilayer is very low (typically 10^{16} to 10^{17} atoms/cm^3) and its junction area is relatively large.

The large junction area provides a greater capacitance than would be tolerable in a mixer diode, usually approximately 0.1 pF for operation near 50 GHz. The large area also facilitates heat dissipation, an important consideration because most of the multiplier's input power is dissipated in the diode's series resistance. Because of the low doping level, the series resistance of the Schottky varactor is greater than that of a mixer diode of the same size and the cutoff frequency is significantly lower.

The equivalent circuit of the mixer diode shown in Figure 2.7 and described by (2.3.3) through (2.3.6) is generally valid for Schottky-barrier varactors, as long as the parameters, especially C_{j0} and γ, are appropriately modified. The diffusion potential ϕ is usually around 1 V, higher than that of a mixer diode, and because of second-order effects, γ is often somewhat lower (approximately 0.45). I_{sat} also differs; however, in normal operation the diode is usually not driven into forward conduction, so the forward *I/V* characteristic $I(V)$ is of secondary concern.

Several figures of merit can be defined for varactors. One is the *dynamic cutoff frequency*, f_{cd}:

$$f_{cd} = \frac{S_{max} - S_{min}}{2\pi R_s} \tag{2.3.16}$$

where S is elastance, or inverse capacitance. S_{min} is the minimum elastance, which occurs as the junction voltage approaches ϕ. S_{min} is very small, and is often neglected in (2.3.16), giving

$$f_{cd} \approx \frac{S_{max}}{2\pi R_s} \tag{2.3.17}$$

where S_{max} is the elastance at reverse breakdown. Thus, varactors are often specified according to their capacitances and cutoff frequencies at a high reverse voltage, rather than to C_{j0}, as are mixer diodes. Another figure of merit is the *dynamic Q*, Q_δ, defined in the same way as in a series RC network:

$$Q_\delta = \frac{S_{max}}{2\pi f_0 R_s} = \frac{f_{cd}}{f_0} \tag{2.3.18}$$

where f_0 is the frequency at which the Q is evaluated. Q_δ is often calculated by using the junction elastance at a reverse voltage other than breakdown as S_{max}; one standard value is -6 V. Infrequently, $S_{max} = 1/C_{j0}$. Clearly, we should always determine precisely how the dynamic Q of a particular varactor is defined.

Schottky-barrier varactors have very high dynamic Qs, allowing good efficiency to be achieved at high frequencies. However, Schottky varactors are limited in power-handling capability; they can be driven only to the point at which the junction begins to conduct. If the input level is increased beyond this point, efficiency suffers, and the output power saturates; this phenomenon will be illustrated in Chapter 7 by Example 1. In the lower microwave range, p^+n junction varactors largely circumvent this problem.

2.3.4 p^+n Junction Varactors

At the lower microwave frequencies, silicon or GaAs p^+n junction varactors are preferred. The dc *I*/*V* characteristic of a p^+n junction has the same general form as that of a Schottky barrier, (2.3.6), but p^+n diodes have greater capacitance variation and, thus, provide greater efficiency at high drive levels. These properties are the result of the long minority carrier lifetimes that obviate the use of *pn* junction diodes in mixers. When the junction is forward-biased during the positive part of a high-frequency RF cycle, the charge is injected into the junction region. Most of that charge (which consists of holes from the p^+ region injected into the *n* region) does not have time to recombine with electrons, so it is stored momentarily and removed when the RF current swings negative. Because almost all of the injected charge is stored and not conducted, the varactor behaves as a nonlinear capacitor even when forward-biased.

The amount of charge stored can be very great, so the forward-bias capacitance is substantial. When the varactor is driven so that the voltage peaks at ϕ and its reverse-breakdown voltage, the varactor is said to be *nominally driven*. If it is driven harder, the positive junction voltage is clamped at ϕ, but the capacitance swing is greater than in the nominally driven varactor; a varactor operated this way is said to be *overdriven*. This charge-storage phenomenon is also used in the step-recovery diode (also called the *SRD* or *snap* diode), described in the next subsection. The main functional difference between the varactor and step-recovery diode is that the SRD obtains its capacitance variation almost entirely by diffusion charge storage, while the p^+n varactor's operation depends much less on high diffusion capacitance than on a gradual capacitance variation over its entire forward and reverse-voltage range.

A disadvantage of the p^+n structure is that it requires a *p*-diffusion step in its fabrication. The diffusion process limits the minimum size of the p^+ region, so the minimum capacitance of the diode is limited as well. The p^+ region also has higher series resistance than the metal anode of a Schottky diode, so p^+n varactors have lower f_{cd} than Schottky varactors. These properties limit p^+n varactors to frequencies below approximately 50 GHz.

A cross section of a p^+n junction varactor is shown in Figure 2.12. The initial part of the varactor's fabrication is much like that of a mixer diode: an *n* epitaxial layer is grown on an n^+ substrate. A p^+ region is then diffused into the epitaxial layer, and ohmic contacts are formed on the p^+ and n^+ regions for the anode and cathode, respectively. An oxide-isolated anode like the structure used in Schottky-barrier diodes is not optimum for the p^+n varactor, because in oxide-isolated diodes the junction's electric field is stronger near the edge of the anode. The nonuniform

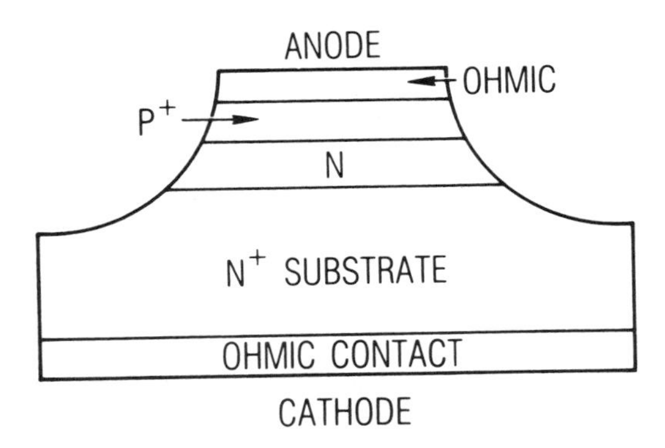

Figure 2.12 Mesa structure of the p^+n varactor.

electric field causes avalanche breakdown to occur near the anode's edge, at a relatively low voltage; much lower than if the field were uniform. In the p^+n varactor, the anode area is formed by etching a mesa, making the electric field more uniform over the junction, and thereby increasing the breakdown voltage. Diodes fabricated in this manner are called *diffused epitaxial varactors*.

A variant of the diffused epitaxial varactor is the *punch-through* varactor. The epitaxial layer of this device is so thin that it is completely depleted at modest reverse voltages, usually a little more than half the breakdown voltage. The varactor has the reverse-bias capacitance characteristic of (2.3.5) at low reverse voltages; at higher voltages, the epi is fully depleted, or *punched through*, and like the Mott diode, the C/V characteristic is nearly flat. The advantage of this structure is that the efficiency of a multiplier using a punch-through diode is less sensitive to changes in input power level than that of one using a conventional varactor. The disadvantage is that the diode's capacitance range is less, so its dynamic Q is lower and consequently the multiplier's efficiency may also be lower.

Most microwave-frequency p^+n varactors are realized in silicon. The minority carrier lifetime in silicon is greater than in GaAs, so for lower-frequency operation (i.e., at output frequencies below about 20 GHz), the charge-storage properties of silicon diodes are better than those of GaAs devices. At higher frequencies, GaAs has the advantage of lower series resistance and consequently higher dynamic Q. Both silicon and GaAs p^+n diodes have lower Qs than comparable Schottky diodes because of the additional series resistance of the p^+ region and its ohmic contact.

2.3.5 Step-Recovery Diodes

Like the varactor, a step-recovery diode, or *SRD*, (also called a *snap* diode) uses capacitance variation to generate harmonics of an applied signal. However, it does so by storing the charge under forward bias and switching very rapidly to a high-impedance state when the diode is discharged. The multiplier is adjusted so that the diode switches at the instant the reverse current is maximum, thus generating a large and very short-lived voltage pulse during each excitation cycle. The resulting pulse train is rich in harmonic content, so it need only be filtered to obtain harmonic output. The SRD multiplier is capable of very high-harmonic multiplication, at high power levels, with remarkably good efficiency and bandwidth. A common application of an SRD is to multiply an input frequency of a few hundred MHz to an output of several GHz.

The SRD must have high charge storage in the forward direction, low capacitance in the reverse direction, low series resistance, and for power applications, high reverse-breakdown voltage. Its switching time must also be adequately short because switching speed establishes its high-frequency limit of operation. To meet these requirements, the SRD must have a relatively long charge-storage time (long recombination time), and the charge that is injected into the junction while it is forward-biased must not travel so far that it cannot be removed during the reverse-bias interval. Finally, the depletion region must not be too wide, or transit-time effects will reduce the efficiency at high frequencies.

The SRD has the *pin* structure shown in Figure 2.13, in which the *i*-region is a layer of undoped or intrinsic semiconductor or is very lightly doped. The *i*-region is formed by the overlap of the *p* and *n* regions, both of which have a steep doping profile. This profile gives a narrow depletion region in the *p* and *n* regions and a strong built-in electric field, which opposes the diffusion of charge into the junction. During forward conduction, holes and electrons are injected into the *i*-region, where they recombine very slowly; the *i*-layer thus becomes a region in which charge is stored. When the SRD is reverse-biased, the *i*-layer is fully depleted; because of the wide depletion width, which includes the entire *i*-layer, reverse capacitance is very low.

The forward *I*/*V* characteristic of the SRD obeys (2.3.6) under dc bias, and the reverse-capacitance characteristic can usually be treated as a constant. Under forward bias, the diode can be modeled as a *pn* junction in parallel with the diffusion capacitance. The stored diffusion charge is

$$Q_s = \tau I \tag{2.3.19}$$

where τ is the recombination time, or minority carrier lifetime, of the material. Thus, unlike a conventional capacitor or the junction capacitance of a Schottky-barrier diode, the SRD's charge storage is a function of current, not of voltage.

The reverse-bias junction capacitance of the SRD is

$$C = \frac{\epsilon_s A}{d} \tag{2.3.20}$$

where A is the area of the junction and d is the depletion width. The largest part of d is the width of the *i*-region, which is large and constant. Consequently, when the SRD is reverse-biased, its capacitance is nearly constant and very low.

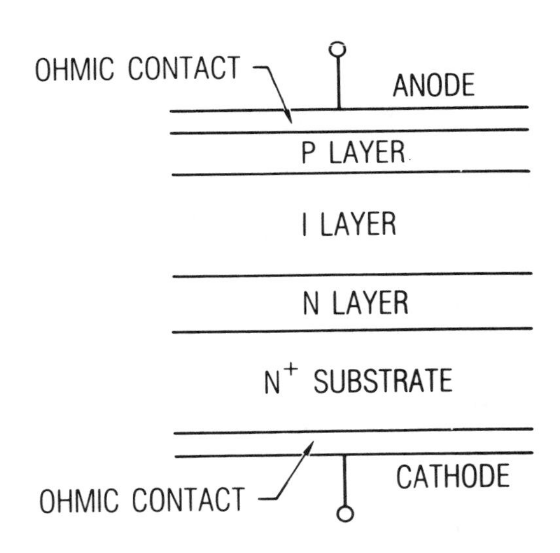

Figure 2.13 Step-recovery diode: a pin structure on an n^+ substrate.

In an SRD frequency multiplier, it is important that all the charge injected into the junction during the positive excursion of the excitation cycle be removed during the negative excursion. Recombination of charge during that time reduces efficiency, because the recombined charge is lost as conduction current. Minimizing charge recombination requires that the minority carrier lifetime τ be long compared to the period of an excitation cycle; because of its longer minority carrier lifetime, silicon is invariably used for SRDs instead of GaAs.

Like other diodes, the SRD has parasitic series resistance. This resistance arises in the ohmic contacts to the *p* and *n* regions, and in the undepleted parts of those regions. Because series resistance causes loss and reduces multiplier efficiency, it is as important to minimize series resistance in an SRD as in any other type of diode.

2.4 GaAs MESFETs

It is no overstatement to say that the GaAs MESFET and its variants, including the high electron-mobility transistor, or *HEMT*, have revolutionized low-noise microwave electronics and linear microwave systems. By contrast, the GaAs MESFET has been accepted more reluctantly in nonlinear circuit applications and an understanding of its quasilinear properties has evolved more slowly. The reasons for this slow acceptance are probably that the procedures for designing and analyzing such components as FET mixers and frequency multipliers are not well known and that the performance achieved in early experimental circuits was not very good.

MESFETs make excellent mixers, however, having low noise figures, broad bandwidths, and conversion gains; and they make frequency multipliers that exhibit high efficiency, gain, and output power. MESFETs are commonly used in quasilinear applications, especially as small-signal and power amplifiers, where an understanding of their nonlinearities is critical in minimizing the less attractive aspects of their performance: intermodulation, saturation, and the generation of spurious outputs.

2.4.1 MESFET Operation

Figure 2.14 shows a cross section of a GaAs MESFET. The MESFET is fabricated by first growing a very pure, semi-insulating buffer layer on a semi-insulating GaAs substrate, then growing an *n*-doped epitaxial layer that is used to realize the FET's active channel. Three connections are made to the channel: the source and drain ohmic contacts and, between them, the Schottky-barrier gate. The epilayer is made thicker than necessary for the channel and is etched to the correct channel thickness in the region where the gate is to be deposited. This *recessed gate* structure allows the layer of epitaxial material under the source and drain ohmic contacts to be quite thick, much thicker than the channel, thus minimizing the parasitic source and drain resistances. Reducing the source resistance

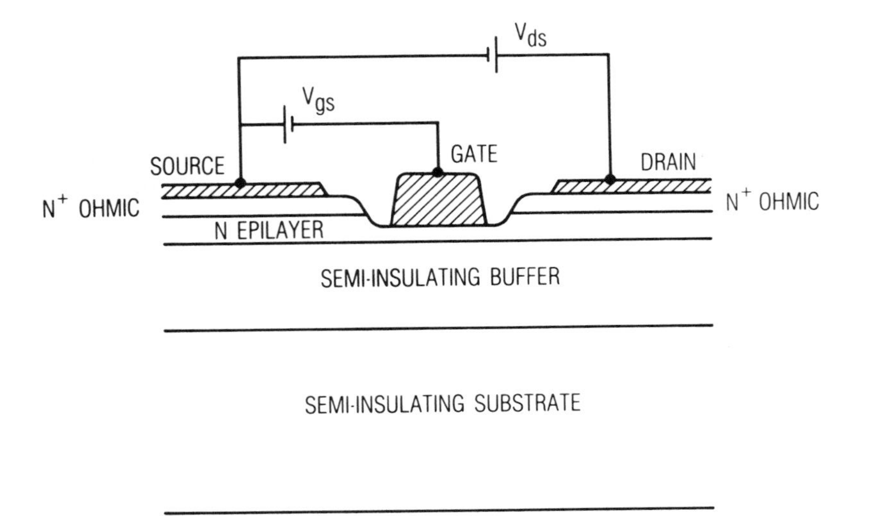

Figure 2.14 GaAs MESFET cross section.

is especially important for low-noise devices; it is also important for achieving good conversion efficiency in FET mixers and frequency multipliers.

The MESFET is biased by the two sources shown in Figure 2.14: V_{ds}, the drain-source voltage, and V_{gs}, the gate-source voltage. These voltages control the channel current by varying the width of the gate-depletion region and the longitudinal electric field. In order to develop a qualitative understanding of MESFET operation, imagine first that $V_{gs} = 0$ and V_{ds} is raised from zero to some low value, as shown in Figure 2.15(a). When $V_{gs} = 0$, the depletion region under the Schottky-barrier gate is relatively narrow, and as V_{ds} is raised, a longitudinal electric field and current are established in the channel. Because of V_{ds}, the voltage across the depletion region is greater at the drain end than at the source end, so the depletion region becomes wider at the drain end. The narrowing of the channel and the increased V_{ds} increase the electric field near the drain, causing the electrons to move faster; although the channel's conductive cross section is reduced, the net effect is increased current. When V_{ds} is low, the current is approximately proportional to V_{ds}. If, however, the gate reverse bias is increased while the drain bias is held constant, the depletion region widens and the conductive channel becomes narrower, reducing the current. When $V_{gs} = V_t$, the *turn-on voltage*, the channel is fully depleted and the drain current is zero, regardless of the value of V_{ds}. Thus, both V_{gs} and V_{ds} can be used to control the drain current. When the FET is operated in this manner (i.e., when both V_{gs} and V_{ds} have a

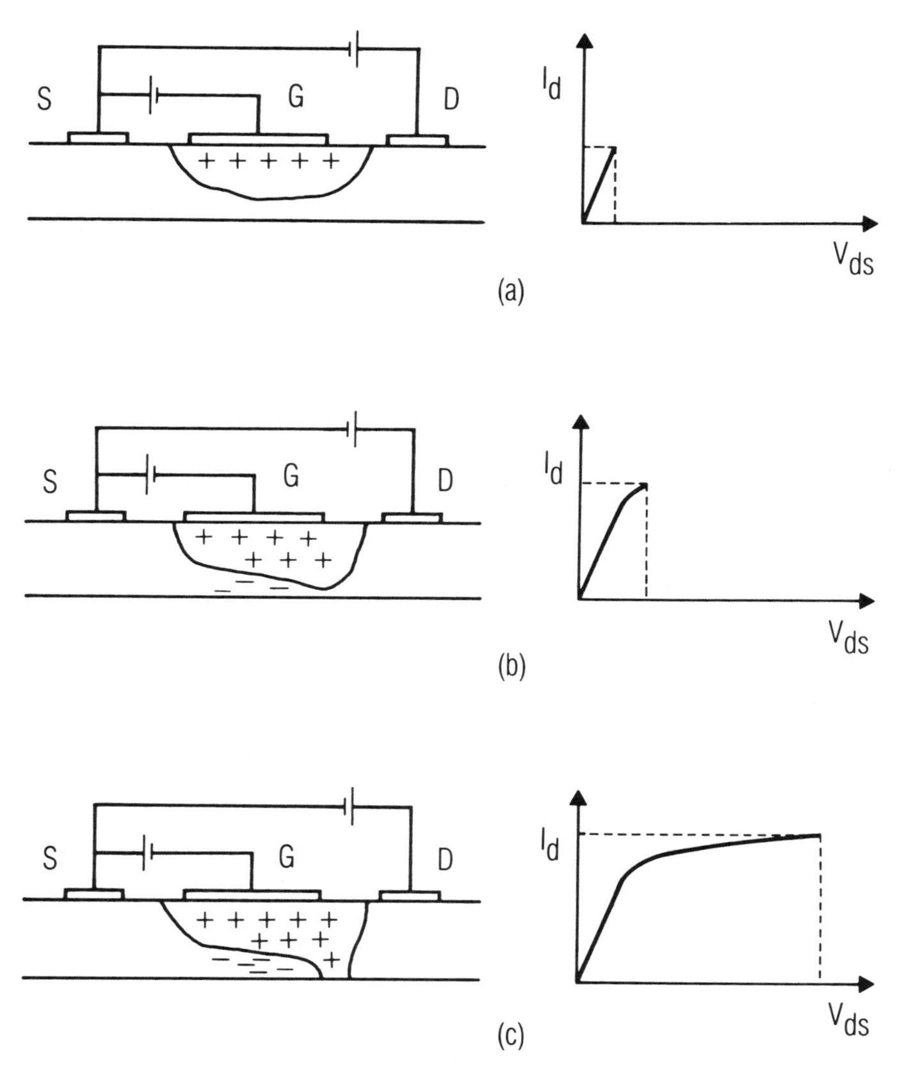

Figure 2.15 GaAs MESFET operation: (a) very low V_{ds} (i.e., a few tenths of a volt); (b) V_{ds} at the saturation point; (c) hard saturation.

strong effect on the drain current), it is said to be in its *linear*, or *voltage-controlled resistor* region.

If V_{ds} is raised further, as in Figure 2.15(b), the channel current increases, the depletion region becomes even wider at the drain end, and the conductive channel becomes narrower. The current clearly must be constant throughout the channel, so as the conductive channel near the drain becomes narrower, the electrons must move faster. However, the electron velocity cannot increase indefinitely; the average velocity of the electrons in GaAs can not exceed a velocity called their *saturated drift velocity*, approximately $1.3 \cdot 10^7$ cm/s. If V_{ds} is increased beyond the value that causes velocity saturation (usually only a few tenths of a volt), the electron concentration rather than velocity must increase in order to maintain current continuity. Accordingly, a region of electron accumulation forms near the end of the gate. Conversely, after the electrons transit the channel and move at saturated velocity into the wide area between the gate and drain, an electron depletion region is formed. That depletion region is positively charged because of the positive donor ions remaining in the crystal. As V_{ds} is increased further (Figure 2.15(c)), progressively more of the voltage increase is dropped across this region, called a *dipole layer, charge domain*, or sometimes a *stationary Gunn domain*, and less is dropped across the unsaturated part of the channel. Eventually, a point is reached where further increases in V_{ds} are dropped entirely across the charge domain and do not substantially increase the drain current; at this point, the electrons move at saturated drift velocity over a large part of the channel length. When the FET is operated in this manner, which is the normal mode of operation for small-signal devices, it is said to be in its *saturation region*, or in *saturated operation*. All FET amplifiers and most FET mixers and frequency multipliers are biased into saturation.

It is important to note that the charge domain begins to form at voltages well below those corresponding to the horizontal portion of the drain *I/V* characteristic. Charge accumulation occurs in silicon MESFETs as well as in GaAs devices, but because silicon's low-field mobility is much lower, the domain's strength (indicated by the magnitude of its voltage drop or the quantity of its stored charge) is much less. Charge-domain formation therefore can be—and usually is—ignored in silicon FETs. In contrast to a silicon FET model, an effective physical model of the GaAs MESFET must account in some way for the charge domain, even at relatively low voltages.

The terms *linear region* and *saturation region* are unfortunately chosen, because they seem to indicate exactly the opposite of their true meaning: a small-signal quasilinear operation takes place in the FET's saturation region, not in its linear region. Further confusion arises because the same terms are used, with opposite meaning, to describe the operating regions of bipolar transistors; that is, the bipolar transistor is said to be in saturation when the collector-emitter voltage is very low. This terminology is widely accepted, however, so little can be done to change it.

The Schottky barrier formed by the gate contact to the channel has a capacitance associated with it; this capacitance exists for the same reasons as that of the Schottky-barrier diode and has similar voltage dependence. At $V_{ds} \approx 0$ the gate-channel capacitance is distributed along the channel, but it can be modeled as two approximately equal capacitors, one between the gate and source, and the other between the gate and drain. These capacitances are related to the change in gate-depletion charge with changes in gate-source voltage V_{gs} and gate-drain voltage V_{gd}, respectively. As V_{ds} is increased and the FET begins saturated operation, however, drain-voltage changes are shielded from the gate depletion region by the dipole layer. Further changes in V_{ds} no longer increase the charge in the depletion region, so the gate-drain capacitance drops to a point where it consists only of stray capacitance between metalizations. In saturation, the gate-source capacitance represents the full gate-depletion capacitance, so the gate-source capacitance rises to approximately twice the value it had in linear operation.

2.4.2 MESFET Modeling

Figure 2.16 shows a lumped-element model of the MESFET that can be used either in a small-signal or a large-signal analysis. R_g is the ohmic resistance of the gate, and R_s and R_d are the source and drain ohmic contact resistances, respectively. R_i is the resistance of the semiconductor region under the gate, between the source and channel. C_{ds} is the drain-source capacitance, which is dominated by metalization capacitance and is therefore often treated as a constant. C_g and C_d are the gate-channel capacitances, which in general are nonlinear, and I_d is the controlled drain current source. If voltages are expected to be great enough to forward-bias or reverse-avalanche the gate, one can include diodes in parallel with C_g and C_d. Such diodes are of limited practical value, however, because operation with gate-channel avalanche breakdown or high values of rectified gate current usually destroys the device. I_d, C_g, and C_d are functions of the gate voltage V_g and the drain voltage V_d. These are called the *internal* gate and drain voltages, to distinguish them from V_{gs} and V_{ds}, the *external* voltages.

Characterizing the linear elements in Figure 2.16 is straightforward; of primary concern is the characterization of the nonlinear capacitances and current source. C_g and C_d account for the displacement current through the gate depletion region; they are large-signal capacitances. These capacitances are logically functions of V_g and the voltage across C_d, which is approximately $V_d - V_g$; however, for simplicity we would like them to have the same control voltages as I_d. Thus,

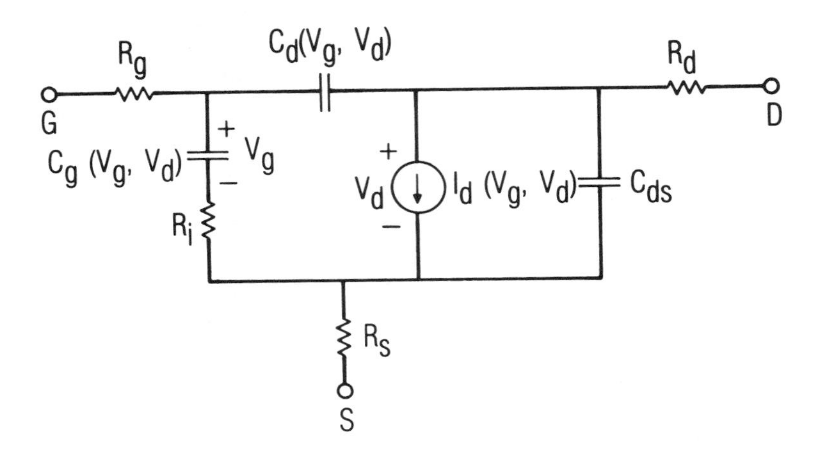

Figure 2.16 GaAs MESFET nonlinear equivalent circuit.

$$C_g(V_g, V_d) = \left.\frac{\partial Q_d}{\partial V_g}\right|_{V_d - V_g = \text{constant}} \tag{2.4.1}$$

$$C_d(V_g, V_d) = \left.\frac{\partial Q_d}{\partial V_d}\right|_{V_g = \text{constant}} \tag{2.4.2}$$

where Q_d is the charge in the gate depletion region. Unlike the small-signal incremental characterization of (2.2.21), this formulation is valid for large-signal operation; it gives the current in C_g and C_d, respectively, as

$$I_{Cg} = \frac{\partial Q_d}{\partial V_g}\frac{\mathrm{d}V_g}{\mathrm{d}t} = C_g(V_g, V_d)\frac{\mathrm{d}V_g}{\mathrm{d}t} \tag{2.4.3}$$

$$I_{Cd} = \frac{\partial Q_d}{\partial V_d}\frac{\mathrm{d}V_f}{\mathrm{d}t} = C_d(V_g, V_d)\frac{\mathrm{d}V_f}{\mathrm{d}t} \tag{2.4.4}$$

where V_f is the voltage across C_d.

Examples of the voltage dependence of C_g and C_d are shown in Figure 2.17. If the FET is in saturation, C_g can usually be modeled with good accuracy as a Schottky-barrier capacitance; however, the *C*/*V* characteristic may deviate significantly from (2.3.5) when the reverse gate voltage exceeds V_t and the channel is fully depleted. At that point, the change in Q_d with gate voltage is significantly reduced, so C_g drops rapidly. C_d can

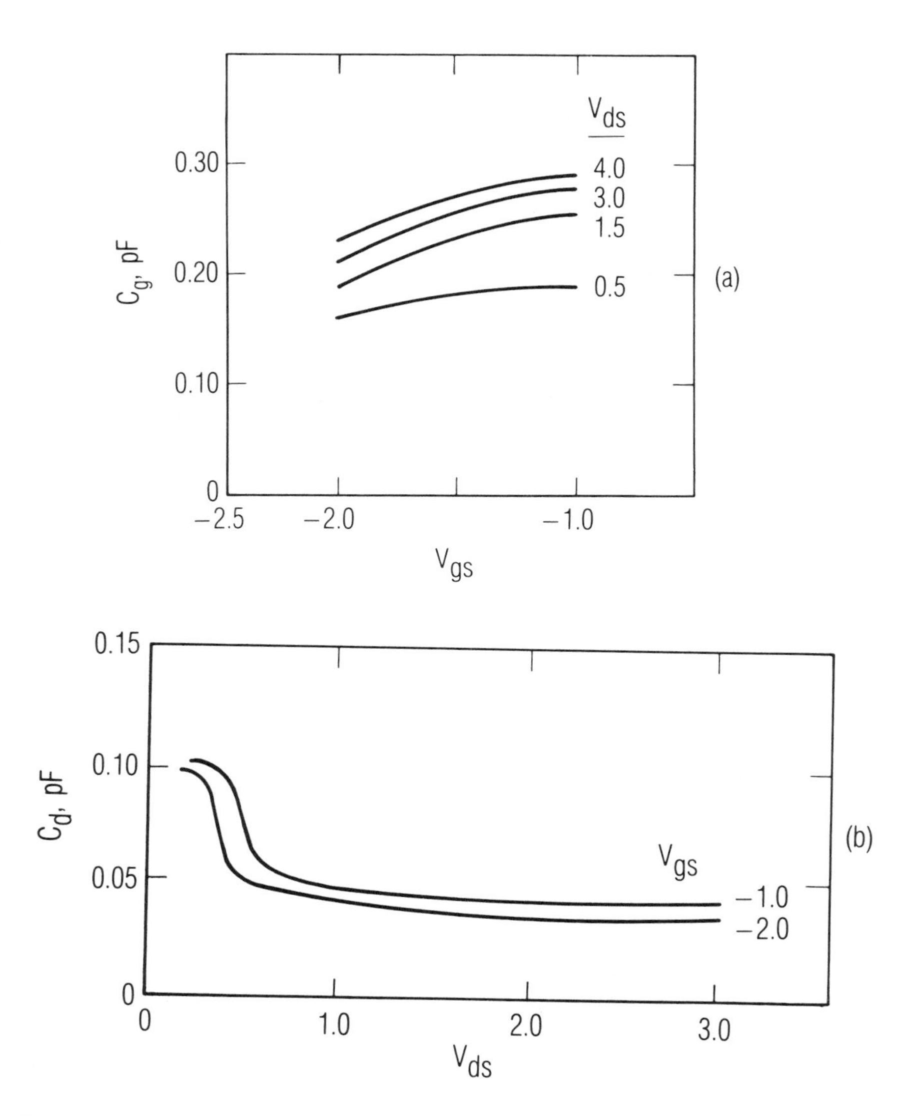

Figure 2.17 (a) Measured gate-source capacitance and (b) gate-drain capacitance of a commercial 0.5μ × 300μ MESFET.

be assumed to be constant as long as the FET remains in its saturation region, as it usually does in most well designed nonlinear or quasilinear components. In saturation, the capacitance between the gate and drain metalizations, rather than depletion layer effects, dominates C_d. The characterization of the controlled current source is the main concern in nonlinear MESFET modeling, because of the strong effect this characterization has upon the accuracy of nonlinear FET analysis.

Many MESFET models have been proposed (References 2.4–2.8). The most successful separate the conductive channel into two regions, the neutral channel between the source end of the gate and the saturation point and the charge domain. The equations describing both regions are solved simultaneously, under the constraints that the channel current in both regions must be the same and the sum of the voltage drops across the regions must equal the drain voltage. Many models make use of Shockley's original junction FET theory (Reference 2.9) to model the neutral part of the channel. The models differ primarily in the way in which they treat the charge domain.

Modeling I_d

The current in the drain current source is a function of internal gate and drain voltages, V_g and V_d. It can be expressed satisfactorily via either a physical model or an empirical expression. The advantage of an empirical expression is that the expression and its derivatives can be in closed form, and usually can be evaluated with less computation, hence less computer time, than a physical model. A physical model (i.e., one in which the current is calculated from the physical parameters and dimensions of the device) usually has a greater range of validity and can be used to relate the device structure directly to the performance of the circuit.

The following empirical expression for drain current has been applied successfully to GaAs MESFETs:

$$I_d = (A_0 + A_1 V_g + A_2 V_g^2 + A_3 V_g^3) \tanh(\alpha V_d) \tag{2.4.5}$$

Equation (2.4.5) is a global representation of the current, not the incremental form, even though the dependence of I_d on V_g is given by a power series. Therefore, it is not correct to identify A_1 as the MESFET's transconductance. The parameters A_0 through A_3 and α must be found by fitting (2.4.5) to measured *I/V* data. The dependence of I_d on V_d is a hyperbolic tangent function. There is no theoretical reason for using the tanh function for V_d; it is used simply because it is coincidentally similar to a drain *I/V* characteristic. Equation (2.4.5) is valid only when $V_d > 0$, because the drain *I/V* characteristic is not symmetrical about the origin of the *I/V* axes and must be used with the constraint $I_d = 0$ when $V_g < V_t$. It is usually adequate to express I_d as a cubic function of V_g, and in most cases, the cubic relationship is sufficient for determining intermodulation products up to third order. A square-law expression for $I_d(V_g)$, as is used in silicon FETs, is almost always inadequate for describing GaAs MESFETs.

It is also possible to derive an *I/V* expression from the physical operation of the device. The usual approach is to use modified Shockley theory to describe the neutral part of the channel, that is, the part between the source end of the gate and the charge domain, and to develop separate expressions for the voltage and current in the domain. Figure 2.18 shows an idealized FET structure, one that does not include R_s, R_i, R_d, or R_g but describes only the intrinsic MESFET. These parasitic resistances can be included in the linear part of the equivalent circuit. The channel's horizontal and vertical coordinates are x and y, respectively; the neutral channel extends from $x = 0$ to L_1, and the charge domain from L_1 to a point, L_2, somewhat beyond the gate length L. We assume that the interface between the electron accumulation and depletion regions is at L, that is, directly under the edge of the gate. This assumption is justified by two-dimensional studies of GaAs MESFETs. We further assume that the electron mobility is constant in the neutral channel and that the electrons move at saturation velocity in the charge domain. This assumption is equivalent to approximating the velocity-field characteristic of GaAs as in Figure 2.19, in which the actual, strongly peaked characteristic is approximated by a piecewise linear curve.

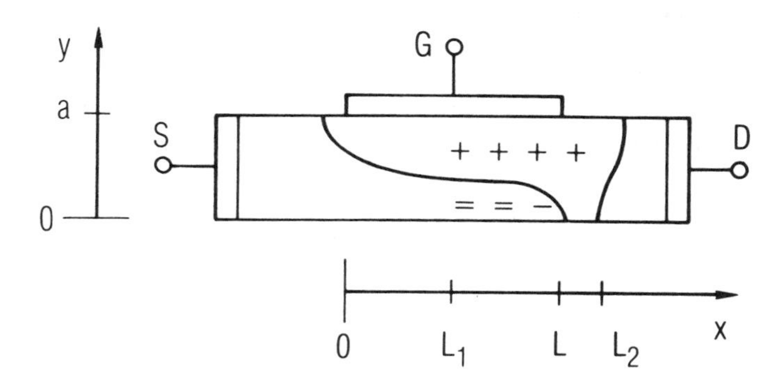

Figure 2.18 An idealized MESFET channel.

Laboratory measurements of pure, lightly doped GaAs give a low-field mobility of 5000 cm^2/V-s and a saturated drift velocity of $1.3 \cdot 10^7$ cm/s. Although these parameters are accurate for describing bulk samples of lightly doped GaAs, they are unreasonably high for modeling GaAs MESFETs for at least four reasons: first, the channels of GaAs MESFETs

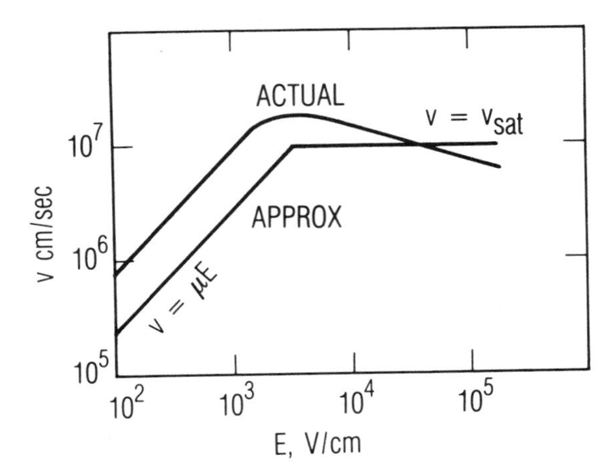

Figure 2.19 Actual velocity-field characteristic of lightly doped GaAs and a piecewise-linear approximation used in the MESFET model.

are more heavily doped than the laboratory samples yielding these parameters, so the actual mobility of the electrons in the FET's channel is lower; second, the electrons in the channel, especially in the charge domain, do not move precisely in the x direction, so the x component of their velocity is lower than their absolute velocity; third, the FET fabrication process subjects the semiconductor to several forms of damage and contamination that affect the purity and structure of the crystal; and fourth, a close fit between the approximate and actual curves in Figure 2.19 generally requires an assumption that mobility is lower. Thus, real devices are better described by mobilities around 2000 cm^2/V-s and saturation velocities of approximately $0.9 \cdot 10^7$ cm/s. Mobilities and saturation velocities in this range generally give much better agreement between calculated and measured I/V characteristics in the following equations than do the higher values.

From Shockley theory, the current in the neutral channel is:

$$I_d = I_1 \{3[u^2(L_1) - u^2(0)] - 2[u^3(L_1) - u^3(0)]\} \tag{2.4.6}$$

where $u(x)$ is the normalized depletion region width:

$$u(x) = \left[\frac{V(x) - V_g + \phi}{V_p}\right]^{1/2} \tag{2.4.7}$$

V_p is the pinch-off voltage, the voltage drop across the depletion region necessary to deplete the full thickness of the channel. V_p is related to V_t as $V_p = -V_t + \phi$ (in this book we will differentiate clearly between V_t and V_p, although many engineers refer loosely to V_t as "pinch-off" voltage). In (2.4.7), V_p and ϕ are defined as positive quantities; V_g is positive if it forward-biases the gate junction. V_t is a specific reverse-bias value of V_g, so it is always negative. $V(x)$ is the channel potential along the line $y = 0$, and ϕ is the diffusion potential of the junction. I_1 is a current parameter. It is derived from the material parameters and dimensions of the FET as

$$I_1 = \frac{W\mu q^2 N_d^2 a^3}{6\epsilon_s L_1} \tag{2.4.8}$$

where μ is the low-field mobility, ϵ_s is the electric permittivity of the GaAs semiconductor, N_d is the doping density, q is the electron charge ($1.6 \cdot 10^{-19}$ coulomb), and W is the gate width. At the edge of the charge domain, where $x = L_1$, the electrons are at saturation velocity, so

$$I_d = I_s[1 - u(L_1)] \tag{2.4.9}$$

where I_s is the open-channel saturation current, the current that would exist in the channel if there were no gate and the electrons moved at saturation velocity. In terms of the device parameters:

$$I_s = qv_sN_dWa \tag{2.4.10}$$

where v_s is the saturation velocity of the electrons. The currents I_d given by (2.4.6) and (2.4.9) must be equal, so equating them at $x = L_1$ gives

$$L_1 = \frac{zL\left\{[u^2(L_1) - u^2(0)] - \frac{2}{3}[u^3(L_1) - u^3(0)]\right\}}{1 - u(L_1)} \tag{2.4.11}$$

where $z = V_p/E_sL$. E_s is the electric field at saturated drift velocity for the piecewise-linear curve in Figure 2.19; it is related to v_s by $v_s = \mu E_s$.

The voltage drop, V_1, in the channel between $x = 0$ and $x = L_1$, from Shockley theory, is

$$V_1 = V(L_1) = V_p[u^2(L_1) - u^2(0)] \tag{2.4.12}$$

and the internal drain voltage, V_d, is

$$V_d = V_1 + V_2 \tag{2.4.13}$$

where V_2 is the voltage drop across the charge domain. If V_2 is known, V_1 can be found from (2.4.13), $u(L_1)$ from (2.4.12), and I_d from (2.4.9). In general, however, V_2 is a function of L_1 and the current I_d, so (2.4.9), (2.4.12), and (2.4.13) must be solved iteratively until all are satisfied. The remaining problem, and the most difficult one, is to find an expression for V_2, the voltage drop across the charge domain.

The difficulty in estimating V_2 arises from the fact that the gradual channel approximation, on which Shockley theory is based, is not valid in the charge domain. Thus, any estimate of V_2 that is based on an explicit or implicit one-dimensional assumption is likely to have a tenuous hold on validity. It is possible to obtain reasonably good results, however, by employing some creative assumptions and empirical corrections to simple theory.

One approach (Reference 2.7) is to integrate over the charge density in the x direction at $y = 0$. First, an estimate of the charge in the domain must be generated. Recognizing that the current must be constant throughout the electron accumulation region between L_1 and L, we obtain an expression for the electron density, $n(x)$:

$$n(x) = N_d \frac{1 - u(L_1)}{1 - u(x)} \tag{2.4.14}$$

in the region $L_1 < x < L$; the charge density is $-q(n(x) - N_d)$. The positively charged area between L and L_2 is assumed to be fully depleted, so its charge density is qN_d. Integrating over the charge region in the x direction gives

$$V_2 = \int_{L1}^{L} \int_{L1}^{x'} \frac{qN_d[u(x) - u(L_1)]}{\epsilon_s[1 - u(x)]}\, dx\, dx' + (L - L_1)E_s + \frac{\epsilon_s E^2(L)}{2qN_d} \tag{2.4.15}$$

where $E(x)$ is the longitudinal electric field in the channel and E_s is the electric field at the saturation point.

It is necessary to integrate (2.4.15) numerically. In (2.4.15), $u(x)$ is found from (2.4.7). However, one may find that, under some circumstances, $V(x)$ is large enough that $u(x) = 1$ and the charge density in

(2.4.14) approaches infinity. This problem can be circumvented by replacing the value of V_p in (2.4.7) by V_{pe}, given by the empirical expression:

$$V_{pe} = \frac{1}{4}\left\{[V(x) - V_g + \phi]^{1/2} + [V(x) - V_g + \phi + 4V_p(1 - u(L_1))]^{1/2}\right\}^2 \quad (2.4.16)$$

This formulation gives good agreement with measured I/V characteristics as long as $L/a > 4$. If L/a is smaller, the MESFET I/V characteristic often has a peak that is not reproduced by these equations.

Other expressions for V_2 are derived from studies of charge domains in Gunn diodes. For example, (Reference 2.8) gives:

$$V_2 = \frac{1.886 L_D E_c^3}{[E(L_1) - E_s]^2} \quad (2.4.17)$$

and the domain width is

$$L_2 - L_1 = \frac{0.707 L_D E_c}{E(L_1) - E_s} \quad (2.4.18)$$

L_D, the Debye length, is

$$L_D = \left(\frac{\epsilon_s KT}{q^2 N_d}\right)^{1/2} \quad (2.4.19)$$

where K is Boltzmann's constant and T is absolute temperature in kelvins. E_c, the *critical field*, is approximately $3.2 \cdot 10^3$ V/cm for the doping levels used in MESFETs. It is found by fitting the expression:

$$v(E) = \frac{\mu E}{1 + E/E_c} \quad (2.4.20)$$

to the velocity-field characteristic in Figure 2.19. The agreement between (2.4.20) and the actual curve in Figure 2.19 is not very good; however, with saturation velocity and mobility reduced to compensate for doping effects and fabrication limitations, (2.4.20) still allows adequate agreement with measured data. These expressions are, of course, much simpler than (2.4.14) through (2.4.16), but they give poorer agreement with measured I/V characteristics.

Both the physical and empirical I/V models describe only the I/V dependence on internal voltages V_g and V_d. Usually, we wish to know the I/V dependence upon the external voltages V_{gs} and V_{ds}, because these are observable. At dc there is no voltage drop across R_i or R_g, so they need not be considered. V_g and V_d are related to V_{gs} and V_{ds} as follows:

$$V_g = V_{gs} - R_s I_d \tag{2.4.21}$$

and

$$V_d = V_{ds} - (R_s + R_d) I_d \tag{2.4.22}$$

Modeling C_g

If the FET remains in saturation, C_g is usually modeled successfully as a Schottky-barrier capacitance, at least up to $V_g = V_t$. Beyond V_t, the capacitance drops because the depletion region cannot expand further. However, it is often acceptable to assume that C_g remains approximately constant at its V_t value or, for best accuracy, decreases gradually according to an empirical expression. Fortunately, most well designed nonlinear or quasilinear FET components are not unduly sensitive to the functional form of C_g, so highly precise modeling is usually not necessary. When V_d drops so low that the FET enters its linear region, C_g must be reduced to approximately half its saturation value.

A by-product of the numerical integration of (2.4.15) is $u(x)$, the normalized depletion width. Knowing $u(x)$ allows us to find C_g accurately in both the saturation and linear regions. The depletion charge Q_d can be found as follows:

$$Q_d = WqN_d a \left\{ \int_0^L u(x) dx + \frac{\pi a}{4} [u^2(0) + u^2(L)] \right\} \tag{2.4.23}$$

in which the depletion charge is found by integrating $u(x)$ between 0 and L and approximating the charge at the source and drain ends of the gate as quarter circles. After Q_d is found from (2.4.23), the capacitances are found from (2.4.1) and (2.4.2).

Modeling C_d

It is almost always valid to assume C_d to be constant in saturation. However, in large-signal operation, the FET's drain voltage waveform may reach low values, so the FET switches periodically between saturated and linear operation. C_d must be measured and modeled empirically in the linear region, or (2.4.23) and (2.4.2) should be used.

2.5 DETERMINING THE MODEL PARAMETERS

Whether a physical or an empirical model is used for the device, a number of parameters must be determined. Some of these are found easily, either because they are physical constants (e.g., electron charge, Boltzmann's constant), are well known properties of the material (e.g., electric permittivity), or are accurately known design parameters of the device (e.g., gate width). Many other parameters that cannot be determined exactly can be established with good enough accuracy for most practical purposes (e.g., epilayer doping density, thickness). Others are not quantifiable a priori, because they depend upon such imponderables as variations in processing cleanliness and defects in the semiconductor or gate junction, so they must be measured (e.g., diffusion potential ϕ, junction ideality factor η, and Schottky-junction current parameter I_{sat}). Of course, in an empirical model, most, if not all, of the parameters must be measured. Most device equivalent circuits include a number of circuit elements that are in this latter category. Therefore, in any model, at least some of the device parameters must be determined by measurements of the actual FET or diode that is to be used or occasionally on a test device fabricated on the same semiconductor wafer.

A solid-state device that is to be used in small-signal quasilinear applications can usually be characterized adequately by S or Y parameters. If the device is to be described by an equivalent circuit, all the circuit elements are assumed to be linear, so no voltage dependence need be determined. In order to model a nonlinear device, however, it is necessary to measure the circuit element values and to determine their dependence upon one or more control voltages or currents (usually voltages) within the circuit. In MESFETs, the channel C/V and drain I/V characteristics are of most interest. In diodes, the junction C/V and I/V characteristics are the strongest nonlinearities.

Parameters can be measured via direct or indirect techniques. Direct techniques involve the measurement of I/V and C/V characteristics by means of a semiconductor curve tracer and a capacitance bridge. Indirect measurements involve determining $C(V)$ and $I(V)$ by measuring an entirely different quantity. For example, the capacitances and resistances in MESFET equivalent circuits are often determined by measuring S parameters of the MESFET over a wide range of frequencies and by adjusting the resistance and capacitance values empirically until the calculated S parameters of the equivalent circuit agree closely with those measured. The process is repeated with a large set of bias voltages, and eventually, a table of *C/V* and *I/V* characteristics is generated. We can then fit an expression such as (2.3.5) or a polynomial to the measured *C/V* characteristics or use the measured data directly with interpolation.

2.5.1 Direct Measurement of Junction *C/V* and *I/V*

The important dc parameters of the Schottky junction, I_0, η, and R_s, can be found very easily from a direct measurement of the *I/V* characteristic. Figure 2.20 shows the measured *I/V* characteristic of a Schottky-barrier diode, plotted on semilog axes. The *I/V* characteristic is a nearly straight line, deviating noticeably at currents above approximately 1 mA because of the voltage drop across the series resistance R_s. Some simple manipulations of (2.3.6) show that at room temperature (72°F) the slope of the straight-line portion of the curve is 58.3η mV per decade of current. The diode's slope parameter can be found by measuring the slope of the closest-fit straight line, in mV/decade of current, and dividing by 58.3. The deviation of the curve from the extrapolated straight line at its high-current end is the voltage dropped across the series resistance. Thus,

$$R_s = \Delta V/I \tag{2.5.1}$$

where ΔV is the voltage deviation and I is the current at which ΔV is determined. Finally, I_{sat} is found from any pair of points (I, V) along the straight-line portion of the curve:

$$I_{sat} = I(V) \exp(-qV/\eta KT) \tag{2.5.2}$$

We should use care in calculating the series resistance of small-diameter GaAs diodes from dc *I/V* measurements because junction heating at high current densities can affect the accuracy of R_s. The thermal resistance of a two-micron GaAs dot-matrix diode is approximately 2°C/mW to 4°C/mW, enough to shift the *I/V* curve slightly toward lower voltages and, thus, to make the series resistance appear to be lower than it is. The series resistance of these small diodes can be measured via indirect methods or at very low currents, approximately 0.1 to 0.2 mA. Pulsed *I/V* measurements usually do not successfully circumvent this problem because the pulse must be much shorter than the thermal time constant of the junction, which is on the order of microseconds.

Direct measurement of the *C/V* characteristic presents some practical difficulties because the junction capacitance of many types of diodes, especially millimeter-wave mixer diodes, is on the order of femtofarads, too

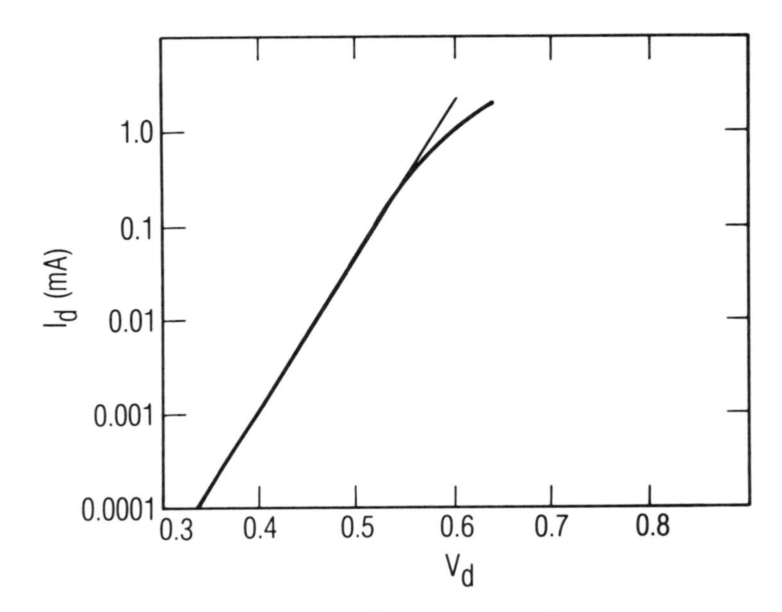

Figure 2.20 Diode *I/V* characteristic. The deviation from a straight line above 1 mA is caused by voltage drop across R_s.

small for direct measurement with a capacitance bridge. One solution to this problem is to fabricate a wafer of diodes, a few of which have large areas, and to measure the capacitance C_{j0} of one of them. The capacitance of the smaller diodes can then be found by scaling C_{j0} according to their area. This process is not highly accurate for very small diodes, those less than 5 microns in diameter, because the capacitive component of the edge of the anode, which does not scale in proportion to area, is significant. For diodes having small C_{j0}, indirect methods of measuring junction capacitance are probably preferable.

If the *C/V* characteristic is known and can be expressed by Equation (2.3.5), the diffusion potential ϕ and exponent γ can be found straightforwardly, by recognizing that a plot of $C^{-1/\gamma}(V)$ is a straight line. Figure 2.21 shows a measured *C/V* characteristic plotted as $C^{-1/\gamma}(V)$. The term γ is adjusted until all the points lie along a straight line. The extrapolation of this line intersects the V axis at ϕ.

Because of package parasitics or metalizations, many diodes have a constant component of capacitance. This component must be subtracted from the *C/V* data or a straight line will not be obtained.

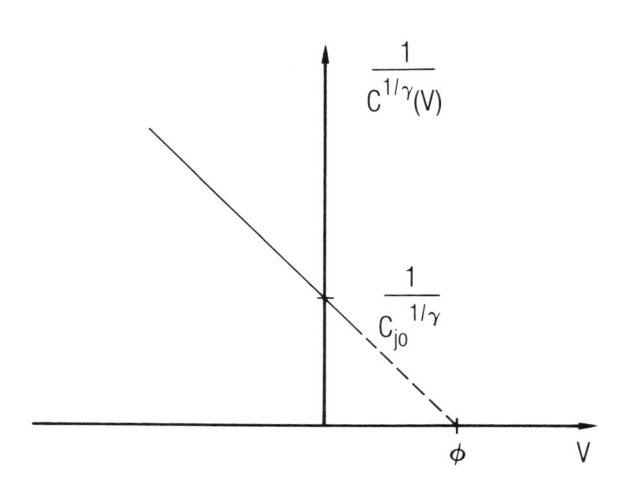

Figure 2.21 Junction capacitance characteristic. Extrapolating the $C^{-1/\gamma}$ characteristic to the V_j axis gives ϕ.

2.5.2 Indirect Measurement of Diode Parameters

Because it is difficult to measure diode capacitance directly, indirect means of measuring diode parameters have been developed. These are based on the microwave-frequency measurement of the cutoff frequency or of Q, which is the ratio of cutoff frequency to measurement frequency. The cutoff frequency is

$$f_c = \frac{1}{2\pi C(V_b) R_{sf}} \tag{2.5.3}$$

where V_b is the junction bias voltage and R_{sf} is the series resistance at the measurement frequency f_0. The measured cutoff frequency is not the same as that given by Equation (2.3.15), because series resistance is increased by the skin effect and capacitance is not necessarily measured at zero bias.

The two most popular indirect methods of measuring diode parameters are those of Houlding (Reference 2.10) and DeLoach (2.11). In addition, diode parameters can be measured by placing the diode in a fixture and measuring input reflection coefficient or S-parameters, as is done with a FET.

Houlding measurements are made with the test setup in Figure 2.22. The diode must be mounted in a fixture that does not introduce any reactance of its own in series with the diode. If a waveguide mount is used,

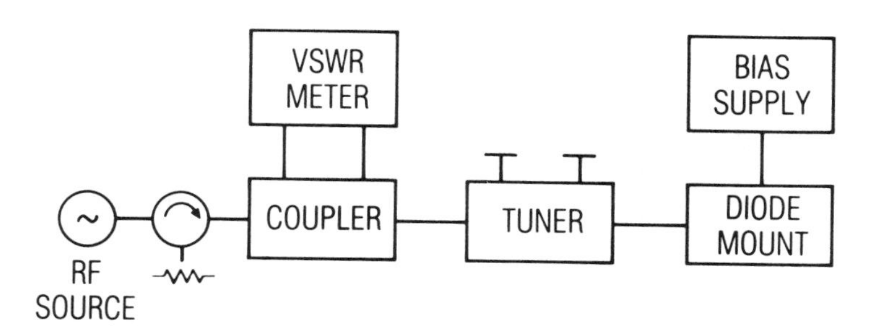

Figure 2.22 Test setup for Houlding measurements.

the backshort must be carefully adjusted so that the open-circuit plane of the short coincides with the position of the diode. This position can be found by first adjusting the backshort so that the shorting plane is at the diode (the proper position is indicated by the absence of change in VSWR with bias) then retracting the short one-quarter wavelength. The diode is then biased to the desired voltage, and the tuner is adjusted until the mount, with the diode in place, is matched. The diode bias is then changed by a few tenths of a volt; because the junction capacitance changes, the mount no longer is matched and the VSWR rises. The diode Q at the original bias voltage is:

$$Q = \frac{2}{\left(\frac{C_{j1}}{C_{j2}} - 1\right)\left(\frac{1}{|\Gamma|^2} - 1\right)^{1/2}} \tag{2.5.4}$$

where Γ is the reflection coefficient corresponding to the increased VSWR. One must have some other way of knowing γ and ϕ in order to find the ratio of the original capacitance to the capacitance that detuned the mount, C_{j1}/C_{j2}; γ and ϕ can be estimated from doping profiles, if known. C_{j1}/C_{j2} is then found directly from (2.3.5).

DeLoach measurements are generally preferred to Houlding because they require no assumptions about the functional form of the C/V characteristic, liberate all diode parameters, and are relatively insensitive to imperfections in the diode mount. DeLoach measurements are made by placing the diode in parallel with a transmission line or waveguide; measuring the series resonant frequency, transmission loss, and Q; and deriving the diode parameters from those measurements. In most diodes the series-resonant frequency is reasonably close to the frequency at which the diode will be used; DeLoach measurements, therefore, give a cutoff-frequency estimate that is relevant to the RF performance of the diode.

Figure 2.23 shows the test setup for DeLoach measurements. The diode is placed in a transmission mount, in parallel with the transmission line, and is zero-biased. Because of its lead inductance and junction capacitance, the diode has a series resonance; the loss L_d through the mount is measured at the resonant frequency f_0. The resonant frequency f_0 and loss L_d are recorded, as are the frequencies at which the loss is 3 dB lower than L_d, f_1 and f_2. The diode series resistance, junction capacitance, and lead inductance are as follows:

$$R_{sf} = \frac{Z_c}{2(L_d^{1/2} - 1)} \tag{2.5.5}$$

$$C_{j0} = \frac{1}{\pi Z_c} \left(\frac{f_2 - f_1}{f_1 f_2} \right) (L_d^{1/2} - 1) \left(1 - \frac{2}{L_d} \right)^{1/2} \tag{2.5.6}$$

$$L_s = \frac{1}{4\pi^2} \left(\frac{1}{f_1 f_2 C_{j0}} \right) \tag{2.5.7}$$

The loss, L_d, is defined as a quantity greater than unity. These relations are strictly valid only if the diode is mounted in a TEM-mode transmission line structure. When a waveguide mount is used, they are accurate if f_0 is well above the waveguide cutoff frequency, and f_1 and f_2 are close to f_0. If these conditions are not met, the precise equations must be used; they are given in References 2.1 and 2.11.

One advantage of the DeLoach measurement is that it provides the contact wire inductance, a parameter that is often important in millimeter-wave mixer design. Whisker inductance is difficult to calculate because the wire often has a complex shape, and its inductance depends not only upon the wire itself but also upon the dimensions of the waveguide or other mount in which the diode is to be used. DeLoach measurements are often made on diodes mounted in the same package or Sharpless wafer as is used in a mixer; this practice prevents inaccuracy caused by differences between the test mount and the one used in the mixer. The ease of making the measurements and calculating the diode parameters from them makes DeLoach measurements very practical for millimeter-wave mixer and frequency-multiplier design.

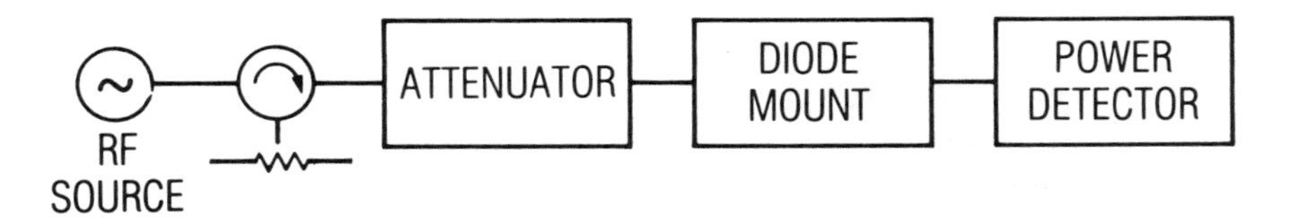

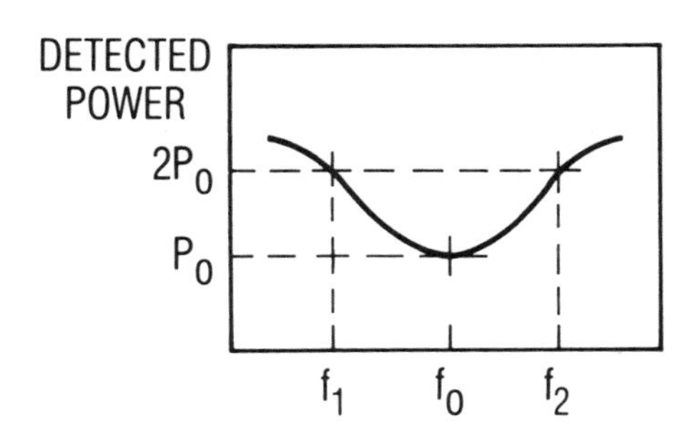

Figure 2.23 Test setup for DeLoach measurements.

REFERENCES

[2.1] S. Maas, *Microwave Mixers*, Artech House, Dedham, MA, 1986.

[2.2] S.M. Sze, *Physics of Semiconductor Devices*, John Wiley and Sons, New York, 1981.

[2.3] E.H. Rhoderick, "Metal-Semiconductor Contacts," *IEEE Proc.*, Part I, Vol. 129, 1982, p. 1.

[2.4] K. Lehovec and R. Zuleeg, "Voltage-Current Characteristics of GaAs JFETs in the Hot Electron Range," *Solid-State Electronics*, Vol. 13, 1970, p. 1415.

[2.5] Leon O. Chua and Y.W. Sing, "Nonlinear Lumped Circuit Model of GaAs MESFET," *IEEE Trans. Electron Devices*, Vol. ED-30, 1983, p. 825.

[2.6] R. Pucel, H.A. Haus, and H. Statz, "Signal and Noise Properties of GaAs Microwave Field Effect Transistors" in L. Martin, ed., *Advances in Electronics and Electron Physics*, Vol. 38, Academic Press, New York, 1975, p. 195.

[2.7] S. Maas, "Theory and Analysis of GaAs MESFET Mixers" Ph.D. dissertation, University of California, Los Angeles, 1984.

[2.8] M.S. Shur and L.F. Eastman, "I/V Characteristics, Small-Signal Parameters, and Switching Times of GaAs FETs," *IEEE Trans. Electron Devices*, Vol. ED-25, 1978, p. 195.
[2.9] W. Shockley, "A Unipolar Field-Effect Transistor," *Proc. IRE*, Vol. 40, 1952, p. 1365.
[2.10] N. Houlding, "Measurement of Varactor Quality," *Microwave J.*, Vol. 3, 1960, p. 40.
[2.11] B.C. DeLoach, "A New Technique to Characterize Diodes and an 800-GHz Cutoff Frequency Varactor at Zero Volts Bias," *IEEE Trans. Microwave Theory Tech.*, Vol. MTT-12, 1964, p. 15.

CHAPTER 3

HARMONIC BALANCE AND LARGE-SIGNAL–SMALL-SIGNAL ANALYSIS

This chapter is concerned with two of the most important techniques for analyzing nonlinear circuits. The first, called *harmonic balance*, is most useful for strongly or weakly nonlinear circuits that have single-tone excitation. Harmonic-balance analysis is applicable to a wide variety of problems in such microwave circuits as power amplifiers, frequency multipliers, and mixers subjected to *local-oscillator* (LO) drive. A beneficial property of harmonic-balance analysis is that it works particularly well in circuits having a mix of long and short time constants and, in fact, was originally proposed to solve the problems inherent in analyzing such circuits (Reference 3.1).

The second technique, *large-signal–small-signal analysis*, is used for nonlinear circuits that are excited by two tones, one of which is very large and the other is vanishingly small. This situation is encountered most frequently in microwave mixers wherein a nonlinear element such as a diode is excited by a large-signal local oscillator and a much smaller received RF signal. The circuit is first analyzed via harmonic balance, under LO excitation alone, and is converted into a small-signal linear, time-varying equivalent. The time-varying circuit is then analyzed as a quasilinear circuit under small-signal RF excitation. The quasilinear assumption is not always necessary, and the small-signal analysis can be extended to include nonlinear effects such as intermodulation.

3.1 HARMONIC BALANCE

3.1.1 Large-Signal, Single-Tone Problems

Figure 3.1 shows a general equivalent circuit that describes many types of nonlinear microwave components. The circuit consists of a nonlinear solid-state device that is connected to a load and a source of large-signal excitation. Initially, we will assume that the excitation source contains only a fundamental frequency component (although any periodic excitation is easily accommodated) and that dc bias is applied to the input and perhaps to the output. Matching networks are used at the input and output to optimize the performance, to couple bias voltages to the device, and to filter and terminate various harmonics appropriately. The solid-state device is most often a transistor or a diode, and we shall assume that it can be described by a quasistatic equivalent circuit that includes both linear and nonlinear elements. The matching circuits are invariably linear and usually have both filtering and impedance-transforming properties.

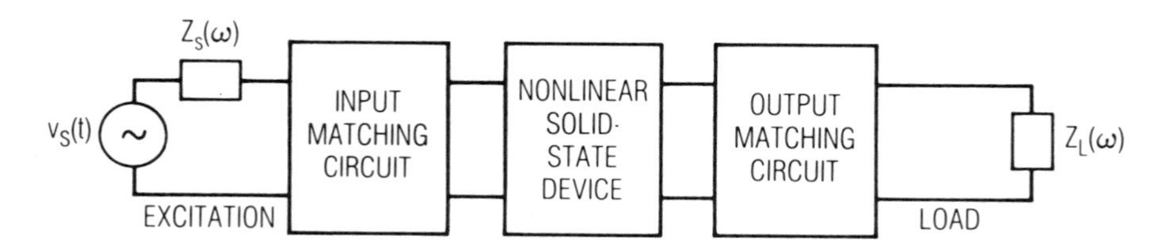

Figure 3.1 Equivalent circuit of a general nonlinear two-port microwave component.

So far, the problem of analyzing this type of circuit does not seem too formidable. One could conceive of writing a set of time-domain differential equations describing the combined nonlinear device equivalent circuit and matching circuits, solving them to obtain the steady-state voltage waveform across the load, and Fourier-transforming to obtain the frequency component corresponding to the desired output harmonic. The differential equations would, of course, be nonlinear and would have to be solved numerically. In fact, many nonlinear circuits are analyzed successfully in this manner.

Three problems often occur that can make such time-domain techniques impractical. The first is that the matching circuits may contain such elements as dispersive transmission lines and transmission-line discontinuities that are difficult, if not impossible, to analyze in the time domain. The best way to characterize these is to use a set of *S*- or *Y*-parameters at each harmonic of the excitation frequency; however, these parameters are frequency-domain quantities and cannot be used to characterize the nonlinear part of the circuit. The second problem is that the circuit may have time constants that are large compared to the inverse of the fundamental excitation frequency. When these exist, it becomes necessary to continue the numerical integration of the equations through many, perhaps thousands, of excitation cycles, until the transient part of the response has decayed and only the steady-state part remains. This long integration is an extravagant use of both computer time and the engineer's patience; furthermore, numerical truncation errors in the long integration may become large and reduce the accuracy of the solution. Although algorithms exist to ameliorate this difficulty (References 3.2 and 3.3), implementing them is an extra complication, and they usually are not included in the general-purpose computer programs most often employed for this task. The third problem with time-domain analysis is that each linear or nonlinear reactive element in the circuit adds a differential equation to the set of equations that describes the circuit. A large circuit can have many reactive elements, so the set of equations that must be solved may be very large.

It is preferable to employ multiport circuit theory to simplify at least part of the circuit by lumping all the linear reactances, impedances, transmission lines, and other linear elements into a single matrix of limited size. By describing the linear part of the circuit as one multiport, it need be evaluated only once at each harmonic, with the results stored as matrices, and no further evaluation is necessary. Harmonic-balance analysis allows this approach to be used.

The circuit elements in Figure 3.1 can be regrouped as shown in Figure 3.2, so that they form a linear subcircuit and a number of nonlinear elements (the nonlinear elements, as a group, are called the *nonlinear subcircuit*). The linear subcircuit can be treated as a multiport and described by its Y-parameters, S-parameters, or some other multiport matrix. The nonlinear elements are modeled by their global I/V or Q/V characteristics, described in Chapter 2, and must be analyzed in the time domain. Thus, the circuit is reduced to an $N+2$–port network, with nonlinear elements connected to N of the ports and voltage sources connected to the other two ports. (The $N+1$th and $N+2$th ports represent, of course, the input and output ports in a two-port network. Usually a sinusoidal source is connected to only one of those ports; however, sources are shown at both ports in Figure 3.2 for complete generality.) $Z_s(\omega)$ and $Z_L(\omega)$, the source and load impedances, respectively, are included, or "absorbed," into the linear subcircuit; they are still in series with the input and output ports, and for some purposes, it will be necessary to resurrect them. The voltages and currents at each port can be expressed in the time or the frequency domain; because of the nonlinear elements, however, the port voltages and currents must have frequency components at harmonics of the excitation. Although in theory an infinite number of harmonics exist at each port, it will be assumed throughout this chapter that the dc component and the first K harmonics (i.e., $k = 0 \ldots K$) describe all the voltages and currents adequately. Consequently all higher harmonics can be ignored.

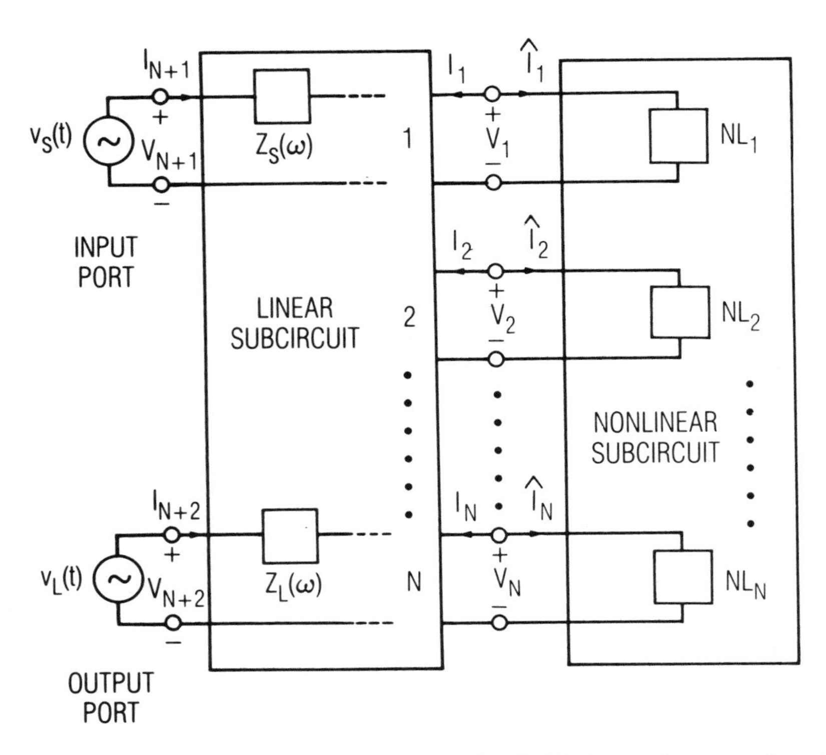

Figure 3.2 A nonlinear microwave circuit, divided into linear and nonlinear subcircuits, with the source and load impedances $Z_s(\omega)$ and $Z_L(\omega)$ absorbed into the linear subcircuit.

The circuit in Figure 3.2 is successfully analyzed when either the steady-state voltage or current waveforms at each port are known. Alternatively, knowledge of the frequency components at all ports constitutes a solution because the frequency components and time waveforms are related by the Fourier series. If, for example, the frequency-domain port voltages are known, one can use the *Y*-parameter matrix for the linear subcircuit to find the port currents. The port currents can also be found by inverse-Fourier-transforming the voltages to obtain their time-domain waveforms, and determining the current waveforms from the nonlinear elements. The idea of harmonic balance is to find a set of port voltage waveforms (or, alternatively, the harmonic voltage components) that give the same currents in both the linear-network equations and the nonlinear-network equations; when that set is found, it must be the solution.

If we express the port-current frequency components as vectors, Kirchhoff's current law requires that:

$$\begin{bmatrix} I_{1,0} \\ I_{1,1} \\ I_{1,2} \\ \cdot \\ \cdot \\ \cdot \\ I_{1,K} \\ I_{2,0} \\ I_{2,1} \\ \cdot \\ \cdot \\ \cdot \\ I_{2,K} \\ \cdot \\ \cdot \\ \cdot \\ I_{N,K} \end{bmatrix} + \begin{bmatrix} \hat{I}_{1,0} \\ \hat{I}_{1,1} \\ \hat{I}_{1,2} \\ \cdot \\ \cdot \\ \cdot \\ \hat{I}_{1,K} \\ \hat{I}_{2,0} \\ \hat{I}_{2,1} \\ \cdot \\ \cdot \\ \cdot \\ \hat{I}_{2,K} \\ \cdot \\ \cdot \\ \cdot \\ \hat{I}_{N,K} \end{bmatrix} = \begin{bmatrix} 0 \\ 0 \\ 0 \\ \cdot \\ \cdot \\ \cdot \\ 0 \\ 0 \\ 0 \\ \cdot \\ \cdot \\ \cdot \\ 0 \\ \cdot \\ \cdot \\ \cdot \\ 0 \end{bmatrix} \tag{3.1.1}$$

where $I_{n,k}$ is a phasor quantity, the kth harmonic component of the current at Port n, calculated via the port voltages and the Y matrix of the linear subcircuit; $\hat{I}_{n,k}$, with the circumflex accent, is the current component calculated via the same port voltages and the nonlinear elements. Equation (3.1.1) shows the general form we will use for the voltage, current, and charge vectors; all such vectors used in this chapter will be of this form unless indicated otherwise. The vectors include only positive-frequency components because the negative-frequency components, being the complex conjugates of the positive-frequency components, can be found immediately if needed. Eliminating the negative-frequency components from (3.1.1) reduces complexity considerably.

First we consider the linear subcircuit. The admittance equations are

$$\begin{bmatrix} \mathbf{I}_1 \\ \mathbf{I}_2 \\ \mathbf{I}_3 \\ \cdot \\ \cdot \\ \cdot \\ \mathbf{I}_N \\ \mathbf{I}_{N+1} \\ \mathbf{I}_{N+2} \end{bmatrix} = \begin{bmatrix} \mathbf{Y}_{1,1} & \mathbf{Y}_{1,2} & \cdots & \mathbf{Y}_{1,N} & \mathbf{Y}_{1,N+1} & \mathbf{Y}_{1,N+2} \\ \mathbf{Y}_{2,1} & \mathbf{Y}_{2,2} & \cdots & \cdot & \cdot & \cdot \\ \mathbf{Y}_{3,1} & \mathbf{Y}_{3,2} & \cdots & \cdot & \cdot & \cdot \\ \cdot & \cdot & \cdots & \cdot & \cdot & \cdot \\ \cdot & \cdot & \cdots & \cdot & \cdot & \cdot \\ \cdot & \cdot & \cdots & \cdot & \cdot & \cdot \\ \mathbf{Y}_{N,1} & \mathbf{Y}_{N,2} & \cdots & \mathbf{Y}_{N,N} & \mathbf{Y}_{N,N+1} & \mathbf{Y}_{N,N+2} \\ \mathbf{Y}_{N+1,1} & \mathbf{Y}_{N+1,2} & \cdots & \cdot & \mathbf{Y}_{N+1,N+1} & \mathbf{Y}_{N+1,N+2} \\ \mathbf{Y}_{N+2,1} & \mathbf{Y}_{N+2,2} & \cdots & \cdot & \mathbf{Y}_{N+2,N+1} & \mathbf{Y}_{N+2,N+2} \end{bmatrix} \begin{bmatrix} \mathbf{V}_1 \\ \mathbf{V}_2 \\ \mathbf{V}_3 \\ \cdot \\ \cdot \\ \cdot \\ \mathbf{V}_N \\ \mathbf{V}_{N+1} \\ \mathbf{V}_{N+2} \end{bmatrix} \tag{3.1.2}$$

The current vector **I**, from (3.1.1), has been written as a set of subvectors, where

$$\mathbf{I}_n = \begin{bmatrix} I_{n,0} \\ I_{n,1} \\ I_{n,2} \\ \cdot \\ \cdot \\ \cdot \\ I_{n,K} \end{bmatrix} \tag{3.1.3}$$

that is, $\mathbf{I}_n$ is the vector of harmonic currents at the nth port, and similarly,

$$\mathbf{V}_n = \begin{bmatrix} V_{n,0} \\ V_{n,1} \\ V_{n,2} \\ \cdot \\ \cdot \\ \cdot \\ V_{n,K} \end{bmatrix} \tag{3.1.4}$$

The elements of the admittance matrix **Y** as given in (3.1.2) are all matrices; each submatrix is a diagonal, whose elements are the values $Y_{m,n}$ at each harmonic $k = 0 \ldots K$ of the fundamental excitation frequency ω_p:

$$\mathbf{Y}_{m,n} = \mathrm{diag}[Y_{m,n}(k\omega_p)], \quad k = 0, 1, 2, \ldots, K \tag{3.1.5}$$

that is,

$$\mathbf{Y}_{m,n} = \begin{bmatrix} Y_{m,n}(0) & 0 & 0 & \cdots & 0 \\ 0 & Y_{m,n}(\omega_p) & 0 & \cdots & 0 \\ 0 & 0 & Y_{m,n}(2\omega_p) & \cdots & 0 \\ \cdot & \cdot & \cdot & \cdot & \cdot \\ \cdot & \cdot & \cdot & \cdot & \cdot \\ \cdot & \cdot & \cdot & \cdot & \cdot \\ 0 & 0 & 0 & \cdots & Y_{m,n}(K\omega_p) \end{bmatrix} \tag{3.1.6}$$

$\mathbf{V}_{N+1}$ and $\mathbf{V}_{N+2}$, the excitation vectors, have the form:

$$\begin{bmatrix} \mathbf{V}_{N+1} \\ \\ \mathbf{V}_{N+2} \end{bmatrix} = \begin{bmatrix} V_{b1} \\ V_s \\ 0 \\ 0 \\ \cdot \\ \cdot \\ \cdot \\ V_{b2} \\ 0 \\ \cdot \\ \cdot \\ \cdot \\ 0 \end{bmatrix} \tag{3.1.7}$$

where V_{b1} and V_{b2} are the dc voltages at Ports $N + 1$ and $N + 2$, respectively, and V_s is the excitation voltage at Port $N + 1$. Equation (3.1.7) implies that the Port $N + 1$ excitation includes a dc and a fundamental frequency source, while the $N + 2$ port includes only dc. This is the usual situation; it corresponds, for example, to a FET amplifier that has gate and drain bias and gate excitation. A two-terminal device would normally have only one bias source, and in this case, the $N + 2$ port could be eliminated. Also, if the excitation were not sinusoidal, the vector on the right in (3.1.7) would include the harmonic components of the excitation instead of zeros. Partitioning the Y matrix in (3.1.2) gives an expression for $\mathbf{I}$, the vector of currents in Ports 1 to N:

$$\begin{bmatrix} \mathbf{I}_1 \\ \mathbf{I}_2 \\ \cdot \\ \cdot \\ \cdot \\ \mathbf{I}_N \end{bmatrix} = \begin{bmatrix} \mathbf{Y}_{1,N+1} & \mathbf{Y}_{1,N+2} \\ \mathbf{Y}_{2,N+1} & \mathbf{Y}_{2,N+2} \\ \cdot & \cdot \\ \cdot & \cdot \\ \cdot & \cdot \\ \mathbf{Y}_{N,N+1} & \mathbf{Y}_{N,N+2} \end{bmatrix} \begin{bmatrix} \mathbf{V}_{N+1} \\ \mathbf{V}_{N+2} \end{bmatrix} + \begin{bmatrix} \mathbf{Y}_{1,1} & \mathbf{Y}_{1,2} & \cdots & \mathbf{Y}_{1,N} \\ \mathbf{Y}_{2,1} & \mathbf{Y}_{2,2} & \cdots & \mathbf{Y}_{2,N} \\ \cdot & \cdot & \cdots & \cdot \\ \cdot & \cdot & \cdots & \cdot \\ \cdot & \cdot & \cdots & \cdot \\ \mathbf{Y}_{N,1} & \mathbf{Y}_{1,2} & \cdots & \mathbf{Y}_{N,N} \end{bmatrix} \begin{bmatrix} \mathbf{V}_1 \\ \mathbf{V}_2 \\ \cdot \\ \cdot \\ \cdot \\ \mathbf{V}_N \end{bmatrix} \tag{3.1.8}$$

or

$$\mathbf{I} = \mathbf{I}_s + \mathbf{Y}_{N \times N} \mathbf{V} \tag{3.1.9}$$

where $\mathbf{Y}_{N\times N}$ is the $N \times N$ submatrix of $\mathbf{Y}$ corresponding to its first N rows and columns. $\mathbf{I}_s$ represents a set of current sources in parallel with the first N ports; the first matrix term in (3.1.8) transforms the input- and output-port excitations into this set of current sources, so the $N+1$th and $N+2$th ports need not be considered further. The equivalent representation is shown in Figure 3.3. This transformation allows us to express the harmonic-balance equations as functions of currents at only the first through Nth ports, the ones connected to nonlinear elements.

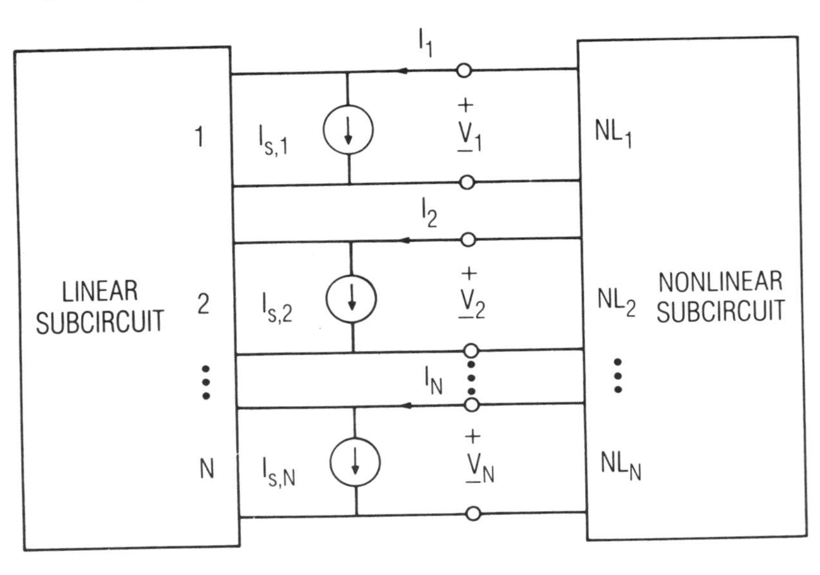

Figure 3.3 The circuit of Figure 3.2, in which the excitation voltage sources at Ports $N + 1$ and $N + 2$ have been transformed into current sources at Ports 1 to N.

The nonlinear-element currents, represented by the current vector $\hat{\mathbf{I}}$ on the right in (3.1.1), can result from nonlinear capacitors or resistor-conductors (because the nonlinear elements all are parts of the solid-state device, nonlinear inductors need not be considered). Inverse-Fourier-transforming the voltages at each port gives the time-domain voltage waveforms at each port:

$$\mathcal{F}^{-1}\{\mathbf{V}_n\} \rightarrow v_n(t) \tag{3.1.10}$$

The nonlinear capacitors will be treated first. Because the port voltages uniquely determine all voltages in the network, a capacitor's charge waveform can be expressed as a function of these voltages:

$$q_n(t) = f_{qn}(v_1(t), v_2(t), \ldots, v_N(t)) \tag{3.1.11}$$

Fourier-transforming the charge waveform at each port gives the charge vectors for the capacitors at each port:

$$\mathcal{F}\{q_n(t)\} \rightarrow \mathbf{Q}_n \tag{3.1.12}$$

and the charge vector $\mathbf{Q}$ is

$$\mathbf{Q} = \begin{bmatrix} \mathbf{Q}_1 \\ \mathbf{Q}_2 \\ \mathbf{Q}_3 \\ \cdot \\ \cdot \\ \cdot \\ \mathbf{Q}_N \end{bmatrix} = \begin{bmatrix} Q_{1,0} \\ Q_{1,1} \\ Q_{1,2} \\ \cdot \\ \cdot \\ \cdot \\ Q_{1,K} \\ Q_{2,0} \\ \cdot \\ \cdot \\ \cdot \\ Q_{2,K} \\ \cdot \\ \cdot \\ \cdot \\ Q_{N,K} \end{bmatrix} \tag{3.1.13}$$

The nonlinear-capacitor current is the time derivative of the charge waveform. Taking the time derivative corresponds to multiplying by $j\omega$ in the frequency domain, so

$$i_{c,n}(t) = \frac{dq_n(t)}{dt} \leftrightarrow jk\omega_p Q_{n,k} \tag{3.1.14}$$

Equation (3.1.14) can be written as

$$\mathbf{I}_c = j\mathbf{\Omega}\mathbf{Q} \tag{3.1.15}$$

where $\mathbf{\Omega}$ is the diagonal matrix:

$$\mathbf{\Omega} = \begin{bmatrix} 0 & 0 & 0 & 0 & \cdots & & & & & 0 \\ 0 & \omega_p & 0 & 0 & \cdots & & & & & 0 \\ 0 & 0 & 2\omega_p & 0 & \cdots & & & & & 0 \\ \vdots & \vdots & \vdots & \vdots & \ddots & & & & & \vdots \\ 0 & 0 & 0 & \cdots & 0 & K\omega_p & 0 & 0 & \cdots & 0 \\ \vdots & \vdots & \vdots & \cdots & 0 & 0 & 0 & 0 & \cdots & 0 \\ \vdots & \vdots & \vdots & & & & & \omega_p & & \\ \vdots & \vdots & \vdots & \cdots & 0 & 0 & 0 & 0 & 2\omega_p \cdots & 0 \\ \vdots & \vdots & \vdots & & \vdots & \vdots & \vdots & \vdots & \ddots & \vdots \\ 0 & 0 & 0 & \cdots & & & & & & K\omega_p \end{bmatrix} \quad (3.1.16)$$

This matrix has N cycles of $(0, \ldots, K)\omega_p$ along the main diagonal.

Similarly, the current in a nonlinear conductance (or in a controlled current source, as described in Chapter 2) is

$$i_{g,n}(t) = f_n(v_1(t), v_2(t), \ldots, v_n(t)) \quad (3.1.17)$$

Fourier transforming these gives

$$\mathcal{F}\{i_{g,n}(t)\} \rightarrow \mathbf{I}_{G,n} \quad (3.1.18)$$

and the vector:

$$\mathbf{I}_G = \begin{bmatrix} \mathbf{I}_{G,1} \\ \mathbf{I}_{G,2} \\ \mathbf{I}_{G,3} \\ \cdot \\ \cdot \\ \cdot \\ \mathbf{I}_{G,N} \end{bmatrix} \quad (3.1.19)$$

Substituting (3.1.9), (3.1.15), and (3.1.19) into (3.1.1) gives the expression:

$$\mathbf{F(V)} = \mathbf{I}_s + \mathbf{Y}_{N \times N}\mathbf{V} + j\mathbf{\Omega Q} + \mathbf{I}_G = \mathbf{0} \quad (3.1.20)$$

Equation (3.1.20) represents a test to determine whether a trial set of port voltage components is the correct one; that is, if $\mathbf{F}(\mathbf{V}) = \mathbf{0}$, then $\mathbf{V}$ is a valid solution. It also represents an equation that can be solved to obtain the port-voltage vector $\mathbf{V}$. Equation (3.1.20) is sometimes called the *harmonic-balance equation*. $\mathbf{F}(\mathbf{V})$, called the *current-error vector*, represents the difference between the current calculated from the linear and nonlinear subnetworks, at each port and at each harmonic, for a trial-solution vector $\mathbf{V}$.

Example 1

We will derive the current-error vector of the circuit in Figure 3.4, which consists of an ideal diode (one having no series resistance or junction capacitance) and a linear *embedding network* described by a Y matrix. As before, the source impedance $Z_s(\omega)$ is absorbed into the linear network. Figure 3.4 might represent, for example, the local-oscillator circuit in a simplified diode mixer model. The embedding network consists of a matching network and a source impedance that had been in series with $v_s(t)$, the excitation source, but in Figure 3.4 has been absorbed into the network. Because only one nonlinear element exists, $N = 1$ and the vector $\mathbf{V} = \mathbf{V}_1$. The Y matrix of the entire embedding network, $\mathbf{Y}_m$, can be written as

$$\mathbf{Y}_m = \begin{bmatrix} \mathbf{Y}_{1,1}\mathbf{Y}_{1,2} \\ \mathbf{Y}_{2,1}\mathbf{Y}_{2,2} \end{bmatrix} \tag{3.1.21}$$

$\mathbf{Y}_{1,1}$ is a submatrix that corresponds to $\mathbf{Y}_{N\times N}$ in (3.1.8) through (3.1.9), and $\mathbf{Y}_{1,2}$ corresponds to the leftmost submatrix in (3.1.8). Then,

$$\mathbf{I}_s = \mathbf{I}_{s,1} = \mathbf{Y}_{1,2}\mathbf{V}_2 \tag{3.1.22}$$

When $\mathbf{V}_2$ is transformed through the Y network, the equivalent circuit of Figure 3.5 results. The linear-circuit equations then depend only upon $\mathbf{I}_{s,1}$ and the admittances seen by the diode at each harmonic, the elements of $\mathbf{Y}_{1,1}$, often called the *embedding admittances*. $\mathbf{V}_2$ consists of only the fundamental source $V_s \cos(\omega_p t)$ and the dc bias source V_b, so

$$\mathbf{V}_2 = \begin{bmatrix} V_b \\ V_s \\ 0 \\ \cdot \\ \cdot \\ \cdot \\ 0 \end{bmatrix} \tag{3.1.23}$$

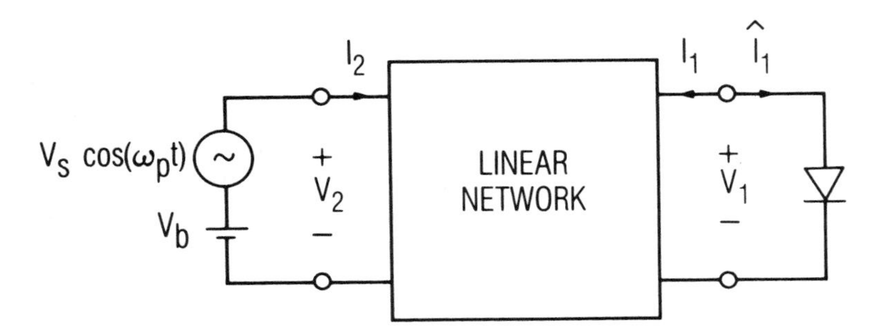

Figure 3.4 Pumped diode circuit of Example 1.

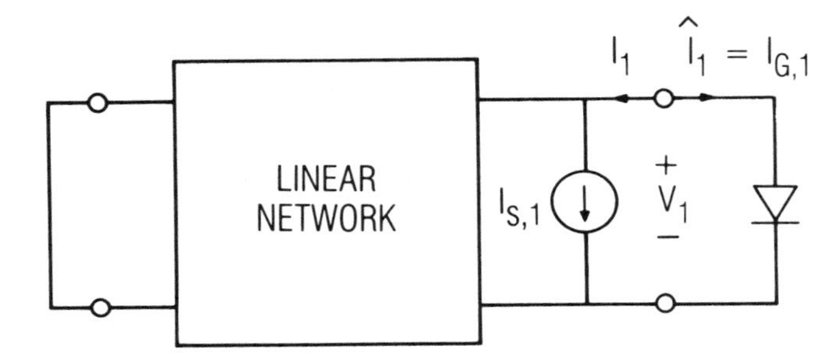

Figure 3.5 Simplified circuit of Example 1, with the two-port linear subcircuit reduced to one port.

We must make an initial estimate of either $v_1(t)$ or $\mathbf{V}_1$; previous experience suggests that a sinusoidal waveform clipped at approximately +0.6 V is often a good initial estimate of $v_1(t)$. Fourier-transforming $v_1(t)$ gives the components of $\mathbf{V}_1$. The diode current is found from the I/V equation of an ideal junction:

$$i_{g,1}(t) = I_{sat}[\exp(\delta v_1(t)) - 1] \tag{3.1.24}$$

where I_{sat} and δ are the diode's junction parameters (these are described in Section 2.3.1). Because there is no capacitance, $q_1(t) = 0$. Fourier-transforming $i_1(t)$ gives the components of the vector $\mathbf{I}_{G,1}$, so the current-error vector is

$$\mathbf{F(V)} = \mathbf{Y}_{1,2}\mathbf{V}_2 + \mathbf{Y}_{1,1}\mathbf{V}_1 + \mathbf{I}_{G,1} \tag{3.1.25}$$

One might wonder if the other Y parameters, $\mathbf{Y}_{2,1}$ and $\mathbf{Y}_{2,2}$, have any use. Indeed they do. Once $\mathbf{V}_1$ is known, they can be used to find the input current from the source, $\mathbf{I}_2$, by

$$\mathbf{I}_2 = \mathbf{Y}_{2,1}\mathbf{V}_1 + \mathbf{Y}_{2,2}\mathbf{V}_2 \tag{3.1.26}$$

$\mathbf{I}_2$, the vector of input currents at all the harmonics, can be used to find the input power:

$$P_{\text{in}} = \mathbf{V}_2\mathbf{I}_2^{*T} \tag{3.1.27}$$

where $^*T$ indicates the conjugate transpose. The power dissipated in the source impedance at $k\omega_p$ is

$$P_k = \frac{1}{2}\,|I_{2,k}|^2\ \text{Re}\{Z_s(k\omega_p)\} \tag{3.1.28}$$

This quantity is more important than it might seem at first, because a nonlinear circuit using a two-terminal device is often modeled in such a way that the linear subcircuit consists only of $Z_s(\omega)$; in this case, $Z_s(\omega)$ is called the *embedding impedance* and represents the impedance seen by the diode at each harmonic, including the source and output impedances. For example, this approach is often used to model a diode frequency multiplier wherein the source impedance is $Z_s(\omega_p)$ and the load at the kth harmonic is $Z_s(k\omega_p)$; then P_k is the output power. Knowing $\mathbf{I}_2$ we can find the fundamental-frequency input impedance:

$$Z_{\text{in}} = \frac{V_s}{I_{2,1}} - Z_s(\omega_p) \tag{3.1.29}$$

$Z_s(\omega_p)$, the source impedance that was previously absorbed into the Y network, is subtracted from the voltage-current ratio to obtain the input impedance of the diode–matching-network combination. As explained in Chapter 1, we cannot define a true input impedance of a nonlinear circuit because an impedance implies a V/I relationship that is independent of voltage or current magnitude. However, the "quasiimpedance" Z_{in} can be used in much the same way as a linear-circuit impedance. For example, Z_{in} is the input impedance to which the source should be matched in order to optimize power transfer at the specific value of V_s; it also relates voltage at one frequency component to the current, as does a linear impedance.

The one remaining problem—and the nastiest part of the whole business—is to solve (3.1.20) to obtain **V**. Each of the $K + 1$ frequency components of **V** at each port is a variable, and each component has a real and imaginary part. Thus, there are $2N(K + 1)$ variables to be determined (we concede that the dc components do not have imaginary parts; however, it is usually easier in the analysis to carry the dc terms' imaginary parts than to try to circumvent them). For example, a FET frequency-multiplier analysis might include nonlinear elements at three ports and have eight significant harmonics plus dc at each port. Thus, $N = 3$, $K = 8$, and there are 54 variables in (3.1.20)! Solving a set of equations having so many variables, especially in view of the circuit's nonlinear nature, is no mean task. In the next section, we examine algorithms for finding that solution.

3.1.2 Solution Algorithms

A number of algorithms have been proposed for solving Equation (3.1.20). Some of these are obvious applications of existing numerical techniques, but others show great ingenuity. The selection of an algorithm depends upon a number of factors, including computational efficiency, computer memory requirements, convergence properties, ease of implementation, and the need for an initial estimate of the solution. Some of these factors are interrelated; for example, if the procedure has good convergence properties, the initial solution estimate need not be very close to the correct solution and can be generated easily.

Other techniques that are not covered in this section have been proposed, most of these are variations on, or improvements in, those that follow. A few of these are described in References 3.4 through 3.6. Reference 3.7 contains an excellent general treatment of considerations involved in solving (3.1.20).

Optimization

At first glance, solving (3.1.20) looks a lot like an optimization problem. Therefore, we might be able to solve it by minimizing the magnitude squared of the current-error function; that is, minimize ϵ where

$$\epsilon = \mathbf{F}(\mathbf{V})\mathbf{F}^{*T}(\mathbf{V}) \tag{3.1.30}$$

An advantage of this approach is that most computer installations have a library of scientific subroutines that includes a general purpose functional optimization routine. Thus, a large and difficult part of the computer programming is completed at the outset. However, most optimization routines are relatively slow and may have convergence problems, especially when a large number of variables must be optimized simultaneously. The popular *gradient-search* technique (sometimes optimistically called the *method of steepest descents*) is particularly bad in this respect; when the number of variables is large, it tends to wander in the vicinity of the global optimum but never quite reaches it. Optimization techniques that do not exhibit this behavior and have better global convergence are much slower. Furthermore, forming the error function in (3.1.30) destroys a lot of information about the individual contribution of each variable to the error. Because of these limitations, optimization is a reasonable approach only for relatively simple problems (e.g., Example 1), in which the ease of programming outweighs the inefficiency. (A discussion of optimization is beyond the scope of this book; a very good and thoroughly readable introduction to optimization can be found in Reference 3.8).

Splitting Methods

A number of relaxation methods, which are both simple to implement and intuitively satisfying, have been proposed. Two of the most popular, which are sometimes called *splitting methods*, are those of Hicks and Khan (Reference 3.9) and Kerr (Reference 3.10). The former is the basis of this section; the latter, called the *reflection algorithm*, is sufficiently distinct to warrant separate examination. This algorithm begins with an initial estimate of the solution vector $\mathbf{V}$; the nonlinear element's current is determined, and the linear circuit's current is assumed to be equal to it. A new voltage vector $\mathbf{V}''$ is calculated from the current and the linear subcircuit, and a new estimate of the solution vector, which is geometrically between $\mathbf{V}$ and $\mathbf{V}''$ is generated. This process continues until convergence is achieved.

The success of this method depends strongly upon the quality of the initial estimate, as well as upon the criteria for generating the new estimate. The initial estimate depends entirely upon both the nature of the circuit and the expected response, so little more of a general nature can be said of it. Reference 3.9 postulates that a good estimate of the update voltage vector is one that reduces the geometrical distance between $\mathbf{V}$ and $\mathbf{V}''$. This idea is valid if the change in $\mathbf{V}$ is not too great.

The algorithm requires an initial estimate of $\mathbf{V}$, $\mathbf{V}^0$ (the superscript is the iteration number). $\hat{\mathbf{I}}^0$ is then calculated from the nonlinear elements by first Fourier-transforming $\mathbf{V}^0$ to obtain $v_n^0(t)$, substituting $v_n^0(t)$ into the nonlinear elements' I/V and Q/V relations to obtain current and charge waveforms, and finally Fourier-transforming again. In some situations, it may be more convenient to make the initial estimate in the time domain; in this case, the nonlinear elements' currents can be calculated directly.

$\mathbf{I}^0$ is then estimated as $-\hat{\mathbf{I}}^0$, and a new voltage vector $\mathbf{V}''$ is found from the linear subcircuit; that is,

$$\mathbf{V}'' = \mathbf{Z}_{N\times N}(\mathbf{I}^0 - \mathbf{I}_s) \tag{3.1.31}$$

where

$$\mathbf{Z}_{N\times N} = \mathbf{Y}_{N\times N}^{-1} \tag{3.1.32}$$

The new estimate of $\mathbf{V}$ is found from

$$\mathbf{V}^1 = s\mathbf{V}'' + (1 - s)\mathbf{V}^0 \tag{3.1.33}$$

where s is a real constant, $0 < s < 1.0$. $\mathbf{V}^0$ is then replaced by $\mathbf{V}^1$, and the process is repeated until minimal change in the voltage vector is observed between iterations.

The variable s in (3.1.33), called the *splitting coefficient*, is a constant that must be determined empirically. Small values of s favor slow but monotonic convergence; increasing s gives faster convergence up to the point at which oscillation begins, then the convergence rate decreases and, eventually, the process becomes unstable. Choosing $s = 1$ almost always causes instability; typically, $s = 0.2$. The splitting coefficient may be even smaller if the initial estimate is not very good, the circuit is strongly nonlinear, or the magnitude of the excitation is large. The following example illustrates the use of the splitting algorithm.

Example 2

The circuit of Example 1, Figures 3.4 and 3.5, will be solved by means of the splitting algorithm. We begin by generating an impedance matrix representing the embedding network; this task is particularly easy, because it requires inverting $\mathbf{Y}_{1,1}$, a diagonal matrix. The circuit model is shown in Figure 3.6; the current-source vector $\mathbf{I}_s$ in Figure 3.6 is the same as that in Figure 3.5. The solution process is as follows:

1. Form an initial estimate of $v_1(t)$. A good initial estimate might be a sinusoid, clipped at approximately $+0.6$ V, with a dc offset equal to the bias voltage.

2. Calculate $i_{g,1}(t) = I_{sat}[\exp(\delta v_1(t)) - 1]$
3. Fourier-transform $i_{g,1}(t)$ and $v_1(t)$ to obtain $\mathbf{I}^0_{G,1}$ and $\mathbf{V}^0_1$, respectively.
4. Assume $\mathbf{I} = -\mathbf{I}^0_{G,1}$ and form $\mathbf{V}'' = \mathbf{Z}_{1,1}(\mathbf{I} - \mathbf{I}_s)$
5. Form the new estimate of $\mathbf{V}_1$ as $\mathbf{V}^1_1 = s\mathbf{V}'' + (1 - s)\mathbf{V}^0_1$.
6. Inverse-Fourier-transform $\mathbf{V}^1_1$ to get a new estimate of $v_1(t)$, and repeat from Step 2 until convergence is achieved.
7. If instability is encountered, reduce s and repeat the procedure from Step 1.

The most important advantages of this algorithm are its conceptual simplicity and the ease with which it can be programmed. Its main disadvantages are the slowness of its convergence, its tendency toward instability, and the numerous iterations required for analyzing nearly linear circuits as well as for analyzing strongly nonlinear ones. Identifying the optimum value of the splitting coefficient a priori is a significant problem, because the optimum value is different for each circuit and set of circuit parameters and may vary throughout the iterative procedure. Some of these problems can be illustrated by the following very simple example.

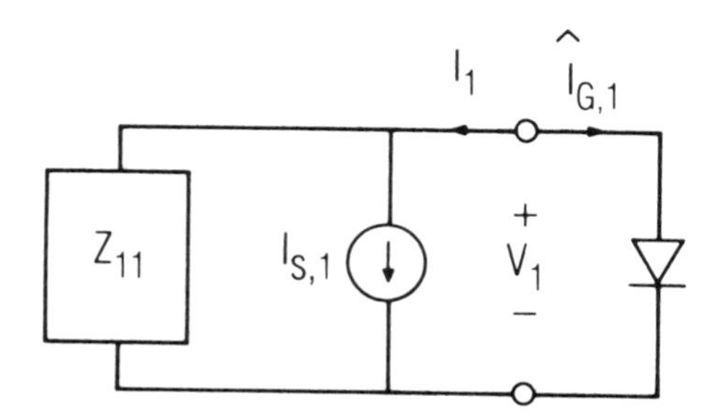

Figure 3.6 Pumped diode circuit for Example 2. This circuit is the same as that of Figure 3.5 except that the linear subnetwork is described as an impedance.

Example 3

The voltage divider network in Figure 3.7 will be analyzed to find the output voltage V. Not content to do things the easy way, we can pretend that this linear circuit is a sort of degenerate nonlinear circuit, and use the splitting algorithm. We begin by estimating $V = 0.75$, and use $s = 0.2$. After the first six iterations (p is the iteration number), we obtain the following:

p	V^p	I	V''
0	0.75	−1.5	0.5
1	0.70	−1.4	0.6
2	0.68	−1.36	0.64
3	0.672	−1.344	0.656
4	0.669	−1.338	0.662
5	0.668	−1.336	0.664
6	0.667		

Six iterations are required to obtain three-digit accuracy, a remarkably large amount of computing in view of the circuit's simplicity. Because of the low value of s, however, convergence is monotonic. We now try to improve the rate of convergence by selecting $s = 0.5$. We obtain

p	V^p	I	V''
0	0.75	−1.5	0.5
1	0.625	−1.25	0.75
2	0.688	−1.375	0.625
3	0.656	−1.313	0.688
4	0.672	−1.344	0.656
5	0.664	−1.328	0.672
6	0.668		

After the same number of iterations, the result is no better. This time the steps toward convergence were larger, but the estimated solution V^p oscillated around the correct value. This result indicates that the optimum value of s is somewhat smaller than 0.5. Finally, just to see what happens, we try $s = 1.0$:

p	V^p	I	V''
0	0.75	−1.5	0.5
1	0.50	−1.0	1.0
2	1.00	−2.0	0.0
3	0.0	0.0	2.0
4	2.0	−4.0	−2.0
5	−2.0		

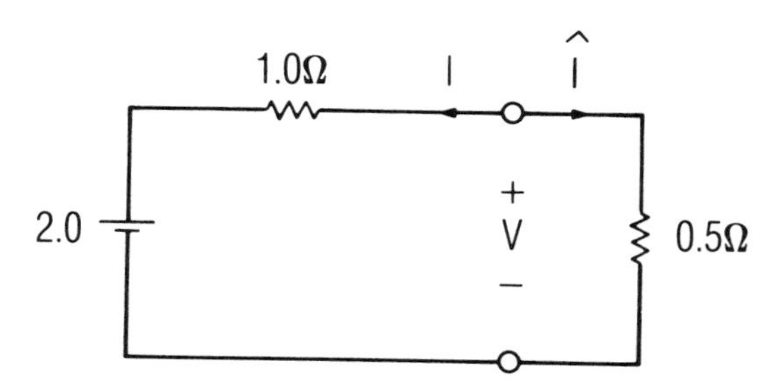

Figure 3.7 Voltage divider circuit of Example 3.

In this case, V^p diverges quite rapidly. Thus, depending on the value of s, convergence may not be achieved even when the circuit is linear and the initial estimate of the solution is very close, less than 15 percent away. In fact, it can be shown (Reference 3.7) that convergence is not assured in some cases, *no matter how close the initial estimate is to the solution!* This example illustrates the fact that the convergence of the splitting algorithm, even when it is applied to linear circuits, is precarious. In nonlinear circuits, it is even more so.

Newton's Method

Newton's method is a very powerful algorithm for finding the zeros of a multivariate function; because the harmonic-balance method involves finding the zeros of $\mathbf{F}(\mathbf{V})$, Newton's method is an obvious choice as a solution algorithm. Newton's method is an iterative technique; it finds the zero of a function by using its first derivative to extrapolate to the axis of the independent variable and repeating the process until the zero is found and has the desired accuracy. Its power comes from its use of the voltage vector and all the derivatives of $\mathbf{F}(\mathbf{V})$ with respect to the voltage components of V in each iteration to estimate a new voltage vector.

This iterative process is most easily illustrated by applying it to a one-dimensional problem. Figure 3.8 shows a function of one variable, $f(x)$, and a Newton estimate of its zero. We can write, for the linear extrapolation:

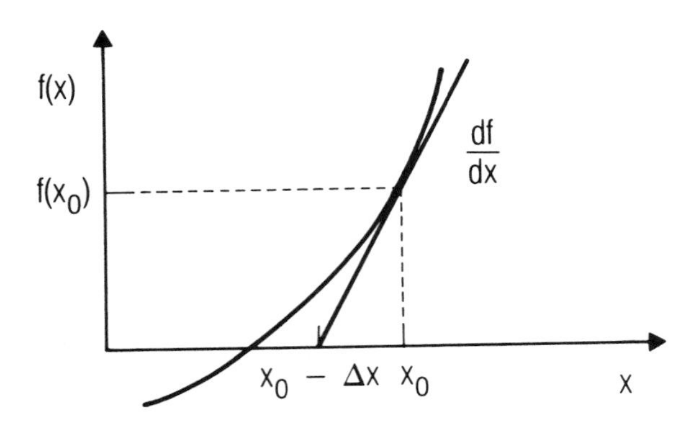

Figure 3.8 Newton's method applied to a one-dimensional problem, estimating the zero of $f(x)$.

$$f(x_0) - \left.\frac{df}{dx}\right|_{x=x_0} \Delta x = 0 \tag{3.1.34}$$

If $f(x_0)$ and its derivative are known, (3.1.34) can be solved easily to obtain Δx, and a new estimate of the zero is found as $x_0 - \Delta x$. The function and its derivative are again evaluated at the estimated zero, and the process is repeated at the new point until the zero is determined with the required accuracy. The analog of (3.1.34) that can be applied to multidimensional functions is

$$\mathbf{F}(\mathbf{V}^p) - \left.\frac{\partial \mathbf{F}(\mathbf{V})}{\partial \mathbf{V}}\right|_{\mathbf{V}=\mathbf{V}^p} \Delta \mathbf{V} = 0 \tag{3.1.35}$$

where $\mathbf{V}^p$ is the pth estimate of the solution vector. With

$$\mathbf{V}^p - \mathbf{V}^{p+1} = \Delta \mathbf{V} \tag{3.1.36}$$

the updated vector, $\mathbf{V}^{p+1}$, is

$$\mathbf{V}^{p+1} = \mathbf{V}^p - \left.\frac{\partial \mathbf{F}(\mathbf{V})}{\partial \mathbf{V}}\right|_{\mathbf{V}=\mathbf{V}^p}^{-1} \mathbf{F}(\mathbf{V}^p) \tag{3.1.37}$$

The trick here is to generate the derivative of $\mathbf{F}(\mathbf{V})$, which is called the *Jocobian of* $\mathbf{F}$. Taking the derivative of (3.1.20) gives

$$\mathbf{J}_F = \frac{\partial \mathbf{F}(\mathbf{V})}{\partial \mathbf{V}} = \mathbf{Y}_{N\times N} + \frac{\partial \mathbf{I}_G}{\partial \mathbf{V}} + j\mathbf{\Omega}\frac{\partial \mathbf{Q}}{\partial \mathbf{V}}$$

$$= \begin{bmatrix}
\frac{\partial F_{1,0}}{\partial V_{1,0}} & \frac{\partial F_{1,0}}{\partial V_{1,1}} & \frac{\partial F_{1,0}}{\partial V_{1,2}} & \cdots & \frac{\partial F_{1,0}}{\partial V_{1,K}} & \frac{\partial F_{1,0}}{\partial V_{2,0}} & \frac{\partial F_{1,0}}{\partial V_{2,1}} & \frac{\partial F_{1,0}}{\partial V_{2,2}} & \cdots & \frac{\partial F_{1,0}}{\partial V_{N,K}} \\
\frac{\partial F_{1,1}}{\partial V_{1,0}} & \frac{\partial F_{1,1}}{\partial V_{1,1}} & \frac{\partial F_{1,1}}{\partial V_{1,2}} & \cdots & \frac{\partial F_{1,1}}{\partial V_{1,K}} & \frac{\partial F_{1,1}}{\partial V_{2,0}} & \frac{\partial F_{1,1}}{\partial V_{2,1}} & \frac{\partial F_{1,1}}{\partial V_{2,2}} & \cdots & \frac{\partial F_{1,1}}{\partial V_{N,K}} \\
\vdots & \vdots & \vdots & & \vdots & \vdots & \vdots & \vdots & & \vdots \\
\frac{\partial F_{1,K}}{\partial V_{1,0}} & \frac{\partial F_{1,K}}{\partial V_{1,1}} & \frac{\partial F_{1,K}}{\partial V_{1,2}} & \cdots & \frac{\partial F_{1,K}}{\partial V_{1,K}} & \frac{\partial F_{1,K}}{\partial V_{2,0}} & \frac{\partial F_{1,K}}{\partial V_{2,1}} & \frac{\partial F_{1,K}}{\partial V_{2,2}} & \cdots & \frac{\partial F_{1,K}}{\partial V_{N,K}} \\
\frac{\partial F_{2,0}}{\partial V_{1,0}} & \frac{\partial F_{2,0}}{\partial V_{1,1}} & \frac{\partial F_{2,0}}{\partial V_{1,2}} & \cdots & \frac{\partial F_{2,0}}{\partial V_{1,K}} & \frac{\partial F_{2,0}}{\partial V_{2,0}} & \frac{\partial F_{2,0}}{\partial V_{2,1}} & \frac{\partial F_{2,0}}{\partial V_{2,2}} & \cdots & \frac{\partial F_{2,0}}{\partial V_{N,K}} \\
\frac{\partial F_{2,1}}{\partial V_{1,0}} & \frac{\partial F_{2,1}}{\partial V_{1,1}} & \frac{\partial F_{2,1}}{\partial V_{1,2}} & \cdots & \frac{\partial F_{2,1}}{\partial V_{1,K}} & \frac{\partial F_{2,1}}{\partial V_{2,0}} & \frac{\partial F_{2,1}}{\partial V_{2,1}} & \frac{\partial F_{2,1}}{\partial V_{2,2}} & \cdots & \frac{\partial F_{2,1}}{\partial V_{N,K}} \\
\vdots & \vdots & \vdots & & \vdots & \vdots & \vdots & \vdots & & \vdots \\
\frac{\partial F_{2,K}}{\partial V_{1,0}} & \frac{\partial F_{2,K}}{\partial V_{1,1}} & \frac{\partial F_{2,K}}{\partial V_{1,2}} & \cdots & \frac{\partial F_{2,K}}{\partial V_{1,K}} & \frac{\partial F_{2,K}}{\partial V_{2,0}} & \frac{\partial F_{2,K}}{\partial V_{2,1}} & \frac{\partial F_{2,K}}{\partial V_{2,2}} & \cdots & \frac{\partial F_{2,K}}{\partial V_{N,K}} \\
\vdots & \vdots & \vdots & & \vdots & \vdots & \vdots & \vdots & & \vdots \\
\frac{\partial F_{N,K}}{\partial V_{1,0}} & \frac{\partial F_{N,K}}{\partial V_{1,1}} & \frac{\partial F_{N,K}}{\partial V_{1,2}} & \cdots & \frac{\partial F_{N,K}}{\partial V_{1,K}} & \frac{\partial F_{N,K}}{\partial V_{2,0}} & \frac{\partial F_{N,K}}{\partial V_{2,1}} & \frac{\partial F_{N,K}}{\partial V_{2,2}} & \cdots & \frac{\partial F_{N,K}}{\partial V_{N,K}}
\end{bmatrix} \tag{3.1.38}$$

The terms of $\mathbf{J}_F$ are

$$\frac{\partial F_{n,k}}{\partial V_{m,l}} = Y_{n,m}\,(k = l) + \frac{\partial I_{G;n,k}}{\partial V_{m,l}} + jk\omega_p \frac{\partial Q_{n,k}}{\partial V_{m,l}} \tag{3.1.39}$$

where, as before, the n and m subscripts refer to the port number in Figure 3.2, and the k and l subscripts indicate the harmonic number. The term $Y_{n,m}\ (k = 1)$ is $Y_{n,m}\ (k\omega_p)$ when $k = 1$, and zero when $k \neq 1$. One can show (References 3.7, 3.11) that

$$\frac{\partial I_{G;n,k}}{\partial V_{m,l}} = \frac{1}{T}\int_{-1/2T}^{1/2T} \frac{\partial i_{g,n}(t)}{\partial v_m(t)} \exp[-j(k - l)\omega_p t]\, dt \tag{3.1.40}$$

and

$$\frac{\partial Q_{n,k}}{\partial V_{m,l}} = \frac{1}{T}\int_{-1/2T}^{1/2T} \frac{\partial q_n(t)}{\partial v_m(t)} \exp[-j(k - l)\omega_p t]\, dt \tag{3.1.41}$$

T is the period of the fundamental excitation frequency. These terms are just the Fourier-series coefficients of the derivative waveforms of the nonlinear elements. Many of these are zero, because some $i_{g,n}$ and q_n are independent of some v_m. The following example clarifies the process.

Example 4

The circuit of Example 1 and Figures 3.4 and 3.5 will be solved by means of Newton's method. $\mathbf{F}(\mathbf{V})$ is given by (3.1.25); differentiating gives the terms of the Jacobian:

$$\mathbf{J}_F = \frac{\partial \mathbf{F}(\mathbf{V})}{\partial \mathbf{V}_1} \tag{3.1.42}$$

or

$$\frac{\partial F_{1,k}}{\partial V_{1,l}} = Y_{1,1}\,(k = l) + \frac{\partial I_{G;1,k}}{\partial V_{1,l}} \tag{3.1.43}$$

where $\mathrm{Y}_{1,1}(k\omega_p)$ is the kth element of $\mathbf{Y}_{1,1}$, given by (3.1.6), and

$$\frac{\partial I_{G;1,k}}{\partial V_{1,l}} = \frac{1}{T}\int_{-1/2T}^{1/2T} \frac{\partial i_{g,1}(t)}{\partial v_1(t)} \exp[-j(k - l)\omega_p t]\, dt \tag{3.1.44}$$

The partial derivative within the integral sign can be interpreted as the incremental junction conductance of the diode; that is:

$$g(t) = \left.\frac{\partial i_g}{\partial v}\right|_{v=v_1(t)} = \frac{d}{dv}\left[I_{sat}\left[\exp(\delta v) - 1\right]\right]\Bigg|_{v=v_1(t)} \tag{3.1.45}$$
$$= \delta I_{sat} \exp(\delta v_1(t))$$
$$\approx \delta i_{g,1}(t)$$

Fourier-transforming $g(t)$ gives the frequency components G_k, $k = -K, \ldots, 0, \ldots, K$. The Jacobian $\mathbf{J}_F$ has the form:

$$\mathbf{J}_F = \begin{bmatrix} \frac{\partial F_{1,0}}{\partial V_{1,0}} & \frac{\partial F_{1,0}}{\partial V_{1,1}} & \frac{\partial F_{1,0}}{\partial V_{1,2}} & \cdots & \frac{\partial F_{1,0}}{\partial V_{1,K}} \\ \frac{\partial F_{1,1}}{\partial V_{1,0}} & \frac{\partial F_{1,1}}{\partial V_{1,1}} & \frac{\partial F_{1,1}}{\partial V_{1,2}} & \cdots & \frac{\partial F_{1,1}}{\partial V_{1,K}} \\ \frac{\partial F_{1,2}}{\partial V_{1,0}} & \frac{\partial F_{1,2}}{\partial V_{1,1}} & \frac{\partial F_{1,2}}{\partial V_{1,2}} & \cdots & \frac{\partial F_{1,2}}{\partial V_{1,K}} \\ \vdots & \vdots & \vdots & & \vdots \\ \frac{\partial F_{1,K}}{\partial V_{1,0}} & \frac{\partial F_{1,K}}{\partial V_{1,1}} & \frac{\partial F_{1,K}}{\partial V_{1,2}} & \cdots & \frac{\partial F_{1,K}}{\partial V_{1,K}} \end{bmatrix} \tag{3.1.46}$$

and from (3.1.43), the Jacobian is

$$\mathbf{J}_F = \begin{bmatrix} Y_{1,1}(0) + G_0 & G_{-1} & G_{-2} & \cdots & G_{-K} \\ G_1 & Y_{1,1}(\omega_p) + G_0 & G_{-1} & \cdots & G_{-K+1} \\ G_2 & G_1 & Y_{1,1}(2\omega_p) + G_0 & \cdots & G_{-K+2} \\ \vdots & \vdots & \vdots & & \vdots \\ G_K & G_{K-1} & G_{K-2} & \cdots & Y_{1,1}(K\omega_p) + G_0 \end{bmatrix}$$

The solution is found by the following process:

1. Form an initial estimate of the waveform $v_1(t)$. As in the previous example, a clipped sinusoid would be a good guess.
2. Fourier-transform $v_1(t)$ to obtain $\mathbf{V}_1^0$, the initial estimate in the frequency domain. The superscript represents the iteration number.

3. Find the conductance waveform $g(t)$ via (3.1.45) and Fourier-transform it.
4. Form $\mathbf{J}_F$ and $\mathbf{F}(\mathbf{V}^0)$ via (3.1.47) and (3.1.25), where $\mathbf{V}^0$ is the combination of $\mathbf{V}_2$ and $\mathbf{V}_1^0$ as shown in (3.1.2) and (3.1.4).
5. Solve (3.1.37) to find the new estimate of the voltage vector $\mathbf{V}_1^1$.
6. Fourier-transform the diode current, which was found in Step 3, and form the vector $\mathbf{I}_{G,1}$.
7. Use (3.1.25) to determine $\mathbf{F}(\mathbf{V}^1)$.
8. If the magnitudes of the components of $\mathbf{F}(\mathbf{V}^1)$ are small enough, the solution has been found. Otherwise, inverse-Fourier-transform to obtain $v_1(t)$, and repeat from Step 3 to obtain $\mathbf{V}_1^2$.

The main advantage of Newton's method is that it makes full use of all the derivatives of the error function with respect to each variable at each port. For this reason it is capable of achieving convergence with a very large number of variables, as long as the nonlinearity is not too strong and the initial estimate is reasonably good (in fact, the algorithm converges in a single step if the circuit is linear). The disadvantage of this algorithm is in the large amount of computer memory and computation time required to generate the Jacobian and to solve the matrix equation (3.1.35). The Jacobian is a square complex matrix of dimension $N(K + 1)$; in our earlier example of a FET frequency multiplier having three nonlinear elements and eight harmonics plus dc, the Jacobian is 27×27. Because the Jacobian is complex, solving (3.1.35) for this case involves solving a 54×54 set of real linear equations, a task that is not intrinsically difficult but is time-consuming and, therefore, expensive. Also, the entire matrix and the solution and update vectors must remain in memory simultaneously; thus, Newton's method requires a large segment of computer memory (many of the matrix entries are zero in large problems, so the use of sparse-matrix techniques can ameliorate this situation somewhat). Finally, generating the Jacobian requires taking a large number of derivatives. This calculation is usually not time-consuming if closed-form expressions are available for those derivatives; however, a closed-form expression is not available for many nonlinear elements, so the derivatives often must be evaluated numerically. This process also consumes a lot of computer time.

One way to minimize the time required to generate the Jacobian is to reuse the Jacobian until the magnitude of the current error function $\mathbf{F}(\mathbf{V})$ begins to rise. If the circuit is not too strongly nonlinear, the Jacobian may be used successfully for a few iterations before it has to be recalculated. Another approach, which also is possible if the nonlinearities are weak, is to eliminate the calculation of some of the Jacobian's more extreme terms, those toward the top right and bottom left corners of the matrix;

these relate low-harmonic error currents to the high-harmonic voltages and are likely to be negligibly small. Carrying this approach to its extreme, we can calculate and use only the main diagonal of $\mathbf{J}_F$. In this case, matrix inversion is very simple, requiring only the inversion of the individual elements of the diagonal, and the computer memory need contain only the terms at one frequency at a time. This approach is attractive because of its high speed and modest memory requirements; however, its convergence properties are often poor.

Even with the best efforts, convergence problems are sometimes encountered in the algorithm, especially in complex circuits having large excitations. One way to circumvent such problems is to begin with the excitation set at a very low level, so that the nonlinearities are weak and a solution is obtained easily. The excitation is then increased, and the previous solution is used as the initial estimate for the next. The process is repeated until the solution is obtained at the desired excitation level. Unless the circuit is inherently unstable, a solution can usually be obtained in this manner, even if the circuit's nonlinearity is strong and convergence is difficult.

Reflection Algorithm

The reflection algorithm solves the harmonic balance equations via a process that mimics the turn-on process of a real circuit. If the circuit is stable, after turn-on it must settle into steady-state operation, so the mathematical process probably will, too.

To implement the algorithm, the equivalent circuit of Figure 3.2 is redrawn as shown in Figure 3.9, so that it is divided into linear and nonlinear subcircuits. According to convenience, the linear subcircuit may be analyzed in either the frequency or time domain. The nonlinear subcircuit must contain all the nonlinear elements, but it may also contain linear elements. The greatest difference between the circuits of Figure 3.2 and Figure 3.9 is the set of transmission lines between the respective ports of the linear and nonlinear networks. These transmission lines are assumed to be ideal, as well as to be a large integer number of wavelengths long at the fundamental excitation frequency; thus, they do not affect the circuit's steady-state response.

When the circuit is turned on, voltage waveforms appear at the N ports and incident waves are impressed on the transmission lines. These propagate toward, and eventually reach, the respective ports of the nonlinear subcircuit. The waves excite the nonlinear elements, and reverse-traveling waves are generated. These return to the ports of the linear

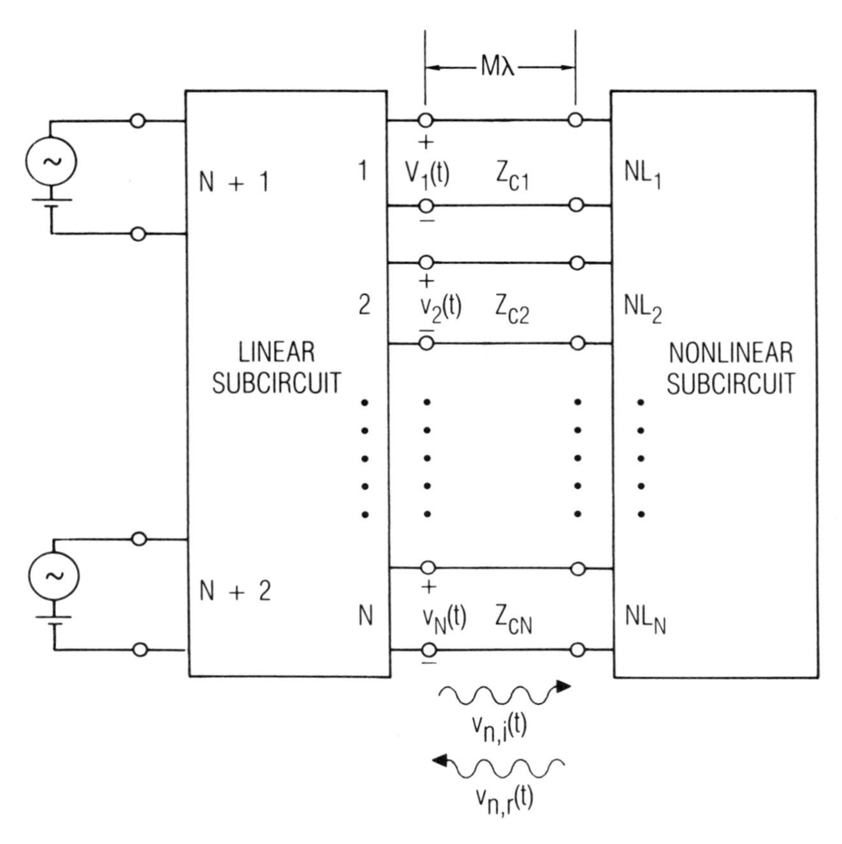

Figure 3.9 A general nonlinear microwave circuit, partitioned into linear and nonlinear subnetworks, with fictitious ideal transmission lines, for analysis via the reflection algorithm.

network, are again reflected, and add to the incident wave; then, the process is repeated. Eventually, steady-state conditions are reached, indicated by minimal change in the incident waves between iterations.

The ruse of the ideal transmission lines allows the circuit to be separated into two parts as shown in Figure 3.10. At the instant the circuit is turned on, no reflected wave exists, so the input impedance of the transmission line is equal to its characteristic impedance. The original incident waves at the N ports, $v_{n,i}^{0}(t)$, $n = 1, \ldots, N$, can be found by a straightforward analysis of the linear subnetwork (as before, the superscript indicates iteration number). When the incident wave reaches the nonlinear subcircuit, the transmission line interface has the equivalent circuit shown

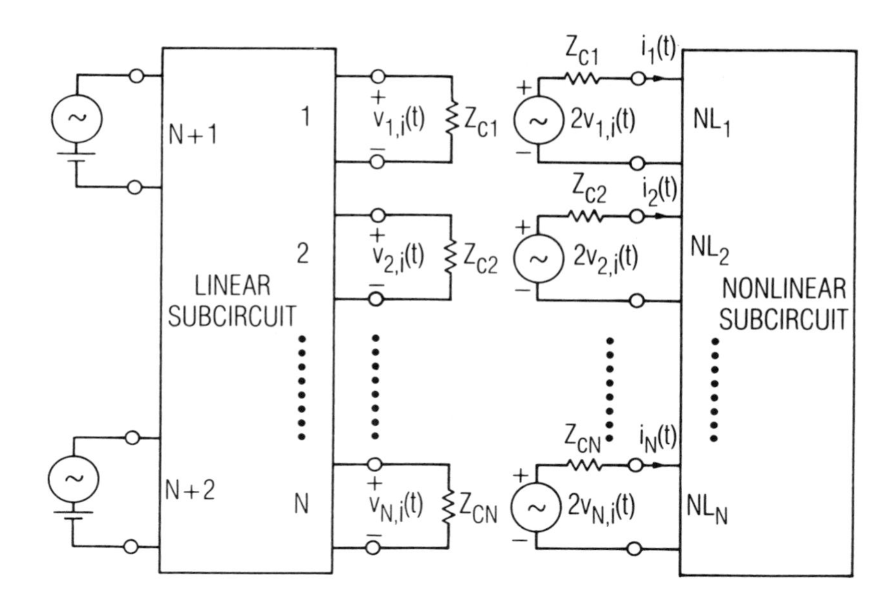

Figure 3.10 Incident-wave circuit for the reflection algorithm, equivalent to that of Figure 3.9.

on the right side of the figure. The reverse wave $v_{n,r}^0(t)$ is found by analyzing the nonlinear network to determine the currents $i_n^0(t)$, and by forming:

$$v_{n,r}^0(t) = v_{n,i}^0(t) - i_n^0(t)\, Z_{cn} \tag{3.1.47}$$

The calculation of $i_n^0(t)$ is often simpler than it looks. Usually, the nonlinear subcircuit consists simply of a set of independent two-terminal nonlinear elements, one connected to each port. In this case, $i_n^0(t)$ can be found algebraically for conductances, and via a simple first-order nonlinear differential equation for capacitances. In the worst case, the nonlinear subcircuit may be a complete solid-state device model, for which a set of nonlinear differential equations must be solved numerically.

To form the new incident wave, the reflected wave must first be Fourier-transformed:

$$\mathcal{F}\{v_{n,r}^0(t)\} \Rightarrow V_{n,r}^0(k\omega_p) \tag{3.1.48}$$

where $V_{n,r}(k\omega_p)$ is the reflected phasor voltage component from Port n at frequency $k\omega_p$, and $V_{n,r}(0)$ is twice the dc value. The reflected wave is multiplied by the source reflection coefficient at the linear-circuit end of the line, and the incident wave in the next iteration, $v_{n,i}^{1}(t)$, is

$$v_{n,i}^{1}(t) = v_{n,i}^{0}(t) + \frac{1}{2} \sum_{k=-K}^{K} \Gamma_n(k\omega_p) V_{n,r}^{0}(k\omega_p) \exp(jk\omega_p t) \tag{3.1.49}$$

After the pth iteration, the new incident wave is

$$v_{n,i}^{p+1}(t) = v_{n,i}^{0}(t) + \frac{1}{2} \sum_{k=-K}^{K} \Gamma_n(k\omega_p) V_{n,r}^{p}(k\omega_p) \exp(jk\omega_p t) \tag{3.1.50}$$

Note that the summation term is always added to the *initial* incident wave, $v_{n,i}^{0}(t)$, not to the previous one. The reflection coefficient $\Gamma_n(k\omega_p)$ is given by

$$\Gamma_n(k\omega_p) = \frac{Z_n(k\omega_p) - Z_{cn}}{Z_n(k\omega_p) + Z_{cn}} \tag{3.1.51}$$

where $Z_n(k\omega_p)$ is the impedance looking into the nth port of the linear subcircuit at frequency $k\omega_p$. When the process has converged, after the Pth iteration, the port-voltage waveform is

$$v_{n}^{P}(t) = v_{n,i}^{P}(t) + v_{n,r}^{P}(t) \tag{3.1.52}$$

Two complications often arise in the implementation of the reflection algorithm. The first concerns the selection of the characteristic impedances Z_{cn}. The algorithm usually is not very sensitive to the choice of Z_{cn}, but it does affect convergence rate somewhat. Equations (3.1.50) and (3.1.51) indicate that Z_{cn} could be selected to minimize the reflection coefficients at the lower harmonics of the excitation frequency. This choice improves the speed of convergence, because the lower harmonics invariably dominate the convergence process. The second complication is that the source impedance at dc at each port, in most circuits, is usually zero. Thus, $\Gamma_n(0) = -1$, and the dc component of the transmission-line waves changes polarity with each iteration, converging only very slowly to its correct value. The wide variation often introduces numerical instability; it can also force some of the voltages in the nonlinear circuit, during intermediate iterations, to exceed the range of the model. The simplest way to avoid this problem is artificially to set the dc source impedance at each port equal to Z_{cn}, making the reflection coefficient zero. This change will, of course, affect the dc values of the voltages at the nonlinear elements, so the bias source value will have to be offset to compensate. Again, the process is clarified by an example.

Example 5

The circuit of Figures 3.4 through 3.6 will be analyzed via the reflection algorithm. The equivalent circuit, including the transmission line, is shown in Figure 3.11. The linear part of the circuit has been converted to a Thevenin equivalent, and the excitation and bias sources have been transformed appropriately, their values distinguished from those in Figures 3.4 through 3.6 by the prime.

Figure 3.12(a) shows the equivalent circuit used for calculating the initial incident wave $v_i^0(t)$. The result is

$$v_i^0(t) = \frac{V_s' Z_c \cos(\omega_p t + \theta)}{[Z_c^2 + |Z_{1,1}(\omega_p)|^2]^{1/2}} + \frac{V_b' Z_c}{Z_c + Z_{1,1}(0)} \tag{3.1.53}$$

$$\theta = \tan^{-1}\left[\frac{\mathrm{Im}\{Z_{1,1}(\omega_p)\}}{\mathrm{Re}\{Z_{1,1}(\omega_p)\} + Z_c}\right]$$

At the right end of the circuit, Figure 3.12(b), the diode current can be found from the equation:

$$i^0(t) = I_{\mathrm{sat}}\{\exp[\delta(2v_i^0(t) - Z_c i^0(t))] - 1\} \tag{3.1.54}$$

Equation (3.1.53) must be solved to obtain $i^0(t)$. The reverse wave is then found:

$$v_r^0(t) = v_i^0(t) - i^0(t) Z_c \tag{3.1.55}$$

and it is Fourier-transformed to find the frequency-domain voltage components $V_r^0(k\omega_p)$. The reflection coefficients are

$$\Gamma(k\omega_p) = \frac{Z_{1,1}(k\omega_p) - Z_c}{Z_{1,1}(k\omega_p) + Z_c} \tag{3.1.56}$$

and the new incident voltage wave is

$$v_i^1(t) = v_i^0(t) + \frac{1}{2}\sum_{k=-K}^{K} \Gamma(k\omega_p) V_r^0(k\omega_p) \exp(jk\omega_p t) \tag{3.1.57}$$

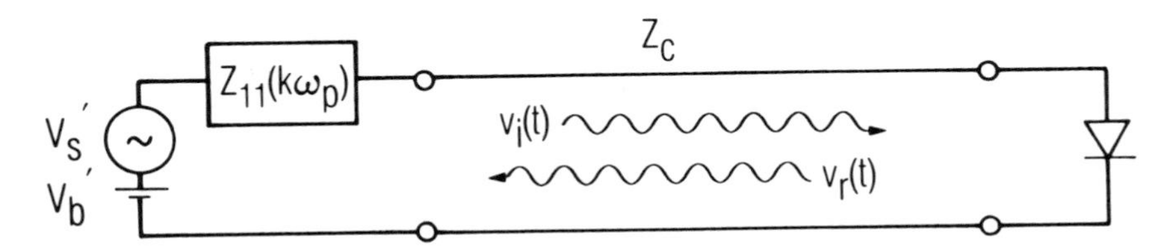

Figure 3.11 Circuit for Example 5. The two-port embedding network has been converted to a series impedance, and the excitation voltage sources have been transformed appropriately.

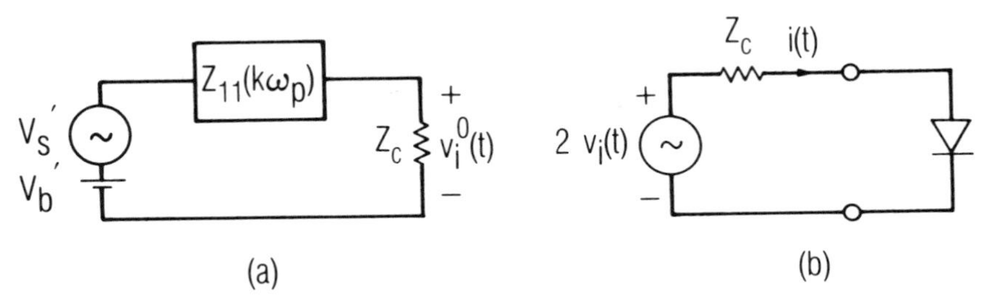

Figure 3.12 Incident-wave equivalent circuit for Example 5: (a) linear subcircuit; (b) nonlinear subcircuit.

The process is repeated until the change between subsequent voltage waves is minimal. After P iterations, the process has converged and the port voltage is

$$v^P(t) = v_i^P(t) + v_r^P(t) \tag{3.1.58}$$

The port current is the final value of $i^P(t)$.

3.1.3 Selecting the Number of Harmonics and Time Samples

In general, the waveforms generated in nonlinear analysis have an infinite number of harmonics, so a complete description of the operation of a nonlinear circuit would appear to require current and voltage vectors of infinite dimension. Fortunately, the magnitudes of frequency components necessarily decrease with frequency; otherwise, the time waveforms would represent infinite power. Accordingly, it is always possible to ignore all harmonics above some maximum number, designated K in Section 3.1.1. An important consideration in implementing a harmonic-balance analysis is the selection of K; selecting K too small results in poor accuracy and often poor convergence. Conversely, selecting K too large slows the solution process, which under the best circumstances is time-consuming, and increases the use of computer memory. Thus, selecting K improperly, whether it is too large or too small, can be expensive.

Perhaps the simplest criterion for selecting K is to consider the magnitudes of the capacitances in the device's equivalent circuit. Above some frequency, the capacitive susceptances are greater than the circuit's conductances, so they effectively are short circuits and their voltage components are negligibly small. This criterion can be applied easily to a diode, for example, where the junction capacitance short-circuits all voltage components across the only other nonlinearity, the resistive junction. It can also be applied to more complex devices, such as FETs, although to do so may require a little more care.

Another important consideration in the selection of K is the strength of the dominant nonlinearity and the magnitude of the excitation. It is often possible to generate a simplified equivalent circuit for the nonlinear device and to approximate the voltages and currents in it well enough to form a rough estimate of the frequency-component magnitudes. For example, in a strongly driven FET, one can often approximate the gate voltage as a sinusoid and the drain current as a rectangular pulse train. The length of each pulse is equal to the length of time that the gate voltage is above V_t, the *threshold* or *turn-on* voltage. A pulse train's Fourier series is found easily, and because the actual drain-current pulse is invariably softer than a rectangular pulse, this series establishes an upper bound to the relative magnitudes of the drain current's harmonic components. In Chapter 1, we saw that an nth-degree nonlinearity generates only n harmonics directly, although higher harmonics are possible as mixing products between these frequencies. These higher harmonics are usually much weaker than those generated directly, so it makes little sense to pick K much larger than the highest degree nonlinearity in the circuit. Conversely, if we wish to determine the levels of high harmonics, we must be careful to model the circuit nonlinearities by using polynomials (or other functions having polynomial expansions) of a degree great enough to generate those harmonics.

The nature of the problem to be solved often places some constraints on K. If the current or voltage at some harmonic k are to be found, $K > k$ is an obvious constraint. It is perhaps less obvious that the errors introduced by harmonic truncation are usually greater at higher harmonics than at the lower ones, so we really should choose $K \gg k$. Calculating the magnitudes of high harmonics accurately also requires that convergence be more complete, so that the errors in the high-harmonic components are small compared to the desired harmonic current or voltage. More precise convergence requires a greater number of harmonics.

The properties of the *fast Fourier-transform* (FFT) algorithm, used to obtain the frequency components from the time waveforms, also places constraints on K. One requirement of the FFT is that the number of harmonics produced must always be an integer power of two. The second, a consequence of the sampling theorem, is that the number of time samples must be twice the number of frequency components. It is not necessary to include all these harmonics in the harmonic-balance equations; it is possible to use, for example, only ten harmonics in the equations but to calculate sixteen via the FFT. It is essential, however, to use all the time samples required by the FFT. Furthermore, there are good reasons to use even more time samples. Using the minimum number of samples required by the sampling theorem may result in aliasing errors, where the neglected high harmonics affect the accuracy of lower-harmonic components. The simplest way to minimize aliasing errors is to oversample, that is, to use a sampling rate at least 25–30 percent greater than the minimum, or 2.5 to 2.6 times the number of required components. In the earlier example, 10 harmonics require a sample rate of 25 or 26 time samples per cycle. The next highest power of two is 32, so 32 samples should be used with 16 harmonics in the FFT. The higher 6 harmonics are simply discarded.

The intended use of the analysis also affects the number of harmonics that must be considered. In Section 3.2 we will show that conversion-matrix analysis, which includes mixing products around the kth local-oscillator harmonic, requires $K = 2k$ harmonics, plus the dc component, in the large-signal analysis. Obtaining good accuracy in the IM analysis of mixers or other time-varying circuits often requires even more harmonics, however; and it is often very difficult to estimate K beforehand. In these problems, we must determine K empirically by increasing it until consistent results, independent of K, are obtained.

What does the process of "discarding" the harmonics $k > K$ really mean? In most cases it implies that the voltage across the nonlinear elements at those frequencies is zero, so the impedance looking into the embedding network from the element terminals is assumed to be a short circuit. In the reflection algorithm, it means instead that the reflection coefficient of the embedding network is zero, or that the element is terminated in the characteristic impedance of the imaginary transmission line. It is sometimes possible, although not always practical, to formulate a dual case for that of Sections 3.1.1 and 3.1.2, wherein the element currents, not the voltages, are the independent variables. In this case, harmonic truncation would set the currents to zero, which implies open circuits at the higher frequencies.

3.1.4 Comparison of Algorithms

A quandary inevitably arises over the selection of one of these algorithms to solve a specific problem. The choice of an algorithm involves the assessment of its speed and reliability of convergence, properties that are not absolute but depend strongly upon the type of circuit, magnitude of excitation, and strength of the nonlinearities. For example, Newton's method has the distinct advantage that it converges much more rapidly in nearly linear circuits than in nonlinear ones, so it might be a good choice for the former. Using the splitting algorithm for nearly linear circuits would not be wise, because it is not much more efficient for nearly linear circuits than for nonlinear ones. Similar considerations apply to the reliability of convergence; for example, the requirement that a close initial estimate of the solution must be made in order to achieve convergence is a distinct disadvantage in a general-purpose circuit analysis program, but it may not be a disadvantage in a program designed to analyze a specific type of circuit.

The political answer to the problem of algorithm selection is fairly clear: a survey of the existing literature shows that most votes are cast for Newton's method and optimization. The reason for their favor is not necessarily inherent superiority but that these algorithms are generally well known and that computer subroutines to implement them are readily available. Indeed, for reasons presented in the previous section, optimization is a relatively inefficient method for solving harmonic-balance equations and Newton's method has frustrating limitations that other techniques may circumvent.

Because it does not require the evaluation of derivatives, the reflection algorithm is much faster per iteration than Newton's method, and at least initially, it approaches the solution about as rapidly. The reflection algorithm requires substantial iteration when applied to nearly linear circuits but can be made to converge in one iteration if the embedding impedances are resistive, regardless of the strength of the nonlinearity: it is necessary only to make the resistances equal to the characteristic impedances of the imaginary transmission lines. Under these conditions, all the reflection coefficients are zero and the solution reduces to a time-domain integration of the nonlinear circuit equations. Additionally, the reflection algorithm is self-starting; that is, it does not require an initial estimate of the solution. This property makes it attractive for use in general-purpose computer programs.

The convergence properties of the reflection algorithm are also very good. It converges rapidly, almost always monotonically, in its initial steps toward a solution; then, as it closes in on the solution, its rate of convergence slows, and it often oscillates with progressively slower convergence around a final set of values, much as does a gradient-search optimization. The point at which this slowing occurs is usually well within the accuracy that is needed for most practical purposes, so it is rarely a serious disadvantage. If serious convergence problems are encountered, they are invariably in the solution of the differential equations that describe the nonlinear subcircuit, not in the algorithm itself. Other advantages of this algorithm are its relative ease of implementation and its low computer memory requirements. Furthermore, it converges easily with a large number of harmonics and nonlinear circuit elements, because it is possible for the number of ports in Figure 3.9 to be much less than the number of nonlinear elements.

One significant disadvantage of the reflection algorithm is the need to use an artificial dc embedding impedance equal to the characteristic impedance of the transmission line. If the nonlinear circuit draws dc current, dc voltage is dropped across this resistance and the dc component of the port voltage is not equal to the bias-source voltage. It is thus necessary to offset the bias source voltage to compensate for the voltage drop, but because the dc current is unknown a priori, the bias-source voltage must be varied iteratively to obtain the required dc components of the port voltages. Furthermore, if there are more ports than sources, it may even be necessary to add fictitious bias sources! The situation can become intractable in multiport circuits; fortunately, there is little difficulty in obtaining the right bias values for the one- or two-port circuits that constitute most nonlinear microwave circuit problems.

The speed and memory use of Newton's method depend strongly on the need to calculate the Jacobian. The size of the Jacobian is increased dramatically by the addition of each harmonic or nonlinear element. Although solving Equation (3.1.37) usually consumes the largest part of the computer time needed to perform each Newton iteration, the time required to generate the derivative waveforms, Fourier-transform them, and set up the matrix is significant. The payoff in using Newton's method results from its strong convergence in a wide variety of applications at a speed that varies directly with the complexity of the problem to be solved. Thus, its speed makes it an attractive method for simple problems, and its robustness makes it attractive for more complex ones as well.

One important property of Newton's method is that its speed and reliability of convergence depend strongly upon the initial estimate of the solution vector. Formulating the initial estimate may not be difficult in analyzing a specific type of circuit, but it may be difficult to conceive of a way to form initial estimates in a general-purpose circuit-analysis program, which must accommodate a wide variety of circuits that have a concomitant variety of possible responses. This property is not always a disadvantage.

Many problems require the generation of a table of output levels as a function of input level; in these, the results of a calculation at one level can be used as a very good initial estimate for the next. Thus, each subsequent calculation begins with a good initial estimate, so it converges rapidly.

An advantage of Newton's method, which may not be obvious before it is used, is that it generates, at each iteration, a measure of the error in the currents at each port and at each harmonic. This feature is particularly useful when the most important information to be obtained from the calculation is a specific harmonic at a particular port (e.g., the drain current at the output frequency in a FET frequency multiplier). Accordingly, it is a simple matter to examine the error current at the desired port and harmonic to ascertain that the error is adequately small. The ability to examine individual error contributors is especially important at high harmonics. The current components at high harmonics are small in magnitude, so they contribute only minimally to the magnitude of the error vector. Thus, the current error at individual high harmonics can be relatively great even when the current-error vector's magnitude is small.

The central advantage of the splitting method is its ease of implementation. Its simplicity, however, comes from the questionable assumption that an adequate estimate of the corrected voltage vector is one that lies on a line between two vectors that are found by applying the same current to the linear and nonlinear subnetworks. By comparison, a gradient-search optimization iteratively corrects the vector in a direction that is known to reduce the magnitude of the error function, but the amount of computation necessary to determine this direction is much greater than in the splitting method. There is no guarantee that either of these methods converges to a global solution, but at least a gradient-search optimization can be trusted to start out in the right direction, and if the splitting algorithm fails, it is a cheap failure. The splitting algorithm is most attractive if the initial solution estimate can be made accurately and if the circuit's nonlinearities are not very strong.

3.2 LARGE-SIGNAL-SMALL-SIGNAL ANALYSIS USING CONVERSION MATRICES

Large-signal–small-signal analysis, or conversion matrix analysis, is useful for a large class of problems wherein a nonlinear device is *pumped* by a single large sinusoidal signal and another signal, which is assumed to be much smaller, is also applied. The most common application of this technique is in the design of microwave mixers, but it is also applicable to such circuits as modulators, parametric amplifiers, and parametric upconverters. The process involves first analyzing the nonlinear device under large-signal excitation only, usually via the harmonic-balance method. The nonlinear element or elements in the device's equivalent circuit are then converted to small-signal, *linear*, time-varying elements and the small-signal analysis is performed, without further consideration of the large-signal excitation. The assumption of small-signal linearity requires that the quasilinear response be of primary interest, but these techniques are also useful, with significant modifications, to circuits wherein quasilinearity is not assumed.

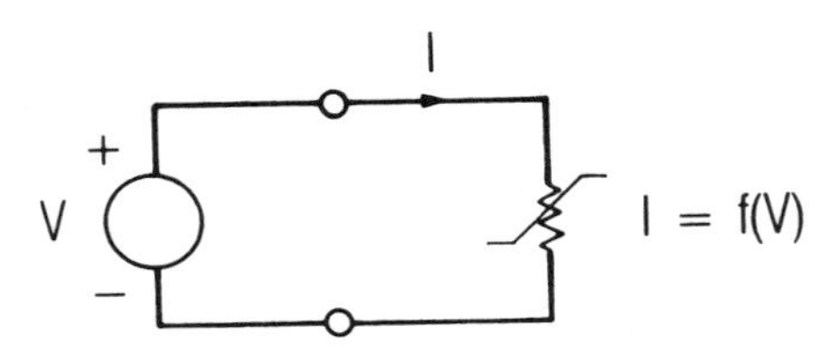

Figure 3.13 Nonlinear conductance driven by a large excitation.

3.2.1 Formulating the Conversion Matrices

Figure 3.13 shows a nonlinear conductance with current I driven by a large-signal voltage, V. It has the I/V relationship $I = f(V)$. Following the process outlined in Chapter 2, we can find the incremental small-signal current by assuming that V consists of the sum of a large-signal component V_0 and a small-signal component v. The current for this excitation can be found by expanding $f(V_0 + v)$ in a Taylor series:

$$
\begin{aligned}
I(V_0 + v) &= f(V_0 + v) \\
&= f(V_0) + \left.\frac{df(V)}{dV}\right|_{V=V_0} v + \left.\frac{1}{2}\frac{d^2f(V)}{dV^2}\right|_{V=V_0} v^2 \\
&\quad + \left.\frac{1}{6}\frac{d^3f(V)}{dV^3}\right|_{V=V_0} v^3 + \ldots
\end{aligned}
\tag{3.2.1}
$$

The small-signal, incremental current is found by subtracting the large-signal component of the current:

$$
i(v) = I(V_0 + v) - I(V_0) \tag{3.2.2}
$$

If $v \ll V_0$ then v^2 and v^3 are negligible, so

$$i(v) = \left. \frac{\mathrm{d}f(V)}{\mathrm{d}V} \right|_{V=V_0} v \tag{3.2.3}$$

V_0 need not be a dc quantity; it can legitimately be a time-varying large-signal voltage $V_L(t)$ (in fact, V_0 and V_L are control voltages). We assume that this is the case and also that $v = v(t)$, a function of time. Then,

$$i(t) = \left. \frac{\mathrm{d}f(V)}{\mathrm{d}V} \right|_{V=V_L(t)} v(t) \tag{3.2.4}$$

Equation (3.2.4) can be expressed as

$$i(t) = g(t)\, v(t) \tag{3.2.5}$$

The time-varying conductance in (3.2.5) is the derivative of the element's I/V characteristic at the large-signal voltage. This is the usual definition of small-signal conductance for static elements. By an analogous derivation, we could have a current-controlled resistor with the V/I characteristic:

$$V = R(I) \tag{3.2.6}$$

and obtain the small-signal v/i relation:

$$v(t) = r(t)\, i(t) \tag{3.2.7}$$

where:

$$r(t) = \left. \frac{\mathrm{d}R(I)}{\mathrm{d}I} \right|_{I=I_L(t)} \tag{3.2.8}$$

Often, the nonlinear element is a function of more than one control voltage. A conductance controlled by two voltages has $I = f_2(V_1, V_2)$. $f_2(V_1, V_2)$ can be expanded in a two-dimensional Taylor series, and after subtracting the large-signal current component and retaining only the linear terms, we have:

$$i(\mathrm{v}_1(t), \mathrm{v}_2(t)) = \frac{\partial f_2(V_1, V_2)}{\partial V_1}\bigg|_{\substack{V_1 = V_{L,1}(t) \\ V_2 = V_{L,2}(t)}} \mathrm{v}_1(t) + \frac{\partial f_2(V_1, V_2)}{\partial V_2}\bigg|_{\substack{V_1 = V_{L,1}(t) \\ V_2 = V_{L,2}}} \mathrm{v}_2(t) \tag{3.2.9}$$

which can be expressed as

$$i(t) = g_1(t)\mathrm{v}_1(t) + g_2(t)\mathrm{v}_2(t) \tag{3.2.10}$$

Equation (3.2.10) shows that the nonlinear conductance that has two control voltages is equivalent to two conductances in parallel. One must be a controlled current source, and the other may be either a controlled source or, in the more common situation where the small-signal voltage is the terminal voltage, a time-varying two-terminal conductance. The equivalent representation is illustrated in Figure 3.14. When the I/V characteristic is a function of more than two voltages, (3.2.10) can be extended in the manner we would expect:

$$i(t) = g_1(t)\mathrm{v}_1(t) + g_2(t)\mathrm{v}_2(t) + g_3(t)\mathrm{v}_3(t) + \ldots \tag{3.2.11}$$

It is, however, unusual to encounter a nonlinear element having more than two control voltages.

The same process can be followed with a capacitor. A nonlinear capacitor has the Q/V (charge/voltage) characteristic $Q = f_Q(V)$, and by an analogous derivation, the incremental small-signal charge is

$$q(t) = \frac{\mathrm{d}f_Q(V)}{\mathrm{d}V}\bigg|_{V = V_L(t)} \mathrm{v}(t) \tag{3.2.12}$$

or

$$q(t) = c(t)\mathrm{v}(t) \tag{3.2.13}$$

the capacitor's current is the time derivative of the charge:

$$i(t) = \frac{\mathrm{d}q(t)}{\mathrm{d}t} = c(t)\frac{\mathrm{d}\mathrm{v}(t)}{\mathrm{d}t} + \mathrm{v}(t)\frac{\mathrm{d}c(t)}{\mathrm{d}t} \tag{3.2.14}$$

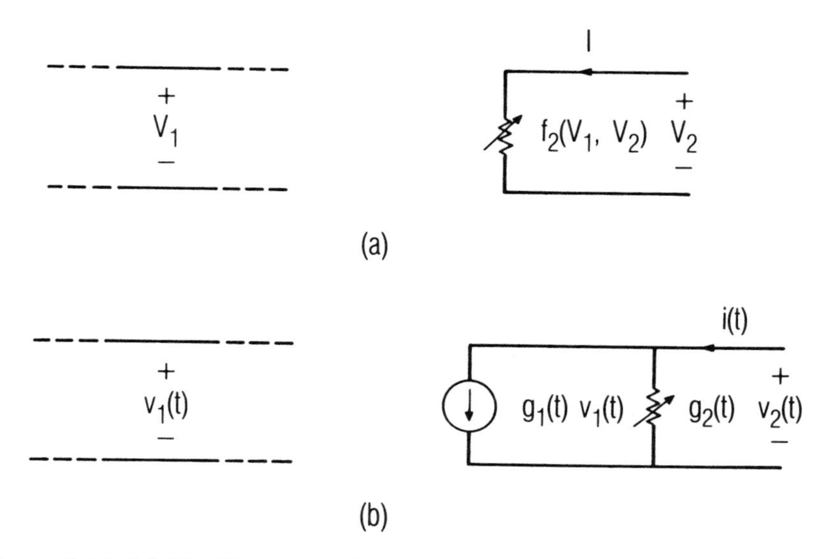

Figure 3.14 (a) Nonlinear conductance having two control voltages; (b) small-signal incremental equivalent circuit.

Like the conductance, the capacitor can have multiple control voltages. In this case, $Q = f_{2,Q}(V_1, V_2)$ and the small-signal charge is

$$q(v_1(t), v_2(t)) = \left.\frac{\partial f_{2,Q}(V_1, V_2)}{\partial V_1}\right|_{\substack{V_1 = V_{L,1}(t) \\ V_2 = V_{L,2}(t)}} v_1(t) + \left.\frac{\partial f_{2,Q}(V_1, V_2)}{\partial V_2}\right|_{\substack{V_1 = V_{L,1}(t) \\ V_2 = V_{L,2}(t)}} v_2(t) \tag{3.2.15}$$

or

$$q(t) = c_1(t)v_1(t) + c_2(t)v_2(t) \tag{3.2.16}$$

and the curent is found by differentiating with respect to time:

$$i(t) = \frac{dq(t)}{dt} = c_1(t)\frac{dv_1(t)}{dt} + v_1(t)\frac{dc_1(t)}{dt}$$
$$+ c_2(t)\frac{dv_2(t)}{dt} + v_2(t)\frac{dc_2(t)}{dt} \qquad (3.2.17)$$

In general, a nonlinear element excited by two tones supports currents and voltages at the mixing frequencies $m\omega_1 + n\omega_2$, where m and n can be any positive and negative integers, including zero. If we assume that one of those tones, ω_1, is at such a low level that it does not generate harmonics, and the other is a large-signal sinusoid at ω_p, the mixing frequencies are $\omega = \pm\omega_1 + n\omega_p$. This equation represents the set of frequency components shown in Figure 3.15, which consists of two tones on either side of each large-signal harmonic frequency, separated by $\omega_0 = |\omega_p - \omega_1|$. A more convenient representation of the mixing frequencies is

$$\omega_n = \omega_0 + n\omega_p \qquad (3.2.18)$$

which is shown in Figure 3.16 and includes only half of the mixing frequencies: the negative components of the lower sidebands and the positive components of the upper sidebands. This set of frequencies is adequate for two reasons: first, the small-signal analysis is linear, so by the superposition principle the results for positive and negative components can be separated; and second, positive- and negative-frequency components are complex conjugate pairs, so knowledge of only one is necessary. We will carry only the components in (3.2.18) in the following analysis, with confidence that the others can be generated when necessary.

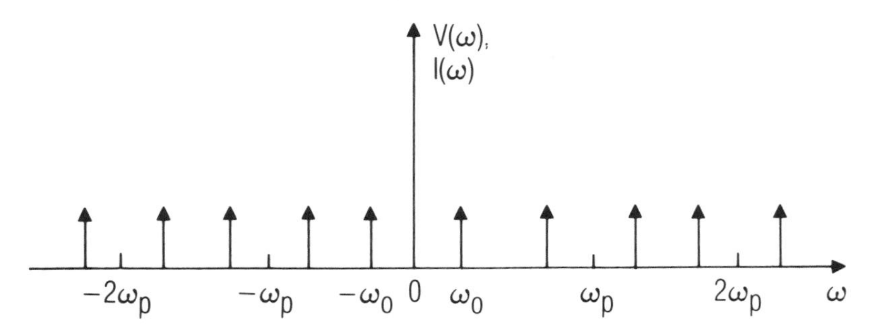

Figure 3.15 Spectrum of small-signal mixing frequencies in the pumped nonlinear element.

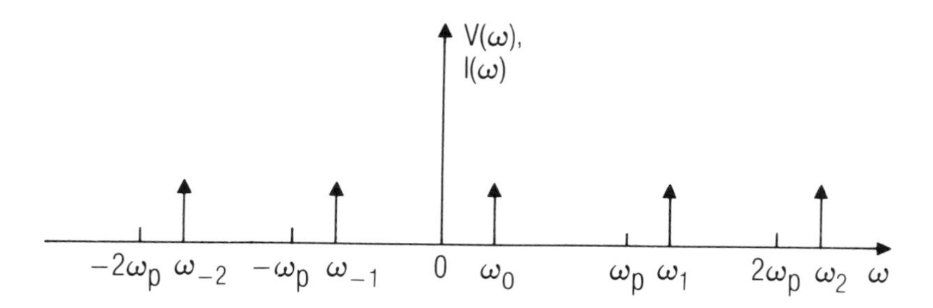

Figure 3.16 Spectrum of small-signal mixing frequencies, illustrating the frequency notation of (3.2.18).

The frequency-domain currents and voltages in a time-varying circuit element are related by a *conversion matrix*. We shall begin by deriving the conversion matrix that represents a time-varying conductance. The small-signal voltage and current can be expressed in the frequency notation of (3.2.18) as

$$v'(t) = \sum_{n=-\infty}^{\infty} V_n \exp(j\omega_n t) \tag{3.2.19}$$

and:

$$i'(t) = \sum_{n=-\infty}^{\infty} I_n \exp(j\omega_n t) \tag{3.2.20}$$

where the primes indicate that $v'(t)$ and $i'(t)$ are sums of the positive- and negative-frequency phasor components in (3.2.18) and are not the complete time waveforms. Above all, (3.2.19) and (3.2.20) are not Fourier series, in spite of their superficial resemblance. The conductance waveform $g(t)$ can be expressed by its Fourier series:

$$g(t) = \sum_{n=-\infty}^{\infty} G_n \exp(jn\omega_p t) \tag{3.2.21}$$

and the voltage and current are related by Ohm's law:

$$i'(t) = g(t)v'(t) \tag{3.2.22}$$

Substituting (3.2.19) through (3.2.21) into (3.2.22) gives the relation:

$$\sum_{k=-\infty}^{\infty} I_k \exp(j\omega_k t) = \sum_{n=-\infty}^{\infty} \sum_{m=-\infty}^{\infty} G_n V_m \exp(j\omega_{m+n} t) \tag{3.2.23}$$

Equating terms on both sides of the equation in (3.2.23) results in a set of equations that can be expressed in matrix form:

$$\begin{bmatrix} I^*_{-N} \\ I^*_{-N+1} \\ I^*_{-N+2} \\ \cdot \\ \cdot \\ \cdot \\ I^*_{-1} \\ I_0 \\ I_1 \\ \cdot \\ \cdot \\ \cdot \\ I_N \end{bmatrix} = \begin{bmatrix} G_0 & G_{-1} & G_{-2} & \ldots & G_{-2N} \\ G_1 & G_0 & G_{-1} & \ldots & G_{-2N+1} \\ G_2 & G_1 & G_0 & \ldots & G_{-2N+2} \\ \cdot & \cdot & \cdot & & \cdot \\ \cdot & \cdot & \cdot & & \cdot \\ \cdot & \cdot & \cdot & & \cdot \\ G_{N-1} & G_{N-2} & G_{N-3} & \ldots & G_{-N-1} \\ G_N & G_{N-1} & G_{N-2} & \ldots & G_{-N} \\ G_{N+1} & G_N & G_{N-1} & \ldots & G_{-N+1} \\ \cdot & \cdot & \cdot & & \cdot \\ \cdot & \cdot & \cdot & & \cdot \\ \cdot & \cdot & \cdot & & \cdot \\ G_{2N} & G_{2N-1} & G_{2N-2} & \ldots & G_0 \end{bmatrix} \begin{bmatrix} V^*_{-N} \\ V^*_{-N+1} \\ V^*_{-N+2} \\ \cdot \\ \cdot \\ \cdot \\ V^*_{-1} \\ V_0 \\ V_1 \\ \cdot \\ \cdot \\ \cdot \\ V_N \end{bmatrix} \tag{3.2.24}$$

Two details in (3.2.24) must be clarified. First, the vectors in (3.2.24) have been truncated to a limit of $n = \pm N$ for I_n and V_n, and $n = \pm 2N$ for G_n to prevent the occurrence of infinite vectors and matrices. We assume that V_n, I_n, and G_n are negligible beyond these limits. The second detail is that the negative-frequency components (V_n, I_n where $n < 0$) are shown as conjugate. The conjugates are caused by a change of definition; according to (3.2.18) ω_n is negative when $n < 0$, so the I_n and V_n are negative-frequency components when $n < 0$. We would rather define them as phasors, which are always positive-frequency components. Positive- and negative-frequency components are related as $V_{-n} = V_n^*$ and $\mathrm{I}_{-n} = I_n^*$, so if we wish V_n, I_n, to represent positive-frequency components, they must be V_n^*, I_n^*. Thus, the conversion matrix relates ordinary phasor voltages to currents at each mixing frequency. The main advantage of making this change is that the conversion matrix is now completely compatible with conventional linear, sinusoidal steady-state analysis.

The dual case, a time-varying resistor, has an unsurprising result. The conversion matrix is

$$\begin{bmatrix} V^*_{-N} \\ V^*_{-N+1} \\ V^*_{-N+2} \\ \vdots \\ V^*_{-1} \\ V_0 \\ V_1 \\ \vdots \\ V_N \end{bmatrix} = \begin{bmatrix} R_0 & R_{-1} & R_{-2} & \cdots & R_{-2N} \\ R_1 & R_0 & R_{-1} & \cdots & R_{-2N+1} \\ R_2 & R_1 & R_0 & \cdots & R_{-2N+2} \\ \vdots & \vdots & \vdots & & \vdots \\ R_{N-1} & R_{N-2} & R_{N-3} & \cdots & R_{-N-1} \\ R_N & R_{N-1} & R_{N-2} & \cdots & R_{-N} \\ R_{N+1} & R_N & R_{N-1} & \cdots & R_{-N+1} \\ \vdots & \vdots & \vdots & & \vdots \\ R_{2N} & R_{2N-1} & R_{2N-2} & \cdots & R_0 \end{bmatrix} \begin{bmatrix} I^*_{-N} \\ I^*_{-N+1} \\ I^*_{-N+2} \\ \vdots \\ I^*_{-1} \\ I_0 \\ I_1 \\ \vdots \\ I_N \end{bmatrix} \tag{3.2.25}$$

where the R_n are the Fourier components of the resistance. As one might expect, the resistance-form conversion matrix of any element is the inverse of its conductance-form matrix, as long as the element can be defined either as a time-varying conductance or resistance.

The conversion matrix of a capacitor is only slightly more complicated. The capacitor's charge is given by:

$$q'(t) = c(t)\, v'(t) \tag{3.2.26}$$

and $c(t)$ has the Fourier series:

$$c(t) = \sum_{n=-\infty}^{\infty} C_n \exp(jn\omega_p t) \tag{3.2.27}$$

The current is

$$i'(t) = \frac{dq'(t)}{dt} \tag{3.2.28}$$

and $q'(t)$ has the form:

$$q'(t) = \sum_{n=-\infty}^{\infty} Q_n \exp(j\omega_n t) \tag{3.2.29}$$

Substituting (3.2.19), (3.2.27), and (3.2.29) into (3.2.26) gives:

$$\sum_{k=-\infty}^{\infty} Q_k \exp(j\omega_k t) = \sum_{n=-\infty}^{\infty} \sum_{m=-\infty}^{\infty} C_n V_m \exp(j\omega_{m+n} t) \tag{3.2.30}$$

The current can be found by differentiating, as in (3.2.28). In the frequency domain, differentiation corresponds to multiplying by $j\omega$, so

$$\sum_{k=-\infty}^{\infty} I_k \exp(j\omega_k t) = \sum_{n=-\infty}^{\infty} \sum_{m=-\infty}^{\infty} j\omega_{m+n} C_n V_m \exp(j\omega_{m+n} t) \tag{3.2.31}$$

Equating terms at the same frequency gives the matrix equation:

$$
\begin{bmatrix} I^*_{-N} \\ I^*_{-N+1} \\ I^*_{-N+2} \\ \cdot \\ \cdot \\ \cdot \\ I^*_{-1} \\ I_0 \\ I_1 \\ \cdot \\ \cdot \\ \cdot \\ I_N \end{bmatrix}
=
\begin{bmatrix}
j\omega_{-N} & 0 & 0 & \ldots 0 \\
0 & j\omega_{-N+1} & 0 & \ldots 0 \\
0 & 0 & j\omega_{-N+2} & \ldots 0 \\
\cdot & \cdot & \cdot & \cdot \\
\cdot & \cdot & \cdot & \cdot \\
\cdot & \cdot & \cdot & \cdot \\
 & & \cdot & \\
0 & 0 & \ldots j\omega_0 & \ldots 0 \\
 & & \cdot & \\
\cdot & \cdot & \cdot & \cdot \\
\cdot & \cdot & \cdot & \cdot \\
\cdot & \cdot & & \\
0 & 0 & \ldots\ldots & j\omega_N
\end{bmatrix}
\begin{bmatrix}
C_0 & C_{-1} & C_{-2} & \ldots & C_{-2N} \\
C_1 & C_0 & C_{-1} & \ldots & C_{-2N+1} \\
C_2 & C_1 & C_0 & \ldots & C_{-2N+2} \\
\cdot & \cdot & \cdot & & \cdot \\
\cdot & \cdot & \cdot & & \cdot \\
\cdot & \cdot & \cdot & & \cdot \\
C_{N-1} & C_{N-2} & C_{N-3} & \ldots & C_{-N-1} \\
C_N & C_{N-1} & C_{N-2} & \ldots & C_{-N} \\
C_{N+1} & C_N & C_{N-1} & \ldots & C_{-N+1} \\
\cdot & \cdot & \cdot & & \cdot \\
\cdot & \cdot & \cdot & & \cdot \\
\cdot & \cdot & \cdot & & \cdot \\
C_{2N} & C_{2N-1} & C_{2N-2} & \ldots & C_0
\end{bmatrix}
\begin{bmatrix} V^*_{-N} \\ V^*_{-N+1} \\ V^*_{-N+2} \\ \cdot \\ \cdot \\ \cdot \\ V^*_{-1} \\ V_0 \\ V_1 \\ \cdot \\ \cdot \\ \cdot \\ V_N \end{bmatrix}
\quad (3.2.32)
$$

Example 6

We shall form the conversion matrix of the circuit shown in Figure 3.17(a). It consists of a conductance in series with a switch; the switch is opened and closed with a duty cycle of 0.5, and thus the combination of the switch and conductance has the waveform shown in Figure 3.17(b). Its Fourier series when $t_0 = 0.5T$ is

$$g(t) = G_p\,(0.5 + 0.318\exp(j\omega_p t) + 0.318\exp(-j\omega_p t) - 0.106\exp(j3\omega_p t) - 0.106\exp(-j3\omega_p t) + 0.064\exp(j5\omega_p t) + 0.064\exp(-j5\omega_p t) + \ldots) \tag{3.2.33}$$

The conversion matrix when $2N = 6$ is:

$$\mathbf{G} = G_p \begin{bmatrix} 0.5 & 0.318 & 0 & -0.106 & 0 & 0.064 & 0 \\ 0.318 & 0.5 & 0.318 & 0 & -0.106 & 0 & 0.064 \\ 0 & 0.318 & 0.5 & 0.318 & 0 & -0.106 & 0 \\ -0.106 & 0 & 0.318 & 0.5 & 0.318 & 0 & -0.106 \\ 0 & -0.106 & 0 & 0.318 & 0.5 & 0.318 & 0 \\ 0.064 & 0 & -0.106 & 0 & 0.318 & 0.5 & 0.318 \\ 0 & 0.064 & 0 & -0.106 & 0 & 0.318 & 0.5 \end{bmatrix} \tag{3.2.34}$$

which relates the mixing products up to ω_3, those close to the third harmonic of the large-signal excitation.

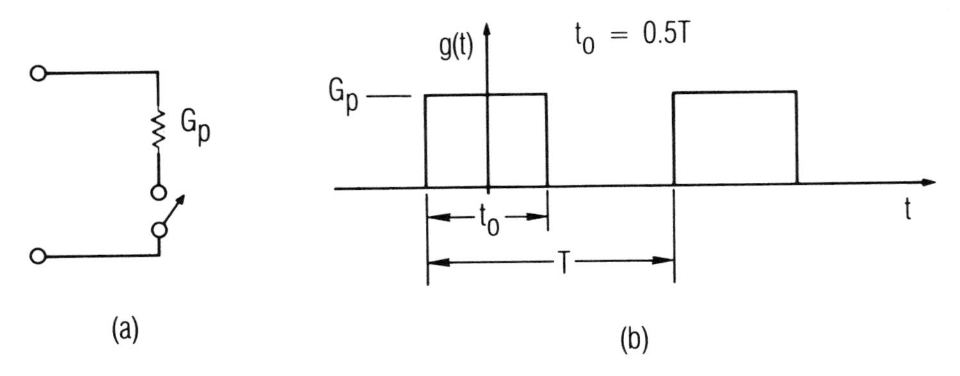

Figure 3.17 (a) Time-varying conductance for Example 6; (b) conductance waveform, $g(t)$.

3.2.2 Applying Conversion Matrices to Time-Varying Circuits

In order to mix fixed-value and time-varying components in the same equations, the fixed elements must have a conversion matrix form. This form is obviously a diagonal matrix, and the element value must occupy all the locations on the main diagonal. The conversion matrix of a frequency-sensitive, time-invariant element, such as a fixed impedance or admittance, is also a diagonal; however, the matrix elements are the impedance or admittance at the frequency corresponding to the location in the matrix. For example, the conversion matrix of a static lumped impedance is

$$
\mathbf{Z} = \begin{bmatrix}
Z^*(-\omega_{-N}) & 0 & 0\ldots & 0\ldots & & & 0 \\
0 & Z^*(-\omega_{-N+1}) & 0\ldots & 0\ldots & & & 0 \\
 & & \vdots & \vdots & & & \vdots \\
0 & 0 \;\cdot & \vdots & \vdots & & & \vdots \\
\vdots & \vdots & \ddots & \vdots & & & \vdots \\
0 & 0\ldots & Z^*(-\omega_{-1}) & 0\ldots & & & 0 \\
0 & 0\ldots & 0 & Z(\omega_0) & 0\ldots & & 0 \\
0 & 0\ldots & 0 & 0 & Z(\omega_1) & 0\ldots & 0 \\
\vdots & \vdots & \vdots & \vdots & & \ddots & \vdots \\
0 & 0\ldots & 0\ldots & 0\ldots & & & 0 \\
0 & 0\ldots & 0\ldots & 0\ldots & & & Z(\omega_N)
\end{bmatrix} \tag{3.2.35}
$$

Remember, when $n < 0$, ω_n is negative and the impedance or admittance in the ω_n position is $V_n^*/I_n^* = Z^*(-\omega_n)$; thus, the entry must be the conjugate of the positive-frequency or admittance at that frequency.

Equations (3.2.24), (3.2.25), and (3.2.32) can be expressed, respectively, as

$$\mathbf{I} = \mathbf{G}\,\mathbf{V} \tag{3.2.36}$$

$$\mathbf{V} = \mathbf{R}\,\mathbf{I} \tag{3.2.37}$$

$$\mathbf{I} = j\boldsymbol{\Omega}\,\mathbf{C}\,\mathbf{V} \tag{3.2.38}$$

These relations have the same form as those that define the I/V relations of linear, time-invariant resistance, conductance, and capacitance in the sinusoidal steady state. The only difference is that these are matrix equations and the latter are scalar. The individual current and voltage components in the $\mathbf{V}$ and $\mathbf{I}$ vectors must satisfy Kirchoff's current and voltage laws in any linear circuit using time-varying elements, just as in time-invariant circuits. Therefore, the matrix equations can be used in exactly the same way as the scalar ones, as long as the requirements of matrix arithmetic are met: the order of multiplication must be preserved, and we must invert and multiply instead of divide.

This realization allows all the tools of conventional sinusoidal, steady-state analysis to be applied to time-varying circuits. Specifically, the conversion matrix for two elements in parallel is the sum of their individual admittance-form matrices, and for two elements in series it is the sum of their impedance-form matrices. One can also generate transfer functions and input-output impedances or admittances in terms of conversion matrices.

A second property of the conversion matrices is that they can be treated in all ways like multiport admittance or impedance matrices; the only difference is that the "ports" in the conversion matrix are currents and voltages at different frequencies instead of physically separate ports. In concept, one could separate the frequency components by using filters and create a physically separate port for each, without changing any of the circuit's properties. Indeed, in designing components that include time-varying elements, such as mixers, we try to separate at least a few of the frequency components in this manner, in order to realize input and output ports, and to terminate other mixing products optimally. This property allows multiport-circuit concepts to be employed in interconnecting time-varying circuits, interfacing them with matching networks, and determining their gain, impedances, and stability. One can even convert a Y or Z matrix to an S-parameter form. These points are illustrated by the following examples.

Example 7

We shall derive the conversion matrix that represents the circuit shown in Figure 3.18. This circuit consists of a time-varying conductance and capacitor in parallel, and a resistor in series (this is a common model of a pumped mixer diode). We assume that a large-signal analysis has been performed, and that the time waveforms and conversion matrices of each circuit element have been determined. $\mathbf{C}_j$ and $\mathbf{G}_j$ are the conversion matrices representing $c_j(t)$ and $g_j(t)$, respectively.

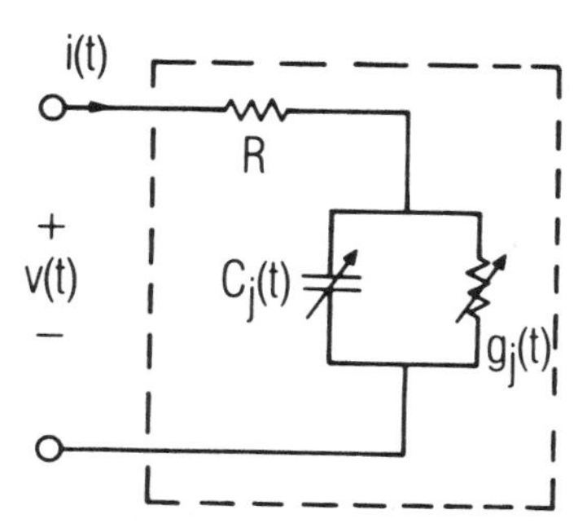

Figure 3.18 Pumped diode equivalent circuit for Example 7.

Because the capacitor and conductance are in parallel, the conversion matrix is the sum of the admittance-form conversion matrices of each component:

$$\mathbf{Y}_j = \mathbf{G}_j + j\mathbf{\Omega}\,\mathbf{C}_j \tag{3.2.39}$$

and their impedance-form conversion matrix is the inverse:

$$\mathbf{Z}_j = \mathbf{Y}_j^{-1} = (\mathbf{G}_j + j\mathbf{\Omega}\,\mathbf{C}_j)^{-1} \tag{3.2.40}$$

The conversion matrix for the resistor is $R\mathbf{1}$, where $\mathbf{1}$ is the $2N + 1 \times 2N + 1$ identity matrix. $R\mathbf{1}$ is in series with $\mathbf{Z}_j$, so the impedance-form conversion matrix of the entire circuit is the sum of $R\mathbf{1}$ and $\mathbf{Z}_j$:

$$\mathbf{Z}_c = R\mathbf{1} + \mathbf{Z}_j = R\mathbf{1} + (\mathbf{G}_j + j\mathbf{\Omega}\,\mathbf{C}_j)^{-1} \tag{3.2.41}$$

The admittance-form matrix, if needed, is just the inverse of $\mathbf{Z}_c$.

Example 8

We shall calculate the conversion matrix that represents the simplified FET equivalent circuit shown in Figure 3.19(a); this circuit could represent a FET mixer. It has two nonlinear circuit elements, $I_d(V_g, V_d)$ and $C_g(V_g)$, and all the remaining elements are linear. The circuit is to be treated as a two-port, so a two-port admittance-form matrix will be needed. It has the form:

$$\begin{bmatrix} \mathbf{I}_1 \\ \mathbf{I}_2 \end{bmatrix} = \begin{bmatrix} \mathbf{Y}_{1,1} & \mathbf{Y}_{1,2} \\ \mathbf{Y}_{2,1} & \mathbf{Y}_{2,2} \end{bmatrix} \begin{bmatrix} \mathbf{V}_1 \\ \mathbf{V}_2 \end{bmatrix} \tag{3.2.42}$$

where $\mathbf{I}_1$, $\mathbf{I}_2$, $\mathbf{V}_1$, and $\mathbf{V}_2$ are current and voltage vectors as shown in (3.2.24), (3.2.25), and (3.2.32), and the $\mathbf{Y}_{m,n}$ submatrices are each complete conversion matrices. Thus, (3.2.42) relates not only the currents and voltages at the mixing frequencies and at each port but also includes transfer terms between ports.

Again, we assume that a large-signal analysis has been performed and that the nonlinear elements have been converted to their incremental, time-varying forms. The drain current source can be split into two elements, $g_m(t)$ and $g_d(t)$, according to (3.2.9) and (3.2.10); the former element is a controlled source, representing the time-varying transconductance, and the latter is a time-varying drain-source conductance. The resulting circuit is shown in Figure 3.19(b).

The submatrices are defined by the relations:

$$\begin{aligned}
\mathbf{I}_1 &= \mathbf{Y}_{1,1}\mathbf{V}_1; \quad \mathbf{V}_2 = \mathbf{0} \\
\mathbf{I}_1 &= \mathbf{Y}_{1,2}\mathbf{V}_2; \quad \mathbf{V}_1 = \mathbf{0} \\
\mathbf{I}_2 &= \mathbf{Y}_{2,1}\mathbf{V}_1; \quad \mathbf{V}_2 = \mathbf{0} \\
\mathbf{I}_2 &= \mathbf{Y}_{2,2}\mathbf{V}_2; \quad \mathbf{V}_1 = \mathbf{0}
\end{aligned} \tag{3.2.43}$$

where $\mathbf{0}$ is the zero vector. The time-varying quantities, $c_g(t)$, $g_m(t)$, and $g_d(t)$, have conversion matrices designated $\mathbf{C}_g$, $\mathbf{G}_m$, and $\mathbf{G}_d$, respectively.

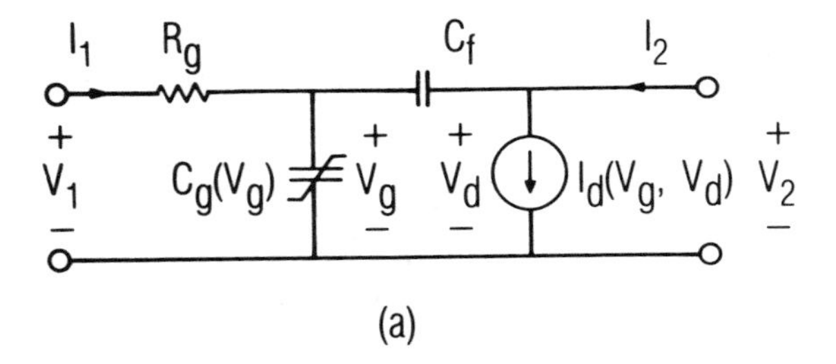

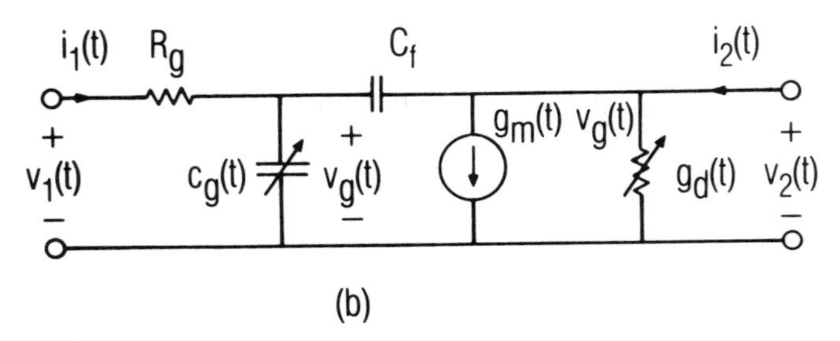

Figure 3.19 (a) FET nonlinear equivalent circuit for Example 8; (b) time-varying linear equivalent circuit.

We begin by finding $\mathbf{Y}_{1,1}$. When Port 2 is shorted at all harmonics, $c_g(t)$ and C_f are in parallel, and we can immediately write

$$\mathbf{I}_1 = \{[j\mathbf{\Omega}(\mathbf{C}_g + C_f\mathbf{1})]^{-1} + R_g\mathbf{1}\}^{-1}\mathbf{V}_1 \tag{3.2.44}$$

and $\mathbf{Y}_{1,1}$ is found by comparing (3.2.44) to (3.2.43). $\mathbf{Y}_{2,1}$ is just a little more trouble. When the output is shorted:

$$\mathbf{V}_g = [j\mathbf{\Omega}(\mathbf{C}_g + C_f\mathbf{1})]^{-1}\mathbf{I}_1 \tag{3.2.45}$$

and:

$$\mathbf{I}_2 = (\mathbf{G}_m - j\Omega C_f)\ \mathbf{V}_g \tag{3.2.46}$$

Substituting the first equation of (3.2.43) into (3.2.45) and the result of that into (3.2.46) gives

$$\mathbf{I}_2 = (\mathbf{G}_m - j\Omega C_f)\ [j\Omega(\mathbf{C}_g + C_f\mathbf{1})]^{-1}\mathbf{Y}_{1,1}\mathbf{V}_1 \tag{3.2.47}$$

$\mathbf{Y}_{2,2}$ and $\mathbf{Y}_{1,2}$ are a little sticky algebraically but straightforward conceptually. Beginning with $\mathbf{Y}_{2,2}$, we note:

$$\mathbf{I}_2 = \mathbf{G}_d\mathbf{V}_2 + \mathbf{G}_m\mathbf{V}_g + \mathbf{I}_f \tag{3.2.48}$$

where $\mathbf{I}_f$ is the current through the capacitor C_f. This time the input port is shorted, and

$$\begin{aligned}\mathbf{I}_f &= \left[\left(\frac{1}{R_g}\mathbf{1} + j\mathbf{\Omega}\ \mathbf{C}_g\right)^{-1} - \frac{j}{C_f}\mathbf{\Omega}^{-1}\right]^{-1}\mathbf{V}_2 \\ &= \mathbf{Y}_f\mathbf{V}_2\end{aligned} \tag{3.2.49}$$

and

$$\mathbf{V}_g = \left(\frac{1}{R_g}\mathbf{1} + j\mathbf{\Omega}\ \mathbf{C}_g\right)^{-1}\mathbf{I}_f \tag{3.2.50}$$

Substituting (3.2.49) and (3.2.50) into (3.2.48) and simplifying gives

$$\mathbf{I}_2 = \left\{\mathbf{G}_d + \mathbf{Y}_f\left[\mathbf{1} + \mathbf{G}_m\left(\frac{1}{R_g}\mathbf{1} + j\mathbf{\Omega}\ \mathbf{C}_g\right)^{-1}\right]\right\}\mathbf{V}_2 \tag{3.2.51}$$

Deriving the submatrix $\mathbf{Y}_{1,2}$ is left to the reader as an exercise. The easy part of the exercise is finding an expression for $\mathbf{Y}_{1,2}$; the hard part is proving that it is equivalent to the author's result:

$$\mathbf{I}_1 = -(\mathbf{1} + jR_g\mathbf{\Omega}\,\mathbf{C}_g)^{-1}\mathbf{Y}_f\mathbf{V}_2 \tag{3.2.52}$$

Deriving (3.2.52) requires the use of the general matrix relation:

$$\mathbf{A}^{-1}\mathbf{B}^{-1} = (\mathbf{BA})^{-1} \tag{3.2.53}$$

where **A** and **B** are any nonsingular square matrices. Equation (3.2.53) is without qualification the most useful relation in the entire field of conversion-matrix manipulation.

Example 9

We will now calculate the input and output impedances, simultaneous conjugate-match impedances, transducer conversion gain, and maximum available conversion gain of the circuit of the previous example, at a specific pair of input and output frequencies. Figure 3.20 shows the circuit to be analyzed, where the two-port is described by the conversion matrix **Y**, given in (3.2.42). It can also be described by $\mathbf{Z} = \mathbf{Y}^{-1}$. The source and load impedances, generally functions of ω, are shown in series with the two-port; shorting either set of terminals loads the input or output port with the appropriate impedance.

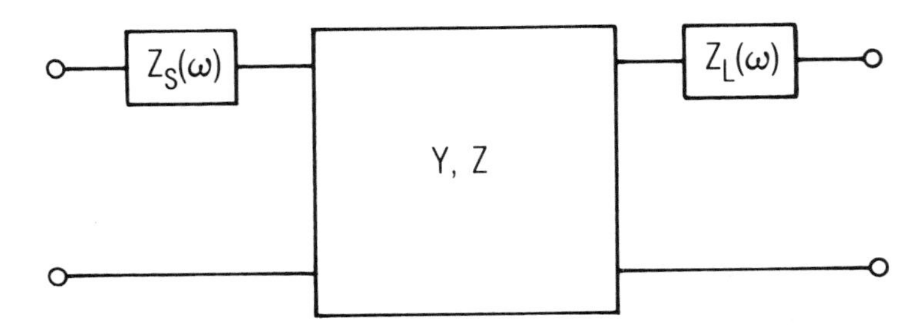

Figure 3.20 Terminated two-port for Example 9.

We wish to calculate this circuit's gain and impedances at specific input and output frequencies. This means that, with the exception of the input port at the input frequency and the output port at the output frequency, we wish to terminate the ports in their source and load impedances at all mixing frequencies. The source and load impedances at the unwanted mixing frequencies are then absorbed into the network, and we are left with a conventional two-port, describable by a simple 2 × 2 Y matrix. The only feature that would distinguish this matrix from the Y matrix of a time-invariant network is that it represents input and output phasors at different frequencies, and if one of those frequencies is a lower sideband, its voltage and current will be conjugate quantities.

We begin by putting the source and load impedances into a compatible two-port conversion matrix representation. This is

$$\mathbf{Z}_t = \begin{bmatrix} \mathbf{Z}_s & 0 \\ 0 & \mathbf{Z}_L \end{bmatrix} \tag{3.2.54}$$

where $\mathbf{Z}_s$ and $\mathbf{Z}_L$ are diagonal matrices, in the form shown in (3.2.35); that is, with $Z^*_{s,L}(-\omega_{-N})$ in the top left corner, $\mathbf{Z}_{s,L}(\omega_0)$ in the center, and $Z_{s,L}(\omega_N)$ in the bottom right. Following the notation for mixing products (3.2.18), we let ω_q be the input frequency and ω_r be the output frequency. The source and load impedances at these frequencies, $Z_s(\omega_q)$ and $Z_L(\omega_r)$, respectively, are set to zero because we want them to remain external to the circuit; the impedances at other frequencies are retained because they will be absorbed into the circuit. Following the rule for conventional two-ports, the impedance-form conversion matrix of the terminated network, $\mathbf{Z}_a$, is

$$\mathbf{Z}_a = \mathbf{Z}_t + \mathbf{Y}^{-1} \tag{3.2.55}$$

The admittance-form matrix for the combination of the FET and the source and load impedances is

$$\mathbf{Y}_a = \mathbf{Z}_a^{-1} \tag{3.2.56}$$

At this point, $\mathbf{Y}_a$ still relates two voltage vectors to two current vectors and has the form:

$$\begin{bmatrix} \mathbf{I}_1 \\ \mathbf{I}_2 \end{bmatrix} = \begin{bmatrix} \mathbf{Y}_{a;1,1} & \mathbf{Y}_{a;1,2} \\ \mathbf{Y}_{a;2,1} & \mathbf{Y}_{a;2,2} \end{bmatrix} \begin{bmatrix} \mathbf{V}_1 \\ \mathbf{V}_2 \end{bmatrix} \tag{3.2.57}$$

We wish to reduce (3.2.57) to a simple 2×2 Y matrix by terminating the ports at all unwanted mixing frequencies. To terminate the output port at all frequencies other than ω_r, we set $\mathbf{V}_2$ to zero by shorting the output terminals at those frequencies; similarly, $\mathbf{V}_1$ is zeroed at all frequencies other than ω_q. Setting these voltage components to zero multiplies all the corresponding columns in $\mathbf{Y}_a$ by zero; therefore, those columns can be eliminated. Furthermore, because the input and output are shorted, the current components at those frequencies are not of interest, so the corresponding rows in $\mathbf{Y}_a$ and $\mathbf{I}$ can also be removed. The only terms left in $\mathbf{Y}_a$ can be put into the 2×2 matrix form:

$$\begin{bmatrix} I_1(\omega_q) \\ I_2(\omega_r) \end{bmatrix} = \begin{bmatrix} y_{1,1} & y_{1,2} \\ y_{2,1} & y_{2,2} \end{bmatrix} \begin{bmatrix} V_1(\omega_q) \\ V_2(\omega_r) \end{bmatrix} \tag{3.2.58}$$

where

$$\begin{aligned} y_{1,1} &= \mathbf{Y}_{a;1,1}(\omega_q,\omega_q) \\ y_{1,2} &= \mathbf{Y}_{a;1,2}(\omega_q,\omega_r) \\ y_{2,1} &= \mathbf{Y}_{a;2,1}(\omega_r,\omega_q) \\ y_{2,2} &= \mathbf{Y}_{a;2,2}(\omega_r,\omega_r) \end{aligned} \tag{3.2.59}$$

The rest is all downhill. Equation (3.2.58) can be used with the usual collection of Y-matrix relations. For example, if the load admittance is $Y_L(\omega_r)$, the input admittance has the familiar relation:

$$Y_{\text{in}}(\omega_q) = y_{1,1} - \frac{y_{1,2}y_{2,1}}{Y_L(\omega_r) + y_{2,2}} \tag{3.2.60}$$

and with a source admittance $Y_s(\omega_q)$, the output admittance is

$$Y_{\text{out}}(\omega_r) = y_{2,2} - \frac{y_{1,2}y_{2,1}}{Y_s(\omega_q) + y_{1,1}} \tag{3.2.61}$$

Note that if $r < 0$ the output admittance is conjugate, and if $q < 0$ the input admittance is conjugate. In these cases, the conjugate of the load or source admittance must also be used in (3.2.60) and (3.2.61), respectively. The equation for transducer conversion gain, in terms of Y parameters, is

$$G_t = \frac{4\,\mathrm{Re}\{Y_s(\omega_q)\}\,\mathrm{Re}\{Y_L(\omega_r)\}\,|y_{2,1}|^2}{|(y_{1,1} + Y_s(\omega_q))(y_{2,2} + Y_L(\omega_r)) - y_{2,1}y_{1,2}|^2} \tag{3.2.62}$$

The Linvill stability factor, c, is

$$c = \frac{|y_{2,1}y_{1,2}|}{2\,\mathrm{Re}\{y_{1,1}\}\,\mathrm{Re}\{y_{2,2}\} - \mathrm{Re}\{y_{2,1}y_{1,2}\}} \tag{3.2.63}$$

If $c < 1$, the circuit is unconditionally stable and no passive source impedance at ω_q or load at ω_r can cause oscillation. If $c < 1$, the *maximum available conversion gain* (MAG) and simultaneous conjugate match impedances $Y_{s,\mathrm{opt}}(\omega_q)$, $Y_{L,\mathrm{opt}}(\omega_r)$ are defined. They are

$$\mathrm{MAG} = \frac{|y_{2,1}|^2}{2\,\mathrm{Re}\{y_{1,1}\}\,\mathrm{Re}\{y_{2,2}\} - \mathrm{Re}\{y_{2,1}y_{1,2}\} + T_y} \tag{3.2.64}$$

and

$$\mathrm{Im}\{Y_{s,\mathrm{opt}}(\omega_q)\} = -\,\mathrm{Im}\{y_{1,1}\} + \frac{\mathrm{Im}\{y_{2,1}y_{1,2}\}}{2\,\mathrm{Re}\{y_{2,2}\}} \tag{3.2.65}$$

$$\mathrm{Re}\{Y_{s,\mathrm{opt}}(\omega_q)\} = \frac{T_y}{2\,\mathrm{Re}\{y_{2,2}\}} \tag{3.2.66}$$

where

$$T_y = \left((2\,\mathrm{Re}\{y_{1,1}\}\,\mathrm{Re}\{y_{2,2}\} - \mathrm{Re}\{y_{2,1}y_{1,2}\})^2 - |y_{2,1}y_{1,2}|^2 \right)^{1/2} \tag{3.2.67}$$

The load impedance $Y_{L,\mathrm{opt}}(\omega_r)$ can be found from (3.2.65) through (3.2.67) by interchanging $y_{1,1}$ with $y_{2,2}$, $y_{2,1}$ with $y_{1,2}$, and $Y_{s,\mathrm{opt}}(\omega_q)$ with $Y_{L,\mathrm{opt}}(\omega_r)$.

As is the case in a time-invariant circuit, unconditional stability at the excitation frequency and large-signal excitation level is not adequate to guarantee that the time-varying circuit will be stable in a practical sense; for the circuit to be stable in practice, it must be unconditionally stable at all possible input frequencies and large-signal excitation levels. Varying the small-signal excitation frequency for which the Y parameters in (3.2.58) are determined also varies the higher-order mixing frequencies and, hence, the embedding impedances at those frequencies. Stability, therefore, is a function of all the parameters that affect the Y parameters; that is, literally all the parameters of the circuit.

It is important to recognize that small-signal and large-signal stability are interrelated. To explain why this is so, we must note that a fundamental assumption in the conversion matrix theory is that small-signal voltages are small variations (in frequency as well as in magnitude and phase) in the large-signal voltage. The conversion matrix is, in fact, nothing more than the large-signal Jacobian, a matrix that relates the current and voltage deviations, evaluated at the mixing frequencies rather than the large-signal harmonics. Small-signal oscillation is a process where these variations build up spontaneously and without bound and eventually become indistinguishable from the large-signal voltage. If they occur at a different frequency from the large signal, they may appear as modulation, "snap" phenomena, parasitic oscillation, or other well known manifestations of instability in nonlinear circuits.

The two-port conversion matrix of Equation (3.2.58) is in Y matrix form only because an admittance-form conversion matrix is usually easiest to derive. However, it need not be expressed in this form; in fact, it can be converted to any two-port matrix form desired, such as an S matrix or even a T matrix (transfer-scattering matrix). The procedure for converting the Y matrix to one of these forms is precisely the same as for any other scalar matrix. For example, the S matrix is found from the Y matrix as

$$\mathbf{S} = (\mathbf{1} + \mathbf{Y}_{\text{norm}})^{-1}(\mathbf{1} - \mathbf{Y}_{\text{norm}}) \tag{3.2.68}$$

where $\mathbf{Y}_{\text{norm}}$ is the Y matrix of Equation (3.2.58) normalized to the admittance for which the S parameters are defined. The interpretation of lower-sideband quantities (q, $r < 0$) in the S matrix may be a little confusing. For example, if $q = -1$ and $r = 0$, a common situation, the S matrix has the form:

$$\begin{bmatrix} b_1^*(\omega_{-1}) \\ b_2(\omega_0) \end{bmatrix} = \begin{bmatrix} s_{1,1} & s_{1,2} \\ s_{2,1} & s_{2,2} \end{bmatrix} \begin{bmatrix} a_1^*(\omega_{-1}) \\ a_2(\omega_0) \end{bmatrix} \tag{3.2.69}$$

where $s_{1,1}$ is the conjugate of the input reflection coefficient:

$$\Gamma_{\text{in}}^* = s_{1,1} = \left. \frac{b_1^*(\omega_{-1})}{a_1^*(\omega_{-1})} \right|_{a_2(\omega_0)=0} \tag{3.2.70}$$

and $|s_{2,1}|^2$ is, as usual, the transducer gain:

$$G_t = |s_{2,1}|^2 = \left| \frac{b_2(\omega_0)}{a_1^*(\omega_{-1})} \right|^2_{a_2(\omega_0)=0} \tag{3.2.71}$$

The fact that a_1 is conjugate in (3.2.71) does not change the magnitude of $s_{2,1}$. Fortunately, the fact that the definitions of $s_{2,1}$ and $s_{1,2}$ include one conjugate and one nonconjugate quantity almost never leads to a meaningless expression for the properties that are usually of most interest: gain, impedances, and stability.

3.2.3 Multitone Excitation and Intermodulation in Time-Varying Circuits

The small-signal analysis of the previous sections was based on the assumption that the excitation was vanishingly small. Accordingly, the nonlinear terms in the incremental Taylor series, (3.2.1), could be ignored, resulting in a linear, small-signal formulation. In this section, that assumption is discarded, and instead it is assumed only that the *incremental* nonlinear element is weakly nonlinear. This is *not* the same as assuming that the nonlinear device is weakly nonlinear; it means instead that the element is weakly nonlinear for any small deviation from its instantaneous large-signal drive voltage. Virtually all nonlinear solid-state devices meet this condition, as long as they are not driven into saturation by the small-signal excitation. Thus, the techniques in this section are most useful for determining intermodulation levels and spurious responses in such heavily pumped circuits as mixers and parametric amplifiers. The method is based on Reference 3.12; it also uses some concepts from the Volterra- and power-series theory in Chapter 4 and could be considered to be a time-varying application of the Volterra series. For these reasons, the reader might do well to become conversant with Chapter 4 and Sections 3.2.1 and 3.2.2 before continuing with this section.

In order to minimize unnecessary complications, the circuit model used in this section includes only a single set of terminals in the nonlinear subcircuit with a resistive or capacitive nonlinearity. The linear circuit can then be described by a Norton equivalent, which consists of a single current source and embedding impedance. The model is shown in Figure 3.21; it is assumed in the figure that the large-signal nonlinear analysis has been performed, and that the currents and voltages shown are the small-signal incremental ones. Except for the excitation source $i_s(t)$, these currents and voltages include intermodulation components as well as linear mixing products. The excitation $i_s(t)$ is a two-tone source:

$$i_s(t) = I_{s1} \cos[(m\omega_p + \omega_1)t] + I_{s2} \cos[(m\omega_p + \omega_2)t] \tag{3.2.72}$$

where, as before, ω_p is the fundamental frequency of the large-signal

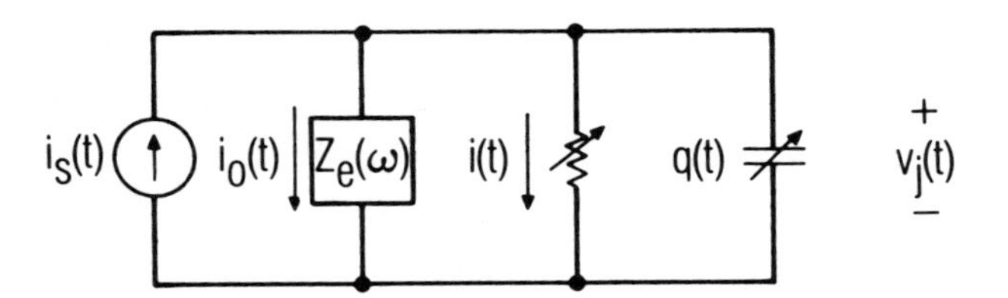

Figure 3.21 Time-varying nonlinear equivalent circuit.

excitation and, in general, $m = 0, \pm 1, \pm 2, \ldots$; for the usual case of an upper-side band input, $m = 1$.

The spectrum of mixing frequencies is shown in Figure 3.22(a), and a detail of those closest to dc is shown in Figure 3.22(b); the spectrum shown in Figure 3.22(b) is mirrored on either side of each large-signal harmonic, at positive and negative frequencies. The terms ω_1 and ω_2 represent the lowest-frequency (usually IF) components of the excitation and the rest are *intermodulation* (IM) products. The IM products shown in the figure are not the only ones possible; they are, instead, those of third or lower order that are closest to ω_1 and ω_2 and are, consequently, of greatest concern in practice.

Following the same process as in (3.2.1) through (3.2.4) and (3.2.12) through (3.2.13) but retaining the terms up to third degree, we have

$$i(t) = \left.\frac{\mathrm{d}f(V)}{\mathrm{d}V}\right|_{V=V_L(t)} v(t) + \frac{1}{2}\left.\frac{\mathrm{d}^2 f(V)}{\mathrm{d}V^2}\right|_{V=V_L(t)} v^2(t) + \frac{1}{6}\left.\frac{\mathrm{d}^3 f(V)}{\mathrm{d}V^3}\right|_{V=V_L(t)} v^3(t) + \ldots \tag{3.2.73}$$

and

$$q(t) = \left.\frac{\mathrm{d}f_Q(V)}{\mathrm{d}V}\right|_{V=V_L(t)} v(t) + \frac{1}{2}\left.\frac{\mathrm{d}^2 f_Q(V)}{\mathrm{d}V^2}\right|_{V=V_L(t)} v^2(t) + \frac{1}{6}\left.\frac{\mathrm{d}^3 f_Q(V)}{\mathrm{d}V^3}\right|_{V=V_L(t)} v^3(t) + \ldots \tag{3.2.74}$$

which can be expressed as

$$i(t) = g_1(t)v(t) + g_2(t)v^2(t) + g_3(t)v^3(t) + \ldots \quad (3.2.75)$$

$$q(t) = c_1(t)v(t) + c_2(t)v^2(t) + c_3(t)v^3(t) + \ldots \quad (3.2.75)$$

Limiting consideration to third-order components, we have

$$v(t) = v_1(t) + v_2(t) + v_3(t) \quad (3.2.77)$$

$$v^2(t) = v_1^2(t) + 2v_1(t)v_2(t) \quad (3.2.78)$$

$$v^3(t) = v_1^3(t) \quad (3.2.79)$$

where $v_n(t)$ is the nth-order voltage, the combination of all nth-order mixing products. Recall that an nth-order mixing product is any combination of n excitation frequencies, including both positive and negative frequencies. The square of the junction voltage obviously creates a second-order product from $v_1^2(t)$ and a third-order product by mixing the first-order $v_1(t)$ and second-order $v_2(t)$.

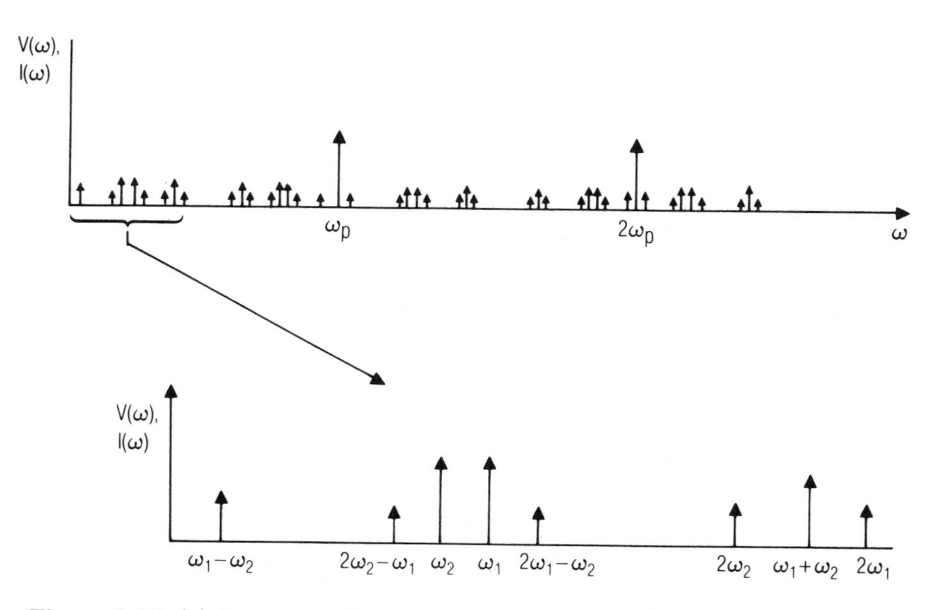

Figure 3.22 (a) Lowest-order mixing frequencies in the nonlinear time-varying circuit; (b) detail of the frequencies closest to dc.

The differential equation describing Figure 3.21 is

$$\frac{\mathrm{d}q(t)}{\mathrm{d}t} + i(t) + i_0(t) = i_s(t) \tag{3.2.80}$$

Substituting (3.2.75) through (3.2.79) into (3.2.80), and separating the equations into first-, second-, and third-order products, we have

$$\frac{\mathrm{d}}{\mathrm{d}t}\left(c_1(t)v_1(t)\right) + g_1(t)v_1(t) + i_{0,1}(t) = i_s(t) \tag{3.2.81}$$

$$\frac{\mathrm{d}}{\mathrm{d}t}\left(c_1(t)v_2(t) + c_2(t)v_1^2(t)\right) + g_1(t)v_2(t) + g_2(t)v_1^2(t) + i_{0,2}(t) = 0 \tag{3.2.82}$$

and

$$\frac{\mathrm{d}}{\mathrm{d}t}\left(c_1(t)v_3(t) + 2c_2(t)v_1(t)v_2(t) + c_3(t)v_1^3(t)\right) + g_1(t)v_3(t) + 2g_2(t)v_1(t)v_2(t) + g_3(t)v_1^3(t) + i_{0,3}(t) = 0 \tag{3.2.83}$$

where $i_{0,n}(t)$ is the nth-order current in $Z_e(\omega)$. These equations imply that a separate circuit can be generated for each mixing product; those circuits are shown in Figure 3.23. Figure 3.23(a), the linear, small-signal circuit, can be used to determine $v_1(t)$ via the conversion matrix techniques of the previous section. The first-order voltage $v_1(t)$ is then used to find the excitation current in Figure 3.23(b), from which the second-order voltage $v_2(t)$ can be found. Note that the circuit in Figure 3.23(b) is *linear*; the only nonlinear process is in the formulation of the excitation current from $v_1(t)$. Therefore, once this current is determined, ordinary linear conversion matrix analysis can be used to find the voltage across and current in $Z_e(\omega)$ at whatever frequency is of interest. Finally, $v_1(t)$ and $v_2(t)$ are used to find the third-order excitation current in Figure 3.23(c). In concept, these currents could be evaluated in the time domain or frequency domain; however, the rest of the circuit uses a frequency-domain characterization, so it is likely to be more convenient to express the source currents in the frequency domain as well. Furthermore, ω_1 and ω_2 are, in general, noncommensurate frequencies, so $v(t)$ is not periodic; this situation introduces

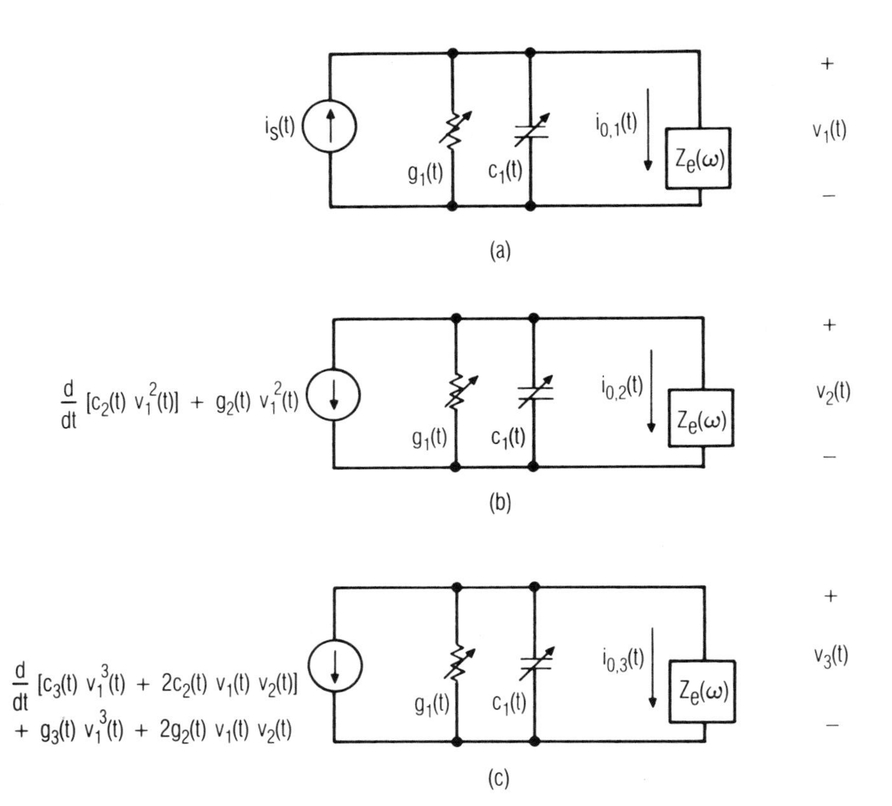

Figure 3.23 Linear circuits for determining the first-order (a), second-order (b), and third-order IM components (c).

further difficulties into a time-domain analysis. We now find frequency-domain expressions for both the junction voltages and the excitation currents.

The voltage, $v_1(t)$, is found from the small-signal linear analysis and has the form:

$$v_1(t) = \frac{1}{2} \sum_{m=-\infty}^{\infty} \sum_{\substack{q=-2 \\ q \neq 0}}^{2} V_{m,q} \exp[j(m\omega_p + \omega_q)t] \tag{3.2.84}$$

so

$$v_1^2(t) = \frac{1}{4} \sum_{m=-\infty}^{\infty} \sum_{n=-\infty}^{\infty} \sum_{q=-2}^{2} \sum_{\substack{r=-2 \\ q,r \neq 0}}^{2} V_{m,q} V_{n,r} \cdot \exp\{j[(m + n)\omega_p + \omega_q + \omega_r]t\} \tag{3.2.85}$$

The second-order terms of most interest are those at $k\omega_p + \omega_1 - \omega_2$ and $k\omega_p + 2\omega_1$. The components at $k\omega_p + \omega_1 + \omega_2$ and $k\omega_p + 2\omega_2$ can be found in a nearly identical manner, if they are of interest, so we will not consider them further. The terms of interest will be designated by subscripts a and b, respectively:

$$v_{1a}^2(t) = \frac{1}{2} \sum_{m=-\infty}^{\infty} \sum_{n=-\infty}^{\infty} V_{m,1} V_{n,-2} \cdot \exp\{j[(m + n)\omega_p + \omega_1 - \omega_2]t\} \tag{3.2.86}$$

and

$$v_{1b}^2(t) = \frac{1}{4} \sum_{m=-\infty}^{\infty} \sum_{n=-\infty}^{\infty} V_{m,1} V_{n,1} \exp\{j[(m + n)\omega_p + 2\omega_1]t\} \tag{3.2.87}$$

The coefficient of (3.2.86) is 1/2 instead of 1/4 because there are two identical terms in the q, r summation in (3.2.85) at this frequency. Also, we should note that $v_{1a}^2(t)$ and $v_{1b}^2(t)$ are complex because they include only some of the terms in (3.2.85); thus, they do not represent real time functions. The Taylor series coefficients can be expressed by their Fourier series' as

$$g_2(t) = \sum_{h=-\infty}^{\infty} G_{2,h} \exp(jh\omega_p t) \tag{3.2.88}$$

and

$$c_2(t) = \sum_{h=-\infty}^{\infty} C_{2,h} \exp(jh\omega_p t) \tag{3.2.89}$$

Substituting (3.2.86) through (3.2.89) into the parts of (3.2.82) that represent the source current gives the current-source components at these two frequencies, $i_{2a}(t)$ and $i_{2b}(t)$:

$$i_{2a}(t) = \frac{1}{2} \sum_{h=-\infty}^{\infty} \sum_{m=-\infty}^{\infty} \sum_{n=-\infty}^{\infty} V_{m,1} V_{n,-2}$$
$$\cdot \{G_{2,h} + C_{2,h}\, j[(h+m+n)\omega_p + \omega_1 - \omega_2]\}$$
$$\cdot \exp\{j[(h+m+n)\omega_p + \omega_1 - \omega_2]t\} \qquad (3.2.90)$$

and

$$i_{2b}(t) = \frac{1}{4} \sum_{h=-\infty}^{\infty} \sum_{m=-\infty}^{\infty} \sum_{n=-\infty}^{\infty} V_{m,1} V_{n,1}$$
$$\cdot \{G_{2,h} + C_{2,h}\, j[(h+m+n)\omega_p + 2\omega_1]\} \qquad (3.2.91)$$
$$\cdot \exp\{j[(h+m+n)\omega_p + 2\omega_1]t\}$$

$i_{2a}(t)$ and $i_{2b}(t)$ have the form:

$$i_{2a}(t) = \frac{1}{2} \sum_{k=-\infty}^{\infty} I_{k,2a} \exp[j(k\omega_p + \omega_1 - \omega_2)t] \qquad (3.2.92)$$

and

$$i_{2b}(t) = \frac{1}{2} \sum_{k=-\infty}^{\infty} I_{k,2b} \exp[j(k\omega_p + 2\omega_1)t] \qquad (3.2.93)$$

Equating terms in (3.2.90) and (3.2.91) with those in (3.2.92) and (3.2.93), respectively, gives

$$I_{k,2a} = \sum_{h=-\infty}^{\infty} \sum_{m=-\infty}^{\infty} \sum_{\substack{n=-\infty \\ h+m+n=k}}^{\infty} V_{m,1} V_{n,-2} [G_{2,h}$$
$$+ C_{2,h} j(k\omega_p + \omega_1 - \omega_2)] \qquad (3.2.94)$$

and

$$I_{k,2b} = \frac{1}{2} \sum_{h=-\infty}^{\infty} \sum_{m=-\infty}^{\infty} \sum_{\substack{n=-\infty \\ h+m+n=k}}^{\infty} V_{m,1} V_{n,1} [G_{2,h}$$
$$+ C_{2,h} j(k\omega_p + 2\omega_1)] \qquad (3.2.95)$$

Limiting k to the range $(-K, \ldots, K)$ allows $I_{k,2a}$ and $I_{k,2b}$ to be expressed as column vectors:

$$\mathbf{I}_{2a} = (I^*_{-K,2a} I^*_{-K+1,2a} \ldots I^*_{-1,2a} I_{0,2a} I_{1,2a} \ldots I_{K,2a})^T \tag{3.2.96}$$

and similarly for $\mathbf{I}_{2b}$. Finally, conversion-matrix analysis gives the vectors of second-order output currents:

$$\mathbf{I}_{0,2a} = -(\mathbf{1} + \mathbf{Y}_j \mathbf{Z}_{e,2a})^{-1} \mathbf{I}_{2a} \tag{3.2.97}$$

$$\mathbf{I}_{0,2b} = -(\mathbf{1} + \mathbf{Y}_j \mathbf{Z}_{e,2b})^{-1} \mathbf{I}_{2b} \tag{3.2.98}$$

where $\mathbf{Y}_j$ is the conversion matrix that represents the parallel combination of the time-varying conductance and capacitance, and $\mathbf{Z}_{e,2a}$ and $\mathbf{Z}_{e,2b}$ are the diagonal embedding impedance matrices (3.2.35) at their respective sets of IM frequencies. The second-order voltages are

$$\mathbf{V}_{2a} = \mathbf{Z}_{e,2a} \mathbf{I}_{0,2a} \tag{3.2.99}$$

$$\mathbf{V}_{2b} = \mathbf{Z}_{e,2b} \mathbf{I}_{0,2b} \tag{3.2.100}$$

The third-order components are found analogously. The components of greatest interest are those at $2\omega_1 - \omega_2$ and $2\omega_2 - \omega_1$; both are derived identically, so only the former will be considered. The $v_1(t)v_2(t)$ terms in (3.2.83) have two components that generate $2\omega_1 - \omega_2$: $v_1(t)$ at ω_1 mixing with $v_{2a}(t)$ at $\omega_1 - \omega_2$, and $v_1(t)$ at $-\omega_2$ mixing with $v_{2b}(t)$ at $2\omega_1$. The components of $v_1^3(t)$ and $v_1(t)v_2(t)$ at this frequency are

$$v_1^3(t) = \frac{3}{8} \sum_{m=-\infty}^{\infty} \sum_{n=-\infty}^{\infty} \sum_{p=-\infty}^{\infty} V_{m,1} V_{n,1} V_{p,-2} \cdot \exp\{j[(m+n+p)\omega_p + 2\omega_1 - \omega_2]t\} \tag{3.2.101}$$

$$v_1(t)v_{2a}(t) = \frac{1}{4} \sum_{m=-\infty}^{\infty} \sum_{n=-\infty}^{\infty} V_{m,2a} V_{n,1} \cdot \exp\{j[(m+n)\omega_p + 2\omega_1 - \omega_2]t\} \tag{3.2.102}$$

and

$$v_1(t)v_{2b}(t) = \frac{1}{4} \sum_{m=-\infty}^{\infty} \sum_{n=-\infty}^{\infty} V_{m,2b} V_{n,-2} \cdot \exp\{j[(m + n)\omega_p + 2\omega_1 - \omega_2]t\} \tag{3.2.103}$$

The Fourier series representations for the time-varying Taylor series coefficients are

$$g_3(t) = \sum_{h=-\infty}^{\infty} G_{3,h} \exp(jh\omega_p t) \tag{3.2.104}$$

and

$$c_3(t) = \sum_{h=-\infty}^{\infty} C_{3,h} \exp(jh\omega_p t) \tag{3.2.105}$$

and the resulting third-order components of the source current are

$$\begin{aligned} I_{k,3} = \frac{3}{4} & \sum_{h=-\infty}^{\infty} \sum_{\substack{m=-\infty \\ h+m+n+p=k}}^{\infty} \sum_{n=-\infty}^{\infty} \sum_{p=-\infty}^{\infty} V_{m,1} V_{n,1} V_{p,-2} \\ & \cdot [G_{3,h} + C_{3,h} j(k\omega_p + 2\omega_1 - \omega_2)] \\ & + \sum_{h=-\infty}^{\infty} \sum_{\substack{m=-\infty \\ h+m+n=k}}^{\infty} \sum_{n=-\infty}^{\infty} (V_{m,2a} V_{n,1} + V_{m,2b} V_{n,-2}) \\ & \cdot [G_{2,h} + C_{2,h} j(k\omega_p + 2\omega_1 - \omega_2)] \end{aligned} \tag{3.2.106}$$

The third-order current in $Z_e(\omega)$ is

$$\mathbf{I}_{0,3} = -(\mathbf{1} + \mathbf{Y}_j \mathbf{Z}_{e,3})^{-1} \mathbf{I}_3 \tag{3.2.107}$$

where $\mathbf{I}_{0,3}$ is the vector of output currents and $\mathbf{I}_3$ is the vector having the form of (3.2.96}, whose components are $I_{k,3}$ from (3.2.106). $\mathbf{Z}_{e,3}$ is the diagonal matrix of embedding impedances at the third-order mixing frequencies. Finally, the power of the third-order IM component dissipated in the embedding network at the frequency $k\omega_p + 2\omega_1 - \omega_2$ is

$$P_{k,3} = 0.5|I_{0;k,3}|^2 \operatorname{Re}\{Z_{e;k,3}\} \tag{3.2.108}$$

Equation (3.2.108) is the output power if the embedding network is lossless. If it is not (e.g., if the diode series resistance has been included in it), it is necessary to subtract the part of $\text{Re}\{Z_{e;k,3}\}$ representing the loss from the impedance in (3.2.108).

3.3 GENERALIZED HARMONIC-BALANCE ANALYSIS

We saw that harmonic-balance analysis was applicable to large-signal, single-tone problems, and that large-signal–small-signal analysis could be used to solve problems that involved multitone small-signal excitations and a single large-signal excitation. We shall see, in the next chapter, that power-series and Volterra-series techniques are very useful in analyzing weakly nonlinear circuits having multiple small-signal excitations.

Although these cases cover a wide range of practical problems, one remaining class of problems still has not been addressed, large-signal multitone excitation of strongly nonlinear circuits. Examples of this type of problem are the calculation of intermodulation levels in power amplifiers and of large-signal intermodulation in mixers. This type of problem cannot be solved by large-signal–small-signal analysis or by Volterra-series techniques because both methods require that at least one signal, and sometimes all of them, be very small. These problems can, however, be handled by a modified type of "harmonic" balance, which we call *generalized harmonic-balance analysis*.

3.3.1 Generalizing the Harmonic-Balance Concept

The concept of harmonic balance is illustrated by Figure 3.2, which shows a nonlinear circuit partitioned into linear and nonlinear subcircuits. The voltages at the interconnections between the two subcircuits are variables that, when determined, define all the voltages and currents in the network. In the case of single-tone excitation, the only case examined so far, the voltages and currents are periodic and, thus, have a fundamental-frequency component and a number of harmonics.

We now consider the case where the excitation may have two or more noncommensurate frequencies and the frequency components of the currents and voltages are no longer harmonically related. In general, the voltages and currents at each port in Figure 3.2 have a set of K frequency components:

$$\omega_k = \omega_0, \omega_1, \omega_2, \ldots, \omega_{K-1} \tag{3.3.1}$$

Usually $\omega_0 = 0$. These frequency components are mixing products, not harmonics; each mixing frequency, ω_k, arises as a linear combination of the excitation frequencies. In the case of two-tone excitation (usually the most important one):

$$\omega_k = m\omega_{p1} + n\omega_{p2}; \quad m, n = \ldots, -2, -1, 0, 1, 2, \ldots \tag{3.3.2}$$

where ω_{p1} and ω_{p2} are the frequencies of the two excitations. Each ω_k represents one mixing frequency and corresponds to one (m,n) pair; all mixing frequencies up to some maximum value of m or n are included in the set of frequencies described by (3.3.1) and (3.3.2), although only positive ω_k need be included (negative-frequency components are simply the complex conjugates of the positive ones, so voltage or current components at these frequencies are redundant). The number of frequency components retained in the set is subject to the same considerations as those that applied to harmonics; these considerations are discussed in Section 3.1.3.

The goal of the harmonic-balance analysis, as before, is to find a set of voltage components $V_{n,k}$ at the frequencies ω_k that satisfies (3.1.1). In this case, however, the components $I_{n,k}$ of the current vector and $Q_{n,k}$ of the charge vector represent the components at Port n and at mixing frequency ω_k, where ω_k is not necessarily a harmonic of a single excitation frequency. The harmonic-balance equation (3.1.20) is still valid in the multitone case; however, in forming (3.1.20), it is necessary to replace the harmonics $k\omega_p$ in $\mathbf{Y}_{m,n}$, (3.1.6), and in $\mathbf{\Omega}$, (3.1.16), with the term ω_k given by (3.3.1) and (3.3.2). It is necessary also to include both tones in the excitation voltage vectors; that is, (3.1.7) becomes

$$\begin{bmatrix} \mathbf{V}_{N+1} \\ \mathbf{V}_{N+2} \end{bmatrix} = \begin{bmatrix} V_{b1} \\ V_{s,1} \\ V_{s,2} \\ 0 \\ \cdot \\ \cdot \\ \cdot \\ V_{b2} \\ 0 \\ \cdot \\ \cdot \\ \cdot \\ 0 \end{bmatrix} \tag{3.3.3}$$

where $V_{s,1}$ and $V_{s,2}$ are the voltages of the two excitations. The positions of these components in (3.3.3) imply that $\omega_0 = 0$, $\omega_1 = \omega_{p1}$, and $\omega_2 = \omega_{p2}$; this is the usual situation.

As in the single-tone case, the central problem in harmonic-balance analysis is to solve (3.1.20) to obtain a voltage vector $\mathbf{V}$ for which $\mathbf{F(V)} = \mathbf{0}$. It is possible to solve (3.1.20) in the multitone case by using Newton's method, optimization, or a splitting technique; in some cases, the reflection algorithm may be applicable. For essentially the same reasons as in the single-tone case, good convergence properties and efficiency, Newton's method is usually preferred for use in a generalized harmonic-balance analysis.

In implementing any of the convergence algorithms, we immediately encounter a very serious problem: We have no valid transformation between the frequency and time domains. In the case of single-tone excitation, the response is periodic, so the transformation is simply a Fourier series. However, in the multitone case ω_{p1} and ω_{p2} are noncommensurate, so the currents and voltages are not periodic, and thus the Fourier series is not applicable. It is necessary either to reformulate the situation so the waveforms are periodic or to find some form of a trigonometric series that describes the time waveforms adequately. The former approach is the "brute force" method; it is not elegant but it has been used successfully. The latter approach requires the generation of an "almost-periodic" Fourier transform. Several methods of doing so have been proposed; we will describe one of the more elegant ones.

3.3.2 Reformulation and Fourier Transformation

In order to use a Fourier transformation, the voltage and current waveforms must be periodic. This will be the case if and only if the excitation frequencies are commensurate; that is,

$$q\omega_{p1} = r\omega_{p2} \tag{3.3.4}$$

for some nonzero positive integers, q and r. Then, the waveforms have a period T, where

$$T = \frac{2\pi}{\omega_{p2} - \omega_{p1}} \tag{3.3.5}$$

In (3.3.5), we have assumed that ω_{p2} is the higher of the two frequencies. In order to avoid aliasing errors, the waveforms must be sampled at a rate equal to twice the highest significant temporal (i.e., not radian) frequency; if that frequency is the Nth harmonic of the higher excitation frequency, $N\omega_{p2}/2\pi$, there must be $N\omega_{p2}/\pi$ samples per second. The number of samples, S, that must be made in each Fourier transformation is therefore the product of this quantity and T, or

$$S = \frac{2N\omega_{p2}}{\omega_{p2} - \omega_{p1}} \tag{3.3.6}$$

If ω_{p1} and ω_{p2} are closely spaced, then S becomes a very large number. Furthermore, because the fast Fourier-transform algorithm requires that S be a power of two, even the large number given by (3.3.6) must be increased to the next power of two. This large number of samples requires a comparably large, and often prohibitive, amount of computation time. It is especially frustrating to note that all but a few of the $S/2$ complex frequency components formed by the FFT are zero and most of the computation time is expended in finding the magnitudes these components rather than the magnitudes of the K components of interest.

The large amount of computation time is not the only problem that this method introduces. The large number of arithmetic operations necessary to form the Fourier transform reduces numerical precision, causing the result to be inaccurate. Furthermore, because of the requirement that ω_{p1} and ω_{p2} be commensurate, it is not possible to use any frequencies of interest; the excitation frequencies that can be used are constrained by (3.3.4) and the need to minimize the number of time samples, S.

3.3.3 Almost-Periodic Fourier Transforms

Although, in the noncommensurate case, the waveforms are not periodic, they are in some sense "almost" periodic with a period given by (3.3.5). It is therefore possible to devise an "almost-periodic" transform that can be used to transform the waveforms between the time and frequency domains; one attractive method for implementing such a transform has been described by Sorkin, Kundert, and Sangiovanni-Vincentelli (Reference 3.13), and their transform will be the basis for this section.

We wish to express the time waveform, $x(t)$, which may represent either a voltage or a current, as

$$x(t) = \sum_{k=0}^{K-1} X_{c,k} \cos(\omega_k t) + X_{s,k} \sin(\omega_k t) \tag{3.3.7}$$

where ω_k are given by (3.3.1) and (3.3.2). If the function $x(t)$ is sampled at the time intervals $t_1, t_2, t_3, \ldots, t_{2K-1}$, then the samples $x(t_i)$ can be expressed via a set of linear equations:

$$\begin{bmatrix} x(t_1) \\ x(t_2) \\ x(t_3) \\ \cdot \\ \cdot \\ \cdot \\ x(t_S) \end{bmatrix} = \begin{bmatrix} 1 & \cos(\omega_1 t_1) & \sin(\omega_1 t_1) & \ldots & \cos(\omega_{K-1} t_1) & \sin(\omega_{K-1} t_1) \\ 1 & \cos(\omega_1 t_2) & \sin(\omega_1 t_2) & \ldots & \cos(\omega_{K-1} t_2) & \sin(\omega_{K-1} t_2) \\ 1 & \cos(\omega_1 t_3) & \sin(\omega_1 t_3) & \ldots & \cos(\omega_{K-1} t_3) & \sin(\omega_{K-1} t_3) \\ \cdot & \cdot & \cdot & & \cdot & \cdot \\ \cdot & \cdot & \cdot & & \cdot & \cdot \\ \cdot & \cdot & \cdot & & \cdot & \cdot \\ 1 & \cos(\omega_1 t_S) & \sin(\omega_1 t_S) & \ldots & \cos(\omega_{K-1} t_S) & \sin(\omega_{K-1} t_S) \end{bmatrix} \begin{bmatrix} X_0 \\ X_{c,1} \\ X_{s,1} \\ \cdot \\ \cdot \\ \cdot \\ X_{s,K-1} \end{bmatrix} \tag{3.3.8}$$

or, in simpler notation:

$$\mathbf{x} = \mathbf{\Gamma}^{-1}\,\mathbf{X} \tag{3.3.9}$$

that is, $\mathbf{\Gamma}$ describes the transformation from the time to the frequency domain. S, the number of time samples, is $S = 2K - 1$. The *discrete Fourier transform* (DFT) is a special case of (3.3.8) in which the time samples are selected uniformly and the ω_k are harmonically related. The *fast Fourier transform* (FFT), which we use in single-tone harmonic-balance analysis, is just an algorithm that implements a DFT but minimizes the number of repeated multiplications. The DFT or FFT generates very little error in transforming between the time and frequency domains, because the rows of $\mathbf{\Gamma}^{-1}$ are orthogonal. When solving (3.3.8) introduces very little error, we say that the matrix is *well conditioned*. If the frequencies are not harmonically related, the rows of $\mathbf{\Gamma}^{-1}$ are not orthogonal, and in fact, it is possible for some rows to be nearly linearly dependent. In this case, we say that the $\mathbf{\Gamma}^{-1}$ is *ill conditioned*; then, the solution of (3.3.8) is poorly defined and, in practice, has large errors. Thus, we need to find a matrix $\mathbf{\Gamma}^{-1}$ that has nonharmonically related ω_k but is well conditioned.

Because uniform time intervals often result in an ill conditioned matrix when the ω_k are not harmonics, the use of nonuniformly-spaced time samples seems likely to result in better conditioning. In any case, if the ω_k are defined, the choice of time samples is our only remaining degree of freedom. But, how do we select those points? One possibility is to oversample; that is, to select more than $2K - 1$ points. Selecting S according to (3.3.6) is an extreme example of this approach, but it happens that a set of $4K$ to $6K$ points, selected randomly over an interval T given by (3.3.5), usually provides a well conditioned system. However, oversampling has the disadvantage that it increases the amount of computation time required to solve (3.3.8) and the equations describing the nonlinear subcircuit; it also introduces the minor problem of an overspecified system. We would therefore like to find a form of $\mathbf{\Gamma}$ that does not require oversampling.

Reference 3.13 shows that it is possible to create a well conditioned system that is not oversampled by first choosing an excessive number of time points and then reducing the number of points to the minimum. We begin by selecting approximately 1.5 times the minimum necessary sampling points t_i, choosing the approximately $3K$ sample points randomly over an interval T given by (3.3.5). The resulting sine-cosine matrix is tall; that is, it has more rows than columns. We then select $2K - 1$ rows of the matrix to form $\mathbf{\Gamma}^{-1}$, and note the corresponding time points; these will be used as the sample points in $\mathbf{\Gamma}$ and $\mathbf{x}$. The rows we retain are rows of the matrix that, as closely as possible, form an orthogonal set of vectors.

The set of nearly orthogonal rows are chosen by a variation of the Gram-Schmidt orthogonalization procedure. Let $\boldsymbol{\gamma}_n$ represent the nth row of the matrix $\mathbf{\Gamma}^{-1}$. We select one row arbitrarily, for example, $\boldsymbol{\gamma}_1$, and remove the components in the direction of $\boldsymbol{\gamma}_1$ of all the other vectors by forming:

$$\boldsymbol{\gamma}_n' = \boldsymbol{\gamma}_n - \frac{\boldsymbol{\gamma}_1^T \boldsymbol{\gamma}_n}{\boldsymbol{\gamma}_1^T \boldsymbol{\gamma}_1} \boldsymbol{\gamma}_1, \; n = 2, 3, \ldots, 2N \tag{3.3.10}$$

The set of vectors $\boldsymbol{\gamma}_n'$ are all orthogonal to $\boldsymbol{\gamma}_1$; because the vectors originally were the same length and had the same norm, the largest remaining vector (the one having the greatest norm) must have been the one most nearly orthogonal to $\boldsymbol{\gamma}_1$. This row is retained and $\boldsymbol{\gamma}_1$ is replaced by it in the next iteration. The process continues until the required number of vectors are selected.

The harmonic-balance equations can be solved via any appropriate algorithm; Sorkin et al. use Newton's method. In order to form the Jacobian, we still need to find expressions for the derivative terms in (3.1.38) and (3.1.39). Because of the nonperiodic character of the voltage and current waveforms, the expressions (3.1.40) and (3.1.41) are no longer valid in describing the derivative terms of the Jacobian matrix, (3.1.38), even if $\mathbf{Y}$ and $\mathbf{\Omega}$ are modified as indicated earlier. An expression for $\partial \mathbf{I}_{G,n}/\partial \mathbf{V}_m$ can be found in the nonperiodic case by using $\mathbf{\Gamma}$:

$$\frac{\partial \mathbf{I}_{G,n}}{\partial \mathbf{V}_m} = \mathbf{\Gamma}\left(\frac{\partial i_{g,n}(t)}{\partial v_m(t)}\right)\mathbf{\Gamma}^{-1} \tag{3.3.11}$$

The variables in (3.3.11) are those defined in Section 3.1, and the derivative in the parentheses represents a diagonal matrix having the derivative at point t_i in the ith position. The expression for $\partial \mathbf{Q}_n / \partial \mathbf{V}_m$ is analogous.

This method is not the only one that has been proposed for implementing an almost periodic Fourier transform or a generalized harmonic-balance analysis. The primary advantage of this method is that it is compatible with the harmonic-balance formulation developed in this chapter, and it appears to work especially well when Newton's method is used as the convergence algorithm. Other methods that have been accepted more or less widely are described in References 3.14 through 3.16.

REFERENCES

[3.1] M.S. Nakhla and J. Vlach, "A Piecewise Harmonic Balance Technique for Determination of Periodic Response of Nonlinear Systems," *IEEE Trans. Circ. Syst.*, Vol. CAS-23, 1976, p. 85.

[3.2] S.W. Director and K.W. Current, "Optimization of Forced Nonlinear Periodic Currents," *IEEE Trans. Circ. Syst.*, Vol. CAS-23, 1976, p. 329.

[3.3] F.R. Colon and T.N. Trick, "Fast Periodic Steady-State Analysis for Large-Signal Electronic Circuits," *IEEE J. Solid-State Circ.*, Vol. SC-8, 1973, p. 260.

[3.4] P.W. Van der Walt, "Efficient Technique for Solving Nonlinear Mixer Pumping Problem," *Electron. Lett.*, Vol. 21, 1985, p. 899.

[3.5] C. Camacho-Peñalosa, "Numerical Steady-State Analysis of Nonlinear Microwave Circuits with Periodic Excitation," *IEEE Trans. Microwave Theory Tech.*, Vol. MTT-31, 1983, p. 724.

[3.6] F. Filicori and V.A. Monaco, "Simulation and Design of Microwave Class-C Amplifiers Through Harmonic Balance Analysis," *IEEE MTT-S Int. Microwave Symp. Digest*, 1975, p. 362.

[3.7] K.S. Kundert and A. Sangiovanni-Vincentelli, "Simulation of Nonlinear Circuits in the Frequency Domain," *IEEE Trans. Computer-Aided Design*, Vol. CAD-5, 1986, p. 521.

[3.8] K.C. Gupta, R. Garg, and R. Chadha, *Computer-Aided Design of Microwave Circuits*, Artech House, Norwood, MA, 1981.

[3.9] R.G. Hicks and P.J. Khan, "Numerical Analysis of Nonlinear Solid-State Device Excitation in Microwave Circuits," *IEEE Trans. Microwave Theory Tech.*, Vol. MTT-30, 1982, p. 251.

[3.10] A.R. Kerr, "A Technique for Determining the Local Oscillator Waveforms in a Microwave Mixer," *IEEE Trans. Microwave Theory Tech.*, Vol. MTT-23, 1975, p. 828.

[3.11] S. Egami. "Nonlinear, Linear Analysis and Computer-Aided Design of Resistive Mixers," *IEEE Trans. Microwave Theory Tech.*, Vol. MTT-22, 1974, p. 270.

[3.12] S. Maas, "Two-Tone Intermodulation in Diode Mixers," *IEEE Trans. Microwave Theory Tech.*, Vol. MTT-35, 1987, p. 307.

[3.13] G.B. Sorkin, K.S. Kundert, and A. Sangiovanni-Vincentelli, "An Almost-Periodic Fourier Transform for Use with Harmonic Balance," *IEEE MTT-S Int. Microwave Symp. Digest*, 1987, p. 717.

[3.14] A. Ushida and L.O. Chua, "Frequency-Domain Analysis of Nonlinear Circuits Driven by Multi-Tone Signals," *IEEE Trans. Circ. Syst.*, Vol. CAS-31, 1984, p. 766.

[3.15] R. Gilmore, "Nonlinear Circuit Design Using the Modified Harmonic-Balance Algorithm," *IEEE Trans. Microwave Theory Tech.*, Vol. MTT-34, 1986, p. 1294.

[3.16] V. Rizzoli, C. Cecchetti, and A. Lipparini, "A General-Purpose Program for the Analysis of Nonlinear Microwave Circuits Under Multitone Excitation by Multidimensional Fourier Transform," *Proc. 17th European Microwave Conf.*, 1987.

CHAPTER 4

VOLTERRA-SERIES AND POWER-SERIES ANALYSIS

The previous chapter was concerned with strongly nonlinear circuits having a single large-signal excitation and sometimes one or more additional excitations that could be assumed to be vanishingly small. In this chapter, we consider the opposite extreme, weakly nonlinear circuits having multiple noncommensurate small-signal excitations. The nonlinearities in these circuits are often so weak that they have a negligible effect on their linear responses. However, the nonlinear phenomena (e.g., intermodulation distortion) in such quasilinear circuits can affect system performance and, thus, may be of great concern. The problem of analyzing such circuits is sometimes called the *small-signal nonlinear problem.*

Two techniques will be examined. The first is *power-series analysis,* which is relatively easy to use but requires a simplifying assumption that is often unrealistic, that the circuit contains only ideal memoryless transfer nonlinearities. The power-series approach is useful in some instances, however, and gives the engineer a good intuitive sense of the behavior of many types of nonlinear circuits. The second technique is *Volterra-series* or *nonlinear transfer-function analysis,* a very powerful technique that does not require such restrictive assumptions. Power-series analysis is equivalent to Volterra-series analysis in the case of memoryless transfer nonlinearities and, therefore, is a good introduction to the Volterra series.

This chapter generally follows the approach of Weiner and Spina (Reference 4.1), but is modified as necessary to be consistent with microwave applications. This book is based on work reported in References 4.2 and 4.3, performed in the early 1970s under US government sponsorship. Reference 4.4 is another excellent source of information on the Volterra series, although it is oriented more toward system applications than circuits. References on practical applications of the Volterra series to microwave circuits are listed at the end of Chapter 8.

4.1 POWER-SERIES ANALYSIS

4.1.1 Power-Series Model and Multitone Response

Many nonlinear systems and circuits are modeled as a filter, or other frequency-selective network, followed by a memoryless, broadband transfer nonlinearity. The model is shown in Figure 4.1, where the frequency-sensitive network has the transfer function $H(\omega)$ and the nonlinear stage has the transfer function:

$$w(t) = f(u(t)) = \sum_{n=1}^{N} a_n u^n(t) = a_1 u(t) + a_2 u^2(t) + a_3 u^3(t) + \ldots \quad (4.1.1)$$

or

$$w(t) = \sum_{n=1}^{N} w_n(t) \quad (4.1.2)$$

where $w_n(t) = a_n u^n(t)$. For practical reasons, the series must be truncated at some value $n = N$.

In Figure 4.1 and (4.1.1), the transfer function variables $w(t)$ and $u(t)$ may be small-signal incremental current or voltage; the stages may represent a nonlinear current, voltage, transresistance, or transconductance amplifier (in most cases, especially those involving transistors, the dominant transfer nonlinearity is a transconductance). It is important that the transfer function $f(u)$ be single-valued and that it be weakly nonlinear, that it express the nonlinearity adequately via a limited number of terms (in practice, N usually must be limited to three or at most five, or the analysis becomes hopelessly laborious). The linear block $H(\omega)$ may represent a filter or a matching network. In order to account for the effects of an output filter or a matching network, we may include another linear network at the output; the inclusion of such a network can be accomplished via ordinary linear circuit theory.

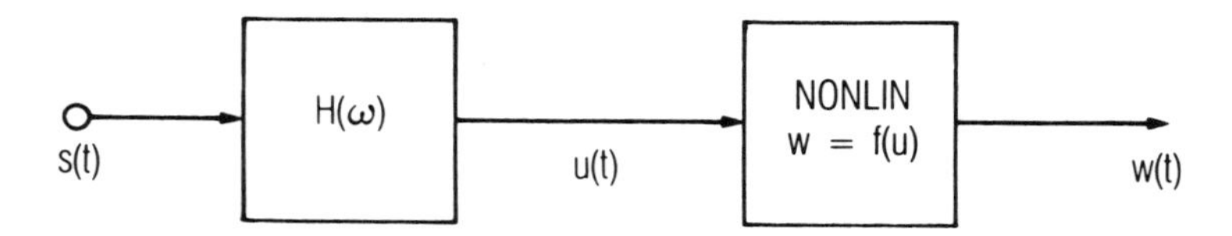

Figure 4.1 Power-series model of a nonlinear system: $H(\omega)$ is a linear circuit, and $f(u)$ is a memoryless nonlinear transfer function.

Figure 4.2 shows a simplified equivalent circuit of a FET, which can be described by this model. The elements of the circuit are readily identified with those of the block diagram in Figure 4.1; $v_s(t)$ corresponds to $s(t)$, $v(t)$ to $u(t)$, and $i(t)$ to $w(t)$. The input linear transfer function $H(\omega) = V(\omega)/V_s(\omega)$, where $V(\omega)$ and $V_s(\omega)$ are the frequency-domain equivalents of $v(t)$ and $v_s(t)$, respectively. Thus, in this example:

$$H(\omega) = \frac{V(\omega)}{V_s(\omega)} = \frac{1}{(R_s + R_i)\, C_i j\omega - L_s C_i \omega^2 + 1} \tag{4.1.3}$$

The only nonlinearity in the circuit is the transfer function $i = f(v)$ between the gate depletion voltage, $v(t)$, and the drain current, $i(t)$. The transfer function, $f(v)$, is the power-series expansion of the current around the bias point; it can be found, as described in Section 2.2.2, by expanding the large-signal drain current–gate voltage characteristic $I_d = F(V)$ in a Taylor series:

$$\begin{aligned} f(v) &= F(V_{g,0} + v) - F(V_{g,0}) \\ &= \left.\frac{dF(V)}{dV}\right|_{V=V_{g,0}} v + \frac{1}{2}\left.\frac{d^2f(V)}{dV^2}\right|_{V=V_{g,0}} v^2 \\ &\quad + \frac{1}{6}\left.\frac{d^3F(V)}{dV^3}\right|_{v=v_{g,0}} v^3 + \dots \end{aligned} \tag{4.1.4}$$

where $V_{g,0}$ is the dc bias voltage across the capacitor. The coefficients a_n are found by comparing (4.1.1) to (4.1.4).

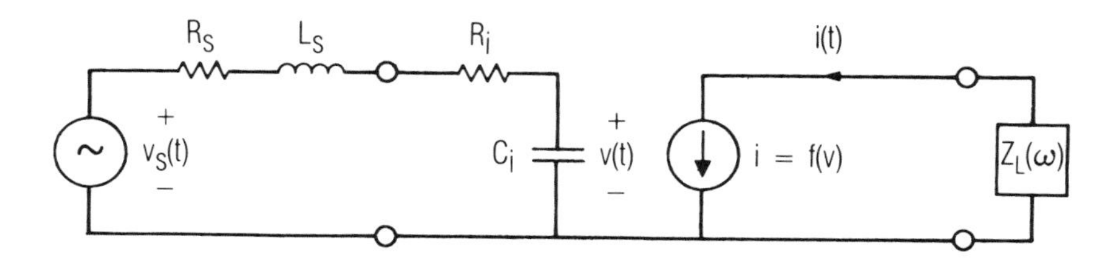

Figure 4.2 Simplified equivalent circuit of a FET, for which the power-series model is applicable.

So far, nothing in this problem precludes the use of time-domain techniques for analyzing the nonlinear circuit of either Figure 4.1 or Figure 4.2. We could easily convert the linear transfer function $H(\omega)$ to an impulse response function $h(t)$, find $v(t)$ by convolving $h(t)$ with $v_s(t)$, and finally substitute $v(t)$ into (4.1.1) to obtain $i(t)$. However, there are two problems inherent in using time-domain techniques to analyze this type of circuit: First, undertaking the solution of such problems is usually motivated by a need for frequency-domain, not time-domain, information, so a frequency-domain approach is the natural first choice. Second, if the excitations are at noncommensurate frequencies, the time-waveforms are not periodic, so obtaining valid frequency-domain data from a numerical time-domain result is often difficult. Finally, we use the frequency-domain approach because of the insight it gives us into a much more powerful technique, the Volterra-series or nonlinear transfer-function approach.

The excitation $s(t)$ in general contains at least two noncommensurate frequencies. In Figure 4.2, $v_s(t)$ corresponds to $s(t)$ and can be expressed as

$$v_s(t) = \frac{1}{2} \sum_{q=1}^{Q} V_{s,q} \exp(j\omega_q t) + V_{s,q}^* \exp(-j\omega_q t) \tag{4.1.5}$$

where the asterisk implies the complex conjugate. Equation (4.1.5) can be written in the following, more compact form:

$$v_s(t) = \frac{1}{2} \sum_{\substack{q=-Q \\ q \neq 0}}^{Q} V_{s,q} \exp(j\omega_q t) \tag{4.1.6}$$

We assume in (4.1.6), and throughout the following analysis, that the excitation and the response have no dc component. In microwave circuits, the excitation invariably has no dc component (other than the bias, which we do not include); however, as shown in Chapter 1, a nonlinear circuit may indeed generate dc components in its output in response to a sinusoidal excitation. In the case of a small-signal excitation and weak nonlinearities (the small-signal assumption is, after all, a foundation of power-series and Volterra-series analysis), the dc components generated by the nonlinearities are very small, and thus invariably negligible. The practical effect of the generation of dc components is that the bias currents and voltages are slightly offset from their quiescent values. In cases where significant bias offset occurs, as in a Class-B amplifier, Volterra- and power-series analyses are usually not applicable.

We shall assume throughout the remainder of this chapter that the excitations do not include dc components, and we will dispense with the notation to that effect ($q \neq 0$) in all the summations.

The output of the linear circuit is $v(t)$, which corresponds to $u(t)$ in Figure 4.1:

$$v(t) = \frac{1}{2} \sum_{q=-Q}^{Q} V_{s,q} \, H(\omega_q) \exp(j\omega_q t) \tag{4.1.7}$$

In (4.1.6) and (4.1.7), $\omega_{-q} = -\omega_q, V_{s,-q} = V^*_{s,q}$, and $H(\omega_{-q}) = H^*(\omega_q)$. The output of the nonlinear stage is found by substituting $v(t)$, expressed by (4.1.7), into (4.1.1) for $u(t)$; the terms are all of the form $a_n v^n$, where

$$\begin{aligned} a_n v^n(t) &= a_n \left[\frac{1}{2} \sum_{q=-Q}^{Q} V_{s,q} H(\omega_q) \exp(j\omega_q t) \right]^n \\ &= \frac{a_n}{2^n} \sum_{q1=-Q}^{Q} \sum_{q2=-Q}^{Q} \cdots \sum_{qn=-Q}^{Q} V_{s,q1} \\ &\quad \cdot V_{s,q2} \ldots V_{s,qn} H(\omega_1) H(\omega_2) \ldots H(\omega_{qn}) \\ &\quad \cdot \exp[j(\omega_{q1} + \omega_{q2} + \ldots \omega_{qn})t] \end{aligned} \tag{4.1.8}$$

The entire response is

$$i(t) = \sum_{n=1}^{N} a_n v^n(t) \tag{4.1.9}$$

where $i(t)$ is equivalent to $w(t)$ in (4.1.1). Equations (4.1.8) and (4.1.9) show that a potentially large number of new frequencies are generated by the nonlinearity; each frequency generated by the nth-degree term is a linear combination of n excitation frequencies, and the total response for each n is the sum of all possible linear combinations of n excitation frequencies. Figure 4.3 shows some of the lowest-order terms, those that are usually of most concern to system designers, when $Q = 2$ (two excitation tones) and $n \leqslant 3$. Furthermore, the amplitude of each frequency component is proportional to the product of the amplitudes of all the contributing excitations.

It is important in the following analysis to distinguish between the concepts of *degree* and *order*. The *degree* of the nonlinearity refers simply to the power of $u(t)$ in the nonlinear transfer characteristic (4.1.1). An nth-*order* mixing frequency is defined as one that arises from the sum of n excitation frequencies. In general, it is not possible to determine the order of a mixing product from its frequency; for example,

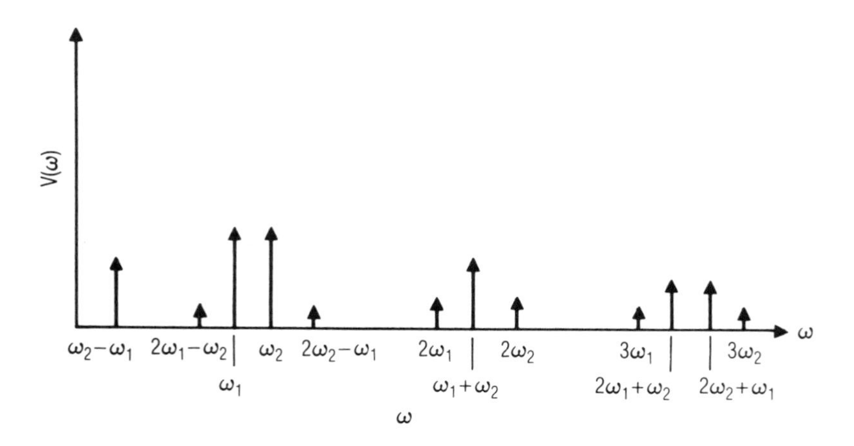

Figure 4.3 Spectrum of intermodulation frequencies resulting from two-tone excitation; the excitation frequencies are ω_1 and ω_2.

the frequency $2\omega_1 - \omega_2$ appears at first to be of third order, that is, $2\omega_1 - \omega_2 = \omega_1 + \omega_1 - \omega_2$, but it could in reality be the fifth-order mixing product $\omega_1 + \omega_1 + \omega_1 - \omega_1 - \omega_2$. In this example, an nth-degree nonlinearity generates nth-order mixing products. This result is a consequence of the fact that our circuit contains only a single ideal transfer nonlinearity. However, in more complicated circuits a nonlinearity of degree n can generate mixing products of order equal to or greater than n. This situation exists because, in reality, the mixing products and the excitation voltages are, in general, not limited to separate parts of the circuit. For example, if the circuit of Figure 4.2 included a feedback capacitance from the top of the current source to the top node of the input capacitor C_i, the control voltage $v(t)$ would include mixing products as well as excitation-frequency components. Thus, $v(t)$ would consist of components at the excitation frequency and at mixing frequencies, and products of order greater than n would arise as nth-order combinations of any of these excitation frequencies and other mixing products. For example, even if $n \leqslant 3$, a $2\omega_1 - \omega_2$ product could mix with a component at $\omega_1 - \omega_2$ to form a fifth-order component at $3\omega_1 - 2\omega_2$.

In order to illustrate power-series analysis, we now consider the case of two-tone excitation ($Q = 2$) and find the part of the response associated with the second- and third-degree components of the output current, $n = 2$ and $n = 3$, respectively. The second-degree component is designated $i_2(t)$ because it corresponds to $w_2(t)$ in (4.1.1) and (4.1.2):

$$
\begin{aligned}
i_2(t) &= a_2 v^2(t) \\
&= \frac{a_2}{4} \sum_{q1=-2}^{2} \sum_{q2=-2}^{2} V_{s,q1} V_{s,q2} H(\omega_{q1}) H(\omega_{q2}) \\
&\quad \cdot \exp[j(\omega_{q1} + \omega_{q2})t]
\end{aligned}
\tag{4.1.10}
$$

The summation in (4.1.8) generates $(2Q)^n$ terms; in this example, $Q = 2$, so there are sixteen terms. The frequencies associated with each term are the following:

$$
\begin{array}{cccc}
-\omega_2 - \omega_2 & -\omega_2 - \omega_1 & -\omega_2 + \omega_1 & -\omega_2 + \omega_2 \\
-\omega_1 - \omega_2 & -\omega_1 - \omega_1 & -\omega_1 + \omega_1 & -\omega_1 + \omega_2 \\
\omega_1 - \omega_2 & \omega_1 - \omega_1 & \omega_1 + \omega_1 & \omega_1 + \omega_2 \\
\omega_2 - \omega_2 & \omega_2 - \omega_1 & \omega_2 + \omega_1 & \omega_2 + \omega_2
\end{array}
$$

We can readily see that these terms include harmonics of the input frequencies (e.g., $\omega_1 + \omega_1$), repeated terms (e.g., $\omega_1 + \omega_2$ and $\omega_2 + \omega_1$), and dc terms (e.g., $\omega_1 - \omega_1$). Also, the terms in (4.1.10) occur in complex conjugate pairs, so the frequency components represent real time waveforms, as they must in any realizable circuit. For example, the $\omega_1 - \omega_2$ component is

$$
\begin{aligned}
i_2'(t) &= a_2 v^2(t) \Big|_{\omega_1 - \omega_2} \\
&= \frac{a_2}{4} 2\{V_{s,1} V_{s,2}^* H(\omega_1) H^*(\omega_2) \exp[j(\omega_1 - \omega_2)t] \\
&\quad + V_{s,1}^* V_{s,2} H^*(\omega_1) H(\omega_2) \exp[-j(\omega_1 - \omega_2)t]\}
\end{aligned}
\tag{4.1.11}
$$

The coefficient of 2 ahead of the brace arises because there are two terms at $\omega_1 - \omega_2$ and two terms at $-(\omega_1 - \omega_2)$ in the double summation: $q1 = 1$, $q2 = -2$ and $q1 = -2$, $q2 = 1$ both give identical terms at $\omega_1 - \omega_2$; $q1 = -1$, $q_2 = 2$ and $q1 = 2$, $q2 = -1$ both give identical terms at $-(\omega_1 - \omega_2)$. Equation (4.1.11) can be expressed in cosine form:

$$
i_2'(t) = a_2 |V_{s,1} V_{s,2} H(\omega_1) H(\omega_2)| \cos[(\omega_1 - \omega_2)t + \phi_2] \tag{4.1.12}
$$

where ϕ_2 is the phase angle associated with the complex coefficients in (4.1.11). The purpose of IM analysis is usually to determine output power at the mixing frequency; thus, ϕ_2 is normally of no interest. Phase angles are important in Volterra analysis and in power-series analysis only when two components of different orders, but having the same frequency, are combined. For example, saturation effects can be analyzed by combining the linear output at ω_1 and the third-order component at $\omega_1 + \omega_1 - \omega_1 = \omega_1$; the phase difference between these components affects the circuit's saturation characteristics.

The current component generated by the third-degree term, $i_3(t)$, can be found in a similar manner. From (4.1.8), with $n = 3$, we have

$$\begin{aligned} i_3(t) &= a_3 v^3(t) \\ &= \frac{a_3}{8} \sum_{q1=-2}^{2} \sum_{q2=-2}^{2} \sum_{q3=-2}^{2} V_{s,q1} V_{s,q2} V_{s,q3} H(\omega_{q1}) H(\omega_{q2}) H(\omega_{q3}) \\ &\quad \cdot \exp[j(\omega_{q1} + \omega_{q2} + \omega_{q3})t] \end{aligned} \tag{4.1.13}$$

The $i_3(t)$ summation has $(2Q)^n = 4^3 = 64$ terms, although not all represent different mixing frequencies. Half of the terms in (4.1.13) are simply conjugates of the others and, as in the second-degree case, many terms are identical. Some of the frequencies in (4.1.13) are

$$\begin{aligned} \omega_2 + \omega_2 - \omega_1 &= 2\omega_2 - \omega_1 \\ \omega_1 + \omega_1 - \omega_2 &= 2\omega_1 - \omega_2 \\ \omega_1 + \omega_1 - \omega_1 &= \omega_1 \\ \omega_2 + \omega_2 - \omega_2 &= \omega_2 \\ \omega_1 + \omega_1 + \omega_1 &= 3\omega_1 \\ \omega_2 + \omega_2 + \omega_2 &= 3\omega_2 \end{aligned} \tag{4.1.14}$$

The first two of these frequencies are of great interest because they often occur at frequencies close to ω_1 and ω_2. There are three identical terms at $2\omega_2 - \omega_1$ and three at $2\omega_1 - \omega_2$; the terms at $2\omega_2 - \omega_1$ occur when:

$$\begin{aligned} &q1 = 2, \quad q2 = 2, \quad q3 = -1 \\ &q1 = 2, \quad q2 = -1, \quad q3 = 2 \\ &q1 = -1, \quad q2 = 2, \quad q3 = 2 \end{aligned} \tag{4.1.15}$$

Because there are three terms at this frequency in (4.1.13), the coefficient 3 is used in the expression for the current component at $2\omega_2 - \omega_1$:

$$\begin{aligned} i_3'(t) &= a_3 v^3(t)\Big|_{2\omega 2 - \omega 1} \\ &= \frac{a_3}{8} 3\Big\{ V_{s,1}^* V_{s,2}^2 H^*(\omega_1) H^2(\omega_2) \\ &\quad \cdot \exp[j(2\omega_2 - \omega_1)t] \\ &\quad + V_{s,1} V_{s,2}^{*2} H(\omega_1) H^{*2}(\omega_2) \\ &\quad \cdot \exp[-j(2\omega_2 - \omega_1)t]\Big\} \end{aligned} \tag{4.1.16}$$

The cosine form of (4.1.16) is

$$i_3'(t) = \frac{3a_3}{4} |V_{s,1} V_{s,2}^2 H(\omega_1) H^2(\omega_2)| \cos[(2\omega_2 - \omega_1)t + \phi_3] \tag{4.1.17}$$

Again, the phase term ϕ_3 represents the combined phases of the complex coefficients in (4.1.16) and may be ignored in determining the power levels of third-order mixing components, the quantities usually of most interest.

4.1.2 Frequency Generation

The new frequencies generated by a transfer nonlinearity, expressed by (4.1.8), are relatively easy to predict. A large number of frequencies are possible; we shall assume that K nth-order frequencies are of interest and any one of them, $\omega_{n,k}$, $k = 1 \ldots K$, can be expressed as follows:

$$\begin{aligned} \omega_{n,k} &= m_{-Q}\omega_{-Q} + \ldots + m_{-2}\omega_{-2} + m_{-1}\omega_{-1} \\ &\quad + m_1\omega_1 + m_2\omega_2 + \ldots + m_Q\omega_Q \end{aligned} \tag{4.1.18}$$

where m_i is the number of times the frequency ω_i occurs in generating the mixing frequency $\omega_{n,k}$. Because exactly n terms are generated by an nth-degree nonlinearity, the set of values of m_i that defines any single mixing frequency is subject to the constraint:

$$\sum_{i=-Q}^{Q} m_i = n \tag{4.1.19}$$

Some of the nth-order terms in (4.1.18) may not appear to be a combination of n frequency components; for example, if $n = 3$ and $Q = 2$, some of those terms are the following:

$$\begin{aligned}\omega_1 + \omega_2 + \omega_{-2} &= \omega_1 + \omega_2 - \omega_2 = \omega_1 \\ \omega_1 + \omega_1 + \omega_{-1} &= \omega_1 + \omega_1 - \omega_1 = \omega_1 \\ \omega_2 + \omega_2 + \omega_{-2} &= \omega_2 + \omega_2 - \omega_2 = \omega_2\end{aligned} \tag{4.1.20}$$

and these seem to involve only one frequency. This is an illustration of the fact, stated in the previous section, that it generally is not possible to determine the order of a mixing product from its frequency. The evidence that these products are indeed third-order is their cubic dependence on $V_{s,q}$ and $H(\omega_q)$ in (4.1.8).

Determining the number of terms at each frequency is clearly necessary in order to obtain the correct magnitude of each IM component. For a mixing frequency given by (4.1.18), the number of terms is given by the multinomial coefficient:

$$t_{n,k} = \frac{n!}{m_{-Q}! \ldots m_{-2}!\, m_{-1}!\, m_1!\, m_2! \ldots m_Q!} \tag{4.1.21}$$

For example, in the second-order case of $\omega_1 - \omega_2 = \omega_1 + \omega_{-2}$ examined earlier, $n = 2$, $m_1 = 1$, $m_{-2} = 1$, so

$$t_{n,k} = \frac{2!}{1!\,1!} = 2 \tag{4.1.22}$$

as we had determined by "brute force." Similarly, for the $n = 3$ case, $\omega_{n,k} = 2\omega_2 - \omega_1$ so $m_2 = 2$, $m_{-1} = 1$, and

$$t_{n,k} = \frac{3!}{2!\,1!} = 3 \tag{4.1.23}$$

which agrees with the coefficient in (4.1.17).

4.1.3 Intercept Point and Power Relations

A commonly used method of determining the intermodulation properties of a nonlinear or quasilinear circuit is to perform a two-tone test, in which two excitations of equal amplitude and separated only slightly in frequency are applied to the circuit, and the powers of the resulting output IM components are measured. Figure 4.4 shows the test setup for such

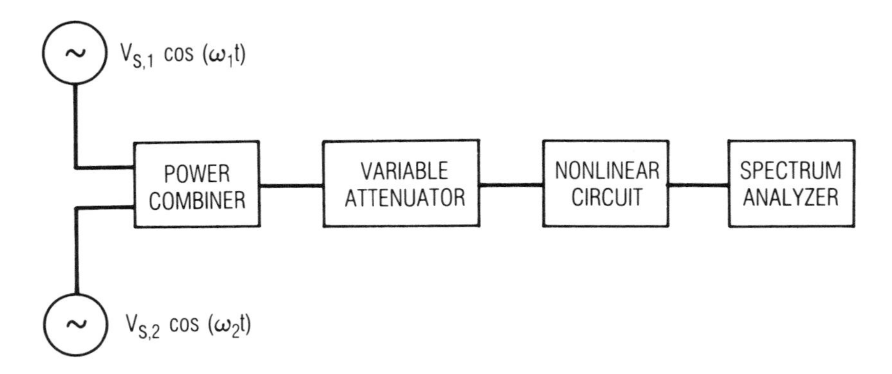

Figure 4.4 Two-tone test circuit. In general, $V_{s,1}$ and $V_{s,2}$ are equal, and ω_1 and ω_2 are nearly equal.

measurements; the two excitations are combined at the input of the nonlinear component, a variable attenuator is used to adjust the input level, and the output-frequency components are observed on the display screen of a spectrum analyzer. A spectrum similar to that shown in Figure 4.3 is normally observed. In the case of a two-tone test, $V_{s,1} = V_{s,2} = V_s$, $\omega_1 \approx \omega_2 = \omega$, and $H(\omega_1) \approx H(\omega_2) = H(\omega)$; V_s can be assumed to be real without loss of generality. These approximations are almost always valid for third-order IM at $2\omega_2 - \omega_1$ and $2\omega_1 - \omega_2$, but they may be somewhat strained when applied to second-order products, which can easily be out of band. We justify them by recognizing that second-order products are of primary concern when they are in band, and thus $H(\omega)$ is approximately constant and, in any case, the qualitative results of the following analysis are more important than the quantitative ones.

When these approximations are made and the phase angle ϕ_2 is ignored, (4.1.12) becomes

$$i_2'(t) = a_2 V_s^2 \, |H(\omega)|^2 \cos[(\omega_1 - \omega_2)t] \tag{4.1.24}$$

The second-order IM output power, the power dissipated in the real part of $Z_L\,(\omega_1 - \omega_2)$, is

$$P_{\text{IM2}} = \frac{1}{2} \, |i_2'\,(t)|^2 \, \text{Re}\,\{Z_L\,(\omega_1 - \omega_2)\} \tag{4.1.25}$$

We assume for simplicity that Re $\{Z_L(\omega)\} = R_L$, a constant. Then,

$$P_{IM2} = \frac{1}{2} a_2^2 V_s^4 |H(\omega)|^4 R_L \tag{4.1.26}$$

The available power of each input tone is

$$P_{\text{av}} = \frac{V_s^2}{8R_s} \tag{4.1.27}$$

and the output IM power can be written in terms of the available input power:

$$P_{\text{IM2}} = 32 a_2^2 |H(\omega)|^4 R_s^2 R_L P_{\text{av}}^2 \tag{4.1.28}$$

The same can be done with the third-order IM component at $2\omega_2 - \omega_1$. The output current at that frequency is

$$i_3'(t) = \frac{3}{4} a_3 V_s^3 |H(\omega)|^3 \cos[(2\omega_2 - \omega_1)t] \tag{4.1.29}$$

and the IM output power is

$$P_{\text{IM3}} = \frac{9}{32} a_3^2 V_s^6 |H(\omega)|^6 R_L \tag{4.1.30}$$

As with the second-order component, the third-order IM output power can be expressed as a function of available input power P_{av}:

$$P_{\text{IM3}} = 144 a_3^2 |H(\omega)|^6 R_s^3 R_L P_{\text{av}}^3 \tag{4.1.31}$$

It is normally most convenient to express the IM powers P_{IM2} and P_{IM3} in dBm, not linear quantities. Thus, with P_{IM2}, P_{IM3}, and P_{av} expressed in terms of dBm, (4.1.28) and (4.1.31) become

$$P_{\text{IM2}} = 10 \log(32 a_2^2 |H(\omega)|^4 R_s^2 R_L) + 2P_{\text{av}} - 30 \tag{4.1.32}$$

$$P_{\text{IM3}} = 10 \log(144 a_3^2 |H(\omega)|^6 R_s^3 R_L) + 3P_{\text{av}} - 60 \tag{4.1.33}$$

The IM output power given by (4.1.32) and (4.1.33), along with the linear output power, are graphed in Figure 4.5. At low levels the second- and

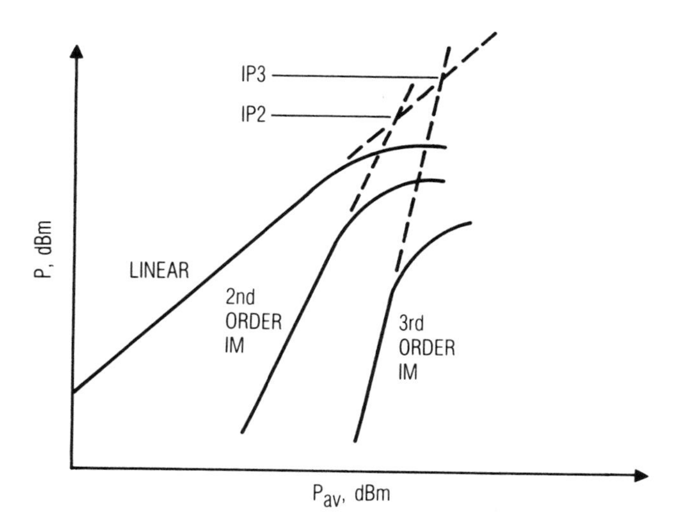

Figure 4.5 Input-output power curves for linear and intermodulation components. By tradition, the power shown in these curves is the power in each tone of the linear or IM product.

third-order IM powers vary by 2 dB/dB and 3 dB/dB, respectively, with input power level, while the linear output varies at the expected 1 dB/dB rate. In fact, further analysis would show that nth-degree IM products always vary by n dB/dB with input power level. At some point, the linear output power saturates because the output power available from any real component is always finite; the IM characteristics also saturate at approximately the same input level. Below this saturation level, however, the IM power curves, in terms of dBm, are straight lines.

The straight-line behavior of the IM characteristic can be used to predict IM levels at any input power level. It is necessary only to know the output level at one point in order to define the entire curve. A convenient point is the extrapolated point at which the IM and linear output powers are equal; this point is different, in general, for each order of IM product (and, in fact, for each different mixing frequency of a given order); it is called the *intermodulation intercept point*. This point is very useful because it defines not only the IM curve but its relation to the linear curve as well. Therefore, it can be used to find not only the IM output power but the ratio of linear to IM power level, often a more important quantity.

The IM characteristic follows the equation for any straight-line function:

$$P_{\text{IM}n} = nP_{\text{av}} + P_0 \tag{4.1.34}$$

where P_0 is a constant that will be evaluated. In terms of linear output power:

$$P_{\text{IM}n} = n(P_{\text{lin}} - G) + P_0 = nP_{\text{lin}} + P_0' \tag{4.1.35}$$

At the nth-order intercept point IP_n:

$$P_{\text{IM}n} = P_{\text{lin}} = IP_n \tag{4.1.36}$$

Substituting (4.1.36) into (4.1.35) gives

$$P_0' = (1 - n)IP_n \tag{4.1.37}$$

Substituting (4.1.37) into (4.1.35) gives the final result:

$$P_{\text{IM}n} = nP_{\text{lin}} - (n - 1)IP_n \tag{4.1.38}$$

Equation (4.1.38) gives the relationship between linear power P_{lin}, nth-order IM output power $P_{\text{IM}n}$, and nth-order intercept point IP_n at input levels below saturation. The only quantity that must be determined in order to define (4.1.38) is the intercept point IP_n. In order to determine IP_n, an expression for IM level must be found via an analysis of the circuit such as that presented earlier; then, IP_n can be found by comparison. A straightforward analysis of the circuit in Figure 4.2 gives the transducer gain G_t in dB:

$$G_t = 10\log(4\, a_1^2\, |H(\omega)|^2 R_s R_L) \tag{4.1.39}$$

Substituting this expression and $P_{\text{av}} = P_{\text{lin}} - G_t$ into (4.1.32) and (4.1.33), and doing some straightforward algebra, gives the expressions:

$$P_{\text{IM2}} = 10\log\left[\frac{2a_2^2}{a_1^4 R_L}\right] + 2P_{\text{lin}} - 30 \tag{4.1.40}$$

and

$$P_{\text{IM3}} = 10\log\left[\frac{9a_3^2}{4a_1^6 R_L^2}\right] + 3P_{\text{lin}} - 60 \tag{4.1.41}$$

By comparing (4.1.40) and (4.1.41) to (4.1.38), we find $IP2$ and $IP3$ in dBm:

$$IP_2 = 10 \log\left[\frac{a_1^4}{a_2^2}\frac{R_L}{2}\right] + 30 \tag{4.1.42}$$

$$IP_3 = 10 \log\left[\frac{2}{3}\frac{a_1^3}{a_3} R_L\right] + 30 \tag{4.1.43}$$

We again note that (4.1.42) and (4.1.43) apply only to specific second- and third-order intermodulation products $\omega_1 - \omega_2$ and $2\omega_2 - \omega_1$. Although they are also valid for the $\omega_2 - \omega_1$ and $2\omega_1 - \omega_2$ products, they are generally not valid for other products of the same orders, for example, $2\omega_1$ and $3\omega_2$. These latter components have different intercept points: they have the same dependence on a_1, a_2, a_3, and R_L, but the fractional coefficient within the parentheses in (4.1.42) or (4.1.43) is different; also some of the assumptions used in generating (4.1.42) and (4.1.43) may not be valid for other products. Equation (4.1.38) is valid for all IM products as long as the correct intercept point is used for IP_n.

Although the power-series concept is simple to implement and gives a good intuitive sense of the IM performance of a quasilinear circuit, it is severely limited. The most obvious limitation of power-series analysis is in the difficulty or, frequently, the impossibility of applying it to circuits that are not described by a simple transfer nonlinearity; it is likely that most practical circuits are not adequately described by such simple models. A second limitation is that a power-series analysis cannot be formulated to include nonlinearities that have memory; in particular, it cannot include nonlinear capacitances, which are often significant contributors to IM distortion in solid-state devices. An effect caused in part by nonlinear reactances is that the "straight-line" portion of the IM characteristic, shown in Figure 4.5, is not precisely straight; it often includes curvature and small ripples, and in some cases, sharp nulls are observed at certain input power levels. Even so, the IM characteristic of most quasilinear circuits often includes a dominant straight-line component, and an intercept point can be defined in such a way that it describes the component's intermodulation characteristics with reasonable accuracy.

We must also remember that the intercept point concept, as described here, is directly applicable only in the case of two-tone excitation and that the power relations are based upon the assumption that the levels of both excitations vary in tandem. In practice, one signal may vary and the other may not; in this case, the variation in the power of any IM output tone with variation in the power of one excitation tone will differ from the previous case. The variation in output level with variations in a single excitation can be found via the following rule: If the level of a single excitation tone at frequency ω_i is varied while all the other tones remain constant, the IM output power at $\omega_{n,k}$ varies by m_i dB/dB, where $\omega_{n,k}$, ω_i, and m_i are as used in (4.1.18). We consider the third-order IM product at $2\omega_2 - \omega_1$ as an example: The IM frequency component at $2\omega_2 - \omega_1$ varies by 2 dB/dB with variations in the level of the ω_2 excitation and by 1 dB/dB with variations in the level of the ω_1 excitation. This rule can be used to find the level of an IM product when the excitation levels are dissimilar, as long as the two-tone IM intercept point for excitations of identical levels is known.

4.1.4 Interconnections of Weakly Nonlinear Components

Equation (4.1.38) is useful for finding the two-tone intermodulation levels in a single quasilinear circuit. However, microwave systems are composed of a number of such circuits interconnected in a variety of ways, and it is invariably necessary to have an IM characterization of the entire system. In this section, we derive the intercept point of a cascade interconnection of stages; having that intercept point, we can use (4.1.38) to find the IM levels at the output of the cascade. The effect on IP_n of the parallel or hybrid interconnection of identical components, a much simpler subject, is considered in Chapter 5; in most cases, we find that all the intercept points are increased by $10 \cdot \log_{10}(M)$ dB, where M is the number of identical components in the combination.

Figure 4.6 shows the cascade interconnection of a number of two-ports. These two-ports may be amplifiers, mixers, control components, or any other type of weakly nonlinear component. One can accommodate linear components by assigning them intercept points that are much greater than those of the other elements. The stages are designated A_m, $m = 1 \ldots M$, and their transducer gains and nth-order intercept points are G_m and $IP_{n,m}$, respectively. We assume that the gain and the input and output impedances of each stage are constant over a frequency range that includes all IM products of interest, and that the stages operate as independent gain blocks; that is, their nonlinearities do not interact. Under these conditions, the system operates as follows: a two-tone signal is applied to the input of A_1, and A_1 passes the linear signal to the output and generates intermodulation distortion products. These are applied to the input of A_2; A_2 again passes the linear and IM outputs from A_1 to its output and generates some IM distortion of its own. These distortion products

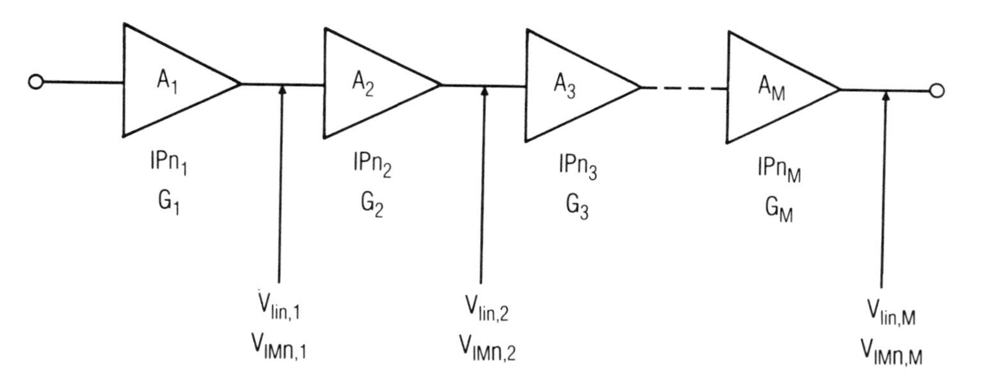

Figure 4.6 Cascade connection of weakly nonlinear components. Although the stages shown here are amplifiers, they can be any type of unilateral two-port.

occur at the same frequencies as those generated by A_1 and, thus, are combined with those from A_1. This process is repeated throughout the rest of the cascade.

In general, the phases of the IM distortion generated in any stage and those passed from its input are unknown. Thus, we generally do not know whether those IM components will be combined in phase (increasing the IM level) or out of phase (decreasing it) or with some other phase. To circumvent this problem, we shall make a worst-case assumption, that all distortion products combine in phase. This assumption gives an upper bound on the intercept point that is close to what is measured in practice because it is inevitable that, somewhere in the system's passband, all the IM components will in fact have nearly the same phase.

The voltage of the linear products at the output of A_1 is designated $V_{\text{lin},1}$ and the IM voltage generated in A_1 is $V_{\text{IM}n,1}$; at the output of A_2, the linear component is $G_2^{1/2} V_{\text{lin},1}$ and the IM voltage is $G_2^{1/2} V_{\text{IM}n,1} + V_{\text{IM}n,2}$. Thus, at the output of the last stage, A_M:

$$V_{\text{lin},M} = V_{\text{lin},1}(G_2 G_3 \ . \ . \ . \ G_M)^{1/2} \tag{4.1.44}$$

and

$$\begin{aligned} V_{\text{IM}n} = {} & V_{\text{IM}n,M} + V_{\text{IM}n,M-1} G_M^{1/2} + V_{\text{IM}n,M-2}(G_M G_{M-1})^{1/2} \\ & + \ . \ . \ . + V_{\text{IM}n,1}(G_2 G_3 \ . \ . \ . G_M)^{1/2} \end{aligned} \tag{4.1.45}$$

Converting (4.1.38) from dBm to units of watts or mW gives

$$IP_n^{(n-1)} = P_{\text{lin}}^n P_{\text{IM}n}^{-1} \tag{4.1.46}$$

We now square $V_{\text{lin},M}$ and $V_{\text{IM}n}$ to obtain $P_{\text{lin},M}$ and $P_{\text{IM}n}$ and substitute the squared forms of (4.1.44) and (4.1.45) into (4.1.46). We also substitute $P_{\text{IM}n,m}^{1/2}$ for $V_{\text{IM}n,m}$ in (4.1.45). This substitution and a little algebra give the result:

$$\begin{aligned} IP_n^{(1-n)/2} = {} & (G_2 \ldots G_M)^{-n/2} P_{\text{lin},1}^{-n/2} [P_{\text{IM}n,M}^{1/2} \\ & + (P_{\text{IM}n,M-1} G_M)^{1/2} \\ & + \ldots + (P_{\text{IM}n,1} G_2 \ldots G_M)^{1/2}] \end{aligned} \tag{4.1.47}$$

Equation (4.1.46) shows that, at the output of any stage:

$$P_{\text{IM},m} = IP_{n,m}^{(1-n)} P_{\text{lin},1}^n (G_2 \ldots G_m)^n \tag{4.1.48}$$

Finally, we substitute (4.1.48) into (4.1.47) and, again, grind through the algebra. The result is the cascade relation for intercept point:

$$\begin{aligned} IP_n^{(1-n)/2} = {} & IP_{n,M}^{(1-n)/2} + (G_M IP_{n,M-1})^{(1-n)/2} \\ & + (G_M G_{M-1} IP_{n,M-2})^{(1-n)/2} \\ & + \ldots + (IP_{n,1} G_2 \ldots G_M)^{(1-n)/2} \end{aligned} \tag{4.1.49}$$

Equation (4.1.49) shows that the amount each stage contributes to the output intercept point of the cascade is a function of the intercept point of that stage multiplied by the gain of all the stages following it. It is important to note that the gain, G_m, and nth-order intercept point, $IP_{n,m}$ in (4.1.49) are in units of watts or mW, not dBm.

4.2 VOLTERRA-SERIES ANALYSIS

4.2.1 Introduction to the Volterra Series

The power-series analysis of Section 4.1 was based upon the system model shown in Figure 4.1, wherein the frequency-sensitive linear part of the circuit and the memoryless nonlinear elements were clearly separate from each other. The model used for the Volterra-series analysis, shown in Figure 4.7, is essentially the same except that the separation between the memoryless and the reactive parts of the circuit has been eliminated, the nonlinear block contains a mix of linear and nonlinear elements. In

Figure 4.7 A weakly nonlinear circuit model for Volterra-series analysis. The circuit may have both reactive and resistive nonlinearities.

this case, the nonlinear elements may be either resistive or reactive, and they are characterized by power series having the same form as (4.1.1). The power-series characterization of each reactive or resistive element can be found via the methods described in Sections 2.2.2 and 2.2.3. As in the previous section, the excitation $s(t)$ contains, in general, a number of individual sinusoidal excitation components having noncommensurate frequencies.

In the previous section, we showed that the response $w(t)$ to the excitation $s(t)$, found via the power-series model, could be expressed as follows:

$$\begin{aligned} w(t) &= \sum_{n=1}^{N} a_n \left[\frac{1}{2} \sum_{q=-Q}^{Q} V_{s,q} H(\omega_q) \exp(j\omega_q t) \right]^n \\ &= \sum_{n=1}^{N} \frac{a_n}{2^n} \sum_{q1=-Q}^{Q} \sum_{q2=-Q}^{Q} \cdots \sum_{qn=-Q}^{Q} V_{s,q1} V_{s,q2} \ldots \\ &\quad \cdot V_{s,qn} H(\omega_{q1}) H(\omega_{q2}) \ldots \\ &\quad \cdot H(\omega_{qn}) \exp[j(\omega_{q1} + \omega_{q2} + \ldots \omega_{qn})t] \end{aligned} \tag{4.2.1}$$

where $H(\omega)$ was the transfer function of the linear part of the circuit, and a_n, $n = 1 \ldots N$, was the coefficient of the nth-degree term in the power series that characterized the memoryless nonlinear block. The excitation $s(t)$ was a small-signal, incremental voltage:

$$s(t) = \frac{1}{2} \sum_{q=-Q}^{Q} V_{s,q} \exp(j\omega_q t) \tag{4.2.2}$$

In the case of Volterra series analysis, we assume that the excitation has the same form as (4.2.2), and we will find that the response can be expressed as the following function of the excitation:

$$w(t) = \sum_{n=1}^{N} \frac{1}{2^n} \sum_{q1=-Q}^{Q} \sum_{q2=-Q}^{Q} \cdots \sum_{qn=-Q}^{Q} V_{s,q1} V_{s,q2} \cdots$$
$$\cdot V_{s,qn} H_n(\omega_{q1}, \omega_{q2}, \ldots, \omega_{qn})$$
$$\cdot \exp[j(\omega_{q1} + \omega_{q2} + \ldots \omega_{qn})t] \tag{4.2.3}$$

The only formal difference between (4.2.3) and (4.2.1) is that (4.2.3) contains a single function $H_n(\omega_{q1}, \omega_{q2}, \ldots, \omega_{qn})$, instead of the product of linear transfer functions $a_n H(\omega_{q1}) H(\omega_{q2}) \ldots H(\omega_{qn})$. The function $H_n(\omega_{q1}, \omega_{q2}, \ldots, \omega_{qn})$ is called the *nth-order nonlinear transfer function.* Knowing $H_n(\omega_{q1}, \omega_{q2}, \ldots, \omega_{qn})$, we can find individual mixing and distortion products in a manner identical to that used in the power-series analysis.

Volterra-series analysis is based on the same assumptions and is subject to many of the same limitations as power-series analysis: that the circuit is weakly nonlinear and that the multiple excitations are small and noncommensurate. In some cases, the two approaches are equivalent; by comparing (4.2.1) and (4.2.3), we can see that power-series analysis is a special case of Volterra analysis, one in which the nonlinear transfer function can be expressed as

$$H_n(\omega_{q1}, \omega_{q2}, \ldots, \omega_{qn}) = a_n H(\omega_{q1}) H(\omega_{q2}) \ldots H(\omega_{qn}) \tag{4.2.4}$$

All the previous work in Section 4.1 regarding frequency mixing in circuits characterized by a power series is valid for the Volterra series (the intercept-point concept is generally valid, as well); the primary difference is in the form of the nonlinear transfer function.

4.2.2 Volterra Functionals and Nonlinear Transfer Functions

A fundamental tenet of linear system and circuit theory is that the output $w(t)$ of a linear system or circuit having excitation $s(t)$ can be expressed by the convolution integral:

$$w(t) = \int_{-\infty}^{\infty} h(\tau) s(t - \tau) \, d\tau \tag{4.2.5}$$

where $h(t)$ is the *impulse response,* the response to a pulse that has infinitesimal width and infinite amplitude but has unit energy. This pulse is called the *Dirac delta function,* $\delta(t)$. It has the property that

$$f(t_0) = \int_{-\infty}^{\infty} f(t)\delta(t - t_0)\, dt \tag{4.2.6}$$

Equation (4.2.5) is valid only for linear circuits and systems. An extension of (4.2.5) to the case of nonlinear systems was proposed by Norbert Wiener (References 4.5 and 4.6), who applied the work of Volterra (Reference 4.7) to the problem of analyzing nonlinear systems. Wiener suggested that, in the case of weakly nonlinear circuits having small excitations, the response may be adequately described by the functional series:

$$\begin{aligned} w(t) = &\int_{-\infty}^{\infty} h_1(\tau_1)\, s(t - \tau_1)\, d\tau_1 \\ &+ \iint_{-\infty}^{\infty} h_2(\tau_1, \tau_2)\, s(t - \tau_1)\, s(t - \tau_2)\, d\tau_1\, d\tau_2 \\ &+ \iiint_{-\infty}^{\infty} h_3(\tau_1, \tau_2, \tau_3)\, s(t - \tau_1)\, s(t - \tau_2) \\ &\cdot s(t - \tau_3)\, d\tau_1\, d\tau_2\, d\tau_3 + \cdots \end{aligned} \tag{4.2.7}$$

In (4.2.7), the multidimensional function, $h_n(\tau_1, \tau_2, \ldots, \tau_n)$, is called the *nth-order kernel* or the *nth-order nonlinear impulse response*. Just as the linear frequency-domain transfer function $H(\omega)$ is the Fourier transform of $h(t)$, the nonlinear transfer function $H_n(\omega_1, \omega_2, \ldots, \omega_n)$ is the n-dimensional Fourier transform of $h_n(t_1, t_2, \ldots, t_n)$. The excitation function, $s(t)$, may be any finite small-signal voltage or current waveform, although in microwave circuits we will be interested exclusively in the case where $s(t)$ is given by (4.2.2), where $s(t)$ is the sum of several sinusoidal components.

Equation (4.2.7) can be expressed in more compact form as

$$w(t) = \sum_{n=1}^{N} w_n(t) \tag{4.2.8}$$

where

$$\begin{aligned} w_n(t) = &\iint \cdots \int_{-\infty}^{\infty} h_n(\tau_1, \tau_2, \ldots, \tau_n)\, s(t - \tau_1) \\ &\cdot s(t - \tau_2) \cdots s(t - \tau_n)\, d\tau_1\, d\tau_2 \cdots d\tau_n \end{aligned} \tag{4.2.9}$$

The frequency-domain form of the response can be found by substituting (4.2.2) into (4.2.9). The result is

$$w_n(t) = \frac{1}{2^n} \iint \ldots \int_{-\infty}^{\infty} h_n(\tau_1, \tau_2, \ldots, \tau_n) \sum_{q1=-Q}^{Q} V_{s,q1}$$
$$\cdot \exp[j\omega_{q1}(t - \tau_1)] \sum_{q2=-Q}^{Q} V_{s,q2}$$
$$\cdot \exp[j\omega_{q2}(t - \tau_2)] \ldots \sum_{qn=-Q}^{Q} V_{s,qn}$$
$$\cdot \exp[j\omega_{qn}(t - \tau_n)]\, d\tau_1\, d\tau_2 \ldots d\tau_n \tag{4.2.10}$$

We assume in (4.2.10) and henceforward that all summations of $qi = -Q \ldots Q$ do not include $qi = 0$, and for simplicity this will not be noted explicitly. Rearranging the terms in (4.2.10) and interchanging the order of integration and summation gives

$$w_n(t) = \frac{1}{2^n} \sum_{q1=-Q}^{Q} \sum_{q2=-Q}^{Q} \ldots \sum_{qn=-Q}^{Q} V_{s,q1} V_{s,q2} \ldots V_{s,qn}$$
$$\cdot \exp[j(\omega_{q1} + \omega_{q2} + \ldots \omega_{qn})t]$$
$$\cdot \iint \ldots \int_{-\infty}^{\infty} h_n(\tau_1, \tau_2, \ldots, \tau_n)$$
$$\cdot \exp[-j(\omega_{q1}\tau_1 + \omega_{q2}\tau_2 + \ldots \omega_{qn}\tau_n)]$$
$$\cdot d\tau_1\, d\tau_2 \ldots d\tau_n \tag{4.2.11}$$

The terms from the integral sign to the end of (4.2.11) can be recognized as a multidimensional Fourier transform:

$$H_n(\omega_{q1}, \omega_{q2}, \ldots, \omega_{qn}) = \iint \ldots \int_{-\infty}^{\infty} h_n(\tau_1, \tau_2, \ldots, \tau_n)$$
$$\cdot \exp[-j(\omega_{q1}\tau_1 + \omega_{q2}\tau_2$$
$$+ \ldots \omega_{qn}\tau_n)]\, d\tau_1\, d\tau_2 \ldots d\tau_n \tag{4.2.12}$$

We could, of course, find the nonlinear impulse response from the frequency-domain nonlinear transfer function by inverse-Fourier transforming:

$$h_n(\tau_1, \tau_2, \ldots, \tau_n) = \frac{1}{(2\pi)^n} \int\int \ldots \int_{-\infty}^{\infty} H_n(\omega_{q1}, \omega_{q2}, \ldots, \omega_{qn}) \cdot \exp[j(\omega_{q1}\tau_1 + \omega_{q2}\tau_2 + \ldots \omega_{qn}\tau_n)] \cdot d\omega_{q1}\, d\omega_{q2} \ldots d\omega_{qn} \tag{4.2.13}$$

Calculating multidimensional Fourier transforms is a nasty business; for this and other reasons, we will work entirely in the frequency domain. Replacing the integrals in (4.2.11) with the nonlinear transfer function $H_n(\omega_{q1}, \omega_{q2}, \ldots, \omega_{qn})$ gives the following expression for $w_n(t)$:

$$w_n(t) = \frac{1}{2^n} \sum_{q_1=-Q}^{Q} \sum_{q_2=-Q}^{Q} \ldots \sum_{q_n=-Q}^{Q} V_{s,q1} V_{s,q2} \ldots V_{s,qn} \cdot H_n(\omega_{q1}, \omega_{q2}, \ldots, \omega_{qn}) \cdot \exp[j(\omega_{q1} + \omega_{q2} + \ldots \omega_{qn})t] \tag{4.2.14}$$

and summing this expression for $w_n(t)$ over n, $n = 1 \ldots N$, to obtain $w(t)$ gives the expected result, (4.2.3).

It is worthwhile at this point to examine (4.2.14) and (4.2.3) and to note some of their important implications. First, as in the power series analysis, the total response in the Volterra case is simply the sum of the individual nth-order responses. In the power-series case this result was guaranteed by the separation of the linear from the nonlinear parts of the circuit, and by the limitation of the analysis to a single transfer nonlinearity. In the Volterra case, this result is an obvious consequence of the form of (4.2.7), but it is not obvious from the nature of the circuit model, where the nonlinear and linear parts of the circuit are freely intermingled. Our ability to separate even orders of mixing products, as well as different mixing products of the same order, is the key to the practicality of the Volterra series; without that ability, the analysis of weakly nonlinear circuits would be hopelessly laborious.

Although it is beyond the scope of this book, it is possible to show in several different ways that the series (4.2.7) is convergent and that the magnitude of each successive term is smaller than the previous one. Because each of the integral terms in (4.2.7) represents a single order of mixing products $w_n(t)$ in the circuit's total response $w(t)$, the power in the higher-order response components must be less than that in the lower-order response components. This result is consistent with the experimental observation that higher-order nonlinear distortion products are invariably weaker than lower-order ones.

A second property of the nonlinear transfer function is that it must be symmetrical in ω. The reason for this symmetry is obvious from a practical standpoint: there is no order associated with the different tones in the multitone excitation of (4.2.2), so one must be able to permute the frequencies in, say, (4.2.14) without changing the response. Equation (4.2.13) implies that $h_n(\tau_1 \ldots \tau_n)$ must also be symmetrical in τ if $H_n(\omega_1, \omega_2, \ldots, \omega_n)$ is symmetrical in ω_q.

4.2.3 Determining Nonlinear Transfer Functions by the Harmonic-Input Method

The nonlinear transfer function can be found via a technique called the *harmonic-input,* or *probing,* method. This method is not very different in concept from the process of finding the frequency-domain transfer function $H(\omega)$ of a linear circuit: we assume that the circuit has the simplest possible excitation, find the response, substitute both into the input-output equation, in this case (4.2.14), and finally solve algebraically for $H_n(\omega_1 \ldots \omega_n)$. Thus, we can find the $H(\omega)$ of a linear circuit by assuming that the input voltage is $1 \cdot \exp(j\omega t)$ and by manipulating the output into the form $H(\omega)\exp(j\omega t)$. The ratio of these quantities is the linear transfer function $H(\omega)$.

In the case of a nonlinear circuit, the situation is a little more complicated, as usual, but the concepts are much the same. In order to find the nth-order part of the response, we assume that the excitation is

$$s(t) = \exp(j\omega_1 t) + \exp(j\omega_2 t) + \ldots + \exp(j\omega_n t) \tag{4.2.15}$$

that is, the excitation used to find the nth-order transfer function is the sum of n positive-frequency phasors of unit magnitude; the negative-frequency components are not included in the excitation. That $s(t)$ is not a real function of time is of no consequence because this excitation is to be used only to determine the transfer functions. From (4.2.14), the nth-order response component at $\omega_1 + \omega_2 + \ldots \omega_n$ has the form:

$$w_n(t)\Bigg|_{\omega=\omega_1+\omega_2+\ldots+\omega_n} = n!\, H_n(\omega_1, \omega_2, \ldots, \omega_n) \cdot \exp[j(\omega_1 + \omega_2 + \ldots \omega_n)t] \tag{4.2.16}$$

This expression for $w_n(t)$ is substituted into the circuit equations; only the terms of nth order are retained (terms that are not of nth order do not contribute to the nth-order response, so they can be ignored), and the nonlinear transfer function $H_n(\omega_1, \omega_2, \ldots, \omega_n)$ is found algebraically.

In all cases, the nth-order nonlinear transfer function is found to be a function of the transfer functions of order less than n. Thus, it is first necessary to use $s(t) = \exp(j\omega_1 t)$ to find $H_1(\omega_1)$, the linear transfer function, then to use $s(t) = \exp(j\omega_1 t) + \exp(j\omega_2 t)$ to find the second-order transfer function $H_2(\omega_1, \omega_2)$ as a function of $H_1(\omega_1)$ and $H_1(\omega_2)$, and to continue until transfer functions of all desired orders have been determined. When all n transfer functions have been found, (4.2.14) and (4.2.8) can be used to find the levels of the frequency components of interest in the total response. It is not necessary in evaluating (4.2.14) to find the levels of all possible frequency components, only those of interest at each order.

Example 1

Figure 4.8 shows a simple weakly nonlinear circuit consisting of a nonlinear capacitor, a linear resistor, and a voltage source. We shall find the transfer function between the excitation, the voltage, $v_s(t)$, and the response, the current $i(t)$.

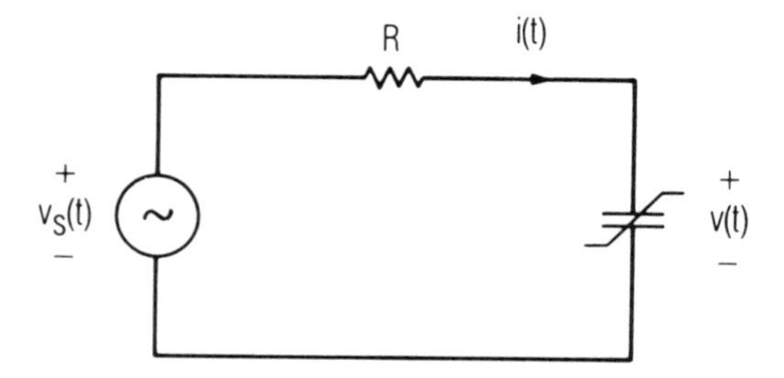

Figure 4.8 Circuit of Example 1, including a nonlinear capacitor.

We assume that the capacitor can be characterized adequately by a second-degree polynomial, so the small-signal incremental voltage, v, across the capacitor can be expressed as follows:

$$v = S_1 q + S_2 q^2 \tag{4.2.17}$$

where S_1 is the capacitor's linear small-signal incremental elastance. It is usually most convenient to represent the capacitor's charge as a function of voltage; however, if the voltage can be expressed as a single-valued function of charge over the range of voltages the capacitor will encounter, the charge-voltage function can be inverted to obtain (4.2.17) (if the voltage is not a single-valued function of charge, the circuit is not amenable to Volterra series analysis and may be unstable).

The loop voltage equation of the circuit is

$$v_s(t) = R\, i(t) + S_1 q(t) + S_2 q^2(t) \tag{4.2.18}$$

where the charge waveform $q(t)$ is the time-integral of the current:

$$q(t) = \int_{-\infty}^{t} i(\tau)\, d\tau \tag{4.2.19}$$

and τ is a variable of integration. We assume in all cases that $q(t)$, $t \to -\infty$, is zero. The nth-order current component $i_n(t)$ is given by (4.2.14), where $i_n(t) \equiv w_n(t)$:

$$\begin{aligned} i_n(t) = \frac{1}{2^n} \sum_{q_1=-Q}^{Q} \sum_{q_2=-Q}^{Q} \cdots \sum_{q_n=-Q}^{Q} & V_{s,q1} V_{s,q2} \cdots \\ & \cdot V_{s,qn} H_n(\omega_{q1}, \omega_{q2}, \ldots, \omega_{qn}) \\ & \cdot \exp[j(\omega_{q1} + \omega_{q2} + \ldots \omega_{qn})t] \end{aligned} \tag{4.2.20}$$

and the current $i(t)$ is

$$i(t) = \sum_{n=1}^{N} i_n(t) \tag{4.2.21}$$

We begin by finding the first-order transfer function. Following the form of (4.2.2), we set

$$v_s(t) = \exp(j\omega_1 t) = \frac{1}{2} \sum_{q=1}^{1} V_{s,q} \exp(j\omega_q t) \tag{4.2.22}$$

and (4.2.22) implies that

$$V_{s,1} = 2.0 \tag{4.2.23}$$

Substituting (4.2.23) into (4.2.20), and (4.2.20) into (4.2.21), we obtain

$$i(t) = \sum_{n=1}^{N} \frac{1}{2^n} 2^n H_n(\omega_1, \omega_1, \ldots, \omega_1) \exp(jn\omega_1 t) \tag{4.2.24}$$

The only first-order component in $i(t)$ is the term $H_1(\omega_1)\exp(j\omega_1 t)$, and the integration of this term in (4.2.19) to form $q(t)$ is a linear process; therefore, if $i(t)$ is limited to first order, $S_1 q(t)$ must be of first order. The term $q^2(t)$ generates components of second order and above but no first-order terms. Accordingly,

$$i_1(t) = H_1(\omega_1)\exp(j\omega_1 t) \tag{4.2.25}$$

We now substitute (4.2.22) and (4.2.24) into (4.2.18) and equate terms of first order; terms other than first-order do not affect the first-order transfer function, so they can be ignored. Because $q^2(t)$ contains no such terms, it is eliminated, and the remaining terms in (4.2.18) contain only $i_1(t)$. Equation (4.2.18) becomes

$$\exp(j\omega_1 t) = RH_1(\omega_1)\exp(j\omega_1 t) + S_1 \int_{-\infty}^{t} H_1(\omega_1)\exp(j\omega_1 \tau)\, d\tau \tag{4.2.26}$$

Performing the integration and solving for $H_1(\omega_1)$ gives the first-order transfer function:

$$H_1(\omega_1) = \frac{j\omega_1}{Rj\omega_1 + S_1} \tag{4.2.27}$$

Equation (4.2.27) is, of course, nothing more than the linear admittance of the series-connected resistor and capacitor. Thus, we find that the first-order transfer function is equivalent to the linear transfer function.

The second-order transfer function is found by setting:

$$v_s(t) = \exp(j\omega_1 t) + \exp(j\omega_2 t) \tag{4.2.28}$$

and by finding the component of $i_2(t)$ at $\omega_1 + \omega_2$. By comparing (4.2.28) to (4.2.2) we can see that

$$V_{s,q} = 2.0, \qquad q = 1, 2 \tag{4.2.29}$$

and we note that $i(t) = i_1(t) + i_2(t)$; because we need only terms of second order to form the second-order function, no current components of order greater than two need be included. Because the excitation has two frequency components, the first-order current $i_1(t)$ also has two frequency components:

$$i_1(t) = \frac{1}{2}\sum_{q=1}^{2} V_{s,q}\, H_1(\omega_q)\exp(j\omega_q t) \tag{4.2.30}$$

Substituting for $V_{s,q}$ gives

$$i_1(t) = H_1(\omega_1)\exp(j\omega_1 t) + H_1(\omega_2)\exp(j\omega_2 t) \tag{4.2.31}$$

$H_2(\omega_1, \omega_2)$ relates the excitation voltages to the second-order current $i_2(t)$:

$$i_2(t) = \frac{1}{4}\sum_{q1=1}^{2}\sum_{q2=1}^{2} V_{s,q1}V_{s,q2}\, H_2(\omega_{q1}, \omega_{q2})\exp[j(\omega_{q1} + \omega_{q2})t] \tag{4.2.32}$$

There are two identical terms in the summation at $\omega_1 + \omega_2$: $q1 = 1$, $q2 = 2$; and $q1 = 2$, $q2 = 1$. Substituting for $V_{s,q}$ via (4.2.29) and performing the summation gives

$$i_2'(t) = 2H_2(\omega_1, \omega_2)\exp[j(\omega_1 + \omega_2)t] \tag{4.2.33}$$

where the prime indicates that only the components at a single frequency of (4.2.32), the ones at $\omega_1 + \omega_2$, are included.

We now substitute $i(t)$ into (4.2.18) in order to find the second-order transfer function. As before, all terms in the equation of order other than two and at frequencies other than $\omega_1 + \omega_2$ do not contribute to $H_2(\omega_1, \omega_2)$, so they can be ignored. Equation (4.2.28) shows that $v_s(t)$ contains only first-order components, so it is ignored; only the second-order current components $i_2(t)$ contribute to second-order terms in $Ri(t)$ and $S_1q(t)$, so in these terms $i_1(t)$ is ignored. However, only $i_1(t)$ contributes to second-order terms in $S_2q^2(t)$. Thus,

$$\begin{aligned} S_1q(t) &= S_1\int_{-\infty}^{t} i_2'(\tau)\,d\tau \\ &= S_1\int_{-\infty}^{t} 2H_2(\omega_1, \omega_2)\exp[j(\omega_1 + \omega_2)\tau]\,d\tau \end{aligned} \tag{4.2.34}$$

and carrying out the integration in (4.2.34) gives

$$S_1q(t) = \frac{2S_1}{j(\omega_1 + \omega_2)} H_2(\omega_1, \omega_2)\exp[j(\omega_1 + \omega_2)t] \tag{4.2.35}$$

The squared term, $S_2q^2(t)$, is

$$S_2q^2(t) = S_2\left[\int_{-\infty}^{t} i_1(\tau)\,d\tau\right]^2 \tag{4.2.36}$$

Substituting (4.2.31) into (4.2.36) gives

$$S_2 q^2(t) = S_2 \left[\int_{-\infty}^{t} \sum_{q=1}^{2} H_1(\omega_q) \exp(j\omega_q \tau) \, d\tau \right]^2 \tag{4.2.37}$$

Changing the order of integration and summation in (4.2.37), performing the integration, and then squaring (4.2.37) gives the result:

$$S_2 q^2(t) = S_2 \sum_{q1=1}^{2} \sum_{q2=1}^{2} \frac{1}{j\omega_{q1}} \frac{1}{j\omega_{q2}} H_1(\omega_{q1}) H_1(\omega_{q2}) \cdot \exp[j(\omega_{q1} + \omega_{q2})t] \tag{4.2.38}$$

The summation in (4.2.38) has two identical terms at $\omega_1 + \omega_2$; therefore,

$$S_2 q^2(t) \Big|_{\omega_1+\omega_2} = S_2 \frac{-2}{\omega_1 \omega_2} H_1(\omega_1) H_1(\omega_2) \exp[j(\omega_1 + \omega_2)t] \tag{4.2.39}$$

Substituting (4.2.33), (4.2.35), and (4.2.39) into (4.2.18) gives the circuit equation for the second-order components:

$$0 = 2RH_2(\omega_1,\omega_2) \exp[j(\omega_1 + \omega_2)t] + \frac{2S_1}{j(\omega_1+\omega_2)} H_2(\omega_1, \omega_2) \exp[j(\omega_1+\omega_2)t] - \frac{S_2}{\omega_1\omega_2} H_1(\omega_1) H_1(\omega_2) \exp[j(\omega_1 + \omega_2)t] \tag{4.2.40}$$

Solving for $H_2(\omega_1, \omega_2)$ gives the expression for the second-order transfer function:

$$H_2(\omega_1, \omega_2) = \frac{j(\omega_1 + \omega_2)}{\omega_1\omega_2} \frac{S_2 H_1(\omega_1) \, H_1(\omega_2)}{[j(\omega_1 + \omega_2)R + S_1]} \tag{4.2.41}$$

The third- and higher-order transfer functions are found in an analogous manner. To find the third-order transfer function, we set

$$v_s(t) = \exp(j\omega_1 t) + \exp(j\omega_2 t) + \exp(j\omega_3 t) \tag{4.2.42}$$

that is:

$$V_{s,q} = 2.0, \qquad q = 1, 2, 3 \tag{4.2.43}$$

and find the component of $i_3(t)$ at $\omega_1 + \omega_2 + \omega_3$. It has the form:

$$i_3(t) = \frac{1}{8} \sum_{q1=1}^{3} \sum_{q2=1}^{3} \sum_{q3=1}^{3} V_{s,q1} V_{s,q2} V_{s,q3} H_3(\omega_{q1},\omega_{q2},\omega_{q3}) \cdot \exp[j(\omega_{q1} + \omega_{q2} + \omega_{q3})t] \tag{4.2.44}$$

Again, because they are linear terms, $Ri(t)$ and $S_1q(t)$ generate third-order mixing products from $i_3(t)$ only. From (4.1.21), there are $n = 3! = 6$ components at $\omega_1 + \omega_2 + \omega_3$ in (4.2.44), so

$$i_3'(t) = 6H_3(\omega_1,\omega_2,\omega_3) \exp[j(\omega_1 + \omega_2 + \omega_3)t] \tag{4.2.45}$$

and

$$\begin{aligned} S_1 q(t) &= S_1 \int_{-\infty}^{t} i_3(\tau)\, d\tau \\ &= \frac{6}{j(\omega_1 + \omega_2 + \omega_3)} H_3(\omega_1, \omega_2, \omega_3) \cdot \exp[j(\omega_1 + \omega_2 + \omega_3)t] \end{aligned} \tag{4.2.46}$$

The excitation $v_s(t)$ has only first-order terms, so it is again eliminated from consideration. The mixing products generated by $S_2q^2(t)$ are not obvious, however, as they were in the second-order case. Evaluating $q^2(t)$ is conceptually straightforward but algebraically messy:

$$q^2(t) = \left[\int_{-\infty}^{t} \sum_{n=1}^{N} \sum_{q1=1}^{3} \cdots \sum_{qn=1}^{3} H_n(\omega_{q1} \ldots \omega_{qn}) \cdot \exp[j(\omega_{q1} + \ldots \omega_{qn})\tau]\, d\tau \right]^2 \tag{4.2.47}$$

Interchanging the order of summation and integration and performing the integration gives

$$q^2(t) = \left[\sum_{n=1}^{N} \sum_{q1=1}^{3} \cdots \sum_{qn=1}^{3} \frac{H_n(\omega_{q1} \ldots \omega_{qn})}{j(\omega_{q1} + \ldots \omega_{qn})} \cdot \exp[j(\omega_{q1} + \ldots \omega_{qn})t] \right]^2 \tag{4.2.48}$$

Squaring (4.2.48) produces the nasty expression:

$$q^2(t) = \sum_{n=1}^{N} \sum_{m=1}^{M} \sum_{q1=1}^{3} \cdots \sum_{qn=1}^{3} \sum_{p1=1}^{3} \cdots \sum_{pm=1}^{3} \frac{1}{j(\omega_{q1} + \ldots \omega_{qn})} \frac{1}{j(\omega_{p1} + \ldots \omega_{pm})} \cdot H_n(\omega_{q1} \ldots \omega_{qn}) H_m(\omega_{p1} \ldots \omega_{pm}) \cdot \exp[j(\omega_{q1} + \ldots \omega_{qn} + \omega_{p1} + \ldots \omega_{pm})t] \tag{4.2.49}$$

A careful inspection of (4.2.49) shows that third-order terms exist only when $m = 1$, $n = 2$ and $m = 2$, $n = 1$, and because of the symmetry in (4.2.49), both these cases give identical results. Thus, we need only consider one of them, say, $n = 1$, $m = 2$, and double the result. Evaluating (4.2.49) and retaining the terms at $\omega_1 + \omega_2 + \omega_3$, we have

$$\begin{aligned} q^2(t)\Big|_{\omega_1+\omega_2+\omega_3} = {} & 2\left\{\frac{1}{j\omega_1} \frac{2}{j(\omega_2 + \omega_3)} H_1(\omega_1) H_2(\omega_2, \omega_3) \cdot \exp[j(\omega_1 + \omega_2 + \omega_3)t]\right\} \\ & + 2\left\{\frac{1}{j\omega_2} \frac{2}{j(\omega_1 + \omega_3)} H_1(\omega_2) H_2(\omega_1, \omega_3) \cdot \exp[j(\omega_1 + \omega_2 + \omega_3)t]\right\} \\ & + 2\left\{\frac{1}{j\omega_3} \frac{2}{j(\omega_1 + \omega_2)} H_1(\omega_3) H_2(\omega_1, \omega_2) \cdot \exp[j(\omega_1 + \omega_2 + \omega_3)t]\right\} \end{aligned} \tag{4.2.50}$$

Substituting (4.2.45), (4.2.46), and (4.2.49) into (4.2.18), we obtain

$$0 = 6RH_3(\omega_1, \omega_2, \omega_3) + \frac{6S_1H_3(\omega_1, \omega_2, \omega_3)}{j(\omega_1 + \omega_2 + \omega_3)}$$
$$- 4S_2\left[\frac{H_1(\omega_1)\,H_2(\omega_2, \omega_3)}{\omega_1(\omega_2 + \omega_3)} + \frac{H_1(\omega_2)H_2(\omega_1, \omega_3)}{\omega_2(\omega_1 + \omega_3)}\right.$$
$$\left. + \frac{H_1(\omega_3)H_2(\omega_1, \omega_2)}{\omega_3(\omega_1 + \omega_2)}\right] \qquad (4.2.51)$$

Solving (4.2.51) for the third-order transfer function, we have

$$H_3(\omega_1, \omega_2, \omega_3) = \frac{2}{3}\frac{j(\omega_1 + \omega_2 + \omega_3)\,S_2\{H_1H_2\}}{j(\omega_1 + \omega_2 + \omega_3)R + S_1} \qquad (4.2.52)$$

where $\{H_1H_2\}$ represents the term in brackets multiplied by $4S_2$ in (4.2.51).

It is important to note that in Example 1 we obtained a third-order transfer function, indicating the presence of third-order mixing products, even though the nonlinearity was only of second degree. This result was predicted in Section 4.1.1 and illustrated in simplified form in Chapter 1.

4.2.4 Applying Nonlinear Transfer Functions

When the nonlinear transfer functions $H_n(\omega_1, \omega_2, \ldots, \omega_n)$, $n = 1 \ldots N$ have been determined, the frequency components of interest can be found in a straightforward manner via (4.2.3). It is important to recognize that there are many identical terms in (4.2.3), as explained in Section 4.1.2, and to find the number of identical terms via (4.1.21). We must also recognize that each mixing frequency may have significant current or voltage components at more than one order. We illustrate these points and the application of the nonlinear transfer functions by the following examples.

Example 2

We wish to find the current component of the third-order mixing frequency $\omega_{3,k} = 2\omega_1 - \omega_2$ in the previous example. The excitation is

$$v_s(t) = V_{s,1}\cos(\omega_1 t) + V_{s,2}\cos(\omega_2 t) \qquad (4.2.53)$$

or

$$v_s(t) = \frac{1}{2} \sum_{q=-2}^{2} V_{s,q} \exp(j\omega_q t) \tag{4.2.54}$$

We begin by recognizing that the product of interest occurs as a mixing frequency of all odd orders of three or greater; that is,

$$\begin{aligned} 2\omega_1 - \omega_2 &= \omega_1 + \omega_1 - \omega_2 && (n = 3) \\ &= \omega_1 + \omega_1 + \omega_1 - \omega_1 - \omega_2 && (n = 5) \\ &= \omega_1 + \omega_1 + \omega_2 - \omega_2 - \omega_2 && (n = 5) \end{aligned}$$

and so on. However, we assumed in Example 1 that orders above three contribute negligibly to this component, a safe assumption when the non-linearity is of degree less than three and the excitation is very weak.

Equation (4.2.20) gives

$$i_3(t) = \frac{1}{8} \sum_{q1=-2}^{2} \sum_{q2=-2}^{2} \sum_{q3=-2}^{2} V_{s,q1} V_{s,q2} V_{s,q3} \cdot H_3(\omega_{q1}, \omega_{q2}, \omega_{q3}) \exp[j(\omega_{q1} + \omega_{q2} + \omega_{q3})t] \tag{4.2.55}$$

Where $H_3(\omega_{q1}, \omega_{q2}, \omega_{q3})$ is given by (4.2.52). Equation (4.1.21) gives the number of terms at $\omega_{3,k} = 2\omega_1 - \omega_2$, where $n = 3$, $m_1 = 2$, and $m_{-2} = 1$:

$$t_{n,k} = \frac{n!}{m_1!\, m_{-2}!} = \frac{3!}{2!\, 1!} = 3 \tag{4.2.56}$$

Evaluating (4.2.55) at the frequency $\omega_{3,k}$ gives

$$\begin{aligned} i_3'(t) = \frac{3}{8} \{ & V_{s,1}^2 V_{s,-2} H_3(\omega_1, \omega_1, -\omega_2) \\ & \cdot \exp[j(2\omega_1 - \omega_2)t] \\ & + V_{s,-1}^2 V_{s,2} H_3(-\omega_1, -\omega_1, \omega_2) \\ & \cdot \exp[-j(2\omega_1 - \omega_2)t]\} \end{aligned} \tag{4.2.57}$$

The prime indicates that (4.2.57) represents only a single frequency component, not all of $i_3(t)$. We note that

$$H_3(-\omega_1, -\omega_1, \omega_2) = H_3^*(\omega_1, \omega_1, -\omega_2) \tag{4.2.58}$$

In this case $V_{s,1}$ and $V_{s,2}$ have arbitrary phases, so without losing generality we can set the phase of both equal to zero; then,

$$V_{s,-1} = V_{s,1}^* = V_{s,1} \tag{4.2.59}$$

and

$$V_{s,-2} = V_{s,2}^* = V_{s,2} \tag{4.2.60}$$

Equation (4.2.57) now becomes

$$i_3'(t) = \frac{3}{8} V_{s,1}^2 V_{s,2} |H_3(\omega_1, \omega_1, -\omega_2)| \cos[(2\omega_1 - \omega_2)t + \phi_3] \tag{4.2.61}$$

where ϕ_3 is the phase of $H_3(\omega_1, \omega_1, -\omega_2)$.

From (4.2.51) and (4.2.52), we see that

$$H_3(\omega_1, \omega_1, -\omega_2) = \frac{2}{3} \frac{\mathrm{j}(2\omega_1 - \omega_2) S_2\{H_1 H_2\}}{\mathrm{j}(2\omega_1 - \omega_2) R + S_1} \tag{4.2.62}$$

and at this mixing frequency:

$$\{H_1 H_2\} = \frac{2H_1(\omega_1)H_2(\omega_1, -\omega_2)}{\omega_1(\omega_1 - \omega_2)} - \frac{H_1(\omega_1)H_2(\omega_1, -\omega_2)}{2\omega_1\omega_2} \tag{4.2.63}$$

Example 3

We wish to find the current in the circuit at ω_1 when the excitation is

$$v_s(t) = V_1 \cos(\omega_1 t) = \frac{1}{2}[V_{s,1} \exp(\mathrm{j}\omega_1 t) + V_{s,-1} \exp(-\mathrm{j}\omega_1 t)] \tag{4.2.64}$$

This is, of course, a trivial problem unless we include the effects of the capacitor's nonlinearity. The effect of this nonlinearity is to add a third-order component at ω_1, the term $\omega_1 = \omega_1 + \omega_1 - \omega_1$.

Equations (4.2.20) and (4.2.21) are evaluated as in Example 2, and we obtain

$$\begin{aligned} i(t) = &\frac{1}{2}[V_{s,1}H_1(\omega_1)\exp(j\omega_1 t) \\ &+ V_{s,-1}H_1(-\omega_1)\exp(-j\omega_1 t)] \\ &+ \frac{3}{8}[V_{s,1}^3 H_3(\omega_1, \omega_1, -\omega_1)\exp(j\omega_1 t) \\ &+ V_{s,-1}^3 H_3(-\omega_1, -\omega_1, \omega_1)\exp(-j\omega_1 t)] \end{aligned} \tag{4.2.65}$$

Equations (4.2.58) through (4.2.60) are used to put (4.2.65) into cosine form:

$$\begin{aligned} i(t) = &V_{s,1}|H_1(\omega_1)|\cos(\omega_1 t + \phi_1) \\ &+ \frac{3}{4}V_{s,1}^3|H_3(\omega_1, \omega_1, -\omega_1)|\cos(\omega_1 t + \phi_3) \end{aligned} \tag{4.2.66}$$

The phase angles ϕ_1 and ϕ_3 in (4.2.66) are the phase angles of $H_1(\omega_1)$ and $H_3(\omega_1, \omega_1, -\omega_1)$, respectively. Because both terms in (4.2.66) are at the same frequency, these phases cannot be ignored; they are important in establishing the behavior of $|i(t)|$ with changes in $V_{s,1}$. Equation (4.2.67) shows that, when $V_{s,1}$ is very small, the current depends upon the linear transfer function only, but as $V_{s,1}$ increases, the third-order transfer function rapidly becomes more significant. The result, in general, is that $|i(t)|$ does not increase linearly with $V_{s,1}$ at large values of $V_{s,1}$. If $\phi_3 \approx \phi_1 + \pi$, then the current saturates; that is, it increases progressively less rapidly with $V_{s,1}$ and, at some point, may even decrease. If $\phi_3 \approx \phi_1$, $|i(t)|$ increases more rapidly than $V_{s,1}$, and gain enhancement is observed (gain enhancement is encountered infrequently and, then, only over a limited range of input levels). We see that saturation effects can be attributed to the progressively greater significance of high-order mixing components, at the fundamental excitation frequency, as excitation level is increased. This phenomenon, which is observed in a wide variety of circuits, especially amplifiers, is also predicted by power-series analysis.

4.2.5 Circuit Analysis via the Method of Nonlinear Currents

Another approach to Volterra series analysis is called the *method of nonlinear currents*. In this technique, current components are calculated from voltage components of lower order, like transfer functions in the harmonic-input method. Voltage components of the same order are then determined from those currents, and the next higher-order currents are found. It is necessary only to calculate the frequency components that are of interest or that contribute to a higher-order component of interest.

The method of nonlinear currents is, in many cases, easier to use than the transfer function approach: it is a little easier to apply to circuits having multiple nodes and probably is more amenable to computer-aided design techniques. However, the nonlinear-current method is not based on transfer functions (although it can be used to numerically generate the nonlinear transfer functions of circuits), so it is not directly useful for system analysis.

Figure 4.9 shows a simple nonlinear circuit consisting of a voltage source, a linear resistor, and a nonlinear conductance. The conductance has voltage $v(t)$ across it and current $i(t)$; its I/V relation is

$$i = g_1 v + g_2 v^2 + g_3 v^3 + \ldots \tag{4.2.67}$$

As in the previous cases (e.g., Section 4.1.1), $i(t)$ and $v(t)$ in (4.2.67) represent the small-signal incremental current and voltage in the nonlinear conductance, that is, the ac deviation around a bias point. The voltage, $v(t)$, consists of all orders of mixing products:

$$v(t) = v_1(t) + v_2(t) + v_3(t) + \ldots \tag{4.2.68}$$

where $v_n(t)$ represents the sum of all nth-order mixing products.

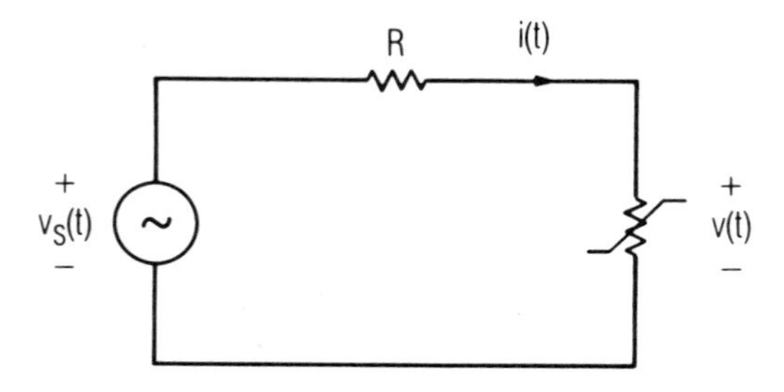

Figure 4.9 Circuit including a weakly nonlinear resistor.

By using the substitution theorem (Section 2.2.1), we can redraw the circuit of Figure 4.9 as shown in Figure 4.10, in which the nonlinear conductance has been replaced by a linear conductance and several current sources. The linear conductance represents the linear part of (4.2.67), and each current source represents a nonlinear term in (4.2.67). If we limit (4.2.67) to the third degree and limit consideration to third-order mixing products, then,

$$v(t) = v_1(t) + v_2(t) + v_3(t) \tag{4.2.69}$$

$$v^2(t) = v_1^2(t) + 2v_1(t)v_2(t) \tag{4.2.70}$$

$$v^3(t) = v_1^3(t) \tag{4.2.71}$$

The $v_1^2(t)$ term on the right side of (4.2.70) generates only second-order mixing products, and the second term, $2v_1(t)v_2(t)$, represents third-order products. The circuit of Figure 4.10 can be rearranged as shown in Figure 4.11, so that each current source represents the same order of mixing frequency; then,

$$i(t) = i_{\text{lin}}(t) + i_2(t) + i_3(t) \tag{4.2.72}$$

where

$$i_{\text{lin}}(t) = g_1 v(t) = g_1[v_1(t) + v_2(t) + v_3(t)] \tag{4.2.73}$$

$$i_2(t) = g_2 v_1^2(t) \tag{4.2.74}$$

$$i_3(t) = 2g_2 v_1(t)v_2(t) + g_3 v_1^3(t) \tag{4.2.75}$$

The current sources $i_2(t)$ and $i_3(t)$ in Figure 4.11 represent all the second- and third-order current components in the nonlinear element that arise from the terms in (4.2.67) of degree greater than one. The linear part of (4.2.67), expressed by (4.2.73), accounts for all the other first- and higher-order current components. Rearranging the circuit this way has two remarkable results: first, the circuit in Figure 4.11 is linear, although the current sources are nonlinear functions of the various-order voltage components. Second, the first-order voltage components, $v_1(t)$, are generated by the first-order source, $v_s(t)$; the second-order current, $i_2(t)$, is a function of the first-order voltages; and the third-order current, $i_3(t)$, is a function of the first- and second-order voltages. We find that the currents of each

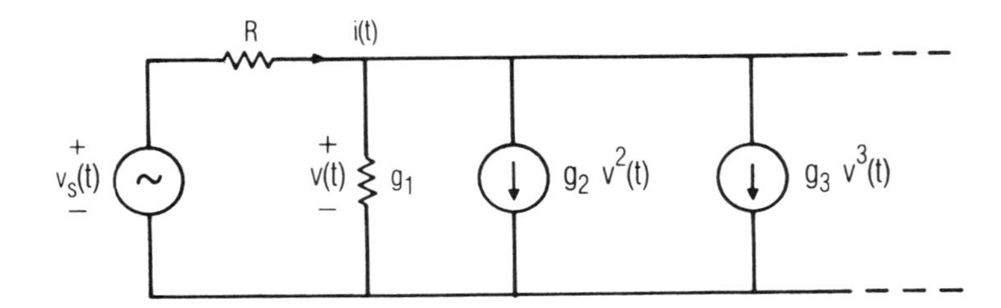

Figure 4.10 The circuit of Figure 4.9, in which the nonlinear resistor has been converted via the substitution theorem to a linear resistor and a set of current sources.

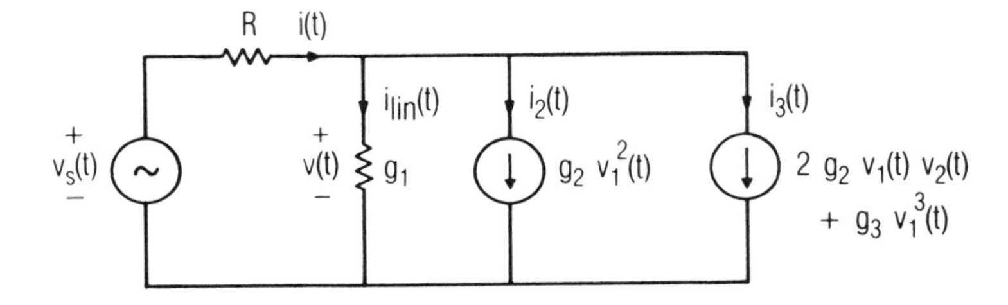

Figure 4.11 A circuit equivalent to that of Figure 4.10, except the current sources have been rearranged so that each represents a single order of mixing products.

order greater than one are always functions of lower-order voltages. These facts suggest a method of solution:

1. Find the first-order components by setting the current sources equal to zero and finding $v_1(t)$ under $v_s(t)$ excitation; this is the ordinary linear analysis.
2. Find the second-order current, $i_2(t) = g_2 v_1^2(t)$, from the voltages $v_1(t)$ found in the previous step. Then, set $v_s(t)$ equal to zero, with $i_2(t)$ the only excitation, and find the second-order voltages, $v_2(t)$, via a linear analysis of the circuit.
3. Find the third-order current $i_3(t)$ from $v_1(t)$, $v_2(t)$, $g_2(t)$, and $g_3(t)$, and with $v_s(t)$ and $i_2(t)$ set to zero, find the third-order voltage components.

Because the circuit in Figure 4.11 is linear, we can find the total voltage, $v(t)$, as the superposition of the responses to each individual excitation source.

Example 4

We use the nonlinear current method to find the response of the circuit of Figures 4.9 to 4.11 to the multitone excitation:

$$v_s(t) = V_{s,1}\cos(\omega_1 t) + V_{s,2}\cos(\omega_2 t) + \ldots V_{s,Q}\cos(\omega_Q t) \tag{4.2.76}$$

or

$$v_s(t) = \frac{1}{2}\sum_{q=-Q}^{Q} V_{s,q}\exp(j\omega_q t) \tag{4.2.77}$$

We set the current sources equal to zero, and find

$$v_1(t) = \frac{1}{Rg_1 + 1} v_s(t) \tag{4.2.78}$$

From Figure 4.11, (4.2.74), (4.2.77), and (4.2.78), the current source $i_2(t)$ is

$$\begin{aligned} i_2(t) &= g_2 v_1^2(t) \\ &= g_2 \frac{1}{(Rg_1 + 1)^2}\frac{1}{4}\sum_{q1=-Q}^{Q}\sum_{q2=-Q}^{Q} V_{s,q1}V_{s,q2} \\ &\quad \cdot \exp[j(\omega_{q1} + \omega_{q2})t] \end{aligned} \tag{4.2.79}$$

Setting all sources except $i_2(t)$ to zero, we find easily that

$$\begin{aligned} v_2(t) &= \frac{-R}{Rg_1 + 1} i_2(t) \\ &= \frac{-g_2 R}{(Rg_1 + 1)^3}\frac{1}{4}\sum_{q1=-Q}^{Q}\sum_{q2=-Q}^{Q} V_{s,q1}V_{s,q2} \\ &\quad \cdot \exp[j(\omega_{q1} + \omega_{q2})t] \end{aligned} \tag{4.2.80}$$

The third-order current, $i_3(t)$, consists of two separate terms:

$$\begin{aligned} i_3(t) &= 2g_2 v_1(t) v_2(t) + g_3 v_1^3(t) \\ &= i_{3a}(t) + i_{3b}(t) \end{aligned} \tag{4.2.81}$$

$$\begin{aligned} i_{3a}(t) = {} & \frac{2g_2}{Rg_1 + 1} \frac{1}{2} \sum_{q1=-Q}^{Q} V_{s,q1} \exp(j\omega_{q1} t) \frac{-g_2 R}{(Rg_1 + 1)^3} \\ & \cdot \frac{1}{4} \sum_{q2=-Q}^{Q} \sum_{q3=-Q}^{Q} V_{s,q2} V_{s,q3} \exp[j(\omega_{q2} + \omega_{q3})t] \end{aligned} \tag{4.2.82}$$

which can be simplified to

$$\begin{aligned} i_{3a}(t) = {} & \frac{-g_2^2 R}{4(Rg_1 + 1)^4} \sum_{q1=-Q}^{Q} \sum_{q2=-Q}^{Q} \sum_{q3=-Q}^{Q} V_{s,q1} V_{s,q2} V_{s,q3} \\ & \cdot \exp[j(\omega_{q1} + \omega_{q2} + \omega_{q3})t] \end{aligned} \tag{4.2.83}$$

Substituting (4.2.78) into the second term of (4.2.81), $i_{3b}(t)$, gives

$$\begin{aligned} i_{3b}(t) = {} & \frac{g_3}{8(Rg_1 + 1)^3} \sum_{q1=-Q}^{Q} \sum_{q2=-Q}^{Q} \sum_{q3=-Q}^{Q} V_{s,q1} V_{s,q2} V_{s,q3} \\ & \cdot \exp[j(\omega_{q1} + \omega_{q2} + \omega_{q3})t] \end{aligned} \tag{4.2.84}$$

The third-order voltage is

$$v_3(t) = -i_3(t) \frac{R}{Rg_1 + 1} \tag{4.2.85}$$

and the explicit form of $v_3(t)$ can be found by substituting (4.2.83) and (4.2.84) into (4.2.85).

The expressions (4.2.78), (4.2.80), and (4.2.85) can be evaluated in order to determine any mixing product of interest. In some cases, however, it is easier to perform an ad hoc analysis to determine these products: In this approach, only the minimum number of frequency components necessary to obtain a particular mixing product are evaluated rather than general expressions for all mixing products. This approach is possible because only a limited number of lower-order mixing products contribute to any specific higher-order product. The analysis is illustrated by the following example.

Example 5

We wish to find the $2\omega_1 - \omega_2$ third-order current in the previous example, where the excitation source has two tones. The excitation is given by (4.2.77) with $Q = 2$. From (4.2.77) and (4.2.78):

$$v_1(t) = \frac{1}{2} \sum_{q=-2}^{2} V_q \exp(j\omega_q t) \tag{4.2.86}$$

where

$$V_q = \frac{V_{s,q}}{Rg_1 + 1} \tag{4.2.87}$$

Clearly, we need only find the positive-frequency component at $2\omega_1 - \omega_2$; the negative-frequency component is just its complex conjugate. Therefore, we need only find the lower-order mixing products that contribute to this positive-frequency component.

The third-order component we wish to find is generated by both terms in (4.2.81), $i_{3a}(t)$ and $i_{3b}(t)$. The term $i_{3b}(t)$ is an obvious contributor, generating the mixing product $\omega_1 + \omega_1 - \omega_2$, but i_{3a} also contributes to $2\omega_1 - \omega_2$ via two mixing products: the second-order frequency $2\omega_1$ in $v_2(t)$ mixing with the first-order frequency $\omega_{-2} = -\omega_2$ in $v_1(t)$; and the second-order frequency $\omega_1 - \omega_2$ mixing with the first-order frequency ω_1. All other mixing products of order three or lower that contribute to $2\omega_1 - \omega_2$ are just the negative-frequency components of these or are repeated, identical terms. Thus, in order to find $2\omega_1 - \omega_2$, we need only find the first-order components at ω_1 and $-\omega_2$, the second-order components at $2\omega_1$ and $\omega_1 - \omega_2$, and the desired third-order component from $i_{3b}(t)$ at $2\omega_1 - \omega_2$.

The second-order current is $i_2 = g_2 v_1^2(t)$; we designate the two second-order current components of interest at $2\omega_1$ and $\omega_1 - \omega_2$ in $i_{2a}(t)$ and $i_{2b}(t)$, respectively. From (4.2.79) and (4.2.87):

$$i_{2a}(t) = \frac{g_2}{4} V_1^2 \exp(j2\omega_1 t) \tag{4.2.88}$$

and

$$i_{2b}(t) = \frac{g_2}{2} V_1 V_2^* \exp[j(\omega_1 - \omega_2)t] \tag{4.2.89}$$

We note in (4.2.89) that $V_{-2} = V_2^*$ (which equals V_2 in this example because, in this purely resistive circuit, all voltages are real) and that there are two terms in (4.2.79) at $\omega_1 - \omega_2$ but only one at $2\omega_1$; thus the difference in the coefficients. The voltage components at these frequencies, which we designate $v_{2a}(t)$ and $v_{2b}(t)$, are

$$v_{2a}(t) = \frac{-R}{Rg_1 + 1} i_{2a}(t) \tag{4.2.90}$$

$$v_{2b}(t) = \frac{-R}{Rg_1 + 1} i_{2b}(t) \tag{4.2.91}$$

The third-order current $i_3'(t)$ at $2\omega_1 - \omega_2$ is

$$\begin{aligned} i_3'(t) &= i_{3a}(t) + i_{3b}(t) \\ &= 2g_2 v_1(t) v_2(t) + g_3 v_1^3(t) \end{aligned} \tag{4.2.92}$$

where $v_1(t)$ is given by (4.2.86), $v_2(t) = v_{2a}(t) + v_{2b}(t)$, and only the terms at $2\omega_1 - \omega_2$ are retained in $i_3'(t)$. Then,

$$\begin{aligned} i_{3a}(t) = 2g_2 \Bigg\{ & \frac{V_1}{2} \exp(j\omega_1 t) \frac{-Rg_2}{2(Rg_1 + 1)} V_1 V_2^* \exp[j(\omega_1 - \omega_2)t] \\ & + \frac{V_2^*}{2} \exp(-j\omega_2 t) \frac{-Rg_2}{4(Rg_1 + 1)} V_1^2 \exp(j2\omega_1 t) \Bigg\} \end{aligned} \tag{4.2.93}$$

or, by simplifying (4.2.93):

$$i_{3a}(t) = \frac{3}{4} \frac{-Rg_2^2}{(Rg_1 + 1)} V_1^2 V_2^* \exp[j(2\omega_1 - \omega_2)t] \tag{4.2.94}$$

The contribution from the cubic term is

$$i_{3b}(t) = \frac{3}{8} g_3 V_1^2 V_2^* \exp[j(2\omega_1 - \omega_2)t] \tag{4.2.95}$$

This results in

$$i_3'(t) = \frac{3}{4}\left[\frac{-Rg_2^2}{(Rg_1 + 1)} + \frac{g_3}{2}\right] V_1^2 V_2^* \exp[j(2\omega_1 - \omega_2)t] \qquad (4.2.96)$$

The cosine form of (4.2.96) is found by including the negative-frequency part of $i_3'(t)$; it is simply the conjugate of (4.2.96). Finally,

$$i_3'(t) = \frac{3}{2}\left[\frac{-Rg_2^2}{(Rg_1 + 1)} + \frac{g_3}{2}\right] |V_1^2 V_2| \cos[(2\omega_1 - \omega_2)t] \qquad (4.2.97)$$

We note that (4.2.94) and (4.2.95) are equivalent to (4.2.83) and (4.2.84) when (4.2.87) is used to substitute $V_{s,1}$ and $V_{s,2}$ for V_1, V_2 and when (4.1.21) is used to calculate the number of identical terms in the triple summation at $2\omega_1 - \omega_2$. As earlier,

$$v_3'(t) = i_3'(t)\,\frac{R}{Rg_1 + 1} \qquad (4.2.98)$$

Example 6

To show how the nonlinear-current method can be applied to nonlinear capacitances, we shall find an expression for the current in the circuit of Figure 4.8 through the second order. In Example 1, the capacitor was characterized (4.2.17) as

$$v = S_1 q + S_2 q^2$$

We now need an expression of the form $q = f(v)$; this can be found, as shown in Reference 1.1, by series reversion:

$$q = C_1 v + C_2 v^2 + C_3 v^3 + \ldots \qquad (4.2.99)$$

where

$$C_1 = \frac{1}{S_1} \qquad (4.2.100)$$

$$C_2 = \frac{-S_2}{S_1^3} \qquad (4.2.101)$$

and

$$C_3 = \frac{2S_2^2}{S_1^5} \tag{4.2.102}$$

The current $i(t)$ is

$$i(t) = \frac{\mathrm{d}q(t)}{\mathrm{d}t} = C_1 \frac{\mathrm{d}v(t)}{\mathrm{d}t} + C_2 \frac{\mathrm{d}[v^2(t)]}{\mathrm{d}t} + C_3 \frac{\mathrm{d}[v^3(t)]}{\mathrm{d}t} \tag{4.2.103}$$

and, after separating the current components of different orders, as in (4.2.69) through (4.2.75), we can write the first- through third-order current components:

$$i_1(t) = C_1 \frac{\mathrm{d}v_1(t)}{\mathrm{d}t} \tag{4.2.104}$$

$$i_2(t) = C_2 \frac{\mathrm{d}[v_1^2(t)]}{\mathrm{d}t} \tag{4.2.105}$$

$$i_3(t) = C_2 \frac{\mathrm{d}[2v_1(t)v_2(t)]}{\mathrm{d}t} + C_3 \frac{\mathrm{d}[v_1^3(t)]}{\mathrm{d}t} \tag{4.2.106}$$

Again we express $v_s(t)$ by (4.2.77). Then,

$$i_1(t) = \frac{1}{2} \sum_{q=-Q}^{Q} \frac{C_1 \mathrm{j}\omega_q}{RC_1 \mathrm{j}\omega_q + 1} V_{s,q} \exp(\mathrm{j}\omega_q t) \tag{4.2.107}$$

and

$$v_1(t) = \frac{1}{2} \sum_{q=-Q}^{Q} \frac{1}{RC_1 \mathrm{j}\omega_q + 1} V_{s,q} \exp(\mathrm{j}\omega_q t) \tag{4.2.108}$$

Substituting (4.2.108) into (4.2.105) gives the second-order current, $i_2(t)$:

$$i_2(t) = \frac{C_2}{4} \sum_{q1=-Q}^{Q} \sum_{q2=-Q}^{Q} \frac{\mathrm{j}(\omega_{q1} + \omega_{q2}) V_{s,q1} V_{s,q2}}{(RC_1 \mathrm{j}\omega_{q1} + 1)(RC_1 \mathrm{j}\omega_{q2} + 1)} \cdot \exp[\mathrm{j}(\omega_{q1} + \omega_{q2})t] \tag{4.2.109}$$

The second-order voltage is found from the linear circuit, in which $i_2(t)$ is the only excitation:

$$v_2(t) = \frac{-C_2}{4} \sum_{q1=-Q}^{Q} \sum_{q2=-Q}^{Q} \cdot \frac{R\,j(\omega_{q1}+\omega_{q2})\, V_{s,q1}\, V_{s,q2} \exp[j(\omega_{q1}+\omega_{q2})t]}{[RC_1 j(\omega_{q1}+\omega_{q2})+1](RC_1 j\omega_{q1}+1)(RC_1 j\omega_{q2}+1)} \tag{4.2.110}$$

The astute reader will recognize parts of (4.2.107) and (4.2.109) as first- and second-order nonlinear transfer functions; the fractional quantity in (4.2.107) is clearly equivalent to (4.2.27) if (4.2.100) is employed to replace C_1 with S_1. Similarly, the terms in (4.2.109) are equivalent to those in (4.2.41), the second-order transfer function, after $H_1(\omega)$ from (4.2.27) is substituted. If we desired third-order current or voltage components, we could use (4.2.106) and follow the same process as was used to obtain the second-order components. The procedure is almost identical to that employed in the conductance examples; the only difference is that it is necessary to differentiate the multiple summation. Because we are limited to sinusoidal steady-state excitations, this differentiation simply involves multiplication by $j(\omega_{q1} + \omega_{q2} + \ldots \omega_{qn})$.

4.2.6 Application to Circuits Having Multiple Nodes

By now you are probably astounded by the enormous amount of effort devoted to the analysis of thoroughly trivial circuits. At this point, you might begin to suspect that Volterra-series analysis of large circuits would be prohibitively laborious. Fortunately, the complexity of the analysis increases approximately in proportion to the number of nonlinear elements, not in proportion to the overall complexity of the circuit, so with a little careful bookkeeping, you can apply the Volterra series successfully to remarkably large circuits. For example, the multitone analysis of mixerlike circuits in Section 3.2.3 is really just an application of the nonlinear-current method to a time-varying, weakly nonlinear circuit.

We now consider the circuit of Figure 4.12(a), which consists of a linear network; an excitation source, $v_s(t)$; a load admittance, $Y_L(\omega)$; and $P - 1$ nonlinear elements (the source impedance is treated as a part of the linear network). The nonlinear elements are separated from the linear network and terminals are added, so that each element is in parallel with a port. As with the single-element circuit of Example 4, we can use the substitution theorem to replace each nonlinear element by a linear element

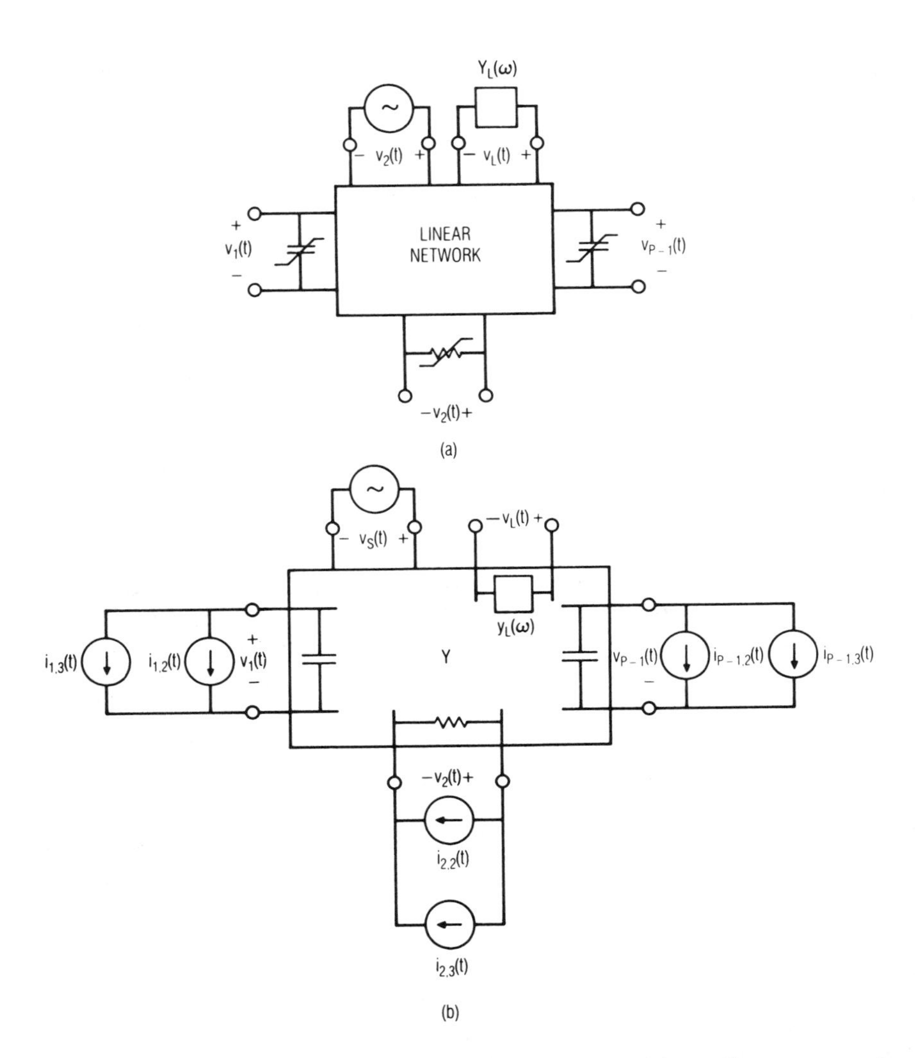

Figure 4.12 (a) A nonlinear circuit divided into a multiport linear network and a set of nonlinear elements, each in parallel with a separate port. (b) The same circuit converted into a linear circuit and a set of current sources.

and a set of current sources representing the current components of order greater than one. The linear elements, including the load admittance, $Y_L(\omega)$, can then be included in the linear part of the network. As earlier, we are left with an equivalent circuit of the nonlinear network, one that consists of a linear network and current sources for each order of the mixing products greater than one; these currents are nonlinear functions of the lower-order mixing voltages. The linear network has $P + 1$ ports, where $v_L(t)$ and $v_s(t)$ are observed at the Pth and $P+1$th ports, respectively, and is described by its Y-matrix.

The port voltages and currents can be expressed by the admittance equations, in matrix form:

$$-\begin{bmatrix} I_{1,n} \\ I_{2,n} \\ \cdot \\ \cdot \\ \cdot \\ I_{P,n} \end{bmatrix} = \begin{bmatrix} Y_{1,1} & Y_{1,2} & \cdots & Y_{1,P} \\ Y_{2,1} & Y_{2,2} & \cdots & Y_{2,P} \\ \cdot & \cdot & & \cdot \\ \cdot & \cdot & & \cdot \\ \cdot & \cdot & & \cdot \\ Y_{P,1} & \cdots & & Y_{P,P} \end{bmatrix} \begin{bmatrix} V_{1,n} \\ V_{2,n} \\ \cdot \\ \cdot \\ \cdot \\ V_{P,n} \end{bmatrix} + \begin{bmatrix} Y_{1,P+1} \\ Y_{2,P+1} \\ \cdot \\ \cdot \\ \cdot \\ Y_{P,P+1} \end{bmatrix} [V_{s,n}] \tag{4.2.111}$$

where $I_{p,n}$ is the phasor representation of the current at some specific nth-order mixing frequency in $i_{p,n}(t)$, the nth-order current source at Port p in Figure 4.12(b), and $V_{p,n}$ is the corresponding voltage (there are, of course, many such mixing frequencies at each order; we have left off the frequency subscript in (4.2.111) for simplicity). The Y-matrix is evaluated at the mixing frequency of interest. We note that $V_{s,n} = 0$ when $n > 1$ and $I_{p,n} = 0$ when $n = 1$; that is, $V_{s,n}$ represents $v_s(t)$, the first-order excitation, and the current sources all represent mixing products. Also, in general, $I_{P,n} = 0$ for all n unless the output port includes a nonlinear element.

The first-order voltages at all the ports can be found readily by setting **I**, the current vector in (4.2.111), to zero, setting $n = 1$, and forming:

$$\begin{bmatrix} V_{1,1} \\ V_{2,1} \\ \cdot \\ \cdot \\ \cdot \\ V_{P,1} \end{bmatrix} = -\begin{bmatrix} Y_{1,1} & Y_{1,2} & \cdots & Y_{1,P} \\ Y_{2,1} & Y_{2,2} & \cdots & Y_{2,P} \\ \cdot & & & \cdot \\ \cdot & & & \cdot \\ \cdot & & & \cdot \\ Y_{P,1} & \cdots & & Y_{P,P} \end{bmatrix}^{-1} \begin{bmatrix} Y_{1,P+1} \\ Y_{2,P+1} \\ \cdot \\ \cdot \\ \cdot \\ Y_{P,P+1} \end{bmatrix} [V_{s,1}] \tag{4.2.112}$$

In general, $v_s(t)$ is a multitone excitation, so **Y** must be formulated and (4.2.112) must be evaluated at each frequency. For orders $n \geqslant 2$, we set $V_{s,n} = 0$ and $\mathbf{I} \neq \mathbf{0}$; then,

$$\begin{bmatrix} V_{1,n} \\ V_{2,n} \\ \cdot \\ \cdot \\ \cdot \\ V_{P,n} \end{bmatrix} = - \begin{bmatrix} Y_{1,1} & Y_{1,2} & \ldots & Y_{1,P} \\ Y_{2,1} & Y_{2,2} & \ldots & Y_{2,P} \\ \cdot & & & \cdot \\ \cdot & & & \cdot \\ \cdot & & & \cdot \\ Y_{P,1} & \ldots & & Y_{P,P} \end{bmatrix}^{-1} \begin{bmatrix} I_{1,n} \\ I_{2,n} \\ \cdot \\ \cdot \\ \cdot \\ I_{P,n} \end{bmatrix} \tag{4.2.113}$$

In order to find the output power at any mixing frequency, we need only evaluate the currents $I_{p,n}$; we can then use (4.2.113) to obtain $V_{P,n}$, the voltage across the load admittance at each mixing frequency of interest. For simplicity we will use the ad hoc evaluation of mixing products described in Example 5 and apply it to a specific port in Figure 4.12(b). In order to minimize confusion, we will streamline the notation somewhat in the following derivation by eliminating the port subscript, *p,* but we will retain the order subscript, *n*; you should recognize that the voltage and current variables in the following derivation refer to any one port.

The excitation $v_s(t)$ is given by (4.2.77), and the first-order voltages at each port are found by applying (4.2.112) at each of the Q excitation frequencies. The first-order voltage at any one port is

$$v_1(t) = \frac{1}{2} \sum_{q=-Q}^{Q} V_{1,q} \exp(j\omega_q t) \tag{4.2.114}$$

We have included the additional subscript 1 in $V_{1,q}$, indicating first order, to distinguish it from higher-order voltages (this subscript was not necessary earlier). If the nonlinear element at that port is a conductance, its incremental I/V characteristic is

$$i = g_1 v + g_2 v^2 + g_3 v^3 + \ldots \tag{4.2.115}$$

and if it is a capacitor, its Q/V characteristic is

$$q = C_1 v + C_2 v^2 + C_3 v^3 + \ldots \tag{4.2.116}$$

The second-order current, in the case of a conductance, is

$$i_2(t) = g_2 v_1^2(t) = \frac{g_2}{4} \sum_{q1=-Q}^{Q} \sum_{q2=-Q}^{Q} V_{1,q1} V_{1,q2} \exp[j(\omega_{q1} + \omega_{q2})t] \tag{4.2.117}$$

and, in the case of a capacitor:

$$i_2(t) = C_2 \frac{d[v_1^2(t)]}{dt}$$

$$= \frac{C_2}{4} \sum_{q1=-Q}^{Q} \sum_{q2=-Q}^{Q} j(\omega_{q1} + \omega_{q2}) V_{1,q1} V_{1,q2}$$

$$\cdot \exp[j(\omega_{q1} + \omega_{q2})t] \tag{4.2.118}$$

We now limit the summation in (4.2.117) and (4.2.118) to the current components in $i_2(t)$ at frequencies of interest. The components of interest are not only those whose levels we wish to know but those that contribute to third- and higher-order mixing products of interest. Current components in $i_2(t)$ at other frequencies may be ignored. Thus, there may be several components in (4.2.117) and (4.2.118) that must be evaluated.

The components of $i_2(t)$ that are retained from (4.2.117) and (4.2.118) each can be put in the form:

$$i_{2,k}(t) = \frac{1}{2} [I_{2,k} \exp(j\omega_{2,k}t) + I_{2,k}^* \exp(-j\omega_{2,k}t)] \tag{4.2.119}$$

and the total current that these terms represent is

$$i_2'(t) = \sum_{k=1}^{K} i_{2,k}(t) \tag{4.2.120}$$

As before, the prime indicates that not all the terms in (4.2.117) or (4.2.118) are represented in (4.2.120), although in this case $i_2'(t)$ is real. There are K second-order mixing frequencies of interest in (4.2.117) and (4.2.118) where

$$\omega_{2,k} = \omega_{q1} + \omega_{q2}; \quad k = 1 \ldots K \tag{4.2.121}$$

that is, each $\omega_{2,k}$ is the sum of some two excitation frequencies ω_{q1} and ω_{q2}. We let $t_{2,k}$, given by (4.1.21), represent the number of terms in (4.2.118) at frequency $\omega_{2,k}$; then, in the case of a conductance:

$$i_{2,k}(t) = \frac{g_2 t_{2,k}}{4} \{V_{1,q1} V_{1q2} \exp[j(\omega_{q1} + \omega_{q2})t]$$

$$+ V_{1,q1}^* V_{1,q2}^* \exp[-j(\omega_{q1} + \omega_{q2})t]\} \tag{4.2.122}$$

and in a capacitor:

$$i_{2,k}(t) = \frac{C_2 t_{2,k}}{4} \{\mathrm{j}(\omega_{q1} + \omega_{q2})\, V_{1,q1} V_{1,q2} \cdot \exp[\mathrm{j}(\omega_{q1} + \omega_{q2})t] - \mathrm{j}(\omega_{q1} + \omega_{q2})\, V_{1,q1}^* V_{1,q2}^* \cdot \exp[-\mathrm{j}(\omega_{q1} + \omega_{q2})t]\} \tag{4.2.123}$$

Equating (4.2.122) and (4.2.123) with (4.2.119) at each frequency $\omega_{2,k}$ gives an expression for $I_{2,k}$:

$$I_{2,k} = \frac{g_2 t_{2,k}}{2} V_{1,q1} V_{1,q2} \tag{4.2.124}$$

for conductances, and

$$I_{2,k} = \frac{C_2 t_{2,k}}{2} \mathrm{j}\omega_{2,k} V_{1,q1} V_{1,q2} \tag{4.2.125}$$

for capacitors.

This process must be repeated at all the ports having nonlinear elements (i.e., all except the Pth port). The second-order currents at all the ports, and at the first mixing frequency ($\omega_{2,1}$), are then substituted into (4.2.113) and the port voltages at that frequency are found. The admittance matrix is then reformulated at the next mixing frequency ($\omega_{2,2}$) and (4.2.113) is invoked to determine all the port voltages at that mixing frequency. This process is repeated K times, until all the port voltages at all the K second-order mixing frequencies $\omega_{2,1}$ to $\omega_{2,K}$ are determined.

We now have the second-order voltage components of interest; at each port:

$$v_{2,k}(t) = \frac{1}{2} [V_{2,k} \exp(\mathrm{j}\omega_{2,k} t) + V_{2,k}^* \exp(-\mathrm{j}\omega_{2,k} t)] \tag{4.2.126}$$

and the second-order voltage at these frequencies is

$$v_2'(t) = \sum_{k=1}^{K} v_{2,k}(t) \tag{4.2.127}$$

We next find the third-order current components in terms of the first- and second-order voltages $V_{1,q}$ and $V_{2,k}$, respectively. These third-order mixing frequencies are designated $\omega_{3,m}$, $m = 1 \ldots M$. We continue to assume that the degree of the nonlinearity in (4.2.115) and (4.2.116) is limited to three; thus, in the case of a conductance, the current at some specific port and frequency $\omega_{3,m}$ is

$$i_{3,m}(t) = 2g_2 v_1'(t) v_2'(t) + g_3 v_1'^3(t) \tag{4.2.128}$$

where $v_1'(t)$ and $v_2'(t)$ include all the first- and second-order frequency components that contribute to $\omega_{3,m}$. Then,

$$\begin{aligned} i_{3,m}(t) = & \sum_{\omega_q + \omega_{2,k} = \omega_{3,m}} 2g_2 \frac{1}{2} [V_{1,q} \exp(j\omega_q t) + V_{1,q}^* \exp(-j\omega_q t)] \\ & \cdot \frac{1}{2} [V_{2,k} \exp(j\omega_{2,k} t) + V_{2,k}^* \exp(-j\omega_{2,k} t)] \\ & + \frac{g_3}{8} \sum_{\substack{q1=-Q \\ \omega_{q1} + \omega_{q2} + \omega_{q3} = \omega_{3,m}}}^{Q} \sum_{q2=-Q}^{Q} \sum_{q3=-Q}^{Q} V_{1,q1} V_{1,q2} V_{1,q3} \\ & \cdot \exp[j\omega_{q1} + \omega_{q2} + \omega_{q3})t] \end{aligned} \tag{4.2.129}$$

The first term in (4.2.129) is summed over all k and q that give the desired mixing frequency, $\omega_{3,m}$, where

$$\omega_{3,m} = \omega_{2,k} + \omega_q = \omega_{q1} + \omega_{q2} + \omega_{q3} \tag{4.2.130}$$

and the triple summation is evaluated only at these same frequencies. Again we designate $t_{3,m}$, given by (4.1.21), as the number of terms in the triple summation at frequency $\omega_{3,m}$, and

$$\begin{aligned} i_{3,m}(t) = & \Bigg[\sum_{\omega_q + \omega_{2,k} = \omega_{3,m}} \frac{g_2}{2} V_{1,q} V_{2,k} \\ & + \frac{g_3 t_{3,m}}{8} V_{1,q1} V_{1,q2} V_{1,q3} \Bigg] \exp(j\omega_{3,m} t) \\ & \quad (\omega_{q1} + \omega_{q2} + \omega_{q3} = \omega_{3,m}) \\ & + \Bigg[\sum_{\omega_q + \omega_{2,k} = \omega_{3,m}} \frac{g_2}{2} V_{1,q}^* V_{2,k}^* \\ & + \frac{g_3 t_{3,m}}{8} V_{1,q1}^* V_{1,q2}^* V_{1,q3}^* \Bigg] \exp(-j\omega_{3,m} t) \\ & \quad (\omega_{q1} + \omega_{q2} + \omega_{q3} = \omega_{3,m}) \end{aligned} \tag{4.2.131}$$

In the case of a nonlinear capacitor:

$$i_{3,m}(t) = 2C_2 \frac{d[v_1'(t)v_2'(t)]}{dt} + C_3 \frac{d[v_1'^3(t)]}{dt} \tag{4.2.132}$$

Via the same approach, we obtain

$$\begin{aligned} i_{3,m}(t) = j\omega_{3,m} \Bigg[& \sum_{\omega_q + \omega_{2,k} = \omega_{3,m}} \frac{C_2}{2} V_{1,q} V_{2,k} \\ & + \frac{C_3 t_{3,m}}{8} \underset{(\omega_{q1}+\omega_{q2}+\omega_{q3}=\omega_{3,m})}{V_{1,q1} V_{1,q2} V_{1,q3}} \Bigg] \exp(j\omega_{3,m}t) \\ & - j\omega_{3,m} \Bigg[\sum_{\omega_q + \omega_{2,k} = \omega_{3,m}} \frac{C_2}{2} V_{1,q}^* V_{2,k}^* \\ & + \frac{C_3 t_{3,m}}{8} \underset{(\omega_{q1}+\omega_{q2}+\omega_{q3}=\omega_{3,m})}{V_{1,q1}^* V_{2,q1}^* V_{3,q1}^*} \Bigg] \exp(-j\omega_{3,m}t) \end{aligned} \tag{4.2.133}$$

The third-order current at $\omega_{3,m}$ can be expressed in the form:

$$i_{3,m}(t) = \frac{1}{2} [I_{3,m} \exp(j\omega_{3,m}t) + I_{3,m}^* \exp(-j\omega_{3,m}t)] \tag{4.2.134}$$

By comparing (4.2.131) and (4.2.133) to (4.2.134), we obtain

$$I_{3,m} = \sum_{\omega_q + \omega_{2,k} = \omega_{3,m}} g_2 V_{1,q} V_{2,k} + \frac{g_3 t_{3,m}}{4} \underset{(\omega_{q1}+\omega_{q2}+\omega_{q3}=\omega_{3,m})}{V_{1,q1} V_{1,q2} V_{1,q3}} \tag{4.2.135}$$

for the conductance, and

$$I_{3,m} = j\omega_{3,m} \Bigg[\sum_{\omega_q + \omega_{2,k} = \omega_{3,m}} C_2 V_q V_{2,k} + \frac{C_3 t_{3,m}}{4} \underset{(\omega_{q1}+\omega_{q2}+\omega_{q3}=\omega_{3,m})}{V_{q1} V_{q2} V_{q3}} \Bigg] \tag{4.2.136}$$

for the capacitor.

Equations (4.2.135) and (4.2.136) represent a single mixing frequency $\omega_{3,m}$ at a single port. Either (4.2.135) or (4.2.136) must be evaluated for each nonlinear element, and then (4.2.113) must be used to find the voltage components $V_{3,m}$, $m = 1 \ldots M$. The Y-matrix is then reformulated at the next third-order mixing frequency of interest, and the currents and voltages are again determined.

4.2.7 Controlled Sources

In all the previous sections, we ignored the possibility that the circuit might include controlled sources. The reason for this neglect is that the Volterra-series modeling of controlled sources is not significantly different from the cases examined earlier. When a controlled source is included in Figure 4.12(b), the circuit's topology does not change but the current is simply a function of a voltage at another port instead of the voltage at the current source's terminals. Thus, equations such as (4.2.135) and (4.2.136) remain valid as long as the voltages are those of the appropriate port, the one that defines the source's control voltage.

REFERENCES

[4.1] D.D. Weiner and J.F. Spina, *Sinusoidal Analysis and Modeling of Weakly Nonlinear Circuits*, Van Nostrand, New York, 1980.

[4.2] J.W. Graham and L. Ehrman, *Nonlinear System Modeling and Analysis with Applications to Communications Receivers*, Rome Air Development Center Technical Report No. RADC-TR-73-178, 1973.

[4.3] J.J. Bussgang, L. Ehrman, and J.W. Graham, "Analysis of Nonlinear Systems with Multiple Inputs," *Proc. IEEE*, Vol. 62, 1974, p. 1088.

[4.4] M. Schetzen, *The Volterra and Wiener Theories of Nonlinear Systems*, John Wiley and Sons, New York, 1980.

[4.5] N. Wiener, *Nonlinear Problems in Random Theory*, Technology Press, New York, 1958.

[4.6] N. Wiener, "Response of a Nonlinear Device to Noise," MIT Radiation Lab. Rpt. V-16S, April 6, 1942.

[4.7] V. Volterra, *Theory of Functionals and of Integral and Integro-Differential Equations*, Dover, New York, 1959.

Chapter 5

Balanced and Multiple-Device Circuits

Single solid-state devices have limitations that may be troublesome in certain applications. One of these is output power; a single device is not always adequate to supply sufficient power or dynamic range. In other cases, a circuit generates potentially troublesome harmonics or intermodulation products or has spurious responses, and these cannot be eliminated by filtering. Some of these problems can be solved by creating a single unit, called a *balanced* circuit. Balanced circuits connecting two or more solid-state devices or two-port components have many attractive properties beyond their improved power and dynamic range; in some cases, balanced circuits can be used to improve bandwidth and input or output VSWR.

Solid-state devices can be combined in many ways. The simplest technique is to connect one or more transistors or diodes in parallel or series. Direct interconnection is often impractical, however, because it changes impedance levels and requires nearly identical devices. Sometimes it is preferable to employ power-combining components such as hybrids and power dividers; these components isolate the individual devices from each other and preserve input and output impedance levels. In some cases, devices can be combined via hybrids at the input or output and directly connected at the opposite port. This chapter examines the properties of hybrid interconnections and the trade-offs involved in the design of balanced and multiple-device circuits.

5.1 BALANCED CIRCUITS USING MICROWAVE HYBRIDS

5.1.1 Properties of Ideal Hybrids

A microwave hybrid coupler is a lossless, four-port passive component. Each port is matched, and the power applied to any input port is split equally between a pair of output ports. The remaining port is isolated;

that is, none of the input power is transferred to it. It is possible to show, by using the properties of the S matrix, that only two types of ideal hybrids are possible: the 180-degree hybrid, in which one path between ports has a phase reversal, and the 90-degree or *quadrature* hybrid, which has two 90-degree phase shifts.

Figure 5.1 shows, schematically, both types of hybrids. The lines between ports show the possible power transfers and phase shifts; for example, power applied to Port 1 of the 180-degree hybrid emerges 3 dB lower and with identical phase at Ports 3 and 4, and Port 2 is isolated. If Port 4 is excited, the outputs are Ports 1 and 2, and the voltages at those ports differ in phase by 180 degrees. Similarly, power applied to Port 1 of the quadrature hybrid emerges at Ports 3 and 4, with 90-degree phase difference, and Port 2 is isolated.

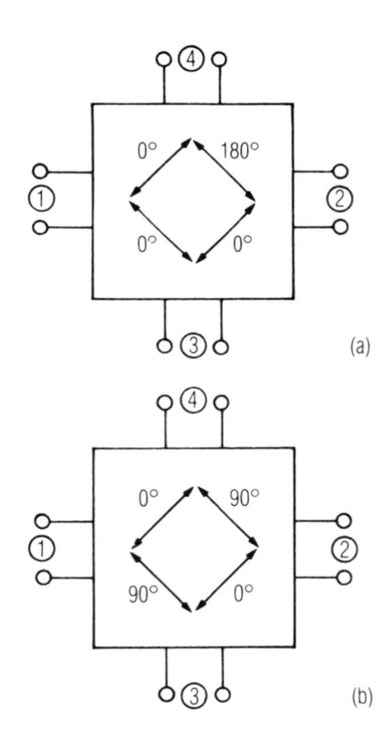

Figure 5.1 Ideal (a) 180-degree and (b) 90-degree, or *quadrature,* hybrids.

The S matrices for ideal 180-degree and quadrature hybrids, $\mathbf{S}_{180}$ and $\mathbf{S}_{90}$, are as follows:

$$\mathbf{S}_{180} = \frac{1}{\sqrt{2}} \begin{bmatrix} 0 & 0 & 1 & 1 \\ 0 & 0 & 1 & -1 \\ 1 & 1 & 0 & 0 \\ 1 & -1 & 0 & 0 \end{bmatrix} \tag{5.1.1}$$

and

$$\mathbf{S}_{90} = \frac{1}{\sqrt{2}} \begin{bmatrix} 0 & 0 & -j & 1 \\ 0 & 0 & 1 & -j \\ -j & 1 & 0 & 0 \\ 1 & -j & 0 & 0 \end{bmatrix} \tag{5.1.2}$$

Equations (5.1.1) and (5.1.2) imply an absolute phase shift of 90 or 180 degrees between ports. However, in real hybrids the difference in phase shift between output ports, not the shift between input and output ports, is of most importance. For example, in Figure 5.1(a) the difference in phase between Ports 3 and 4, when driven at Port 2, must be 180 degrees; the phase difference between Ports 2 and 4, and between 2 and 3, is rarely of concern. The S matrices verify that, in both hybrids, transmission between Ports 1 and 2, or between Ports 3 and 4, is impossible; these ports are called *mutually isolated* pairs.

Hybrids can be used as power combiners as well as power dividers if inputs are applied to mutually isolated ports. If, for example, waveforms are applied to Ports 1 and 2 of the 180-degree hybrid in Figure 5.1(a), the output at Port 3 is proportional to the sum of the inputs, and the output at Port 4 is proportional to their difference. When the 180-degree hybrid is used this way, Port 3 is called the *sum,* or *sigma,* port and Port 4 is called the *difference,* or *delta,* port. If the other ports, 3 and 4, are used as inputs, then Port 1 is the sigma port and Port 2 is the delta port.

Practical hybrid couplers do not exhibit these ideal characteristics and have only a limited bandwidth over which they approximate the ideal response. The nonidealities of greatest concern are isolation, phase balance, amplitude balance, loss, and port VSWR. *Phase balance* is the deviation from the ideal phase difference at a pair of output ports; *amplitude balance,* usually expressed in dB, is the ratio of the amplitude levels at the output ports. *Isolation,* also expressed in dB, is the loss between a pair of mutually isolated ports; and the *loss* is the ratio of available input power to the sum of the powers at the two output ports. The loss accounts for the dissipation and reflection loss in the hybrid, including power delivered to the termination of the isolated port; it does not include the unavoidable

3-dB minimum power split loss. The port VSWRs are invariably imperfect, not only because of manufacturing limitations, but also because VSWR, isolation, and loss are not independent quantities; for example, in all hybrids, if even one port VSWR is imperfect, isolation cannot, in theory, be perfect. The VSWRs of the individual ports (as functions of frequency) generally are not the same unless the hybrid is symmetrical. A hybrid's balance, isolation, and VSWR, as a function of frequency, usually establish its bandwidth.

5.1.2 Practical Hybrids

Transformer Hybrid

The transformer hybrid is a practical structure for use at frequencies between a few MHz and approximately 500 MHz, although careful design occasionally allows for operation near 1 GHz. This hybrid uses the symmetry properties of a transformer to achieve 180-degree hybrid operation. Its simplest form is shown in Figure 5.2, in which the ports are numbered in a manner that corresponds to Figure 5.1(a). In this configuration, the impedance at Ports 3 and 4 is not the same as that of Ports 2 and 3; however, it is possible to devise more complex transformer hybrid circuits that have equal port impedances.

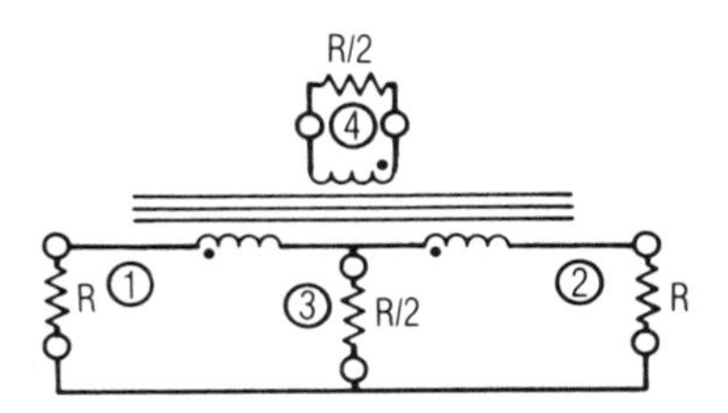

Figure 5.2 The transformer hybrid: all three windings have the same number of turns, and the port numbering follows Figure 5.1(a).

Figure 5.3 illustrates the operation of the transformer hybrid. In Figure 5.3(a), power is applied to Port 4 and is split between the load resistors at Ports 1 and 2. Because of the symmetry of the structure, no voltage appears across the resistor at Port 3, so it is isolated from Port 4. In Figure 5.3(b), Port 3 is excited. The currents in the secondary windings (those connected to Ports 1 and 2) must be equal and opposite because of

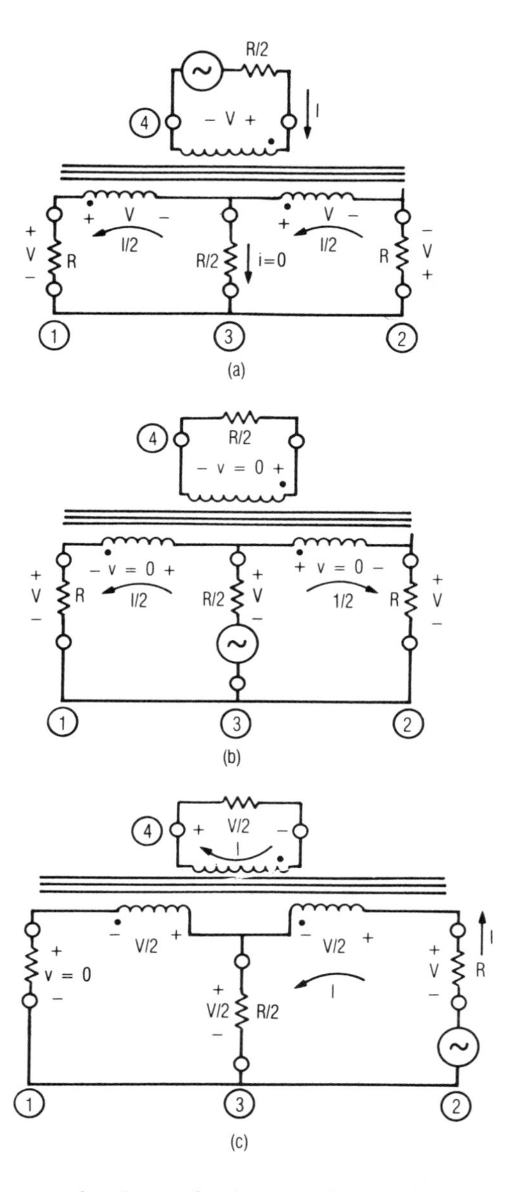

Figure 5.3 Currents and voltages in the transformer hybrid when different ports are excited: (a) Port 4 excited; (b) Port 3 excited; and (c) Port 2 excited.

the symmetry of the structure, so no voltage is generated across any of the windings, and the loads at Ports 1 and 2 are effectively in parallel with Port 3. Figure 5.3(c) shows the operation of the hybrid with Port 2 excited. Because the windings all have an equal number of turns, the current generated in the primary (Port 4) winding is equal to that generated in the winding connected to Port 2, causing a power division between those ports. These currents also induce equal but opposite currents in the remaining winding, isolating Port 1. Note that the voltage polarities in Figure 5.3(a) and 5.3(c) imply that the 180-degree phase division is between Ports 2 and 4.

Transformer hybrids are often realized as shown in Figure 5.4, via a set of trifilar windings on a toroidal core. The core is usually made of a ferrite material. This structure is favored because it confines the magnetic field within the windings and thus provides very good coupling over a wide bandwidth.

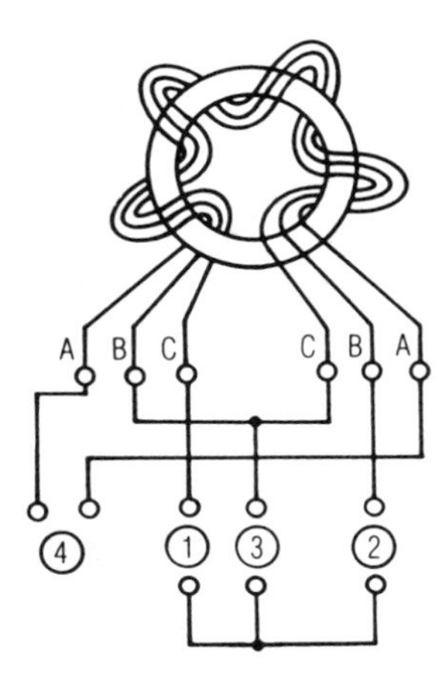

Figure 5.4 The transformer hybrid realized via a trifilar winding on a toroidal core.

An important property of the transformer hybrid is that its phase and amplitude balance are determined by the structure of the circuit and not by frequency-sensitive elements such as half-wavelength transmission lines. Accordingly, the hybrid's balance is usually very good over a broad frequency range, and its bandwidth is generally limited by loss and degradation of isolation. This degradation occurs at high frequencies because of stray inductance and capacitive coupling between the transformer windings. Operation is limited at low frequencies by a standard requirement for transformers that the self inductances of the windings have reactances much greater than the load and source impedances.

Ring (Rat-race) Hybrid

Figure 5.5 shows the ring, or *rat-race*, hybrid. Unlike the transformer hybrid, the ring hybrid requires frequency-sensitive elements, namely transmission lines of a precise length, that make it a narrow-band component. Figure 5.5 shows a ring hybrid in a form that can be realized in microstrip or stripline; ring hybrids have also been realized in a wide variety of other transmission media, including waveguide.

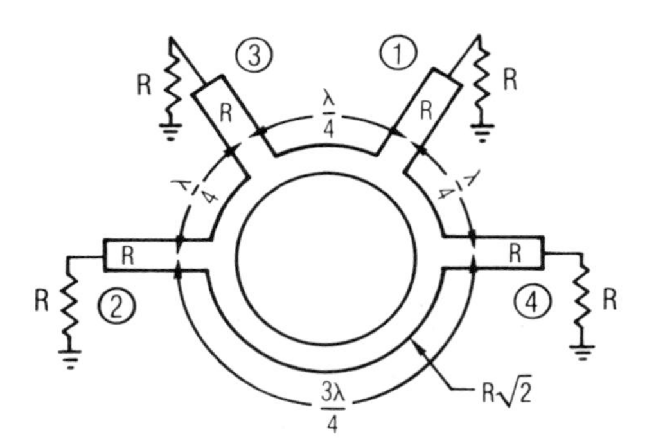

Figure 5.5 The ring hybrid in microstrip or stripline form.

Power applied to any port of the ring hybrid is divided equally between the two adjacent ports. The remaining port is isolated because there are always two paths between the input port and the isolated port: Going around the ring in one direction leads from the input to the isolated port over a 0.5-wavelength path; in the other direction, the path is 1.0 wavelength, or 0.5-wavelength longer. The longer path introduces a phase reversal that cancels the voltage at the isolated port and creates a virtual ground at its point of connection to the ring. Because of the extra half wavelength of transmission line, the path from Port 4 to Port 2 has the 180-degree phase shift.

Because of its relatively low loss and the simplicity of its design and fabrication, the ring hybrid is a very popular design. All the ports have the same impedance, and the ring's characteristic impedance is $\sqrt{2}$ times the port impedance. If transmission line dispersion and junction effects are negligible, the VSWR of each port is less than 2.0 over a nearly 100 percent bandwidth; however, the transmission bandwidth is much narrower than the VSWR bandwidth, 10–20 percent at most, depending upon the criticality of the application.

Wilkinson Hybrid

The Wilkinson hybrid of Figure 5.6 is another simple but popular design. It is usually used to combine or split power, but it actually is a type of 180-degree hybrid that has a built-in termination, the 2R resistor, on Port 2. The Wilkinson hybrid uses quarter-wavelength transmission lines, but its phase and amplitude balance depend primarily upon circuit symmetry and thus are broadband. Port 2 is almost never used for input or output and is, therefore, terminated with a resistor whose value is twice the port impedance. Ports 1 and 2 are mutually isolated, as are Ports 3 and 4.

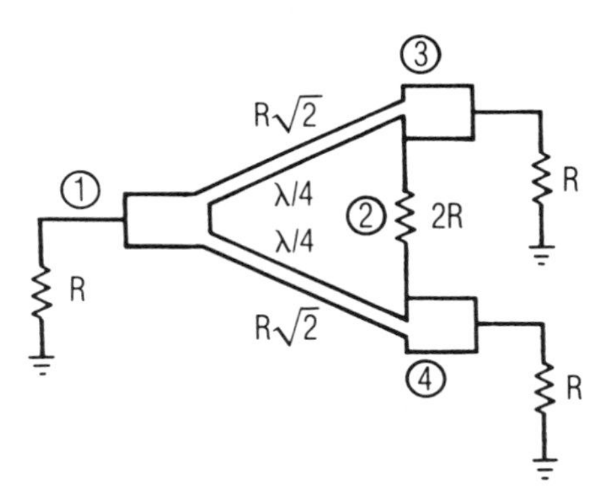

Figure 5.6 The Wilkinson hybrid or power divider.

The Wilkinson hybrid is often used as a power combiner for individual amplifier stages. Its advantages over other hybrids in this application are that its port termination, the 2R resistor, need not be connected to ground, and because of its excellent balance, large trees of Wilkinson combiners and dividers can be made with good overall balance and low loss. Furthermore, it is possible to make broadband, well balanced, Wilkinson-like dividers having multiple outputs.

Branch-Line Quadrature Hybrid

A branch-line quadrature hybrid is shown in Figure 5.7. It consists of two quarter-wave transmission lines connected by quarter-wave

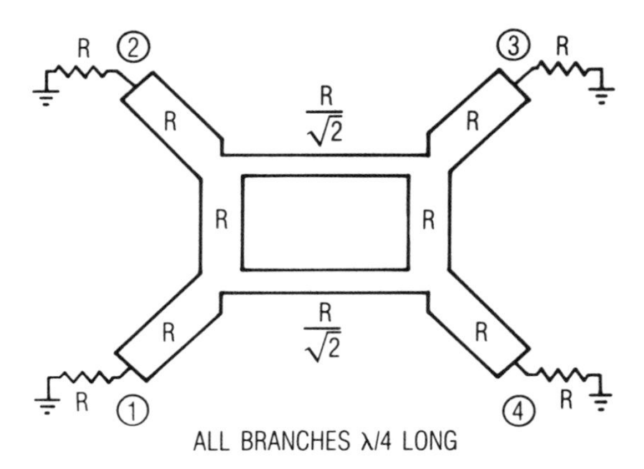

Figure 5.7 The branch-line quadrature hybrid in microstrip or stripline form; all branches are one-quarter wavelength long.

branches; the series lines have characteristic impedances $R/\sqrt{2}$, and the branch impedances are simply R. This hybrid is simple to design and frabricate, and it has very low loss. Because it does not require bond wires or narrow microstrip lines, it can be fabricated successfully on soft substrates to realize low-cost circuits.

The branch-line hybrid has a relatively narrow bandwidth, approximately 10 percent, and the VSWRs and transmission bandwidths of different pairs of ports are not the same. Branch-line hybrids are practical only at relatively low frequencies; at frequencies at which the branch lengths are on the order of the line widths, performance is often very poor. The hybrid's bandwidth can be improved by using multisection designs, however, and the careful selection of materials and geometry can be helpful in achieving good high-frequency operation.

Coupled-Line Quadrature (Lange) *Hybrid*

One of the most popular types of quadrature directional couplers consists of a pair of coupled microstrip lines, 0.25-wavelength long, as shown in Figure 5.8. This coupler is designed by selecting the even- and odd-mode characteristic impedances of the coupled lines so that

$$Z_{oe} = R\left(\frac{1 + c}{1 - c}\right)^{1/2} \tag{5.1.3}$$

and

$$Z_{oo} = R\left(\frac{1 - c}{1 + c}\right)^{1/2} \tag{5.1.4}$$

where R is the port impedance and c is the voltage coupling ratio, the square root of the power coupling ratio. If such a coupler is designed to achieve a 3-dB power division, it is a quadrature hybrid. In order to achieve 3-dB coupling ($c = 0.707$), (5.1.3) and (5.1.4) imply that $Z_{oe} = 2.41R$ and $Z_{oo} = 0.414R$.

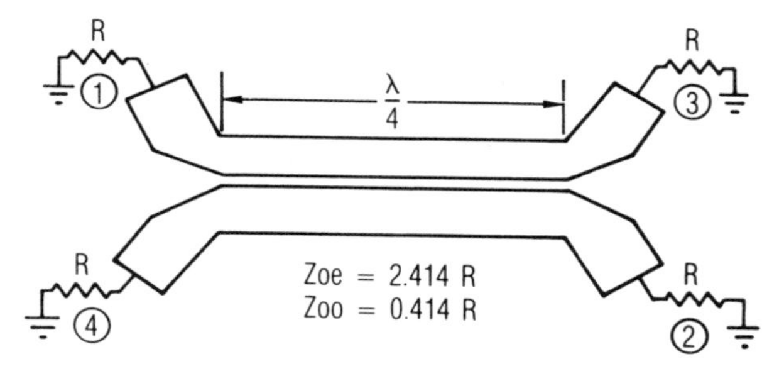

Figure 5.8 The simplest form of the coupled-line hybrid. This structure is used to realize directional couplers having low coupling coefficients; it cannot provide sufficient coupling to be used as a 3-dB hybrid.

Unfortunately, it is virtually impossible in practice to obtain 3-dB coupling from a single pair of coupled lines, even on substrates having high dielectric constants because the required spacing between the microstrips is too small. Furthermore, the coupler has the practical disadvantage that the output ports are always on opposite sides of the structure, so a symmetrical circuit is impossible.

Both of these problems can be solved in a remarkably simple and elegant manner. A solution to the coupling problem is to split the two coupled lines into four, as shown in Figure 5.9. The four strips now have three pairs of adjacent edges instead of only one in the two-strip case, so the capacitance between the strips is approximately tripled. This modification allows the even- and odd-mode characteristic impedances required to realize a 3-dB coupler to be achieved successfully. The output ports can be interchanged by dividing one of the outer strips and moving half of it to the other side of the coupler; this modification moves the port connected to that strip to the desired location. It is necessary to interconnect the separated strips via wires. This hybrid, named after its inventor, J. Lange

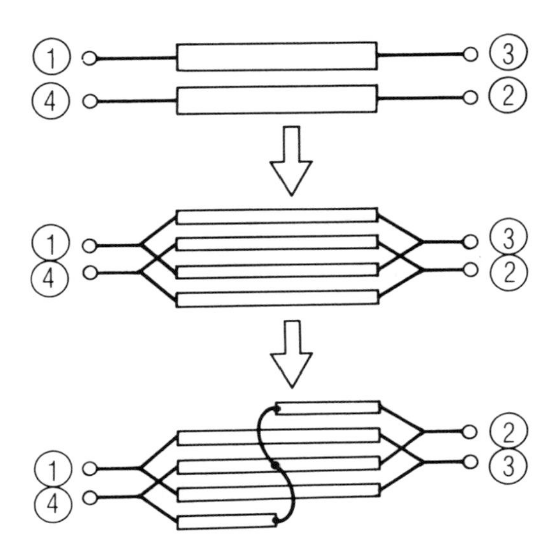

Figure 5.9 Evolution of the Lange hybrid: The simple coupler of Figure 5.8 is split into four strips to increase its coupling, and the strips are rearranged to place both outputs on the same side of the structure.

(Reference 5.1), is one of the most popular and broadband quadrature couplers in use.

The ideal coupled-line hybrid has a 0.5-dB coupling bandwidth of approximately 50 percent. Furthermore, the phase difference between output ports, the isolation, and the port VSWRs of a coupled-line hybrid are theoretically perfect and independent of frequency. Perfect isolation implies that any two output ports are complementary, that the sum of the powers at the output ports equals the input power. However, the balance is not frequency-independent. If Port 1 is excited with voltage V_1, the voltages V_3 and V_4 at the terminated outputs, Ports 3 and 4, are, respectively,

$$\frac{V_4}{V_1} = \frac{jc \sin(\theta)}{(1 - c^2)^{1/2} \cos(\theta) + j \sin(\theta)} \tag{5.1.5}$$

and

$$\frac{V_3}{V_1} = \frac{(1 - c^2)^{1/2}}{(1 - c^2)^{1/2} \cos(\theta) + j \sin(\theta)} \tag{5.1.6}$$

The electrical length θ of the coupler is

$$\theta = \frac{\pi}{2}\frac{\omega}{\omega_0} \tag{5.1.7}$$

where ω_0 is the hybrid's center frequency.

If $c = 0.707$ (a 3-dB hybrid), dividing (5.1.5) by (5.1.6) gives

$$\frac{V_4}{V_3} = j\sin(\theta) \tag{5.1.8}$$

showing that V_4 leads V_3 by 90 degrees at all frequencies. The balance is

$$\frac{|V_4|^2}{|V_3|^2} = \sin^2(\theta) \tag{5.1.9}$$

Practical Lange couplers are, of course, not ideal. The parasitic inductances of the wires needed for the crossover connections and the unequal phase velocities of even and odd modes on microstrip coupled lines are the dominant effects that limit the coupler's performance. These effects are especially severe at high frequencies. Even so, it is relatively easy to minimize the effects of these factors and to realize Lange hybrids having remarkably good performance over broad bandwidths. In applications that are not sensitive to amplitude balance, Lange hybrids can be used over very wide bandwidths, often greater than one octave.

5.1.3 Properties of Hybrid-Coupled Components

Figure 5.10 shows a pair of two-ports connected in parallel via microwave hybrids. The two-ports can be of any type, and the hybrids can be 180-degree or quadrature designs. The requirements for a successful interconnection are (1) that the ports of each hybrid that are connected to the two-ports be mutually isolated pairs; (2) that the phase shifts from the input to the output through each branch (i.e., through the input hybrid, one of the two-ports, and the output hybrid) be equal; and (3) that the two-ports be identical. If these conditions are met, the coupled pair of two-ports has the same gain as either two-port but twice the output-power capability. The coupled pair may have other useful properties, depending upon the type of hybrid used for the interconnection.

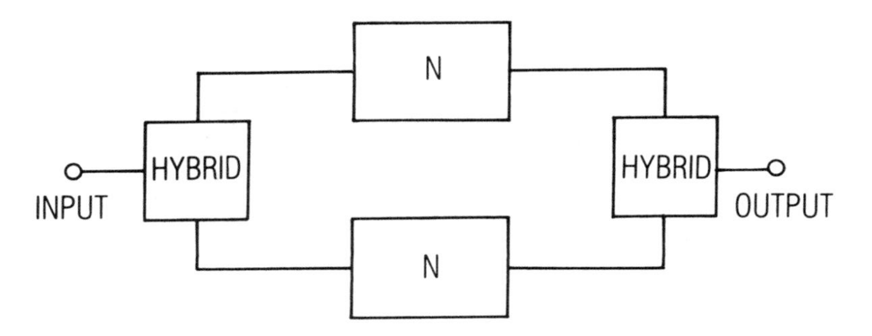

Figure 5.10 General configuration of hybrid-coupled two-ports.

180-Degree Hybrid-Coupled Components

Figure 5.11 shows a pair of identical two-port components connected in parallel via 180-degree hybrids. The components can be connected to in-phase or out-of-phase pairs of the hybrid's ports; however, if an out-of-phase pair is selected, obtaining good balance over even a moderate bandwidth usually necessitates connecting the input of one two-port and the output of the other two-port to identically numbered ports (see Figure 5.1).

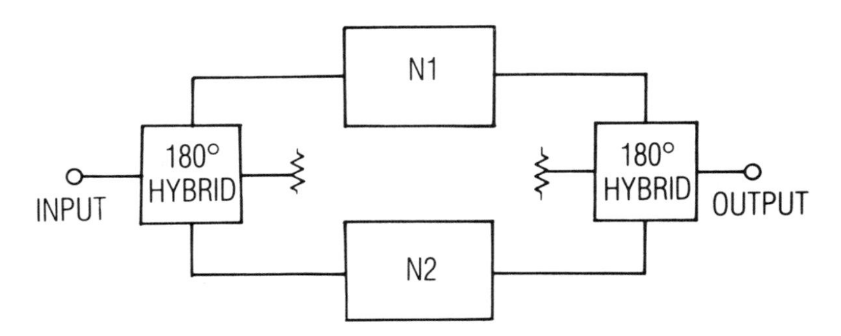

Figure 5.11 Hybrid-coupled two-ports using 180-degree hybrids or power dividers.

This configuration is frequently used to combine power of amplifier stages. Because of their structural simplicity and excellent balance, Wilkinson hybrids are a natural choice for combining power; they are configured with Port 1 as the input or output and the amplifiers connected to Ports 3 and 4 (see Figure 5.6). In order to achieve high output power, Wilkinson-like power dividers having multiple outputs are often used to combine a large number of amplifiers.

Figure 5.12 illustrates the operation of the input and output hybrids. In Figure 5.12(a), the input port has voltage V_s and current I_s; the available power at the input is divided in half by the hybrid, so the voltage V_i and current I_i at the two outputs of the hybrid, the inputs of N_1 and N_2, is

$$|V_i| = \frac{|V_s|}{\sqrt{2}} \tag{5.1.10}$$

and

$$|I_i| = \frac{|I_s|}{\sqrt{2}} \tag{5.1.11}$$

We have assumed that the hybrid is ideal, the two-ports are identical, and both two-ports have the same input impedance Z_i. We can show from (5.1.1) that the input impedance of the hybrid under these conditions must also be Z_i; that is, the input reflection coefficient of the hybrid-coupled components is that of the individual components. The voltages and currents at the outputs of N_1 and N_2 (i.e., those at the inputs of the output hybrid), V_o and I_o, are identical. The 3-dB power division in the output hybrid reduces V_o by $\sqrt{2}$, but two input voltages are combined in phase, so the output voltage and current V_L and I_L are

$$|V_L| = \frac{2|V_o|}{\sqrt{2}} \tag{5.1.12}$$

and

$$|I_L| = \frac{2|I_o|}{\sqrt{2}} \tag{5.1.13}$$

Unsurprisingly, the output power is 3 dB greater than the powers at either input. Thus, the hybrid introduces a 3-dB loss in available power at its inputs, but reclaims that loss at its output by combining the voltages in

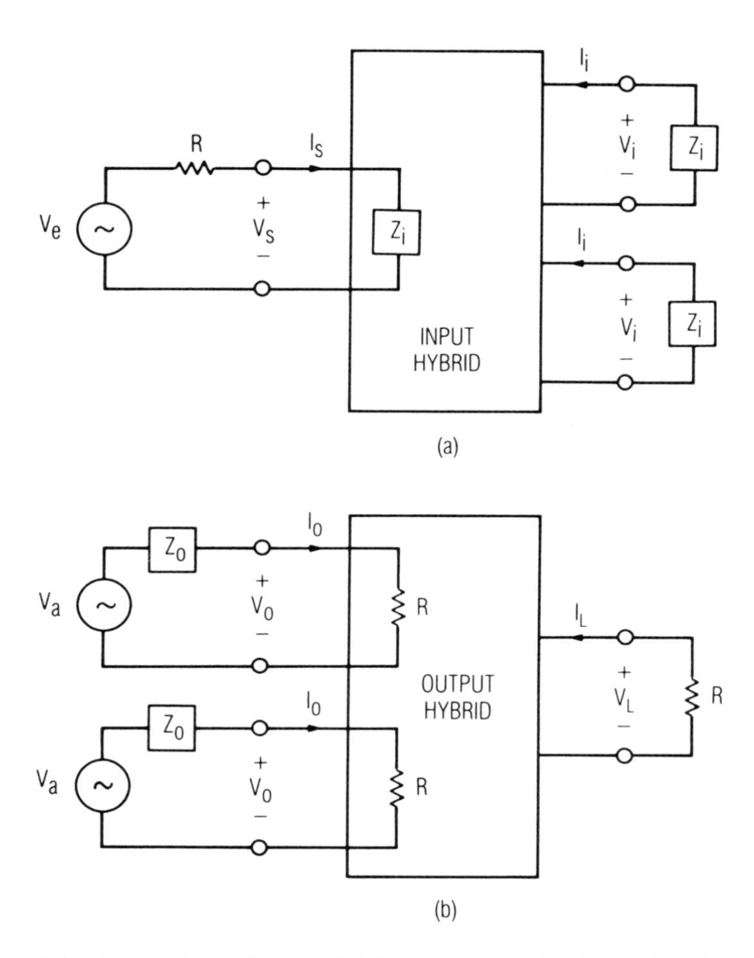

Figure 5.12 Equivalent circuit of (a) the input hybrid and (b) the output hybrid. The isolated ports are not shown.

phase. The transducer gain of the hybrid-coupled pair is, therefore, the same as that of the individual stages, but the available output power is 3 dB greater.

A consequence of coupling the components via a 180-degree hybrid is a modest improvement in the worst-case VSWR. It can be shown that the reflection coefficient at the input of the ideal hybrid is

$$\Gamma_{in} = 0.5(\Gamma_3 + \Gamma_4) \tag{5.1.14}$$

where Γ_{in} is the input impedance looking into Port 1, and Γ_3 and Γ_4 are the reflection coefficients of the terminations on the output ports, Ports 3 and 4, respectively, in Figure 5.1. Thus, if the two-ports are not precisely identical, and at some frequency the input reflection coefficient of one two-port is much poorer than that of the other, the averaging effect of the 180-degree hybrid reduces the worst-case input reflection coefficient. The same property is evident at the output.

Quadrature-Coupled Components

Figure 5.13 shows a pair of two-port components coupled via quadrature hybrids. The crossover paths between ports in the ideal hybrids have identical phase delays of 90 degrees, and the straight-through paths have no phase delay; the components are connected to the hybrids' ports in such a way that the phase shift through each branch of the balanced structure is the same.

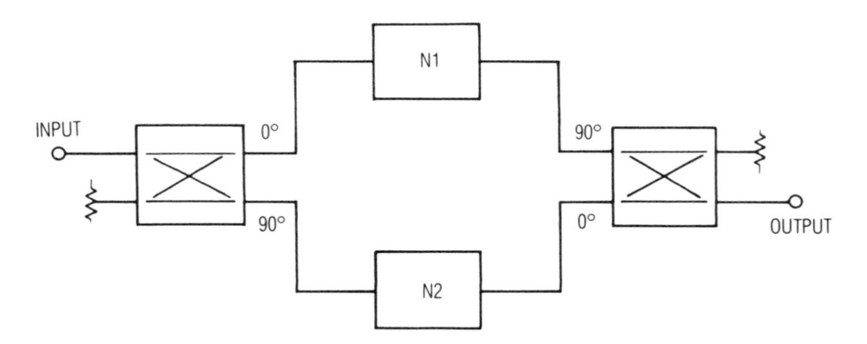

Figure 5.13 Hybrid-coupled two-ports using quadrature hybrids. The paths in the hybrid that have the 90-degree phase shift are the crossover paths.

The magnitudes of the port voltages in the quadrature hybrids are the same as those of the 180-degree hybrid; that is, (5.1.10) through (5.1.13) apply to Figure 5.13 as well as to Figure 5.12. The only consequence of the different phase shifts is that the quadrature hybrid's ports must be configured as shown in Figure 5.13, so that the output voltages of the individual components *N*1 and *N*2 combine in phase. Thus, the gain of the quadrature-coupled components is equal to that of the individual two-ports, and the output power capability is 3 dB greater.

The most attractive property of the quadrature-coupled configuration is that, in the ideal case, the input reflection coefficient is zero, regardless of the input reflection coefficients of the individual two-ports. We can derive this property by using (5.1.2) and by assuming that Port 1 is the input and that Ports 3 and 4 have terminations with reflection coefficient Γ. The terminations constrain the *a* and *b* waves at Ports 3 and 4 as follows:

$$a_3 = \Gamma b_3 \tag{5.1.15}$$

and

$$a_4 = \Gamma b_4 \tag{5.1.16}$$

Substituting (5.1.15) and (5.1.16) into (5.1.2) gives

$$\begin{bmatrix} b_1 \\ b_2 \end{bmatrix} = \Gamma \begin{bmatrix} 0 & -\mathrm{j} \\ -\mathrm{j} & 0 \end{bmatrix} \begin{bmatrix} a_1 \\ a_2 \end{bmatrix} \tag{5.1.17}$$

Equation (5.1.17) implies that all the input power reflected from the individual components is dissipated in the load at Port 2, and none emerges from Port 1; that is, the input port is matched. The same considerations apply to the output port; it, too, is matched. An intuitive explanation of this phenomenon is that energy reflected from Port 4 returns to Port 1 without phase shift, but energy reflected from Port 3 passes through the hybrid's 90-degree path twice, returning to Port 1 with 180-degree phase shift. Thus, reflected waves cancel at Port 1. However, the reflected waves undergo identical phase shifts between the input and Port 4 and, therefore, combine in phase. Similarly, one can show that, when the terminations on Ports 3 and 4 are unequal and the termination on Port 2 is ideal:

$$\Gamma_{\text{in}} = 0.5(\Gamma_3 - \Gamma_4) \tag{5.1.18}$$

where $\Gamma_{\text{in}} = b_1/a_1$, and Γ_3, Γ_4 are the reflection coefficients of the terminations at Ports 3 and 4, respectively. Thus, even when the port terminations are not precisely equal, the input VSWR may still be very low.

This property of quadrature-coupled components is indeed delightful; equally delightful is the fact that, if the quadrature hybrid is an ideal coupled-line hybrid, good gain is achieved over a very broad bandwidth. The reason for the broadband operation is that the coupled-line hybrid's phase balance, port VSWR, and isolation are theoretically perfect and frequency-independent. Its amplitude balance is imperfect, varying as

$\sin^2(\omega/\omega_0)$, but this imperfection is not as important as it may seem at first. Because the hybrid's isolation and input VSWR are perfect, all the available input power must appear at the output ports; thus, if the loss from the input to one port of the hybrid is L ($L < 1$), the loss from the input to the other port must be $1 - L$. Figure 5.13 shows that the coupled stages are connected to the hybrids in such a way that a signal must experience loss L through one hybrid and loss $1 - L$ in the other. Therefore, the gain through the input hybrid, either component, and the output hybrid is $L(1 - L)G_t$, where G_t is the transducer gain of the identical two-port components. The gain of the coupled pair of components is $4L(1 - L)G_t$; even if the imbalance is fairly large, this gain is very close to the ideal gain G_t. For example, at 0.5 and 1.5 times the center frequency, the coupling of an ideal hybrid drops to 0.33 (i.e., $L = 0.33$) and its amplitude balance, $L/(1 - L)$, is a seemingly horrific 3 dB; however, the gain reduction of the coupled pair of components over this 3 : 1 frequency range is only 0.8 dB. Even greater bandwidth can be achieved by designing the hybrid to have a bandcenter power division other than 3 dB: if the coupling between Port 4 and Port 1 is made greater than 3 dB at center frequency, the band-edge balance of the hybrid is improved, as is the worst-case imbalance over the entire band. Similarly, one can show that the effect of imperfect amplitude balance is to raise the magnitude of the input reflection coefficient to $|(2L - 1)\Gamma|$, where Γ is the component's input reflection coefficient.

Effect of Imperfect Balance

In the previous sections, we generally assumed that the hybrids were ideal and the two-ports were identical. Such perfection never occurs in practice, of course, so it is important to be able to estimate the effects of imperfection. Estimating the effects of phase, amplitude, and gain imbalance is not difficult as long as the hybrids are not too far from ideal.

In any balanced circuit, two voltage components combine in phase after traveling through different paths between the circuit's input and output, each path consisting of the input hybrid, one of the two parallel two-port components, and the output hybrid. The effect of dissimilar gains in the two-ports and amplitude imbalance in the hybrids is that the amplitudes of those voltage components are not identical. Similarly, phase imbalance arising in either the components or the hybrids causes the two output voltage components to have different phases.

The two voltage components at the output of the output hybrid can be described as phasors, as shown in Figure 5.14. The phase difference

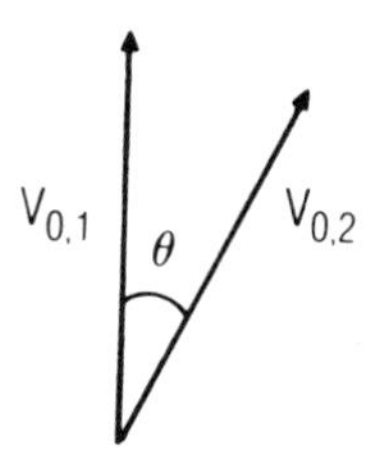

Figure 5.14 Voltage phasors at the output of the hybrid-coupled circuit. $V_{o,1}$ is the voltage of the signal that passed through the upper path in Figure 5.11 or 5.13, that is, through the input hybrid, $N1$, and the output hybrid; $V_{o,2}$ is the signal that followed the lower path.

between the two components is θ, and the amplitudes of the voltage components, $V_{o,1}$ and $V_{o,2}$, generally are also different. If the voltage components had equal amplitude and $\theta = 0$, then the output power would be

$$P_{o,e} = \frac{1}{2}\frac{|V_{o,1} + V_{o,2}|^2}{R} \tag{5.1.19}$$

or

$$P_{o,e} = \frac{2|V_{o,1}|^2}{R} \tag{5.1.20}$$

where R is the hybrid's output termination resistance and $V_{o,1} = V_{o,2}$ is the amplitude of either component. When they are unequal and $\theta \neq 0$:

$$P_{o,u} = \frac{1}{2}\frac{|V_{o,1} + V_{o,2}\exp(j\theta)|^2}{R} \tag{5.1.21}$$

or, letting $\delta = V_{o,2}/V_{o,1}$ with $V_{o,2} < V_{o,1}$:

$$P_{o,u} = \frac{1}{2}\frac{|V_{o,1}|^2[1 + 2\delta\cos(\theta) + \delta^2]}{R} \tag{5.1.22}$$

If we assume that $V_{o,1}$ is a reference voltage, that it has the same value for the cases of perfect and imperfect balance, then,

$$\frac{P_{o,u}}{P_{o,e}} = \frac{1}{4}\,[1 + 2\delta\cos(\theta) + \delta^2] \tag{5.1.23}$$

Equation (5.1.23) indicates that a 20-degree phase imbalance and 1-dB gain imbalance between the two branches reduces the overall gain of the circuit by 0.6 dB. This degree of balance is not particularly difficult to maintain in most cases, even over broad bandwidths, so we may conclude that the penalty, in terms of gain, for phase and amplitude imbalance is not particularly severe.

Harmonics and Spurious Signals

Hybrid-coupled circuits may provide limited rejection of spurious signals and harmonics generated in the parallel two-ports. Such rejection is by no means guaranteed, because it depends upon the type of hybrid used in the balanced ciruit and that hybrid's properties at the harmonic or spurious frequency. The spurious signals of greatest concern are usually close to the frequency of the desired signal; harmonics invariably are not close to the desired signal but may still be of concern in broadband systems. If the phase and amplitude balance of the hybrid are uniform over a wide frequency range (an acceptable assumption for certain hybrid types, such as the transformer hybrid), then it is possible for the balanced structure to have significant spurious rejection.

The only balanced structure having significant spurious- and harmonic-rejection properties is shown in Figure 5.15. In this figure, the two-ports are combined via 180-degree hybrids and are connected to mutually isolated out-of-phase ports. If a two-tone signal is applied to the circuit, the input voltage $v_{i,1}(t)$ at $N1$ is

$$v_{i,1}(t) = V_1\cos(\omega_1 t) + V_2\cos(\omega_2 t) \tag{5.1.24}$$

and $v_{i,2}(t)$, the input voltage at $N2$, is

$$v_{i,2}(t) = V_1\cos(\omega_1 t + \pi) + V_2\cos(\omega_2 t + \pi) \tag{5.1.25}$$

For simplicity, we can model the networks via the power-series approach of Section 4.1 and Figure 4.1, wherein each network consists of a linear two-port having the transfer function $H(\omega)$, followed by a nonlinear frequency-independent element having the transfer function:

$$f(V) = a_1 V + a_2 V^2 + a_3 V^3 + \ldots \tag{5.1.26}$$

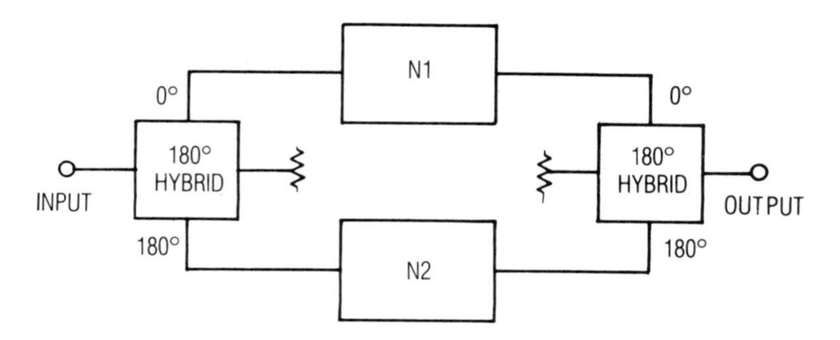

Figure 5.15 180-degree hybrid-coupled components: this configuration has limited second-order harmonic and spurious rejection as long as the output hybrid's amplitude and phase balance are maintained at the spurious output frequency.

The second-order voltage components at the output of $N1$, $v_{o,1}(t)$, are

$$\begin{aligned} v_{o,1}(t) = \; & a_2|H(\omega_1)H(\omega_2)|V_1V_2 \cos[(\omega_1 + \omega_2)t] \\ & + a_2|H(\omega_1)H(\omega_2)|V_1V_2 \cos[(\omega_1 - \omega_2)t] \\ & + 0.5a_2|H(\omega_1)|^2V_1^2 \cos(2\omega_1 t) \\ & + 0.5a_2|H(\omega_2)|^2V_2^2 \cos(2\omega_2 t) \end{aligned} \tag{5.1.27}$$

The second-order outputs at N_2, $v_{o,2}(t)$, are

$$\begin{aligned} v_{o,2}(t) = \; & a_2|H(\omega_1)H(\omega_2)|V_1V_2 \cos[(\omega_1 + \omega_2)t + 2\pi] \\ & + a_2|H(\omega_1)H(\omega_2)|V_1V_2 \cos[(\omega_1 - \omega_2)t] \\ & + 0.5a_2|H(\omega_1)|^2V_1^2 \cos(2\omega_1 t + 2\pi) \\ & + 0.5a_2|H(\omega_2)|^2V_2^2 \cos(2\omega_2 t + 2\pi) \end{aligned} \tag{5.1.28}$$

which is clearly the same as $v_{o,1}(t)$. The signal $v_{o,2}(t)$ undergoes an additional 180-degree phase shift in the output hybrid, but $v_{o,1}(t)$ does not; thus, all the second-order voltages cancel in the output. Similar analysis shows that all even-order mixing products and harmonics are rejected, as long as the hybrid's balance and phase properties are the same at the mixing or harmonic frequency as they were at the excitation frequency. Conversely, all odd-order harmonics and mixing products differ in phase by 180 degrees at the outputs of $N1$ and $N2$ and combine in phase after the 180-degree phase shift in the output hybrid. Thus, odd-order products are not rejected.

The spurious-rejection properties of the quadrature-coupled circuit are significantly different from those of the 180-degree hybrid-coupled circuit. Applying the same analysis to the quadrature-coupled circuit shows that second-order mixing products are 180 degrees out of phase at the outputs of $N1$ and $N2$. These voltage components are applied to the quadrature output hybrid, so the second-order mixing products would be expected to have a 90-degree phase difference when they are combined, providing 3-dB rejection. However, most quadrature hybrids do not have the same amplitude or phase characteristics at the second harmonic as at the intended operating frequency, so it is usually not possible to make general statements about their second-order rejection properties. Third harmonics at the output of $N2$ are delayed by 270 degrees, and the third-harmonic properties of most quadrature hybrids are the same as the fundamental-frequency properties. Thus, the voltage components have a 180-degree phase difference when they are combined in the output hybrid, so third harmonics are rejected. Some, but not all, third-order mixing products are rejected: The third-order intermodulation products at $2\omega_1 - \omega_2$ and $2\omega_2 - \omega_1$ are not rejected, but those at $2\omega_1 + \omega_2$ and $2\omega_2 + \omega_1$ are rejected.

Intermodulation Intercept Point

Section 4.1 introduced the intermodulation (IM) intercept point, the point at which the extrapolated two-tone IM levels and linear output levels are identical. Because of the balanced circuit's power-combining effect, interconnecting two components in a balanced structure gives the combination a greater intercept point than that of the individual components.

The level of nth-order IM products $P_{\mathrm{IM}n}$ can be found from the linear output level and the intercept point as follows:

$$P_{\mathrm{IM}n} = nP_{\mathrm{lin}} - (n - 1)\mathrm{IP}_n \tag{5.1.29}$$

where P_{lin} is the level of the linear output power and IP_n is the nth-order intercept point. All power levels are in dBm. Equation (5.1.29) can be rearranged to express IP_n:

$$\mathrm{IP}_n = \frac{nP_{\mathrm{lin}} - P_{\mathrm{IM}n}}{n - 1} \tag{5.1.30}$$

If the two-ports are operated at identical output levels P'_{lin}, they have identical IM output levels $P'_{\text{IM}n}$. At the output of the balanced circuit, both the IM and linear output levels are 3 dB higher than those of the individual two-ports. The IP_n of the balanced circuit must differ from that of the individual components, so for the balanced circuit (5.1.30) becomes

$$\text{IP}'_n = \frac{n(P'_{\text{lin}} + 3) - (P'_{\text{IM}n} + 3)}{n - 1} \tag{5.1.31}$$

Equation (5.1.31) can be rearranged to give

$$\text{IP}'_n = \text{IP}_n + 3 \tag{5.1.32}$$

that is, the intercept point of the balanced structure is 3 dB greater than that of the individual two-ports, regardless of order. Furthermore, we can show via the same approach that combining m two-ports via any power-combining technique increases the intercept point by $10 \cdot \log_{10}(m)$ dB.

5.2 DIRECT INTERCONNECTION OF MICROWAVE COMPONENTS

It is not always necessary or desirable to use some type of hybrid to interconnect solid-state devices or components. Although the previous section was concerned primarily with ideal hybrids, real, nonideal hybrids must be used in practical circuits, and real hybrids are not perfect. Hybrids introduce additional loss, which may not be tolerable in power circuits, and their imperfect balance and VSWR may also degrade circuit performance. They also increase the circuit's size and weight and make it more expensive to design and fabricate. The direct interconnection of components circumvents some of these problems, but at the expense of losing the inherent isolation between stages that hybrids provide.

The primary purpose of connecting solid-state devices directly is to increase output power without adding complexity; for example, high-power microwave bipolar and field-effect transistors are realized by directly connecting many smaller devices, called *cells*, in parallel. A second purpose is to eliminate undesired harmonics and intermodulation products; even- or odd-order products sometimes can be eliminated by appropriately connecting devices together. These rejection properties are often exploited in the design of frequency converters such as frequency multipliers and mixers.

5.2.1 Harmonic Properties of Two-Terminal Device Interconnections

Nonlinear two-terminal circuit elements such as diodes are often connected in parallel or series in order to eliminate certain harmonics or mixing products. Because spurious signals often are in-band and cannot be removed by filtering, the ability to eliminate such products without resorting to filters is often valuable. In circuits designed to have very wide bandwidths, harmonics may be within the output passband, but even in narrow-band circuits, certain spurious mixing products may be in-band. Two of the most important interconnections of nonlinear devices having spurious-rejection properties are called the *antiparallel* and the *antiseries* or, more commonly but less elegantly, the *push-push* interconnection. A third type of interconnection is called the *series* interconnection; it is a variation on the antiparallel interconnection but has different properties.

Antiparallel Interconnection

Figure 5.16(a) shows a two-terminal nonlinear conductance that has the I/V characteristic:

$$I = f(V) = aV + bV^2 + cV^3 + dV^4 + eV^5 + \dots \quad (5.2.1)$$

The element's I/V characteristic is generally not symmetrical, so it is marked with a + sign at one terminal in order to indicate its polarity. In Figure 5.16(b), the applied voltage is reversed. In this case,

$$I = f(-V) = -aV + bV^2 - cV^3 + dV^4 - eV^5 + \dots \quad (5.2.2)$$

that is, the odd-degree components of the power series are negative. Finally, if the element is reversed, but the voltage and current conventions remain as in Figure 5.16(a):

$$I = -f(-V) = aV - bV^2 + cV^3 - dV^4 + eV^5 + \dots \quad (5.2.3)$$

This case, illustrated in Figure 5.16(c), is the converse of the previous one: The even-degree currents components are negative. We conclude from (5.2.3) that reversing the terminals of a nonlinear circuit element changes the sign of the even-degree terms in its power series.

Figure 5.17 shows the *antiparallel* interconnection of two identical conductive nonlinear elements described by (5.2.1) through (5.2.3). The current in element A, I_A, is found from (5.2.1):

$$I_A = f(V) = aV + bV^2 + cV^3 + dV^4 + \ldots \tag{5.2.4}$$

I_B, the current in element B, is found from (5.2.3):

$$I_B = -f(-V) = aV - bV^2 + cV^3 - dV^4 + \ldots \tag{5.2.5}$$

and finally the total external current I is

$$I = I_A + I_B = f(V) - f(-V) = 2aV + 2cV^3 + \ldots \tag{5.2.6}$$

I

+

V

−

$I = f(V) = aV + bV^2 + cV^3 + dV^4 + \cdots$

(a)

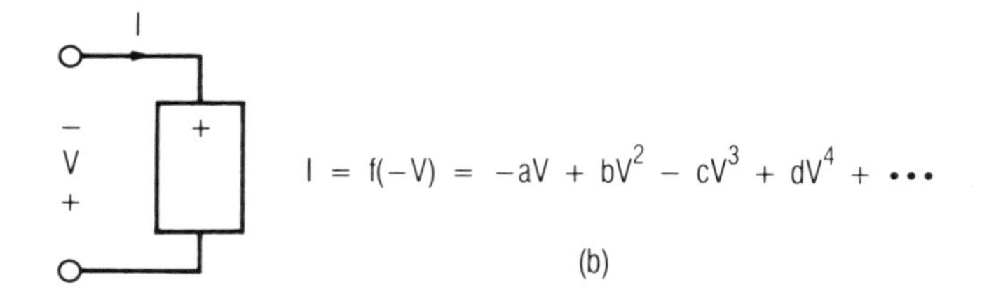

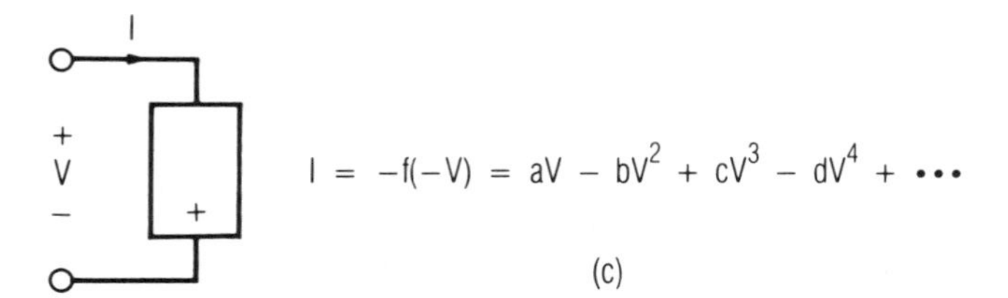

Figure 5.16 Voltage-current relations in a conductive nonlinearity with three different voltage and current polarities.

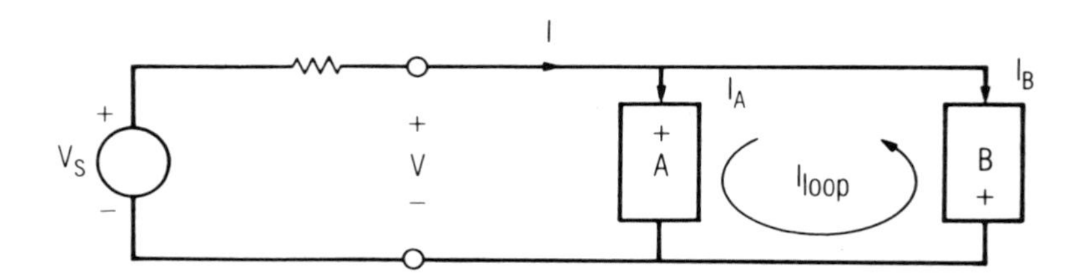

Figure 5.17 Antiparallel connection of two identical nonlinear elements.

From (5.2.6), we can see that the external current does not include any even-degree components. Therefore, the antiparallel pair of nonlinear elements operates as if it were a single element with only odd-degree nonlinearities. It was shown in Chapter 1 that even-degree nonlinearities generate even-order mixing products, and odd-degree nonlinearities generate odd-order mixing products. Thus, the antiparallel-connected nonlinear elements generate no even-order mixing products from the frequencies in their terminal voltages.

This result may at first seem impossible because, from (5.2.4) and (5.2.5), the even-degree current components still exist in I_A and I_B. In order to examine this mystery further, we consider the loop current, I_{loop}, in Figure 5.17. The loop current must consist only of the components for which

$$I_{\text{loop}} = I_A = -I_B = bV^2 + dV^4 + \ldots \tag{5.2.7}$$

The odd-degree components in I_{loop} must be zero because it is impossible to have $-aV = aV$, $-cV^3 = cV^3, \ldots$, regardless of V. Equation (5.2.7) shows that I_{loop} contains the missing even-degree current components and does not include any of the odd-degree terms. We now can see clearly what has happened: The even- and odd-order mixing components have been separated; the even-order current components circulate in the loop, while the odd-order current components circulate in the external circuit.

The lack of even-order components circulating in the external circuit can be used to advantage. For example, antiparallel diodes can be used to realize a frequency tripler having inherently low second- and fourth-harmonic output (it will, of course, have no inherent rejection of fifth-harmonic output, but fifth- and higher-order components are usually weak because of the high multiplication factor). The antiparallel pair can also be employed as a mixer that has no mixing response between the RF input frequency and the fundamental component of the local oscillator (LO), because this mixing product is of second order. A mixer using antiparallel diodes achieves efficient mixing between the RF and the second LO harmonic because this is a third-order mixing product. Such *subharmonically pumped* mixers are in wide use; they are particularly valuable at millimeter wavelengths where fundamental-frequency LO power may be difficult or expensive to obtain.

The separation of the even- and odd-order frequency components of the current has another implication, one that is subtle but very important. Because no even-order currents circulate in the external circuit, no even-order voltages are generated between the elements' terminals. Thus, the even-order components of the terminal voltage, as well as the external even-order currents, are zero. Furthermore, each nonlinear element generates even-order currents that are equal to those of the other and are opposite in direction. The existence in each component of a circulating current and zero terminal voltage implies a short circuit at the terminals; each element in effect short circuits the other at all even-order mixing frequencies.

This result (the fact that each element short-circuits the other at even-order harmonics and mixing frequencies) allows considerable simplification of the analysis of such components. It is not necessary to include both nonlinear components in the analysis, or the embedding impedances at even-order harmonics and mixing frequencies. Instead, we need only include one element in the circuit, express the I/V characteristic of the single element as $I = 2f(V)$, and set all the even-order embedding impedances to zero. In this way, we obtain a single-device equivalent circuit that describes the two-device circuit completely. The results will be the same as those of an analysis that includes both nonlinear elements.

Scaling the nonlinear element of $I = 2f(V)$ in such a circuit is not always wise or even possible because, in some cases, the nonlinear element may not be a simple two-terminal conductive nonlinearity; it may have a relatively complex equivalent circuit, one described by a model that is difficult to modify. Furthermore, modifying the model may require modifying the computer program that is used to analyze the circuit; such changes may not be possible if the source program is not available and, in any case, may not be advisable because of the possibility of introducing errors. Instead, it may be preferable to generate a single-device equivalent circuit by modifying the external circuit. If a single-device equivalent circuit can be defined, it is possible to analyze the interconnected circuit via established techniques. For example, the use of a single-device equivalent circuit allows a subharmonically pumped mixer to be analyzed via a single-diode mixer program such as DIODEMX (References 2.1, 5.2).

To scale the external circuit, all odd-order embedding impedances are artificially set to twice their true values, even-order embedding impedances are set to zero, and the antiparallel pair is replaced by a single device. This process is illustrated in Figure 5.18, in which $Z'(\omega) = Z(\omega)$ at odd-order mixing frequencies and $Z'(\omega) = 0$ at even-order frequencies. On completing the analysis of the single-device circuit, all absolute power levels (e.g., LO power of a mixer, input and output powers of a multiplier) must be doubled, but all relative levels (e.g., conversion loss or gain) must remain unchanged.

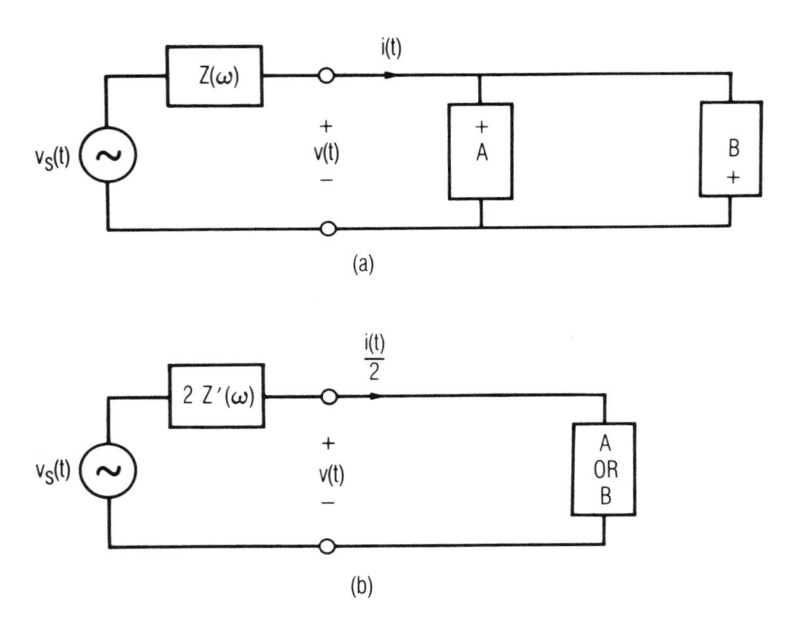

Figure 5.18 Generation of the single-device equivalent circuit of the antiparallel-connected elements: (a) the complete circuit; (b) the single-device equivalent. $Z'(\omega)$ equals $Z(\omega)$ for odd-order frequencies; it is zero for even-order frequencies.

A limitation of the single-device equivalent circuit is that it is strictly valid only in the case of perfect balance and identical nonlinear devices. This limitation is not as severe as it seems, however, because, as with the hybrid-coupled circuits, performance is not highly sensitive to balance. The effects of slight imbalances can be determined in the same manner as for hybrid-coupled components: the output currents are treated as phasors that combine in phase (the desired output) or out of phase (the cancelled output). Amplitude and phase imbalance cause those phasors to have unequal magnitudes and to have a phase difference other than zero or 180 degrees.

Antiseries Interconnection

A dual case of the antiparallel connection is the antiseries connection shown in Figure 5.19. Because of the symmetry of the circuit, $V_A = V_B = V$ and

$$I_L = I_A + I_B \tag{5.2.8}$$

I_A is found from (5.2.1) and I_B from (5.2.2). Adding these gives the output current:

$$I_L = 2bV^2 + 2dV^4 + \ldots \tag{5.2.9}$$

that is, the output current, the current in the load R_L, is an even-degree function of the voltage across the nonlinear elements. Thus, under sinusoidal excitation the load current and voltage contain only even-order mixing frequencies and harmonics. We find the loop current in a manner similar to that of the antiparallel case, by recognizing that $I_A = -I_B$ for current components in I_{loop}:

$$I_{\text{loop}} = I_A = -I_B = aV + cV^3 + \ldots \tag{5.2.10}$$

I_{loop} is an odd-degree function of V, so the loop current must contain the odd-order mixing components and harmonics of the frequencies in its terminal voltage waveform. Like the antiparallel interconnection, the antiseries interconnection separates even- and odd-order frequency components, but in the latter the even-order components circulate in the external circuit.

Figure 5.20 shows how the even- and odd-order voltage components, $v_e(t)$ and $v_o(t)$, respectively, and the even- and odd-order current components, $i_e(t)$ and $i_o(t)$, are distributed in the circuit ($v_o(t)$ does not include the fundamental-frequency component). The load current $i_L(t)$ has only even-order components, so only even-order voltages $v_e(t)$ exist at its terminals. Although element B does not directly short-circuit element A at the odd-order frequencies, as it did in the antiparallel case, the entire lower half of the loop short-circuits the entire upper half of the loop (as evidenced by the lack of odd-order voltage components across R_L). This observation can be used to decompose the antiseries-connected pair of devices into a single-device equivalent circuit; the process is illustrated in Figure 5.21.

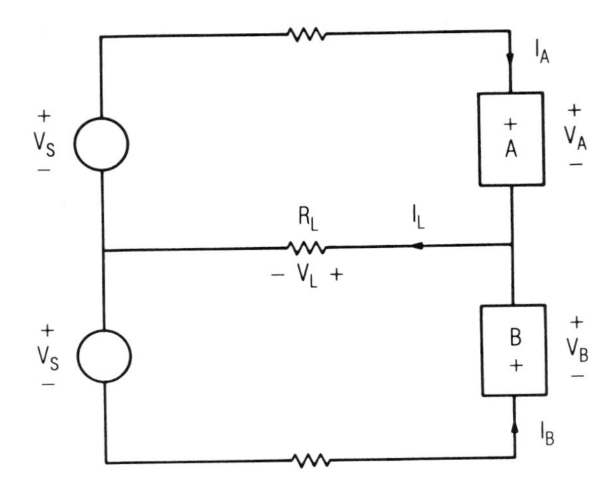

Figure 5.19 Antiseries connection of two identical nonlinear elements. The dual sources must be realized via a transformer or other 180-degree hybrid.

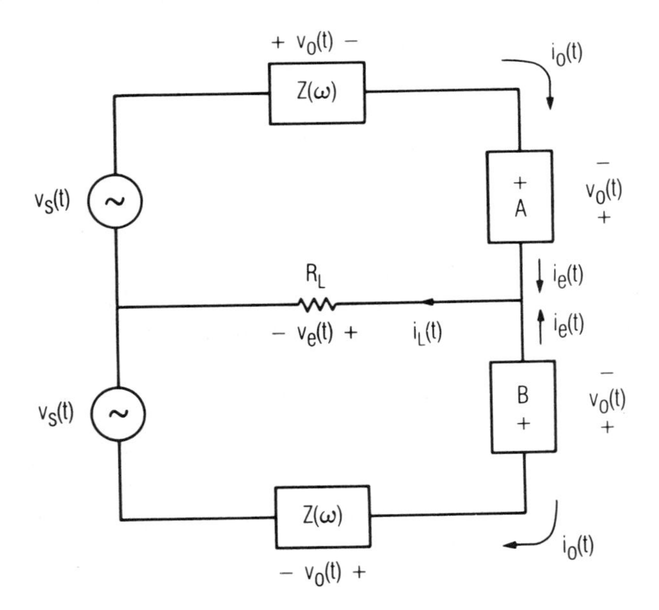

Figure 5.20 Even- and odd-order voltages and currents in the antiseries circuit. Although the odd-order voltage components exist only across *A*, *B*, and both $Z(\omega)$, the voltages across these elements do not consist solely of odd-order components.

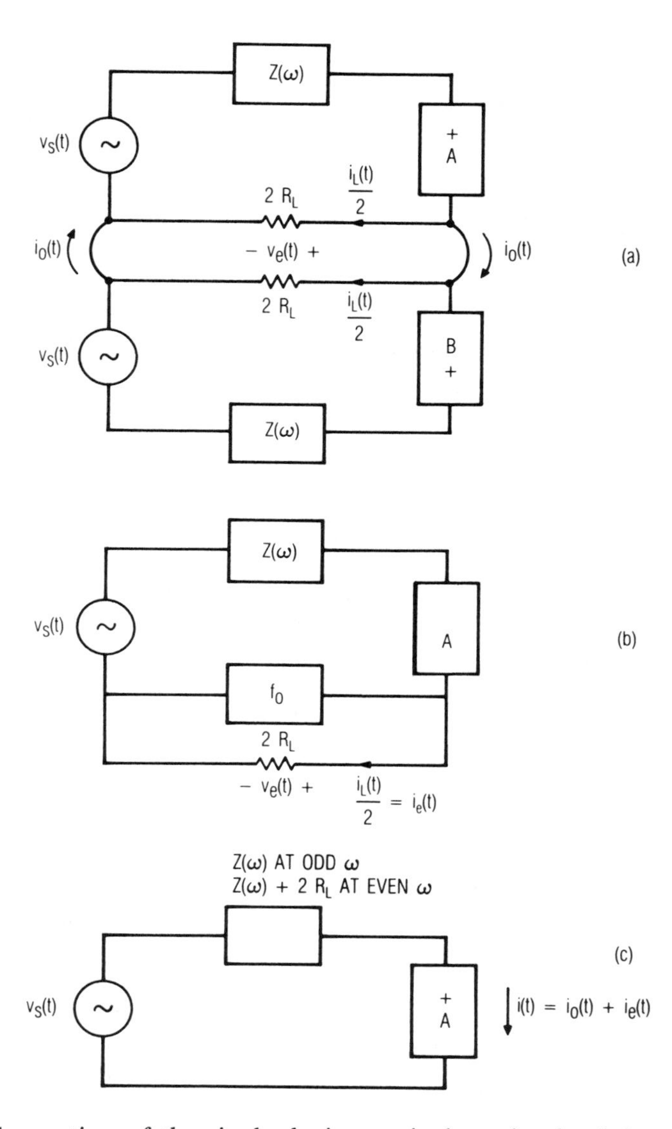

Figure 5.21 Generation of the single-device equivalent circuit of the antiseries circuit: (a) the antiseries circuit is split into two half-circuits; (b) the lower half-circuit is replaced by a short circuit f_o at the odd-order frequencies; (c) the single-device equivalent is formed by including the load resistor $2R_L$ in $Z(\omega)$.

In Figure 5.21(a), the load resistor R_L has been split into two resistors, each having resistance $2R_L$; each resistor also has half the load current, $i_L(t)/2$. In this circuit, $i_L(t)/2 = i_e(t)$, where $i_e(t)$ is the even-order current in either half of the divided circuit. The odd-order current components do not circulate in R_L, but pass through the links between the half-circuits; because the odd-order current in the top half of the circuit passes through the bottom half without generating any voltage across it, the bottom half-circuit effectively shorts the upper half-circuit at the odd-order frequencies. For this reason, the lower half-circuit can be replaced by an ideal filter, f_o, shown in Figure 5.21(b), that short-circuits the load resistor at odd-order frequencies and is an open circuit at even-order frequencies. This circuit is a valid single-device equivalent of the circuit in Figure 5.20.

The circuit in Figure 5.21(b) is not in the form we would prefer, however; it is not in the canonical form that we have assumed to exist in the previous chapters. We would prefer an equivalent circuit that consists of the device, embedding impedance, and voltage source in series. That circuit is shown in Figure 5.21(c), in which the load resistance is absorbed into the embedding network. The embedding impedance of this circuit therefore is $Z(\omega) + 2R_L$ at even-order frequencies and $Z(\omega)$ at odd-order frequencies. The loop current, $i(t)$, in Figure 5.21(c) consists of both the even- and odd-order components; that is, $i(t) = i_e(t) + i_o(t)$. These components can be separated easily in the frequency domain, and the output power is found from the desired frequency component of $i_e(t)$ and $2R_L$. As before, the output power of the single-device equivalent circuit is half that of the complete circuit.

The dual excitation sources shown in Figures 5.19 through 5.21 must be realized by some type of balun or 180-degree hybrid. The transformer hybrid is often employed for this task; Port 4 in Figure 5.2 is typically used for the input, and Ports 2 and 3 are the outputs. Because Port 3 is in series with R_L, Port 3 is usually shorted and connected to ground instead of being terminated. Other types of 180-degree hybrids can also be used to provide the dual sources, as long as an out-of-phase pair of ports is used as the output.

Although the antiseries connection of two-terminal devices has many important uses in microwave electronics, one of the most familiar applications is in a low-frequency circuit, the full-wave rectifier shown in Figure 5.22(a). Fourier analysis of the output current and voltage waveforms, Figure 5.22(b), shows that the waveforms contain no excitation-frequency component; the output frequencies are only dc and even harmonics of the excitation frequency. This same circuit can be scaled to microwave frequencies and used as a second-harmonic frequency multiplier having minimal fundamental and third-harmonic output; one such multiplier is described in Reference 5.3.

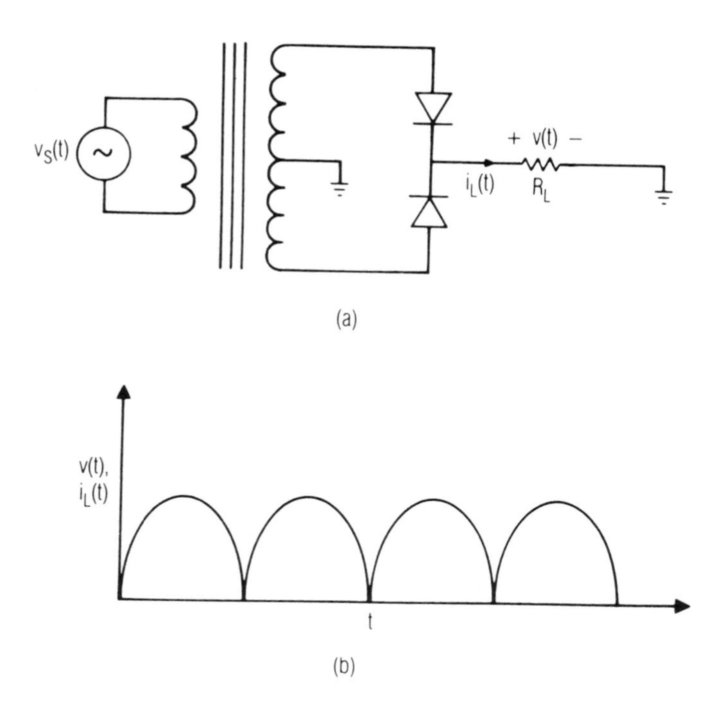

Figure 5.22 A common antiseries circuit, the full-wave rectifier: (a) circuit; (b) voltage and current waveforms.

Series Interconnection

Figure 5.23 shows an interconnection of two nonlinear elements with an output transformer that couples them to the load, R_L; for lack of a better term, we call this a *series interconnection*. From the descriptions of the antiparallel and antiseries circuit, it should be clear that the even- and odd-order currents in the nonlinear elements are as shown in the figure.

The primary circuit (i.e., the tapped side) of the output transformer is excited in phase by the odd-order currents in the nonlinear elements; these currents induce equal but opposing currents in the secondary side, and consequently there is no odd-order current in the transformer's secondary winding. Because there is no secondary current, the odd-order voltage across both the secondary and primary must be zero; thus, the nonlinear elements are connected to ground through the transformer at odd-order mixing frequencies. Furthermore, the even-order currents are equal and opposite at the node connecting A, B, and $Z(\omega)$; thus, this node is a virtual ground point for even-order products.

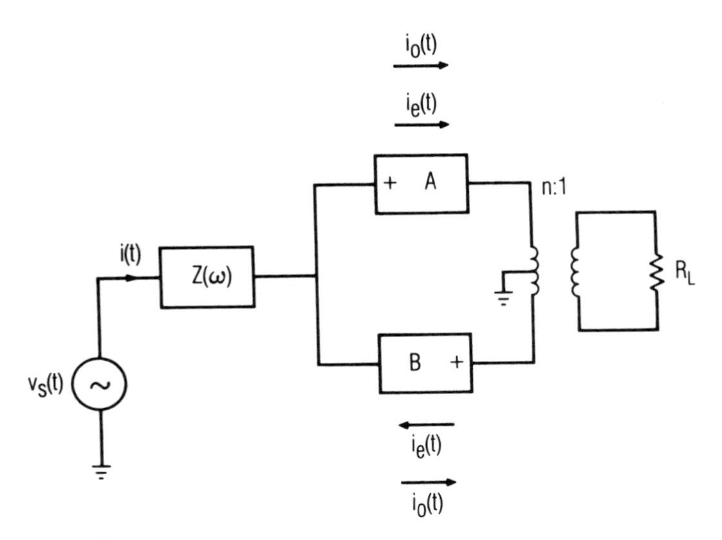

Figure 5.23 Series connection of two nonlinear elements. In a microwave circuit, the transformer is usually realized by a 180-degree hybrid.

The single-device equivalent circuits are found in the usual manner, by splitting $Z(\omega)$ and the transformer into two parallel elements. The circuit can then be separated into two separate circuits, each of which has the form shown in Figure 5.24(a). In this figure, the load impedance has been transferred to the primary side of the transformer, and the elements f_o and f_e, ideal filters that are short circuits at the odd- and even-order mixing frequencies, respectively, have been included to provide the short circuits at the virtual ground points. The elements can be consolidated further as shown in Figure 5.24(b), in which $Z(\omega), f_o, f_e$, and the load have been expressed as a single impedance: $2Z'(\omega) = (n^2/2) R_L$ at even-order frequenciess, and $2Z'(\omega) = 2Z(\omega)$ at odd-order frequencies. Except for the change in impedances and the fact that the even-order products are the output quantities, this circuit is identical to the single-device equivalent circuit of the antiparallel interconnection in Figure 5.18. The series circuit operates in a manner similar to that of the antiparallel circuit; in the former, however, the output is coupled to the devices via the circulating odd-order current instead of the even-order terminal current. Consequently, in the series circuit the even-order products are the output, not the odd-order.

In a microwave realization of the series circuit, some type of 180-degree hybrid would be used in place of the transformer. Depending upon the characteristics of the hybrid, it may be necessary to change the description of the series impedance $2Z'(\omega)$ in Figure 5.24(b) because many

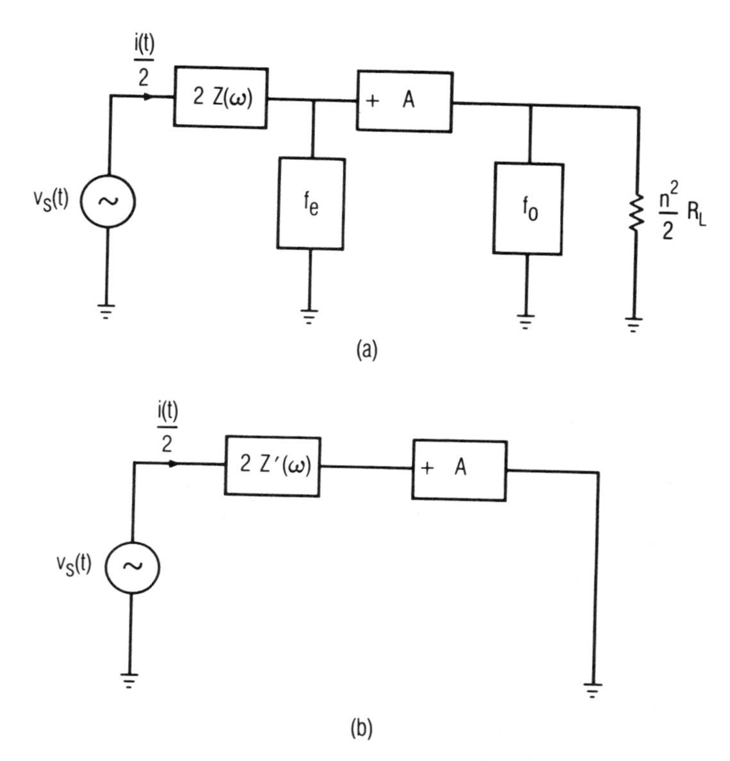

Figure 5.24 Single-device equivalents of the circuit in Figure 5.23: (a) half-circuit representation, where f_o and f_e are ideal series resonators tuned to the odd- and even-order frequencies, respectively; (b) representation in which $Z(\omega)$ has been modified to account for f_o and f_e.

microstrip hybrids do not present short circuits to all odd-order currents, as does the transformer. Many hybrids present a short circuit to the odd-order currents at the excitation frequency, but they present open circuits to other odd-order products. In this case, it is necessary to modify $Z'(\omega)$ in Figure 5.24(b) appropriately.

5.2.2 Properties of Direct Parallel Interconnection of Two-Port Components

Two-port components can be connected in parallel at their input and output ports, as shown in Figure 5.25. Direct parallel interconnection is very common in microwave circuits; power FET and bipolar devices, and even some small-signal devices, are realized as parallel combinations of

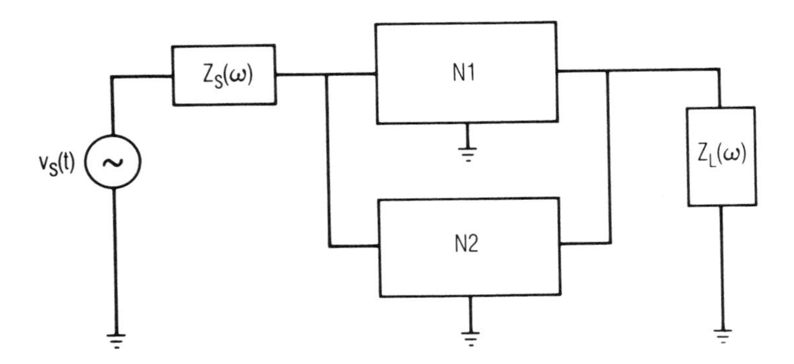

Figure 5.25 Directly connected two-port components.

many low-power devices. Characterizing parallel-connected linear two-ports is straightforward: The Y matrix of parallel two-ports equals the sum of their individual Y matrices. Nonlinear two-ports cannot be described by Y parameters, however, nor by any similar two-port equations because two-port parameters are based upon an assumption of linearity.

The best, and simplest, approach to the analysis of parallel-connected two-ports is by generating a single-component equivalent circuit. Generating the equivalent circuit requires changing only the source and load impedances at all harmonics or mixing frequencies.

Figure 5.26 illustrates the process of generating the single-device equivalent circuit. Because of the symmetry of the structure, it is possible to split $Z_s(\omega)$ and $Z_L(\omega)$ into two parallel impedances of $2Z_s(\omega)$ and $2Z_L(\omega)$. This operation preserves the voltage levels in the circuit but reduces the input and output currents by a factor of two compared to the currents of the combined pair. The input and output power levels in the single-component equivalent circuit are, therefore, half those of the parallel-connected pair of two-ports; however, because both available input power and output power are reduced by the same factor, the gain is the same.

The single-component equivalent circuit in Figure 5.26 shows that each component is effectively terminated by an impedance twice that of the actual load; thus, the impedance levels necessary to match a parallel-coupled pair of two-ports is half that required by a single two-port. This is the major disadvantage of such interconnections and the primary limitation in achieving high power by paralleling individual solid-state devices: The impedance level necessary to match the parallel combination drops by a factor equal to the number of devices. In paralleling devices, we eventually reach a point at which it is no longer possible to match the

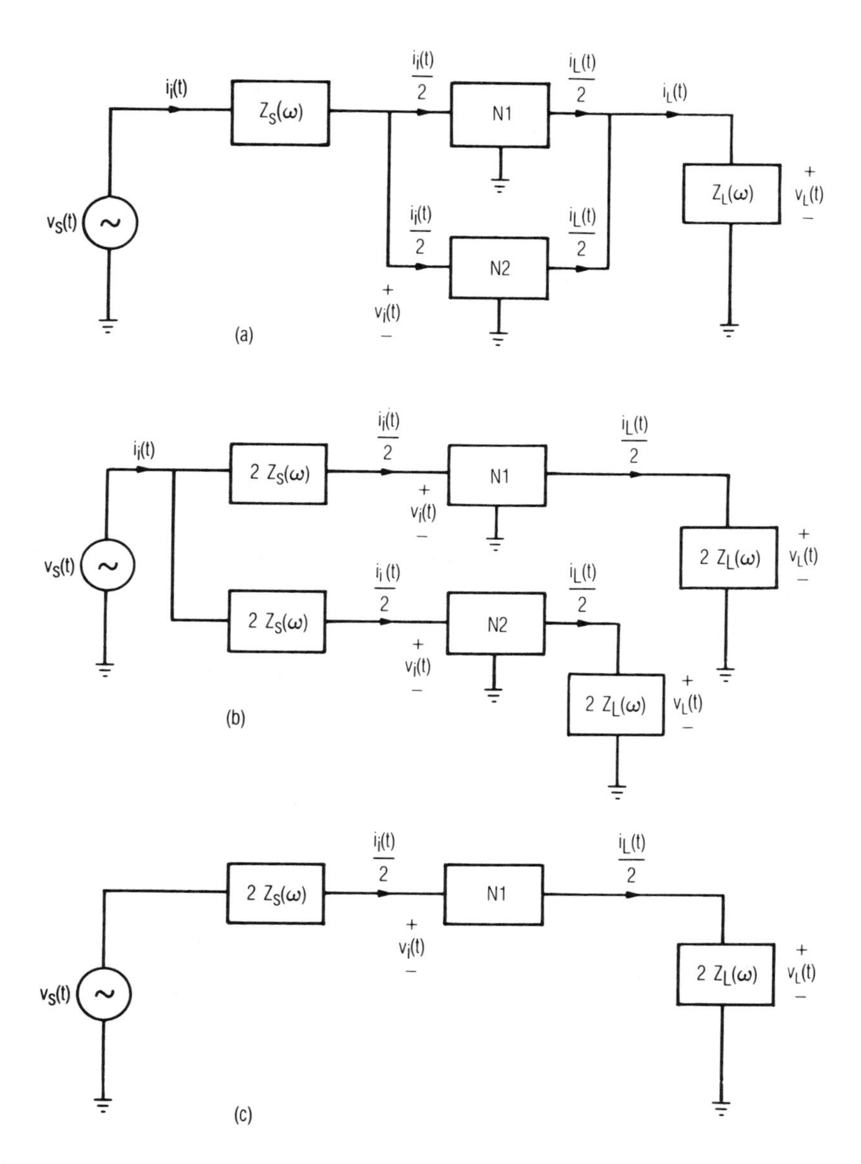

Figure 5.26 Evolution of the single-element equivalent circuit of the directly connected two-ports: (a) the directly connected circuit; (b) splitting the source and load impedances does not change the voltages or total currents; (c) the single-element equivalent.

combination by means of circuits having practical topologies and element values. Furthermore, because of the low impedances, parallel-connected devices have high combined input and output currents, and power dissipation in the matching circuits may be substantial. Other practical problems, such as maintaining phase and amplitude balance in a large number of separate devices, and even thermal problems, also limit the number that can be connected in parallel. We will explore some of these practical matters further in Chapter 9.

Another implication in Figure 5.26, one particularly relevant to the design of mixers and frequency multipliers, is that the embedding impedance presented to each device is twice the actual terminating impedance. Because of this property, it is sometimes possible to achieve optimum terminating impedances more easily in a balanced structure than in a single-device structure. For example, the optimum IF load impedance for each diode in a balanced diode mixer is usually approximately 100 Ω. This impedance can be achieved without transformation by connecting in parallel the IF outputs of the two individual mixers in a balanced pair. The shared 50-Ω IF load is equivalent to each mixer having an individual load of 100 Ω at its IF port.

Because the phase shifts are identical in both two-ports in the parallel-coupled circuit, the circuit does not reject any harmonics or mixing frequencies of any order, even or odd. It does, however, achieve the same 3-dB improvement in intermodulation intercept point as do the hybrid-coupled two-ports. Accordingly, in applications where harmonic or spurious rejection is important, it is best to use circuits that are hybrid-coupled.

REFERENCES

[5.1] J. Lange, "Interdigitated Stripline Quadrature Hybrid," *IEEE Trans. Microwave Theory Tech.*, Vol. MTT-17, 1969, p. 1150.

[5.2] S. Maas, "An Interactive Microwave Mixer Analysis Program," Aerospace Corp. Technical Report No. SD-TR-86-98, 1986, Aerospace Corp., El Segundo, CA.

[5.3] H. Ogawa, A. Minagawa, "Unipolar MIC Balanced Multiplier—A Proposed New Structure for MICs," *IEEE Trans. Microwave Theory Tech.*, Vol. MTT-35, 1987, p. 1363.

CHAPTER 6

DIODE MIXERS

The most common type of microwave-frequency mixer uses a Schottky-barrier diode. Diode mixers are useful over a remarkably broad range of frequencies; inexpensive doubly balanced mixers can be obtained for use at the lower microwave frequencies (below 20 GHz) and mature single-diode mixer designs are available for millimeter-wave applications as well. This chapter is concerned with the practical aspects of designing mixers in both frequency ranges.

For further information on the subject of microwave and millimeter-wave mixers, the author shamelessly suggests his own book on the subject (Reference 2.1). Another good source of information is a volume of reprints of technical papers (Reference 6.1), which includes most of the classic mixer papers, some of which are referenced in this chapter.

6.1 MIXER DIODES

Although a few ancient mixers that use point-contact diodes still exist, virtually all modern diode mixers employ Schottky-barrier diodes as the mixing elements. Inexpensive silicon diodes are used in most prosaic mixer applications; these diodes, available in a wide variety of chip and packaged forms, are adaptable to virtually any transmission medium. In particular, they can be obtained in so-called *dual* and *quad* forms: two or four nearly identical diodes in a single package, connected in a ring or star configuration for use in balanced circuits.

GaAs diodes are considerably more expensive than silicon diodes, but they can provide better conversion loss and noise performance, especially at high frequencies. GaAs diodes are more expensive than those in silicon and at low frequencies their performance advantage is minimal, so they generally are not obtainable in the low-cost (e.g., molded plastic) packages sometimes used for silicon diodes. GaAs diodes are available as chips, as beam-lead devices, and in miniature ceramic and quartz packages.

Schottky-barrier diodes are discussed generally in Section 2.3, and mixer diodes in particular are examined in Section 2.3.2; these sections describe the theory and modeling of the Schottky junction. This section is concerned with the use of such diodes in practical mixer circuits.

6.1.1 Diode Model

Figure 6.1 shows the circuit model of a Schottky-barrier diode. Although this equivalent circuit is applicable to either a GaAs or silicon diode, it describes only the seminconductor chip; that is, it does not include package parasitics or other parasitics not associated directly with the junction. The circuit consists of a nonlinear resistive junction represented by the controlled current source, $I_j(V)$; a nonlinear junction capacitance, C_j; and a series resistance, R_s (which is generally assumed to be linear). The junction current has the characteristic:

$$I_j(V) = I_{\text{sat}}[\exp(qV/\eta KT) - 1] \tag{6.1.1}$$

where I_{sat} is the reverse-saturation current, q is the electron charge ($1.6 \cdot 10^{-19}$ C), K is Boltzmann's constant ($1.37 \cdot 10^{-23}$ J/K), and T is absolute temperature in Kelvins. The constant η, called the *ideality factor*, accounts for differences between the ideal I/V characteristic (in which $\eta = 1.0$) and real diodes (in which η is usually between 1.05 and 1.25). I_{sat} depends strongly upon temperature and is proportional to junction area; for diodes of the same junction area, I_{sat} is much smaller in GaAs than in silicon. This characteristic implies that the knee of the I/V characteristic occurs at higher voltage in GaAs then in silicon.

The incremental conductance $g_j(V)$ of the junction is found (Section 2.2) by differentiating (6.1.1):

$$g_j(V) = \frac{dI_j(V)}{dV} = \frac{q}{\eta KT} I_{\text{sat}} \exp(qV/\eta KT)$$

$$\approx \frac{q}{\eta KT} I_j(V) \tag{6.1.2}$$

The approximation is valid when $V \gg \eta KT/q$; because $\eta KT/q$ is only 25 mV at room temperature, this is rarely a restrictive assumption. The junction capacitance has the characteristic:

$$C_j(V) = \frac{C_{j0}}{\left(1 - \frac{V}{\phi}\right)^{\gamma}} \tag{6.1.3}$$

where C_{j0} is the zero-voltage junction capacitance, ϕ is the diffusion potential, and γ is a constant ($\gamma = 0.5$ in a uniformly doped diode). Although the I/V characteristics of most mixer diodes follow (6.1.1) very closely, (6.1.3) does not describe all Schottky-barrier diodes adequately. Equation (6.1.3) is valid if the epilayer doping is uniform and the epitaxial layer is thick enough so that it is not fully depleted over the range of V; however, the use of nonuniform doping or very thin epilayers (e.g., the Mott diode, Section 2.3.2) may modify this characteristic. The effects of those modifications may be as minor as to reduce γ slightly, or as major as to require a completely different, perhaps empirical expression for $C_j(V)$.

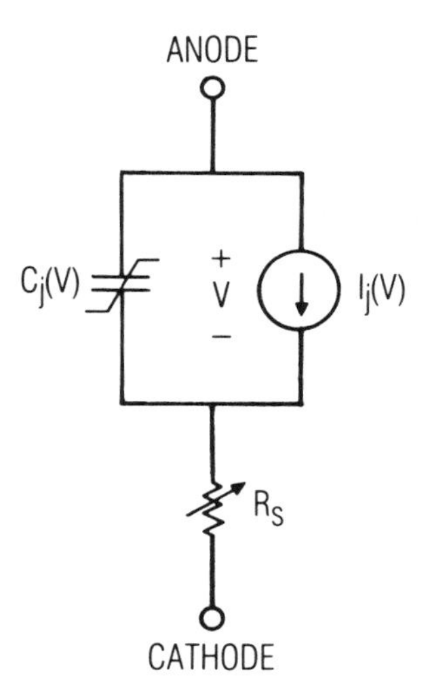

Figure 6.1 Schottky-barrier diode model applicable to an unpackaged chip diode.

R_s, the series resistance, results primarily from the resistance of the undepleted epitaxial region under the diode junction. Because R_s depends strongly upon such characteristics as frequency and the diode's structure, it is not possible to present a general expression for its value. R_s is very weakly nonlinear; its nonlinearity is usually negligible in mixer analyses, so R_s is invariably assumed to be linear.

A package or a complex diode structure introduces additional parasitics for which a circuit model must account. The modeling of these parasitics depends upon the structure of the diode or package. Figure 6.2, for example, shows a simple model of a chip diode mounted in a miniature "pill" package; the package is modeled by a lumped inductor representing the inductance of the connecting wire or ribbon and by a lumped capacitor representing the capacitance of the ceramic sidewalls and the parallel-plate capacitance of the top and bottom caps. In many cases, the transmission medium in which the diode is to be used influences the modeling of parasitics; a model of a packaged diode, for example, used in a microstrip circuit may not be identical to the model of the same diode mounted in a waveguide.

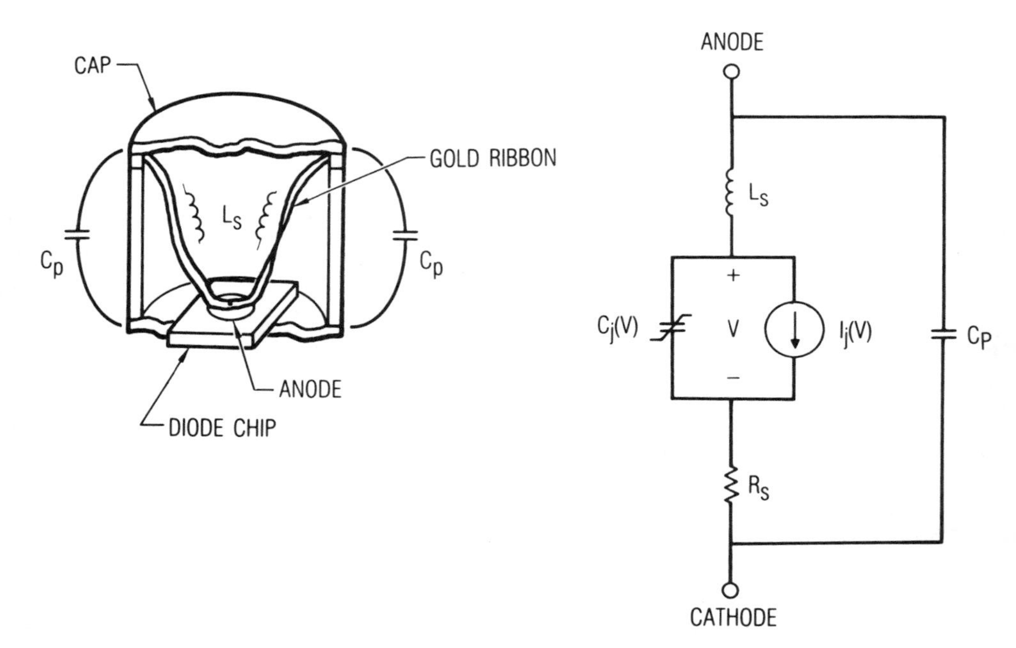

Figure 6.2 Cutaway view and equivalent circuit of a packaged chip diode: L_s is the inductance of the bonding wire or ribbon; C_p is the capacitance of the package, resulting from the metalic top and bottom and ceramic sidewalls.

A figure of merit of the diode is its cutoff frequency, a parameter calculated from dc or low-frequency measurements of the diode's characteristics. Because package parasitics can be absorbed into the matching circuits, they are not included in determining the cutoff frequency. The cutoff frequency, f_c, is

$$f_c = \frac{1}{2\pi R_s C_{j0}} \tag{6.1.4}$$

Minimizing both R_s and C_{j0} is necessary in order to achieve low conversion loss at high frequencies; however, reducing one of these parameters usually increases the other. Thus, a high cutoff frequency shows that the diode designer has done a good job of simultaneously minimizing both R_s and C_{j0}.

Although a first-order analysis indicates that cutoff frequency is independent of junction area, second-order effects cause a diode's cutoff frequency to rise slightly as its area is decreased. Small, well made Schottky-barrier diodes often have cutoff frequencies above 2500 GHz.

6.1.2 Mixer Diode Types

Perhaps the simplest diode commonly used in mixers is an unpackaged chip having the cross section shown in Figure 2.9; such chips can be mounted in ceramic or plastic packages or may be used unpackaged in hybrid circuits. Chip diodes usually have more than one anode; several anodes of different diameters may be defined on a single chip in order to compensate for manufacturing tolerances or to allow one type of diode to be used at widely differing frequencies. However, in order to allow bonding of a wire or ribbon to the anode, the anode must be at least 10 to 15 microns in diameter or have a metal overlay to increase its area. Chip diodes that have such large anodes have relatively high junction capacitance and thus are not optimum for millimeter-wave operation; reducing the capacitance, without reducing junction area, requires low epilayer doping levels that increase R_s and result in a lower cutoff frequency than would be obtained with a smaller anode.

If a different method is used to connect the anode, the diodes' anodes can be smaller. One such method, which facilitates connections to very small anodes, is to use a pointed spring wire, or *whisker,* similar to the whisker used in a point-contact diode. The whisker point is chemically etched to a very small radius, less than 1 micron, allowing anodes as small as 1.5 microns to be used. Connecting a finely pointed whisker to an anode that has a diameter equal to 3 wavelengths of red light is not a simple task, so the diode must be designed to facilitate the anode-contacting process as much as possible. The best way to simplify the contacting process is to cover the top surface of the chip with thousands of closely spaced identical anodes, so the whisker point need only be set on the top surface of the chip, where it will undoubtedly contact an anode. After the contact is made, it can be checked in a scanning electron microscope and with a semiconductor curve tracer to verify that the point is reliably seated and contacts only a single anode. Diodes made this way are called *dot-matrix diodes,* and they are used primarily for millimeter-wave applications.

Because the whisker contact is relatively delicate, the mechanical design of the mount used for a dot-matrix diode is critical to the mixer's reliability. One very old but viable design for mounting a millimeter-wave dot-matrix diode is called a *Sharpless wafer,* after W.M. Sharpless (Reference 6.2). The Sharpless wafer, shown in Figure 6.3, is used in a waveguide structure; the cavity in which the diode is mounted is actually a piece of the waveguide. The chip is mounted on the end of an insulated pin, which serves as part of the IF circuit, and the whisker, mounted on the end of a press-fit post, protrudes into the waveguide from the opposite side. The whisker and the part of the post that protrudes into the waveguide couple the RF and LO power from the waveguide into the chip; the capacitance of the IF pin effectively connects the cathode of the diode to the waveguide topwall at the RF and LO frequencies. Other types of wafer mounts have also been used, but these are usually variations on the Sharpless-wafer theme.

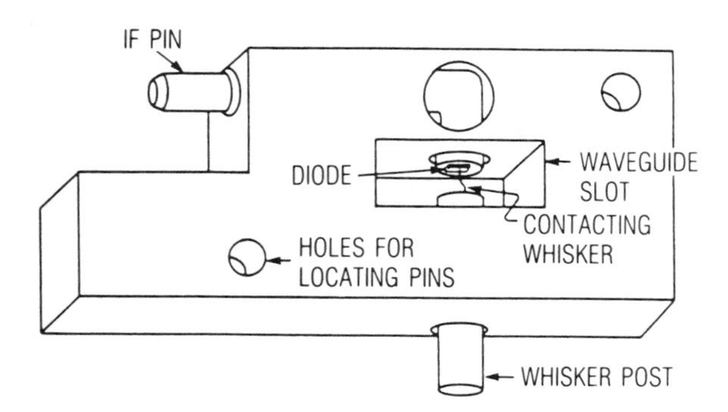

Figure 6.3 Sharpless wafer for millimeter-wave diodes: The diode chip is mounted on the end of the IF pin, and the whisker, attached to the whisker post, contacts the diode from the opposite side of the waveguide. The whisker post is press-fit into the wafer.

An important and very versatile type of diode is the Schottky-barrier *beam-lead* diode, shown in cross section in Figure 6.4. A beam-lead device is a chip diode that has an integral ribbon lead connected to its anode; its cathode contact, which also has an integral ribbon lead, is on the same side of the chip as the anode. The beam-lead diode has many of the desirable characteristics of both packaged and chip devices: it has a very small anode (smaller than chip devices intended to have wire or ribbon connections); it has very low series inductance; and it can be handled and connected into a circuit without the use of whiskers, wire, or special equipment.

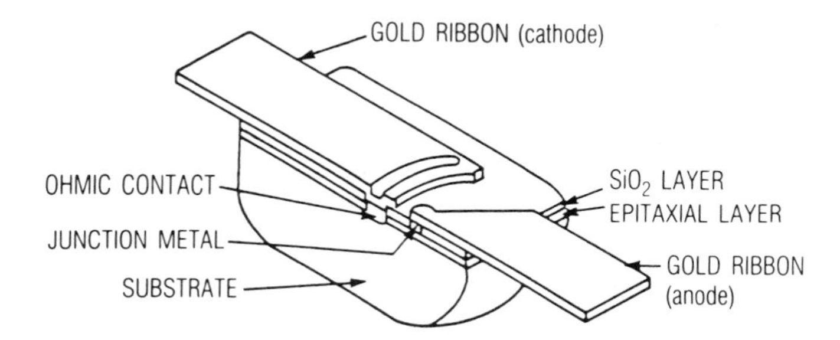

Figure 6.4 Cross section of a beam-lead diode designed for operation below approximately 12 GHz. The diode's most serious limitation for high-frequency operation is the overlay capacitance between the anode ribbon and the epilayer.

The structure of the semiconductor chip used in a beam-lead diode requires two major modifications of the Schottky-barrier diode structure shown in Figure 2.9: first, the top surface must somehow be connected to the substrate; and second, the anode ribbon must be so designed that it does not cover a large area above the epilayer, or a large parasitic "overlay" capacitance results. The cathode connection can be made by etching away the oxide layer and epilayer in a region near the anode, and the most common method of minimizing overlay capacitance is to locate the anode close to the edge of the chip. The resulting structure is not entirely satisfactory for many purposes, however, because the very small anode connection close to the edge of the chip is delicate and the overlay capacitance may still be substantial. The series resistance is also relatively high because the lateral current path through the substrate and epilayer to the cathode is relatively long.

Many ingenious structures have been proposed to circumvent the problems inherent in beam-lead diodes (References 6.3–6.6). The usual approach to reducing overlay capacitance is to separate the anode ribbon from the conductive layers of the diode by a thick insulating layer, which may be air. One simple modification to the generic beam-lead structure in Figure 6.4 is to fabricate the diode on a semi-insulating substrate (the buffer layer, of course, remains heavily doped) and to etch away the epitaxial and buffer layers where the anode ribbon overlays the substrate. Thus, the anode ribbon overlays a relatively wide nonconductive region instead of the conductive epilayer. Other techniques that reduce overlay capacitance include the use of air bridges, proton-bombarded regions in the semiconductor under the anode lead (proton bombardment changes the conductive semiconductor to an insulator), and the use of very thick oxide layers. Such diodes have been generally successful, although they sometimes involve more difficult and expensive processing than does the generic beam-lead diode. Many of these diodes are suitable for use at millimeter wavelengths, and performance of mixers using them is often indistinguishable from that of whisker-contacted dot-matrix diodes. Finally, the ability to mass-produce such diodes, even with relatively expensive processing techniques, usually makes them less expensive than whisker-contacted diodes in Sharpless mounts, which require laborious fabrication and assembly.

6.2 NONLINEAR ANALYSIS OF MIXERS

The analysis of diode mixers requires a straightforward application of the methods described in Chapter 3, especially Section 3.2. The conversion loss and input-output impedances can be found via a method called *large-signal–small-signal analysis,* wherein the diode and its embedding circuit are first analyzed under LO excitation via a harmonic-balance technique to determine the diode's junction conductance and capacitance waveforms. These waveforms are then used to formulate conversion matrices representing the time-varying junction conductance and capacitance, and the small-signal linear analysis, described in Sections 3.2.1 and 3.2.2, is performed. This approach is adequate for determining the single-tone, small-signal properties of the mixer. If multitone intermodulation data is desired, the methods of Section 3.2.3 can be applied. Volterra-series analysis (described in Chapter 4) is normally limited to cases of small-signal excitation and weak nonlinearities and is not applicable to the analysis of diode mixers, which always entail large LO voltages and strong resistive nonlinearities.

The data obtained from the large-signal and small-signal analyses can be used, along with information about the diode's noise sources, to calculate a mixer's noise temperature. Noise analysis is beyond the scope of this book, but the calculation of mixer noise temperatures is examined in References 2.1 and 6.1.

6.2.1 Large-Signal Analysis

Harmonic-balance techniques are ideally suited to the large-signal part of the large-signal–small-signal analysis. In fact, harmonic-balance techniques were introduced into the world of microwave circuits specifically for the purpose of mixer analysis and subsequently were applied to other types of circuits; most of the convergence algorithms described in Section 3.1 were originally developed for mixers.

Figure 6.5 shows a canonical model of the mixer under LO excitation. In this model, the LO excitation source and the matching circuit have been reduced to a voltage source, $V_s(t)$ (which may contain a dc component), and a frequency-dependent impedance, called the *embedding impedance,* $Z_e(\omega)$. In this model, $Z_e(\omega)$ equals the output impedance of the matching circuit at the fundamental frequency and all the harmonics of the LO frequency, ω_p, and $V_s(t)$ is scaled so that the available power at the fundamental frequency from $V_s(t)$ and $Z_e(\omega_p)$ is the same as that from the LO source and matching network. The junction current, $I_j(V)$, and capacitance, $C_j(V)$, are given by (6.1.1) and (6.1.3), respectively. The logical division of this circuit into linear and nonlinear subnetworks is to make the parallel combination of the capacitor and current source the nonlinear subnetwork and to make the series resistance and $Z_e(\omega)$ the linear subnetwork. The goal of the harmonic-balance analysis is to find the small-signal incremental conductance and capacitance waveforms, $g_j(V(t))$ and $C_j(V(t))$, respectively (for simplicity, we shall denote these waveforms as $g_j(t)$ and $C_j(t)$ to distinguish them from the functions $g_j(V)$ and $C_j(V)$ in (6.1.2) and (6.1.3)). The terminal voltage waveform, $V_d(t)$, is sometimes of interest; $V_d(t)$ can be found as $V_d(t) = V(t) + R_s I(t)$, where $I(t)$ is the sum of the resistive junction current $I_j(t) \equiv I_j(V(t))$ and the capacitive current.

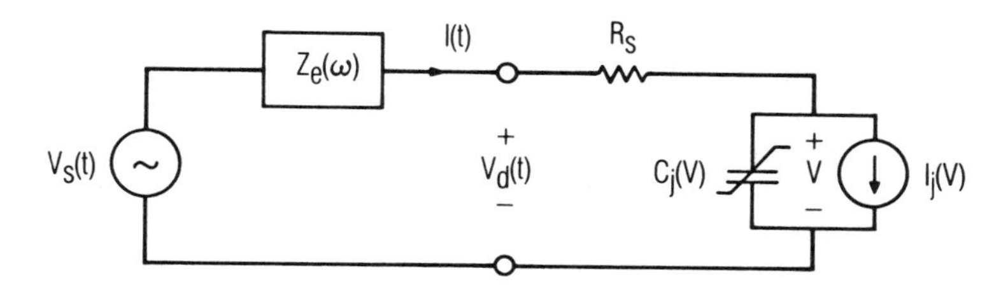

Figure 6.5 Canonical model of a single-diode mixer under LO excitation. $Z_e(\omega)$, the embedding impedance, represents all source and load impedances.

Several other quantities are of interest in the pumped diode. The LO input impedance is defined as

$$Z_{\text{in,LO}} = \frac{V_{d,1}}{I_1} \tag{6.2.1}$$

where $V_{d,1}$ and I_1 are the fundamental-frequency components of the waveforms $V_d(t)$ and $I(t)$, respectively. To be precise, the input impedance of a nonlinear circuit cannot be defined because impedance is a fundamentally linear concept; however, the "quasiimpedance," $Z_{\text{in,LO}}$, given by (6.2.1) is what must be matched by the source impedance to optimize power transfer. $Z_{\text{in,LO}}$ is a function of the magnitude of $V_s(t)$, bias voltage, and $Z_e(\omega)$ at the fundamental and harmonic frequencies. The power dissipation in the diode is

$$P_d = \frac{1}{T}\int_{-T/2}^{T/2} V_d(t)\, I(t)dt \tag{6.2.2}$$

where T is the period of the LO waveforms' fundamental frequency. Because the mixer's RF excitation is assumed to be much smaller than that of the LO, the RF power dissipation is negligible. The part of P_d supplied by the LO source, exclusive of the dc bias, is

$$P_{d,\text{LO}} = P_d - V_{d,0}I_0 \tag{6.2.3}$$

where $V_{d,0}$ and I_0 are the dc components of $V_d(t)$ and $I(t)$, respectively. The LO efficiency is

$$\epsilon_{\text{LO}} = \frac{P_{d,\text{LO}}}{P_{\text{av,LO}}} \tag{6.2.4}$$

where $P_{\text{av,LO}}$ is the available LO power. The term ϵ_{LO} indicates how well the mixer is utilizing available LO power; if $Z_e(\omega_p) = Z^*_{\text{in,LO}}$ and the harmonic embedding impedances are reactive then $\epsilon_{\text{LO}} = 1.0$. LO efficiency is an approximate concept; even if all the LO power is dissipated in the diode, there is no strict guarantee that the power is used efficiently to vary C_j and g_j. However, in practical situations high LO efficiency generally implies that the LO power is used efficiently, and low LO efficiency always implies that LO power is wasted.

The most important considerations in the selection of an algorithm for harmonic-balance analysis are its efficiency (speed) and reliability of convergence. There is, as we might expect, considerable controversy among experts over the merits and flaws of the various convergence algorithms used for analyzing mixers. Even so, two harmonic-balance algorithms used often and successfully are Newton's method and the reflection algorithm. Newton's method is favored because of its strong convergence and its efficiency; it is especially effective in mixers, where one can generate the required initial estimate of $V(t)$ relatively easily. The reflection algorithm is also popular because it converges reliably (if sometimes slowly) and because it does not require an initial estimate of $V(t)$. Although optimization is often used as a convergence algorithm, it should be avoided because of its inefficiency. Likewise, splitting methods, although they are relatively fast, should be avoided because of their poor convergence properties.

A rough but adequate initial estimate of $V(t)$, for use with Newton's method, can be found from Figure 6.5. If R_s is more than a few ohms, C_{j0} is not impractically great, and the mixer is strongly pumped, then the LO junction voltage waveform, $V(t)$, is well approximated by a clipped sinusoid. When the junction is biased positively, $V(t)$ is clipped at a value corresponding to the peak forward current, which can be estimated from $V_s(t)$, R_s, and $Z_e(\omega_p)$. When $V(t)$ reverse-biases the junction, $V(t)$ is approximately sinusoidal; we can find its peak value from the circuit in Figure 6.5 by setting $I_j(V) = 0$.

The convergence properties of Newton's method are good enough that the clipped-sinusoid estimate is adequate to achieve convergence in most cases; nevertheless, Newton's method sometimes fails to converge. In these cases, a simple solution is to set $V_s(t)$ to a very low value, so the circuit can be treated as quasilinear and $V(t)$ can be estimated very accurately. This value of $V(t)$ is used as the initial estimate for a harmonic-balance calculation. The $V(t)$ waveform resulting from that calculation is then used as the initial estimate in another harmonic-balance calculation, this time at an increased level of $V_s(t)$. The process is repeated until the large-signal calculation is performed at the desired $V_s(t)$. This process is time-consuming, but it usually results in a solution.

The reflection algorithm is a good choice for use in computer-aided design programs that must be used by people who may not have a good understanding of the harmonic-balance process and may have difficulty assessing the cause of convergence failure or circumventing it. The algorithm's rate of convergence near the global solution is sometimes slow, but this slowing usually occurs well past the point where the solution is adequate for practical purposes. Further comparisons of the convergence algorithms may be found in Section 3.1.4.

6.2.2 Small-Signal Analysis

The large-signal analysis yields the large-signal voltage and current waveforms in the pumped diode, most importantly $V(t)$ and $I_j(t)$. When these are known, the capacitance waveform, $g_j(t)$, and the conductance waveform, $C_j(t)$, can be found from (6.1.2) and (6.1.3), respectively. The small-signal canonical equivalent circuit of the mixer is as shown in Figure 6.6, where the diode has been replaced by its time-varying equivalent circuit. Again, $v_s(t)$ represents the RF excitation, and $Z_e(\omega)$ is the impedance presented to the diode, from all the mixer's matching circuits, at frequency ω. The embedding impedances of interest in $Z_e(\omega)$ are those at the mixing frequencies:

$$\omega_n = n\omega_p + \omega_0; \quad n = 0, \pm 1, \pm 2, \pm 3, \ldots \qquad (6.2.5)$$

The embedding impedance $Z_e(\omega)$ represents all the source and load impedances at all LO harmonics and mixing frequencies, as measured at the terminals of the diode. Thus, $Z_e(\omega)$ is a canonical representation of all the matching networks and filters in the mixer. In forming $Z_e(\omega)$, we have assumed all those circuits to be lossless at the LO, RF, and IF frequencies (this assumption is not essential, but it simplifies the calculation of input and output impedances and conversion loss). The current $i(t)$ represents the entire current at all the mixing frequencies in (6.2.5).

A conversion matrix for the diode can be formed via the methods of Sections 3.2.1 and 3.2.2; the formulation of this matrix is derived in detail in Example 7 of Section 3.2.2. The impedance-form conversion matrix of the diode is

$$\mathbf{Z}_c = R_s\mathbf{1} + (\mathbf{G}_j + j\mathbf{\Omega}\mathbf{C}_j)^{-1} \qquad (6.2.6)$$

where $\mathbf{1}$ is the identity matrix. The conversion matrix relates the small-signal current and voltage components at the frequencies given by (6.2.5), at the diode terminals, as

$$\mathbf{V}_d = \mathbf{Z}_c\,\mathbf{I} \qquad (6.2.7)$$

An example of the form of this matrix and the current and voltage vectors is given in (3.2.24). The impedance element $Z_e(\omega)$ has the conversion matrix $\mathbf{Z}_e$ in the form of (3.2.35); because the impedance is in series with the diode, the conversion matrix of the combined diode and impedance is

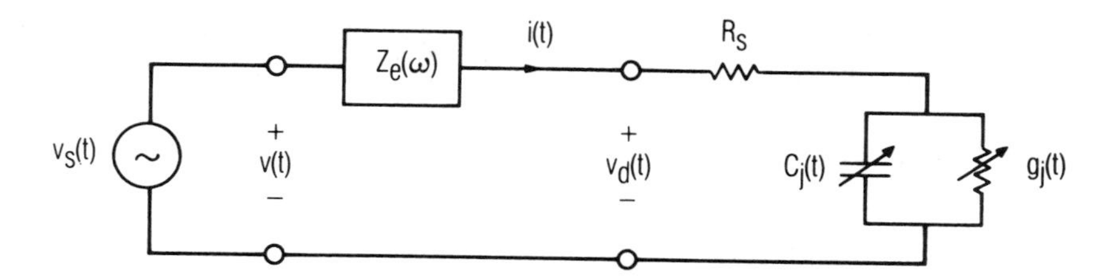

Figure 6.6 Small-signal equivalent circuit of a single-diode mixer.

the sum of $\mathbf{Z}_c$ and $\mathbf{Z}_e$. Thus, the vector of mixing currents and voltage components at the source terminals can be related, as in (3.2.24), by a single admittance-form coversion matrix $\mathbf{Y}$:

$$\mathbf{I} = \mathbf{YV} = (\mathbf{Z}_c + \mathbf{Z}_e)^{-1}\,\mathbf{V} \tag{6.2.8}$$

Because $v(t) = v_s(t)$, the vector $\mathbf{V}$ representing the sinusoidal excitation has only one nonzero component, V_q, where the excitation frequency is $\omega_q = q\omega_p + \omega_0$ (in the most common case of an RF frequency above the LO frequency, $q = 1$). $\mathbf{I}$, therefore, is simply the product of a single column of the conversion matrix, the column corresponding to ω_q, and the phasor V_q. The current component I_r at frequency ω_r is

$$I_r = Y_{r,q} V_q \tag{6.2.9}$$

The small-signal input impedance of the pumped diode at ω_q is

$$Z_{\text{in}}(\omega_q) = \frac{1}{Y_{q,q}} - Z_e(\omega_q) \tag{6.2.10}$$

If we assume that ω_r is the IF frequency (usually $r = 0$), the output impedance is

$$Z_{\text{out}}(\omega_r) = \frac{1}{Y_{r,r}} - Z_e(\omega_r) \tag{6.2.11}$$

It is interesting to note that, although the circuit is time-varying, time-invariant input and output impedances can be defined. We must remember in evaluating (6.2.10) and (6.2.11) that when $q,r < 0$, ω_q and

ω_r are negative frequencies. Consequently, the impedances $Z_e(\omega_q)$ and $Z_e(\omega_r)$ and the input/output impedances $Z_{\text{in}}(\omega_q)$ and $Z_{\text{out}}(\omega_r)$ are conjugates of the positive-frequency impedances at $|\omega_q|$ or $|\omega_r|$. The output power at the IF frequency ω_r is

$$P_{L,\text{IF}} = \frac{1}{2}\,|I_r|^2\,\text{Re}\{Z_e(\omega_r)\} \tag{6.2.12}$$

and the available RF input power at ω_q is

$$P_{\text{av,RF}} = \frac{|V_q|^2}{8\,\text{Re}\{Z_e(\omega_q)\}} \tag{6.2.13}$$

The transducer conversion loss is the ratio of these quantities:

$$L_c = \frac{P_{\text{av,RF}}}{P_{L,\text{IF}}} = \frac{|V_q|^2}{4\,|I_r|^2\,\text{Re}\{Z_e(\omega_q)\}\,\text{Re}\{Z_e(\omega_r)\}}$$

$$= \frac{1}{4\,|Y_{r,q}|^2\,\text{Re}\{Z_e(\omega_q)\}\,\text{Re}\{Z_e(\omega_r)\}} \tag{6.2.14}$$

Equation (6.2.14) illustrates a common misconception about the upconverting and downconverting properties of a mixer. In a purely resistive mixer (i.e., one that has no nonlinear junction capacitance), the upconversion and downconversion gains are equal because $|Y_{r,q}| = |Y_{q,r}|$. This is the case regardless of the shape of the LO waveform. However, the resistive mixer is not a reciprocal component unless $Y_{r,q} = Y_{q,r}$, a condition that exists only if the point $t = 0$ can be selected to make the LO current waveform an even function of time. The statement has often been made, however, that a purely resistive mixer has equal upconversion and downconversion gains only if its LO current waveform can be made an even function of time by a shift in the time axis. It is true that the mixer is reciprocal only under this condition, but even if the condition is not met, the upconversion and downconversion gains are still equal. However, in a mixer that has nonlinear junction capacitance, $|Y_{r,q}| \neq |Y_{q,r}|$ and the upconversion and downconversion gains are not equal.

A more elegant method of finding the conversion loss and input-output impedances is to generate an equivalent two-port representing the diode, as demonstrated in Example 9 of Section 3.2.2. Although Example 9 is concerned with the conversion matrix of a two-port, it is possible to generate the two-port conversion Y-matrix of (3.2.58) from the diode and impedance $Z_e(\omega)$ in the circuit of Figure 6.6. We wish to consider the RF source and IF load impedances apart from the two-port, so they can be varied independently in order to optimize performance. We can separate them by setting $Z_e(\omega_q) = 0$ and $Z_e(\omega_r) = 0$ and by placing the following source and load impedances in series with $v_s(t)$, to the left of the $v(t)$ node in Figure 6.6:

$$\begin{aligned} Z_s(\omega) &= Z_e(\omega) & \omega &= \omega_q \\ &= 0 & \omega &\neq \omega_q \end{aligned}$$

and:

$$\begin{aligned} Z_L(\omega) &= Z_e(\omega) & \omega &= \omega_r \\ &= 0 & \omega &\neq \omega_r \end{aligned}$$

Again, we form a conversion matrix as in (6.2.8), now designated $\mathbf{Y}'$, that represents the elements to the right of the $v(t)$ node, and that relates $\mathbf{I}$ to the mixing product vector for $v(t)$, $\mathbf{V}$:

$$\mathbf{I} = \mathbf{Y}'\,\mathbf{V} \tag{6.2.15}$$

V_q and V_r are the only nonzero components of $\mathbf{V}$. The voltage components of $\mathbf{V}$ are zero at all frequencies other than ω_q and ω_r, so we can decompose $\mathbf{Y}$ in (6.2.15) into a 2×2 matrix by removing all the rows and columns except those corresponding to ω_q and ω_r. The resulting 2×2 matrix has the form:

$$\begin{bmatrix} I_q \\ I_r \end{bmatrix} = \begin{bmatrix} y_{1,1} & y_{1,2} \\ y_{2,1} & y_{2,2} \end{bmatrix} \begin{bmatrix} V_q \\ V_r \end{bmatrix} \tag{6.2.16}$$

Figure 6.7 shows the two-port equivalent circuit of the mixer and the separate $Z_e(\omega_q)$ and $Z_e(\omega_r)$. In effect, we have treated $Z_e(\omega)$ as part of the diode at all mixing frequencies except the RF and IF; this arrangement allows us to select $Z_e(\omega_q)$ and $Z_e(\omega_r)$ independently in order to optimize the mixer's performance, with a given set of embedding impedances at all other mixing frequencies and a given set of LO waveforms. We can now replace the terms $Y_s(\omega_q)$ and $Y_L(\omega_r)$ in (3.2.60) through (3.2.71) by $1/Z_e(\omega_q)$ and $1/Z_e(\omega_r)$, respectively, and use those equations, with the two-port Y-parameters in (6.2.16), to find input-output impedances and conversion loss under both conjugate-matched and -unmatched conditions, as well as stability factors.

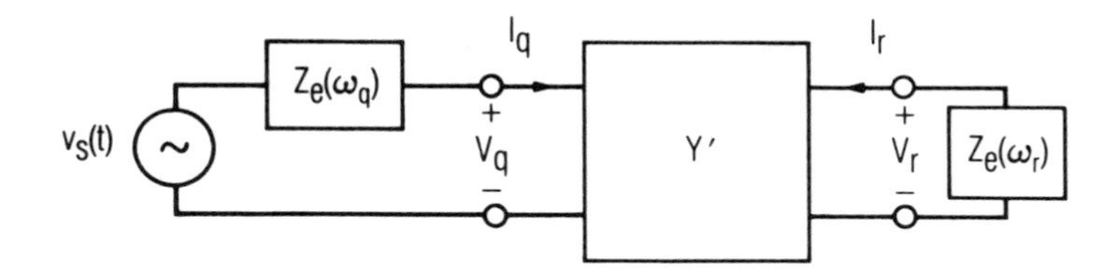

Figure 6.7 Two-port equivalent circuit of the single-diode mixer. The two-port represents the pumped diode and all embedding impedances except those at the RF and IF frequencies, ω_q and ω_r.

6.3 SINGLE-DIODE MIXER DESIGN

Diode mixer design is primarily a process of matching the pumped diode to the RF input and IF output, terminating the diode properly at LO harmonics and unwanted mixing frequencies (i.e., those other than the RF and IF), and adequately isolating the input, LO, and output ports. That isolation, and in some cases the termination, can be provided by filters, a balanced structure, or both; the choice depends on the frequencies and the intended application.

We begin by examining single-diode mixers. Single-diode mixers are important components, both intrinsically and because many types of balanced mixers consist of nothing more than a hybrid-coupled pair of single-diode mixers. A single-diode equivalent circuit can be found for any balanced mixer, so the process for designing a single-diode mixer is fundamental to that of all diode mixers.

6.3.1 Design Approach

In this section, we introduce a partially empirical process for the design of single-diode mixers. The process consists of estimating the source and load impedances and designing matching circuits that present those impedances to the diode and provide appropriate filtering functions. A design produced via this process usually results in respectable performance; however, the design can be further improved and better performance can be achieved if it is refined via the techniques described in Chapter 3 and Section 6.2.

Figure 6.8 shows the circuit of a single-diode mixer. It consists of a diode and three matching circuits that match the RF, IF, and LO terminations to the diode. It is clearly necessary in any mixer that these circuits do not interact, that each circuit does not affect the tuning of the other

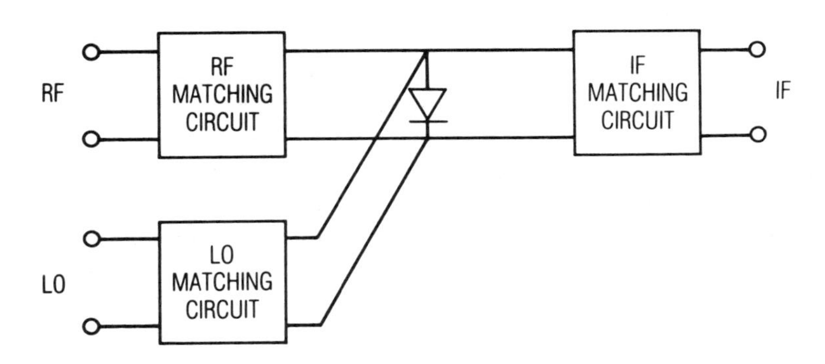

Figure 6.8 Single-diode mixer circuit.

two; they should also provide the appropriate termination (which, for lowest conversion loss, is always reactive) to the diode at LO harmonics and at unwanted mixing frequencies, especially the image frequency. These requirements, termination and noninteraction, imply a filtering function as well as a matching one. Filtering alone, however, is not the only requirement; the impedance presented to the diode over a wide range of frequencies must be controlled by the design.

Some of these requirements may be met automatically by the structure of the mixer. For example, in a mixer that has a waveguide input and a coaxial output, interaction between the RF and IF matching circuits at the IF frequency is usually obviated by the high cutoff frequency of the waveguide; the IF frequency, which is ordinarily below cutoff cannot propagate in the waveguide. The IF circuit, however, must still be designed to reject the RF and LO frequencies. Furthermore, in a single-diode waveguide mixer (and in many balanced mixers), the RF and LO signal are usually diplexed onto the same waveguide, so the RF and LO matching circuits are physcially the same structures.

Before we can design a mixer, we need to know the following:

1. the input impedance of the pumped diode at the RF frequency;
2. the output impedance of the pumped diode at the IF frequency;
3. the LO input impedance;
4. the diode's optimum termination at undesired mixing frequencies and LO harmonics;
5. the required port-to-port isolation.

If these five parameters are known, designing a mixer requires only designing the matching-filtering networks. We can design these networks in the conventional manner, by first making an initial design and then optimizing it via a general-purpose computer-aided circuit-design program. Unfortunately, in attempting to determine these parameters, we are faced with a problem and a paradox. The problem is that there is no simple way to determine the port impedances or optimum terminations without resorting to the formidable theory of Chapter 3. The paradox is that this theory is one of analysis not synthesis, so we must somehow generate a design before the theory can be invoked. The only way around this situation is to develop a simple approximate design approach, use it to obtain an initial design, and finally to check that design via theory. If the predicted performance is not good enough, the design can be modified and rechecked. One or two iterations of this process are usually adequate to optimize a design.

A long history of doing large-signal–small-signal mixer calculations allows us to make some generalizations about the embedding impedances at unwanted mixing frequencies and input-output impedances of the pumped diode. Exceptions to the following observations can be found, of course; nevertheless, they are generally valid for practical mixers.

1. The pumped diode can be modeled at the RF frequency as a resistor and capacitor in parallel. The resistor is usually in the range of 50–150 Ω, and the capacitor is between C_{j0} and $1.5C_{j0}$. The IF ouput impedance is usually between 75 and 150 Ω and may have a small susceptance, either capacitive or inductive, in parallel.
2. The large-signal LO input impedance is usually close to the small-signal RF input impedance, especially if the IF frequency is low.
3. The RF and IF input impedances are close to the high end of the impedance range given in observation 1 if the embedding impedances are open circuits at the unwanted mixing frequencies and LO harmonics and close to the low end of the range if the impedances are short circuits at those frequencies.
4. Open-circuit embedding impedances, especially at the image frequency, give the IF output impedance a surprisingly large reactive part that is often inductive.
5. Short-circuit embedding impedances generally result in a well behaved circuit. Open-circuit impedances can cause unstable operation.
6. Open-circuit embedding impedances give the lowest conversion loss, highest noise temperature, and worst intermodulation performance; short-circuit embedding impedances result in slightly greater conversion loss but generally in lower noise temperature and intermodulation performance.
7. The diode's termination at the image frequency has a strong effect on the conversion loss; however, the image termination that gives

optimum conversion loss does not necessarily give optimum noise temperature. Furthermore, it may result in very poor intermodulation performance.

8. There is an optimal combination of bias, source and load impedance, and LO level for a specific diode and set of embedding impedances. Diode bias and LO level affect input-output impedance strongly, and can be adjusted to help match the input and output if the choice of source and load impedances is not optimum; however, in this case conversion loss and noise temperature will not be optimum.

The source and load impedances of interest are those presented to the terminals of the intrinsic diode; that is, they do not include package or other parasitics. Such parastitics must be included as part of the matching circuits. These observations indicate that a short circuit is probably the best diode termination at unwanted mixing frequencies for the best overall mixer performance. In the case of a short circuit, the IF output impedance is usually close to 100 Ω and the RF and LO input impedances are approximately those of a 100-Ω resistor in parallel with C_{j0}.

The diode's termination at the image frequency is the most critical of all the terminations at unwanted mixing frequencies. In many mixers, the image frequency is close to the RF frequency, so the image termination is the same as the RF source impedance. However, if the IF frequency is relatively high, it is possible to use a filter (or the filtering properties of the RF matching circuit) to terminate the diode in a reactance at the image frequency. This practice, called *image enhancement,* can be used to improve the conversion efficiency of the mixer. Image enhancement must be used with care, however, because it is possible for an image-enhanced mixer to achieve only a modest improvement in conversion loss at the expense of poor intermodulation and noise performance.

It may be surprising that these generalizations apply to any diode. They are possible because the *I*/*V* characteristics of all diodes are fundamentally the same: they are exponentials. The only apparently significant difference in the *I*/*V* characteristics of different diodes is in their values of I_{sat}; however, even differences in I_{sat} are less important than they might appear, because of the current's exponential dependence on voltage. One property of an exponential *I*/*V* characteristic is that the same conductance waveform can be achieved with any value of I_{sat}, simply by appropriately scaling the bias voltage and LO power. The junction capacitance, of course, has a significant effect on the LO waveform, but even here most diodes have the same *C*/*V* dependence, and C_{j0} in most well-designed mixers is selected to have approximately the same reactance. This similarity in *I*/*V* and *C*/*V* characteristics between diodes is responsible for the fact that all well designed mixers have conversion losses within a few dB of each other, regardless of frequency, structure, or intended application.

Because matching circuits are inherently filter circuits, the filtering function may arise more or less automatically. If the inherent rejection of the matching circuit is not adequate, then an additional filter can be added. Because the cascading of filters does not always result in predictable rejection levels, a filter included to provide additional port isolation should be optimized as part of the matching circuit; it should not be designed as a separate unit and simply added to the circuit.

6.3.2 Diode Selection

The selection of an appropriate diode for a specific design is important in achieving good performance and minimizing cost. Most mixers intended for use at the lower microwave frequencies are not designed for high performance; instead, they are designed for well behaved operation (i.e., they have good port VSWRs, flat frequency response, and are stable) and for low-cost manufacture. In these mixers, the important trade-offs are such things as the cost of a purchasing a beam-lead diode compared to the cost of assembling a mixer that uses chips. The electrical parameters of the diode are more critical in millimeter-wave mixers, however, where maintaining good conversion and noise performance is much more difficult.

An initial consideration is the selection of a silicon or a GaAs device. GaAs devices can achieve significantly better conversion performance than silicon, but their advantage at low frequencies is minimal and their cost is much greater. Therefore, GaAs devices probably should be reserved for use at higher frequencies, especially millimeter-wave applications, where their superior performance justifies their cost. GaAs devices generally have higher breakdown voltages than silicon ones and better resistance to ionizing radiation; in addition, mixers using GaAs diodes usually exhibit a wider range of optimum conversion loss and noise figure with LO power variation. The greater I_{sat} of GaAs devices implies that these diodes usually must be operated with dc bias, or LO power requirements will be relatively high.

The diode parameters that most strongly affect mixer performance are C_{j0} and R_s. It is desirable to minimize both of these; however, in the design of a Schottky-barrier diode, reducing C_{j0} is accomplished most directly be reducing the diameter of the anode, inevitably increasing R_s. Other methods of increasing one of these parameters (e.g., adjusting the doping density) also improve that parameter at the expense of degrading the other and may reduce the cutoff frequency, f_c. Therefore, the mixer designer's task is to obtain a diode that has a high cutoff frequency and the most appropriate trade-off between junction capacitance and series resistance.

Several things must be considered in making this trade-off. One is the *conversion-loss degradation factor* δ, which accounts for the loss in the series resistance at the RF frequency. It is

$$\delta = 1 + \frac{R_s}{Z_s} + \frac{Z_s f_{\mathrm{RF}}^2}{R_s f_c^2} \tag{6.3.1}$$

where f_{RF} is the RF frequency, f_c is the cutoff frequency (R_s is evaluated at f_{RF}, not dc), and Z_s is the source impedance at the RF frequency and at the terminals of the resistive junction (i.e., the terminals of the current source $I_j(V)$ in Figure 6.1). In order to match the resistive junction, Z_s must, of course, be real. The cutoff frequency usually remains approximately constant with small changes in R_s, so f_c can be treated as constant and (6.3.1) can be minimized; the value of R_s that minimizes δ is

$$R_s = Z_s \frac{f_{\mathrm{RF}}}{f_c} \tag{6.3.2}$$

For example, a mixer operating at 20 GHz, using a diode that has a cutoff frequency of 1000 GHz and $Z_s \approx 100\ \Omega$, has an optimum R_s of 2 Ω. This is a very low series resistance, and in fact, a diode that has such a low R_s would have a large anode area and, consequently, a high value of C_{j0}, 0.08 pF. We noted earlier that second-order effects (associated with the nonuniform junction electric field near the edge of the anode) generally cause large-area diodes to have lower cutoff frequencies than small diodes. Consequently, such a large diode would not have optimum f_c, and the high junction capacitance might introduce matching difficulties. Thus, a 20 GHz mixer could probably have a higher R_s (perhaps 4 to 6 Ω) and lower C_{j0}, and still achieve close to the optimum δ. In general, considerations of matching and cutoff frequency dictate a lower C_{j0} than the optimum given by (6.3.2); for this reason, the best R_s/C_{j0} trade-off is usually to use a relatively large C_{j0}, consistent with matching limitations.

The parameters in the I/V characteristic are either of secondary importance or are not under the designer's control. Obviously, it is desirable to minimize η; η depends primarily upon the quality of the diode manufacturing process. The parameter I_{sat} in (6.1.1) is proportional to the junction area; in small millimeter-wave diodes, I_{sat} is very small, implying that the diode has very low total current for any given current density, or in other terms, high current density for moderate total current. The conductance waveform, however, is proportional to total junction current, not to current density; thus, in small diodes, current-density limitations in the junction may prevent the peak current from being great enough to achieve a high peak conductance. This situation complicates the design of millimeter-wave mixers by raising impedance levels and increasing conversion loss.

6.3.3 Mixer Design Example

The general principles of the previous section will be illustrated by an example, the design of a 94 GHz single-diode mixer having a 1.5 GHz IF. The mixer is shown in a cutaway view in Figure 6.9; this is one of the most commonly used structures for millimeter-wave mixers. The unpackaged diode is located in the waveguide, mounted in a Sharpless (or similar) mount, on the end of a coaxial IF filter-matching structure. The diode is mounted in a reduced-height section of waveguide, and a quarter-wave transformer is provided to match the reduced-height waveguide to the standard waveguide. Because both the LO and RF propagate in the waveguide, an external diplexer must be used to combine them.

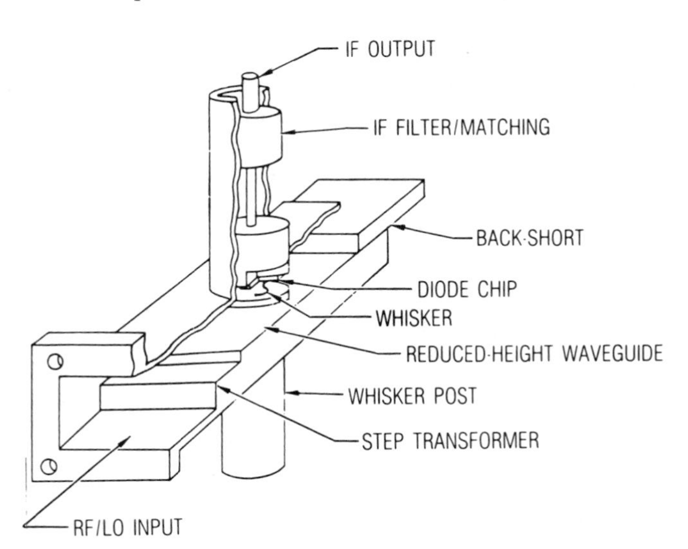

Figure 6.9 Cutaway view of the single-diode waveguide mixer. The diode and IF filter are also partially cut away to show the whisker.

The diode selected for the mixer is a 2.0-micron GaAs dot-matrix device. It has the following parameters:

$R_s = 7.0\ \Omega$
$C_{j0} = 0.008$ pF
$\eta = 1.15$
$I_{\text{sat}} = 2.5 \cdot 10^{-13}$A
$\gamma = 0.5$
$\phi = 0.75\ V$

Its cutoff frequency is 2840 GHz, a relatively high value. We begin by modeling the pumped diode as a 100-Ω resistor in parallel with a capacitor equal to $1.5C_{j0}$, or 0.012 pF. The diode is also in series with the contacting whisker, so the whisker's inductance must be included in the model. Finally, the tuning short in the waveguide behind the diode must be included and the source impedance, which is just the waveguide's characteristic impedance, determined via the power-voltage definition. The RF/LO model of the waveguide-mounted diode is shown in Figure 6.10.

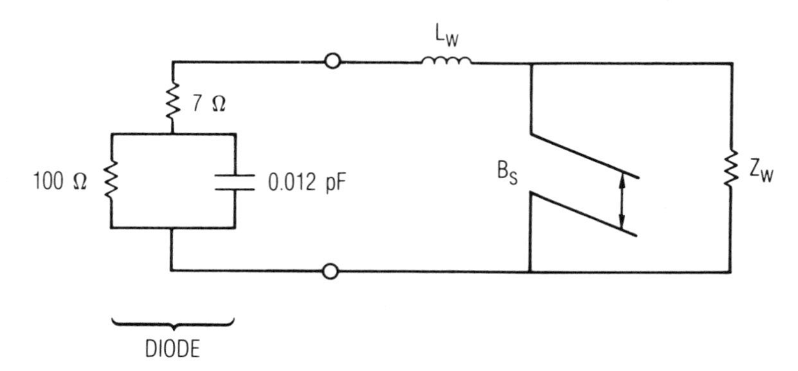

Figure 6.10 Equivalent circuit of the waveguide and pumped diode: Z_w is the waveguide impedance, B_s is the susceptance of the tuning short, and L_w is the whisker inductance.

The IF filter, on which the diode is mounted, is not included in Figure 6.10 because it will be designed to have an input impedance of nearly zero at the RF and LO frequencies; thus, at these frequencies the diode's cathode is effectively connected to the waveguide's topwall. The IF filter, however, is designed to match the standard 50-Ω load to the diode's IF output impedance; this impedance is expected to be real and approximately 100 Ω. The zero input impedance of the IF filter at 94 GHz implies high

RF/LO-to-IF isolation, and the standard 94 GHz waveguide (type WR-10), which has a cutoff frequency of 59 GHz, effectively provides IF-to-RF/LO isolation.

It is almost always necessary to mount a diode in a waveguide whose height is less than the standard value. A reduced-height waveguide has a lower impedance than a standard-height waveguide and allows the use of a whisker that has reasonably low inductance; it is therefore more conducive to effective matching. In practice, the waveguide impedance should be in the range of 150–200 Ω, a value that usually requires reducing the height to one-quarter to one-half its standard value. We are left with two degrees of freedom, the whisker inductance, L_w, and the short susceptance, B_s, to match the diode to a waveguide impedance in this range. A few minutes with a Smith chart gives a whisker reactance of 140 Ω, implying 0.23 nH inductance, and a normalized short susceptance of 1.40, giving a backshort position of 0.402 λ_g behind the diode. The waveguide impedance Z_w is 200 Ω.

The waveguide's dimensions are found from the relation for waveguide impedance (the power-voltage definition):

$$Z_w = 754 \frac{b}{a} \frac{\lambda_g}{\lambda_0} \tag{6.3.3}$$

where λ_0 is the free-space wavelength, λ_g is the guide wavelength, a is the waveguide height, and b is the width. The ratio λ_g/λ_0 is

$$\frac{\lambda_g}{\lambda_0} = \left(1 - \frac{f_c^2}{f^2}\right)^{-1/2} \tag{6.3.4}$$

where f_c is the waveguide's cutoff frequency. From (6.3.3) and (6.3.4) we find that $\lambda_g/\lambda_0 = 1.285$ and, when $Z_w = 200$, $b/a = 0.20$. The width, a, of a WR-10 waveguide is 0.100 in, so the height, b, is 0.020 in. The whisker inductance can be estimated from Schelkunoff's equation for the inductance of a straight wire across a waveguide:

$$L = 2 \cdot 10^{-9}\, d \ln\left(\frac{2a}{\pi r}\right) \tag{6.3.5}$$

where d is the length of the wire in cm, r is the radius of the wire, and L is in henries (H). A straight wire across this waveguide would have approximately 0.4 nH inductance; the inductance of the whisker can be reduced from 0.4 nH to the desired 0.23 nH by making it shorter than 0.020 in. and allowing the diode and the whisker post to protrude into the waveguide. Invariably, some experimentation with the whisker length in millimeter-wave mixers is necessary in order to optimize performance.

We now turn our attention to the IF filter. Because the IF and RF/LO frequencies are well separated, the design is relatively easy. A good approach is to use a filter structure to provide the desired short-circuit termination at 94 GHz, and a matching circuit at the output end of the filter to provide the impedance transformation. After the usual trial-and-error process, the design shown in Figure 6.11 was generated. The angle of the input reflection coefficient of this filter is 180 ± 5 degrees between 90 and 100 GHz, and if the diode's impedance is 100 Ω, its IF VSWR is less than 1.3 between 1.25 GHz and 1.75 GHz. The filter's loss at 94 GHz is greater than 35 dB, giving good isolation. A particularly fastidious designer could optimize the structure to eliminate its spurious passband near 190 GHz, to eliminate from the IF output the LO harmonics generated in the diode.

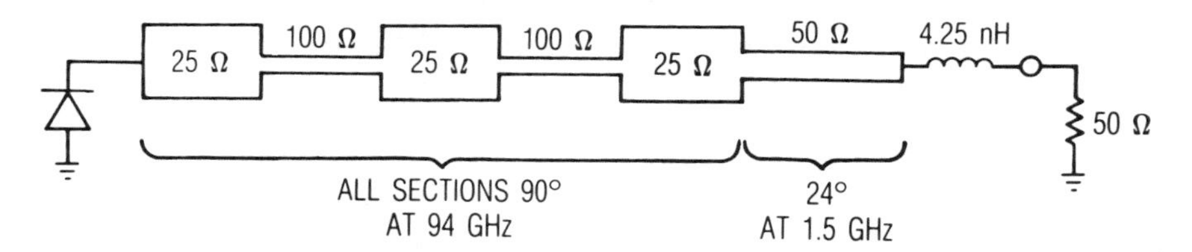

Figure 6.11 IF filter-matching circuit for the 94-GHz mixer.

In order to check the design, we now should calculate the expected conversion loss via the techniques described in Chapter 3. First, we must obtain the embedding impedances at the unwanted, high-order mixing frequencies; these can be estimated from the diode-mount equivalent circuit in Figure 6.10. A problem in using Figure 6.10 is that the waveguide is multimoded at the high-order frequencies and LO harmonics, so the embedding impedances at these frequencies cannot be calculated very accurately. Nevertheless, the embedding impedances calculated from this circuit are probably adequate for several reasons: (1) Although the waveguide is multimoded at frequencies near the second LO harmonic, the symmetry of the structure is such that it will not couple effectively to such modes; (2) at higher harmonics, the whisker's inductance dominates the embedding impedances, making them effectively open circuits regardless of the rest of the waveguide circuit; and (3) mixer performance is less sensitive to the embedding impedances at higher-order mixing frequencies, where the embedding impedances are less easily determined, than at the lower-order frequencies, where the impedances are well known.

Conversion loss is calculated via a program similar to that in Reference 2.1. This calculation predicts a conversion loss of 4.6 dB, an RF input VSWR of 1.6, and an IF VSWR of 1.3. The mixer requires 3 dBm of LO power and 0.3 V of dc bias, the rectified diode current is 3 mA, and the LO efficiency is 93 percent. The SSB noise temperature, in the absence of hot-electron noise and circuit losses, is 210 K. At 94 GHz, we should expect close to 1 dB of parasitic circuit loss in addition to that of an LO diplexer, so the conversion loss of the optimized mixer may be around 5.5 dB. This result agrees well with the published conversion losses of experimental mixers at this frequency.

6.4 BALANCED MIXERS

Most diode mixers used at microwave and the lower millimeter-wave frequencies are balanced. The advantages of balanced mixers over single-diode mixers are (1) the inherent rejection of spurious responses and intermodulation products (Section 1.3); (2) LO/RF and, in some cases, LO/RF-to-IF isolation without the need for filters; and (3) rejection of AM noise in the LO. The disadvantages are (1) greater LO power requirements; (2) generally higher noise figure and conversion loss (because of difficulties in biasing and matching individual diodes); and (3) few types of balanced mixers that exhibit all these characteristics. Commercially available balanced mixers are frequently small, lightweight, inexpensive, broadband components; in many applications, the need for a small, inexpensive mixer that has good spurious-response properties outweighs the need for low conversion loss. Furthermore, in cases where the LO and RF bands overlap, balanced mixers are essential because it is impossible to separate the LO from the RF via filters.

6.4.1 Singly Balanced Mixers

A singly balanced mixer consists of two mixers combined by either a 180-degree or a 90-degree hybrid. The LO and RF are applied to one pair of mutually isolated ports, and single-diode mixers are connected to the other pair of ports. The diodes in the two mixers must be connected to the ports in such a way that their polarities are opposite: If the cathode of one diode is connected to one port, the anode of the other must be connected to its port. The IF outputs of the individual mixers can be connected via another hybrid, or more commonly, they can be connected directly in parallel. The properties of hybrid-connected nonlinear elements are described in detail in Chapter 5.

Figure 6.12 shows a singly balanced mixer that uses a 180-degree hybrid. The RF and LO are connected to one pair of mutually isolated ports; the single-diode mixers, represented by diode symbols in the figure, are connected to the other pair. At frequencies low enough that the diode's parasitics do not severely increase conversion loss, the "mixers" may be nothing more than individual diodes connected directly to the ports of a transformer or ring hybrid; in microstrip circuits, at higher frequencies, some tuning may be used. Balanced millimeter-wave mixers are often realized by two complete single-diode waveguide mixers at the ports of a waveguide hybrid. The IF ports of the mixers are shown connected in parallel, and they have a common low-pass filter; this filter isolates the IF from the RF and LO and grounds the common terminal of the diodes at the RF and LO frequencies.

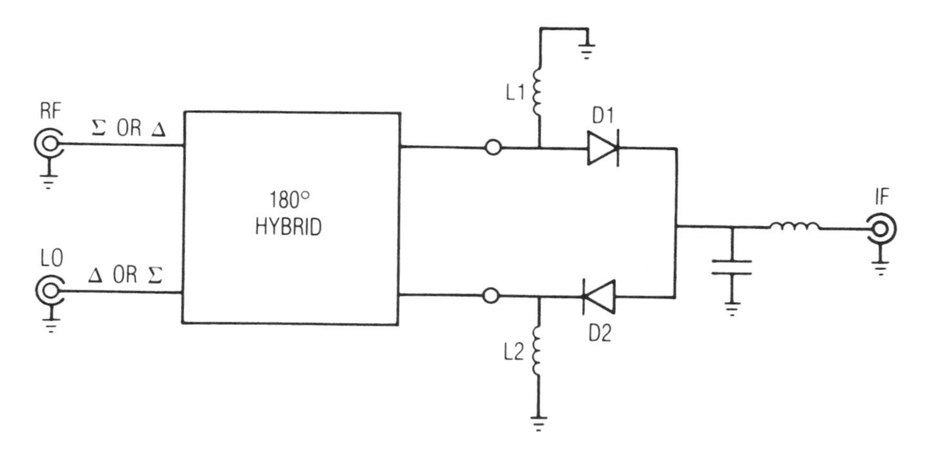

Figure 6.12 A singly balanced mixer: Diodes *D*1 and *D*2 can be unmatched diodes or complete individual single-diode mixers. The mixer can be configured with either the sigma or delta port as the RF; the other is the LO.

The anode of *D*1 and the cathode of *D*2 in Figure 6.12 must be bypassed to ground at dc and the IF frequency. In a waveguide mixer, this bypassing is provided inherently by the structure, but in a microstrip mixer, it must be provided in some specific way. In Figure 6.12, the inductors *L*1 and *L*2 realize the so-called IF return: they ground their respective ends of the diodes at the IF frequency and dc. Dc bias can be supplied to both diodes by means of a voltage source in series with either of these inductors.

Because the IF ports are connected in parallel, the impedance presented to each single-diode mixer at the IF frequency is twice that of the actual IF load. Section 5.2 explains this point in detail; however, we can see that this is the case by thinking of the IF load as two loads in parallel, each having twice the impedance of the actual load. Because of the symmetry of the circuit, the two loads can be separated so that each is connected to only one mixer.

We can calculate the conversion loss of the balanced mixer by analyzing the individual single-diode mixers, each having the doubled IF load impedance, and from Chapter 5 the conversion loss of the balanced mixer must be the same as that of the individual single-diode mixers. In Chapter 5, the single-diode circuit that is analyzed to determine the performance of the balanced mixer is called the *single-diode equivalent circuit* of the mixer; a single-diode equivalent circuit can be found for any balanced mixer. Thus, the mixer analysis and design approaches of Chapter 3 and the previous section are applicable to balanced mixers as well as to single-diode structures. A theoretical constraint on the process of forming a single-diode equivalent circuit is that the circuit must be perfectly balanced: the individual mixers in the balanced circuit must be identical and the hybrid must be ideal. In practice, however, it is necessary only that these requirements be met approximately; Section 5.1.3 describes the effects of imperfect balance.

A singly balanced mixer can also be realized by replacing the 180-degree hybrid in Figure 6.12 by a quadrature hybrid. The individual mixers are the same as those used with the 180-degree hybrid, and as before, they are connected to mutually isolated ports. The bypassing and IF filtering are performed in the same manner as in the 180-degree hybrid mixer.

The main difference in operating characteristics between the 180-degree and quadrature hybrid mixers are in the ports' VSWRs, isolation, and spurious response properties. In the 180-degree mixer, the input VSWR at the LO and RF ports is dominated by the VSWRs of the individual mixers, and the RF-LO isolation is dominated by the isolation of the hybrid. The quadrature hybrid, however, operates in a very different manner (Section 5.1.3). LO power reflected from the individual mixers does not return to the LO port but instead exits the RF port; similarly, reflected RF power exits the LO port. The LO-RF isolation is therefore equal to the input return loss of the individual mixers at the LO and RF frequencies, respectively; the port isolation of the quadrature-hybrid mixer depends primarily on the input VSWRs of the two individual mixers not on the isolation of the hybrid itself. As long as the LO and RF source VSWRs are good, the mixer's LO and RF input VSWRs are also very good. However, if the RF port termination has a poor VSWR at the LO frequency, the circuit's balance can be upset and the LO pumping of the individual mixers can be unequal; similarly, a poor LO port termination at the RF frequency can upset RF balance.

The spurious-response properties of the 180-degree and quadrature-hybrid mixers are also different. If the sigma port of the hybrid is used as the LO port, the 180-degree hybrid mixer rejects mixing products that involve even harmonics of the LO; if the sigma port is the RF port, even harmonics of the RF that mix with any harmonics of the LO are rejected. The quadrature-hybrid mixer, however, does not reject the even harmonics of one signal (either the RF or LO) mixing with the odd harmonics of the other. Both types of mixers, however, reject the even LO harmonics that mix with the even RF harmonics.

It is worth noting that although many seemingly different types of singly balanced mixers have been developed, all are fundamentally only realizations of either the 180-degree or quadrature structure. An example is the crossbar mixer shown in Figure 6.13(a), which, except for its use of two diodes, does not appear at first glance to be a balanced mixer, but is, in fact, a 180-degree hybrid mixer. In the crossbar mixer, two diodes are connected in series across the RF waveguide, and the LO is coupled to the diodes via a metallic strip (the *crossbar*) that acts as a coupling probe in the LO waveguide. The probe is also used for the IF output. The orientation of the probe and the RF and LO waveguides is such that the probe does not couple the LO and RF waveguides.

That the crossbar mixer is a type of 180-degree hybrid balanced mixer is evident from the polarities of the LO and RF voltages at the diode, as shown in Figure 6.13(b). The RF voltage applied to the diodes has the same phase as if the RF signal had been applied to the delta port of a 180-degree hybrid, and the LO voltage pumps the diodes out of phase, as if it had been applied to the sigma port. Except there being two diodes in series instead of one, the same matching considerations as in the single-diode mixer apply to the crossbar mixer, and the same methods of optimizing the diode terminations (especially, image enhancement) can be readily applied.

The main advantage singly balanced mixers have over other types of mixers is that they have many of the desirable properties of balanced mixers, yet can be treated in many ways like single-diode mixers: They can have carefully designed matching circuits and dc bias, giving them conversion performance nearly as good as that of single-diode mixers. The structure of doubly balanced mixers, described in the next section, does not allow for practical matching circuits and dc bias. Accordingly, the conversion and noise performance of doubly balanced mixers is not as good, and they are used primarily in applications where their superior

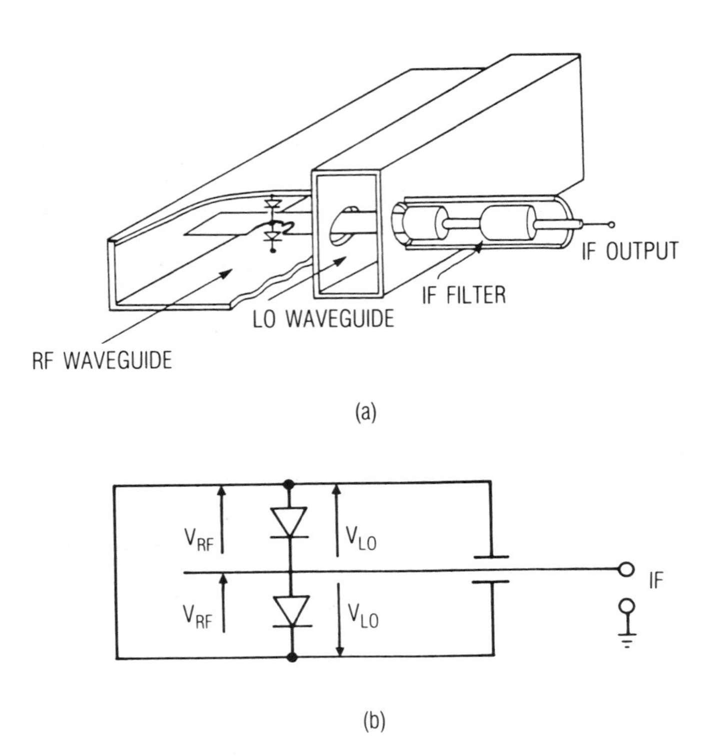

Figure 6.13 (a) The crossbar mixer; (b) polarities of the LO and RF voltages at the diodes.

spurious-response properties are more important than conversion performance.

6.4.2 Doubly Balanced Mixers

The two most common types of doubly balanced mixers are the ring mixer and the star mixer. The ring mixer is more amenable to low-frequency applications, in which transformers can be used. The star mixer is used primarily in microwave applications, as it is more amenable to operation with microwave baluns. There is no significant difference in the properties or performance of both mixer types.

The classic ring mixer circuit, sometimes called a *ring modulator,* is shown in Figure 6.14. The circuit consists of a ring of four diodes, designated $D1$ through $D4$, and two transformers, $T1$ and $T2$. The transformers are identical to the transformer hybrid described in Section 5.1.2 and are

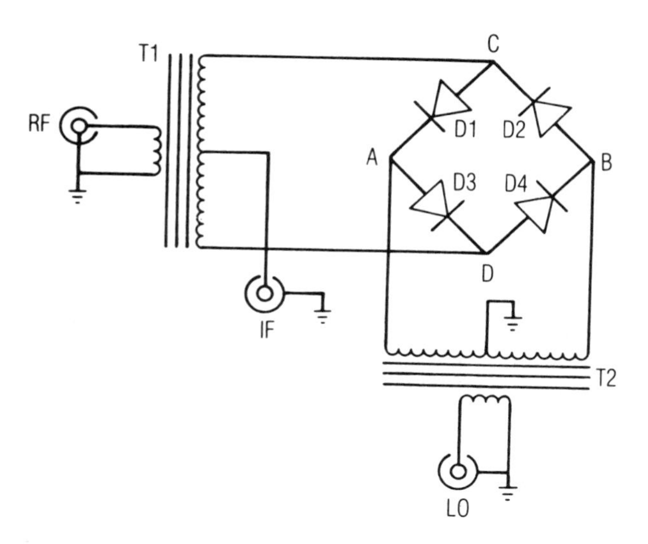

Figure 6.14 The ring mixer: The IF port can be the center tap of either transformer; however, LO-to-IF isolation is usually better if the RF transformer's center tap is used.

often realized as separate trifilar windings on toroidal cores (one winding is used as the primary one, and the other two are connected in series to form the secondary ones). The secondary windings of these transformers are connected to the nodes of the diode ring, labeled A through D.

The operation of the mixer can be described very simply if the diodes are treated as ideal LO-driven switches. When the polarity of the LO voltage is such that the right side of the secondary winding of $T2$ is positive, diodes $D1$ and $D2$ are turned on and $D3$ and $D4$ are turned off. During this half of the LO cycle, $D1$ and $D2$ short-circuit $T2$, so that node C is connected to ground through the center tap of $T2$. The upper half of $T1$ is thus connected through these diodes to the IF port, and the RF port is momentarily connected to the IF port. When the LO voltage reverses, $D3$ and $D4$ are turned on and $D1$ and $D2$ are turned off. Then, the lower half of $T1$ is connected to the IF, so the RF is again connected to the IF but with its polarity now reversed. The mixer therefore acts as a polarity-reversing switch, connecting the RF port to the IF but reversing its polarity every half LO cycle.

The IF voltage is, therefore,

$$v_{\mathrm{IF}}(t) = s(t)\, v_{\mathrm{RF}}(t) \tag{6.4.1}$$

or

$$v_{IF}(t) = \sum_{\substack{n=1 \\ (n \text{odd})}}^{\infty} b_n \sin(n\omega_p t)\, v_{RF}(t) \tag{6.4.2}$$

where $s(t)$, the switching waveform in Figure 6.15, has been represented as a Fourier series. Downconversion occurs via the product of the fundamental-frequency component of $s(t)$ and the sinusoidal $v_{RF}(t)$. Because $s(t)$ is a symmetrical square wave, it has no dc component in its Fourier-series representation, so $v_{IF}(t)$ can have no RF-frequency component; consequently the RF and IF are isolated, even though no filters are used. The waveform $s(t)$ also has no even-harmonic components, so there can be no spurious responses associated with even LO harmonics; because of the symmetry of the circuit, spurious responses associated with the even harmonics of the RF must also be rejected. Furthermore, at all times either $D1$ and $D2$ are shorted or $D3$ and $D4$ are shorted; this short-circuit prevents the coupling of RF voltage to the LO port or LO voltage to the RF port. Even if the diodes are not ideal switches (i.e., if the diodes are not perfect short circuits when turned on), good RF-to-LO isolation is guaranteed by the fact that the instantaneous RF voltage at A and B must always be the same as long as the voltage drops in the pairs ($D1$, $D2$) and ($D3$, $D4$) are identical. If the voltage drops across these diodes are identical, then the symmetry of the circuit causes nodes C and D, the terminals of the $T1$ secondary, to be virtual ground points for the LO. The RF transformer secondary is connected to these LO ground points, so the LO/RF isolation must be very great.

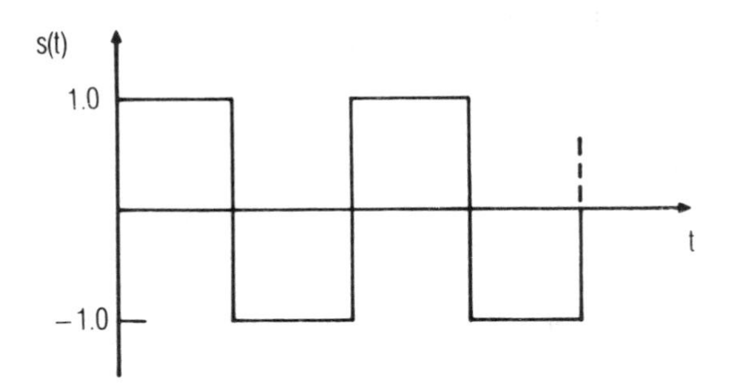

Figure 6.15 The switching waveform of the ideal ring mixer, which also is valid for the star mixer.

Ring mixers are used most commonly in broadband applications at frequencies up to a few hundred MHz. Their bandwidths are limited primarily by the bandwidths of the RF and LO transformers; although they are untuned, the diodes rarely limit performance in this frequency range. With the use of more creative transformer designs, ring mixers are occasionally produced at frequencies in the GHz range. These mixers are sometimes used as modulators, phase detectors, and even voltage-controlled attenuators as well as mixers; they are very versatile components.

In a star mixer, one terminal of each of four diodes is connected to a common node; this node is used as the IF terminal. Figure 6.16 shows a version of a star mixer that uses a high-frequency balun and, therefore, is useful at microwave frequencies. This balun, a variation of a structure known as the *Marchand balun*, is described in detail in References 2.1 and 6.7. The Marchand balun is remarkably broadband and is sometimes used in mixers having decade bandwidths. The balun is equivalent to the transformer circuit shown in Figure 6.17, which, like the ring mixer, uses two transformers. In this case, however, the transformer windings are shown separately; *T2a* and *T2b* represent the secondaries of one transformer, while *T1a* and *T1b* represent the secondaries of another.

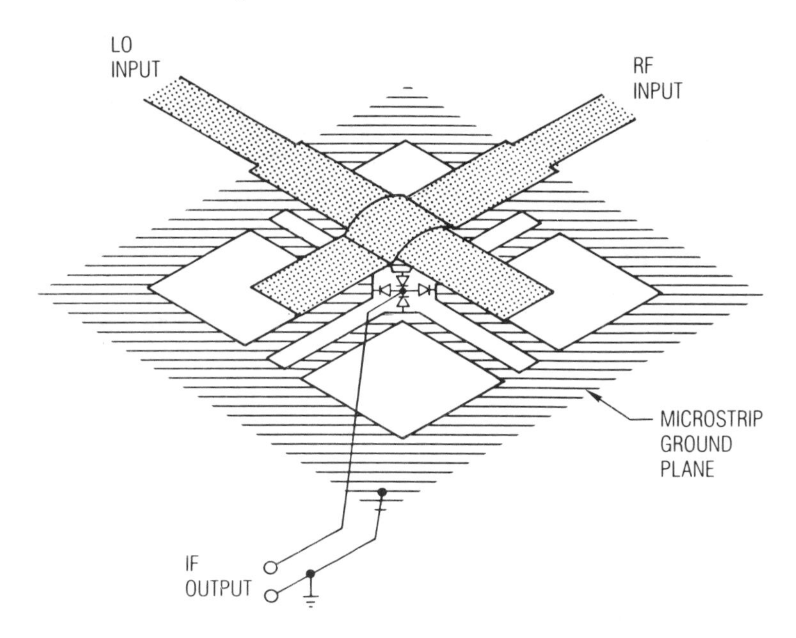

Figure 6.16 A doubly balanced star mixer for use at microwave frequencies; the mixer must be used in a housing that comprises the outer shield of the modified Marchand balun.

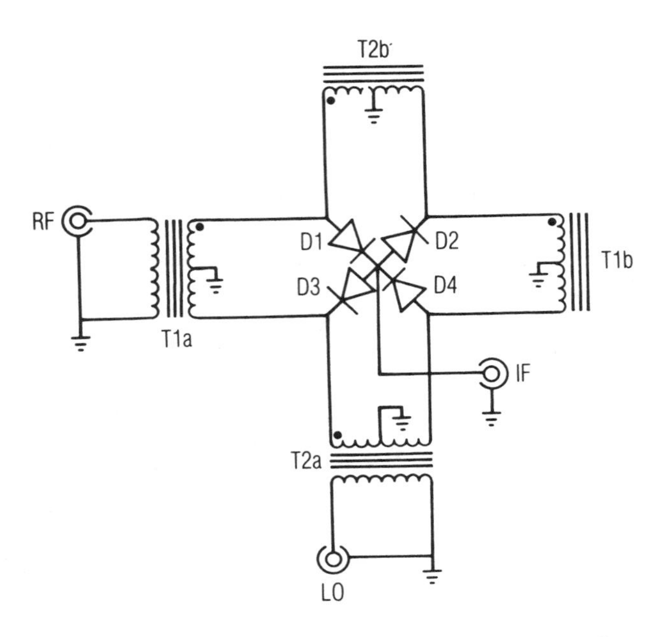

Figure 6.17 A transformer equivalent circuit of the star mixer.

This mixer, like the ring mixer, operates as a polarity-reversing switch. When the dotted sides of the LO transformer secondaries *T2a* and *T2b* are positive, *D*1 and *D*2 are turned on, *D*3 and *D*4 are turned off, and the dotted sides of *T*1*a* and *T*1*b* are connected to the IF port. The RF port is thus connected to the IF through the transformer *T*1 and the diodes. When the LO polarity reverses, *D*1 and *D*2 are off and *D*2 and *D*3 are on; then, the undotted side of both *T*1 windings is connected to the IF port and the RF is again connected to the IF but its polarity is reversed. The RF polarity is therefore applied to the IF port, but its polarity is reversed periodically at the LO frequency. Because the star mixer operates via the same principle as the ring mixer, it should be no surprise that the spurious-response properties of the star mixer are the same as those of the ring mixer.

REFERENCES

[6.1] E.L. Kollberg, *Microwave and Millimeter-Wave Mixers,* IEEE Press, New York, 1984.

[6.2] W.M. Sharpless, "Gallium Arsenide Point-Contact Diodes," *IRE Trans. Microwave Theory Tech.*, Vol. 9, 1961, p. 6.

[6.3] A.G. Cardiasmenos, "New Diodes Cut the Cost of MMW Mixers," *Microwaves*, Sept. 1978, p. 78.
[6.4] W.C. Ballamy and A.Y. Cho, "Planar Isolated GaAs Devices Produced by Molecular Beam Epitaxy," *IEEE Trans. Electron Devices,* Vol. ED-23, 1976, p. 481.
[6.5] B.J. Clifton, "Schottky Diode Receivers for Operation in the 100–1000 GHz Region," *Radio Electron. Eng.*, Vol. 49, 1979, p. 333.
[6.6] W.L. Bishop, K. McKinney, R.J. Mattauch, T.W. Crowe, and G. Green, "A Novel Whiskerless Schottky Diode for Millimeter and Submillimeter Wave Applications," *IEEE MTT-S Int. Microwave Symp. Digest,* 1987, p. 607.
[6.7] N. Marchand, "Transmission Line Conversion Transformers," *Electronics,* Vol. 17, No. 12, 1979, p. 52.

CHAPTER 7

DIODE FREQUENCY MULTIPLIERS

A large part of the electronics in any microwave communication system is devoted to generating signals of specific frequencies. Often signals of high stability and low noise are needed; these are sometimes obtained by generating harmonics from a very stable low-frequency source, such as a crystal oscillator. For better or worse, harmonic generation is one of the things that nonlinear circuits do best, so it should be no surprise that varactor, step-recovery, and Schottky-barrier diodes are widely employed in frequency-generating electronic systems.

Diode circuits that use varactors or *step-recovery diodes* (SRDs) are often employed as harmonic generators at microwave frequencies. These are reactive multipliers that make use of the diode's nonlinear capacitance characteristic. Varactors are used primarily to multiply microwave signals to low harmonics (i.e., rarely over four times the source frequency); in contrast, SRDs are used to multiply signals in the UHF or low microwave range to very high harmonics. Both components are inherently narrowband and, when properly designed, have good efficiency and low noise.

Resistive diodes (invariably Schottky-barrier diodes) are sometimes used in low-order frequency multipliers. Resistive multipliers are less efficient than reactive multipliers, but they can be made very broadband. Furthermore, it is usually easier to develop a resistive multiplier than a reactive one; reactive multipliers are very sensitive to slight mistuning and, therefore, have a well deserved reputation for being difficult to optimize. In contrast, resistive multipliers are relatively easy to tune and are not nearly as sensitive.

7.1 VARACTOR FREQUENCY MULTIPLIERS

7.1.1 Noise Considerations

In the past, it was common to use varactor frequency multipliers to generate moderate to high levels of RF power. It is now possible to generate RF power via solid-state sources (IMPATT and Gunn devices) and power amplifiers (GaAs MESFETs); these generally have greater efficiency, fewer components, and greater bandwidth than varactor multipliers. Furthermore, GaAs MESFET frequency multipliers, described in Chapter 10, are capable of greater efficiency and bandwidth (for low-order multiplication) than diode multipliers. We might wonder why varactor frequency multipliers are still used at all.

The major advantage the varactor frequency multiplier has over other types of multipliers is that, because it is a reactive device, it generates very little noise. This property is particularly valuable in applications where low phase noise is desired, as in local oscillator (LO) sources for radar applications, phase-modulated communications systems, and low-noise millimeter-wave systems. The only noise source in a varactor multiplier is the thermal noise of its series resistance and of its circuit losses, both of which are very small in a well designed device and circuit. As a result, the phase noise of a varactor multiplier is usually that of an ideal frequency multiplier. Frequency multipliers using Schottky-barrier varactors can achieve high efficiency and low noise at output frequencies of several hundred GHz; such multipliers, driven by Gunn sources, can generate adequate LO power for single-diode mixers and can have very low AM noise levels. Reference 7.1 gives an example of one such multiplier.

Even if a multiplier introduces no phase noise of its own, the process of frequency multiplication (even by an ideal, noiseless multiplier) inevitably reduces the *carrier-to-noise ratio* (CNR) of an input signal that has phase noise. The reason for this unfortunate characteristic is that a frequency multiplier is, in fact, a phase multiplier, so it multiplies the phase deviations as well as the frequency of the input signal. The minimum CNR degradation, in dB, in any frequency multiplier (including an ideal, noiseless one), is

$$\Delta \text{CNR} = 20 \log_{10}(n) \tag{7.1.1}$$

when n is the multiplication factor. Thus, a frequency doubler ($n = 2$) degrades the CNR of the input signal by at least 6 dB; a quadrupler degrades the CNR by at least 12 dB. If the multiplier is noisy, it can add even more phase noise to the input signal, and ΔCNR can be even greater.

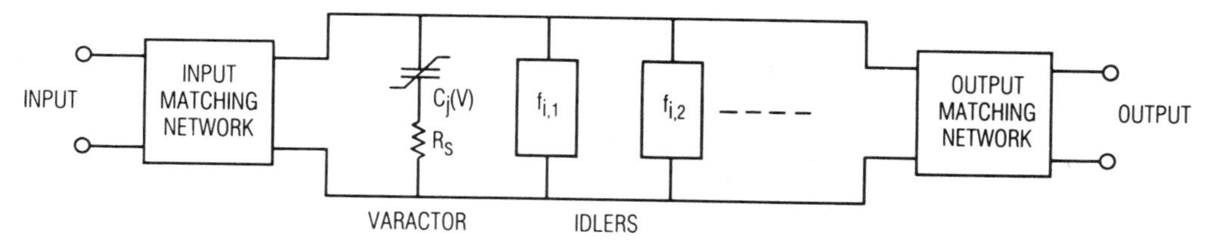

Figure 7.1 A varactor frequency multiplier. The blocks marked $f_{i,n}$ are the idlers at the nth harmonic.

In addition to phase noise, amplitude-modulated (AM) noise is of concern in many types of systems. AM noise in the LO can be an especially serious problem in low-noise receivers: if the LO signal has AM noise sidebands at the RF frequency, that noise can be downconverted to the IF frequency, significantly increasing the noise temperature of the receiver.

Solid-state or vacuum devices used in frequency sources or in multiplier chains are usually driven into saturation, and their limiting effects usually remove much of the signal's AM noise. However, these effects do not completely eliminate AM noise, so sources that are inherently noisy (e.g., IMPATT devices, klystron tubes, and backward-wave oscillators) often generate signals that have high AM noise levels in spite of any limiting that may occur. Using an amplifier to increase the power level of a signal, even if the amplifier is driven into saturation, also can introduce a large amount of AM noise. The use of balanced mixers or narrowband filters at the mixer's LO port can do much to reduce the effects of such noise; however, in some cases (e.g., in many types of low-noise millimeter-wave receivers), balanced mixers or filtering may not be possible. An LO system consisting of a Gunn source followed by a varactor multiplier usually has negligible AM noise and is therefore a preferred configuration for millimeter-wave systems.

7.1.2 Power Relations and Efficiency Limitations

Manley and Rowe (Reference 7.2) developed a set of general relations among the real powers at all mixing frequencies in a nonlinear capacitor. These equations, often called the *Manley-Rowe relations*, are valid for any nonlinear capacitor driven by one or two noncommensurate frequencies. The relations are remarkable in that they do not depend directly upon the capacitor's *Q*/*V* characteristic or the levels of the applied excitations (they do require, however, that currents and voltages exist at at least some of these frequencies; therefore, the nature of the capacitor's

Q/V characteristic and the circuit's embedding impedances must cause those voltage and current components to exist). The Manley-Rowe relations have been applied to parametric amplifiers and upconverters as well as to varactor frequency multipliers, and they can be used to establish limits to the gain or loss of such components.

The two Manley-Rowe relations are as follows:

$$\sum_{m=0}^{\infty} \sum_{n=-\infty}^{\infty} \frac{mP_{m,n}}{mf_1 + nf_2} = 0 \tag{7.1.2}$$

$$\sum_{n=0}^{\infty} \sum_{m=-\infty}^{\infty} \frac{nP_{m,n}}{mf_1 + nf_2} = 0 \tag{7.1.3}$$

where f_1 and f_2 are the frequencies of the two excitation signals and $P_{m,n}$ is the average real power into the capacitor at the frequency $|mf_1 + nf_2|$ (note that in this case the input powers at the excitation frequencies, $P_{1,0}$ and $P_{0,1}$, are the powers absorbed by the network, not available powers from the sources). These relations can be derived from the sole considerations that the capacitor is lossless (i.e., in the case of a varactor, the series resistance is zero) and that the capacitor's Q/V characteristic is single-valued.

In a frequency multiplier, there is only a single excitation, f_1, so $f_2 = 0$, the summation over n can be elimated, and all the terms of (7.1.3) become zero. Equation (7.1.2) becomes:

$$\sum_{m=0}^{\infty} P_m = 0 \tag{7.1.4}$$

where P_m is the power in the diode's reactive junction (i.e., not including power in the series resistance) at the frequency mf_1. Equation (7.1.4) states that all the input power must be converted to output power at the harmonics of f_1, none can be dissipated in the reactive junction (note that (7.1.4) does not say *where* the output power must be dissipated; in practice, much of it may be dissipated in circuit losses or in the series resistance). In an Mth-harmonic multiplier, the highest possible value of P_M occurs when only P_1 and P_M are not zero; then $P_M = -P_1$, that is, the output power at P_M is equal to the input power at P_1 and, if the input power equals the available power of the source, the multiplier has 100 percent efficiency.

For this optimum efficiency to be achieved, there must be no real power in the circuit at any of the unwanted harmonic frequencies. This condition is guaranteed when the diode's junction is terminated in a pure reactance at all harmonics other than that desired. In practice, however, it is not possible to present a pure reactance to the junction because of the diode's series resistance; this resistance is always in series with the terminating impendance and thus dissipates power at all harmonics. In order to eliminate power dissipation in the series resistance, one might be tempted to design the embedding circuit to open-circuit the diode at all unwanted harmonics; in this case, the unwanted harmonic currents in the series resistance would be zero and no power would be dissipated. The next best approach would seem to be to short-circuit the diode at all unwanted harmonics; a short circuit would not eliminate the dissipation in the series resistance but would prevent harmonic power dissipation in the output network.

It happens, however, that in diodes having *C/V* characteristics close to that of the ideal Schottky or *pn* junction and in frequency multipliers that generate harmonics greater than the second, short-circuit terminations at unwanted harmonics are preferred. The reason for this counterintuitive situation is that the diode voltage, as a function of charge, has a square-law characteristic and therefore cannot generate voltage components beyond the second harmonic unless harmonic current components also exist. Let us suppose that the diode is driven at the excitation frequency by an ideal current source and thus has only open-circuit harmonic terminations. The *Q/V* characteristic of an ideal, uniformly doped junction, given by (2.3.3), is

$$Q(V) = -2C_{j0}\,\phi\,(1 - v/\phi)^{1/2} \tag{7.1.5}$$

Equation (7.1.5) can be rearranged to express V as a function of Q:

$$V = \phi\left(\frac{Q_\phi^2 - Q^2}{Q_\phi^2}\right) \tag{7.1.6}$$

where $Q_\phi = -2C_{j0}\phi$, a constant. If the diode is open-circuited at all harmonics, the current can have no harmonic components and thus must be sinusoidal at the fundamental frequency. Because the current is sinusoidal, the charge also varies sinusoidally at the same frequency; if the voltage has a square-law dependence on Q, it must also have a square-law dependence upon the current. Squaring this sinusoid produces only second harmonics; therefore, if the varactor is open-circuited at all harmonics, there can be no voltage components across the junction at any harmonics beyond the second and the multiplier will be limited to second-harmonic operation. In order to have a third-harmonic output, it is necessary to have a large second-harmonic component of junction current; then, the third "harmonic" arises as a second-order mixing product between the fundamental excitation and the second-harmonic current. In order to have this large second-harmonic current, there must be a short circuit across the junction, called a *short-circuit idler,* at the second harmonic. Similarly, for higher-harmonic outputs, idlers must be provided at the intermediate harmonics. For example, a quadrupler could have a second-harmonic and a third-harmonic idler; a quintupler would likely have second- and third-harmonic idlers.

The idea that varactor multipliers can generate only second harmonics is strictly valid only for varactors that have the ideal Q/V characteristic of (7.1.5). The Q/V characteristics of real varactors normally deviate somewhat from (7.1.5), and some second-harmonic current is generated by second-harmonic voltage dropped across the finite embedding impedance. Furthermore, because the charge-storage properties of a p^+n varactor increase its V/Q nonlinearity beyond the second degree, overdriving the diode generates current and voltage components at harmonics greater than the second. An extreme case is that of the SRD, which has a very strong C/V nonlinearity; idlers are not normally needed in SRD multipliers. However, both theory and experimental evidence indicate that the use of idlers improves the efficiency of all reactive frequency multipliers, even those using SRDs.

Idlers are usually realized as short-circuit resonators that are separate from the input and output matching circuits. In practice, idlers are usually realized by a series resonance chosen more for its convenience than for high performance, such as the series resonance of the varactor's package, commonly used as an idler at high frequencies (tuning elements are often included to tune the resonance precisely to the desired harmonic). This technique results in a very compact multiplier that can be realized easily in strip transmission media but probably has a lower Q than would a waveguide-mounted diode having a separate idler cavity. It is important to maximize idler Q in multipliers designed to have high efficiency, because the large idler currents must circulate in the idler resonator's loss resistance; this resistance, like the diode's series resistance, can generate significant power losses. Low series resistance and high unloaded Q are therefore clear requirements of high-efficiency frequency multiplication.

In theory, high-order multipliers are most efficient when they have idlers at all intermediate harmonics. Unfortunately, it is rarely practical at microwave frequencies to provide more than one idler, but harmonics up to the fourth can be generated efficiently in multipliers having only one idler circuit. The difficulty in realizing several idlers is one of the factors that limits the order of multiplication of a varactor multiplier.

Finally, we examine two very important details. The first is that the efficiency limitation established by the Manley-Rowe relations is only part of the story. These relations show that all the input power must be converted to output power at the fundamental and harmonic frequencies; this result is obvious because a reactive element, linear or nonlinear, cannot dissipate power. Thus, the power gain of a frequency multiplier using an ideal diode can be as great as 0 dB. However, we are really interested in the transducer gain of the multiplier, not the power gain, and the Manley-Rowe relations do not prove anything regarding the transducer gain. In order to show that the transducer gain can be as great as the power gain, we must show that it is possible to achieve a conjugate match at the multiplier's input; then, the available power equals the input power and the Manley-Rowe limit is valid for transducer gain as well as power gain.

Proving that the input can be matched is not a simple task, however; indeed, we can show that in many nonlinear circuits, the input can not be matched. For example, a circuit consisting of an ideal diode (one having zero resistance in the "on" state) in series with a load resistor cannot be matched and has the property that the total power at all frequencies delivered to the load can be no more than half the available power. Although no power can be dissipated in the diode and all the input power is delivered to the load, the maximum transducer gain is -3 dB. This loss is reflection loss at the input; the input of this circuit cannot be matched at the fundamental frequency. Fortunately, we can see from harmonic-balance calculations and other evidence that the input of a reactive frequency multiplier can be matched; we will not attempt to prove this point in any general way.

The second detail is that the Manley-Rowe relations do not simply establish a limit to the efficiency of a varactor frequency converter; they describe a fundamental characteristic of any pumped nonlinear capacitance. In the case of a frequency multiplier, that characteristic, expressed by (7.1.4), is consistent with intuition: the sum of all the harmonic output powers must equal the input power. This relationship is precisely valid for the reactive junction of the diode and does not depend in any way upon the excitation level or the external circuit of the multiplier: If the multiplier is badly designed, the input power may be low and the output power is dissipated in the series resistance and wasted in unwanted harmonics; if the circuit is well designed, input power is coupled efficiently to the diode junction, loss in the series resistance is minimized, and significant real power exists in the diode only at the input and output frequencies. In both cases, however, the Manley-Rowe relations are satisfied. Thus, although these relations can be used to find limits to a multiplier's efficiency, they do not guarantee that efficiency; achieving optimum efficiency in a frequency multiplier requires using a varactor that has low series resistance, selecting the varactor that is appropriate for the frequency and power level at which it is to be operated, using idlers, and matching the input and output impedances of the multiplier.

7.1.3 Design of Varactor Frequency Multipliers

Figure 7.1 shows the general structure of a varactor frequency multiplier. It consists of input and output matching circuits, a varactor, and M idler resonators, $f_{i,1}, \ldots, f_{i,M}$. Designing the multiplier requires determining the parameters of a diode that is appropriate for the frequency and power level to be used and finding the source and load impedances. When these parameters are known, the matching circuits and idler resonators can be realized in the conventional manner.

A classic paper by Burckhardt (Reference 7.3) has been the basis for the design of many frequency multipliers. Burckhardt's analysis is not unlike the harmonic-balance analysis described in Chapter 3, but his results are presented in a normalized and tabulated form so they can be used to design a wide variety of multipliers. The assumptions and limitations in Burckhardt's work are that (1) the idlers are lossless series resonators (short-circuit idlers); (2) only input, output, and idler currents in the diode are considered; (3) idlers and input-output circuits resonate with the average diode elastance; (4) the diode junction voltage varies between the reverse-breakdown voltage and ϕ, although the varactor may be overdriven; and (5) the varactor's dynamic Q (2.3.18), evaluated at the output frequency, is greater than 50.

The diode's normalized drive level, D, is defined as

$$D = \frac{q_{\max} - Q_B}{q_\phi - Q_B} \tag{7.1.7}$$

where Q_B is the depletion charge at breakdown and q_ϕ is the charge when the junction voltage just reaches ϕ. The charge $q_{\max}$ is the maximum stored charge; if the junction voltage just barely reaches ϕ, $q_{\max} = q_\phi$ and $D = 1.0$. This is the maximum drive level possible in a Schottky-barrier varactor, although in practice the junction voltage is usually limited by resistive conduction to a value less than ϕ, and thus $D < 1.0$. In a p^+n varactor, $q_{\max}$ may be greater than q_ϕ, so D can be greater than unity; however, in this case, the positive excursion of the junction voltage is clamped at ϕ. Burkhardt gives data for drive levels from 1.0 to 1.6.

Tables 7.1 and 7.2 give the important parameters necessary for designing varactor multipliers. The optimum conversion efficiency ϵ_c and output power P_L are related to two tabulated parameters, α and β, as follows:

$$\epsilon_c = \exp(-\alpha / Q_\delta) \tag{7.1.8}$$

and

$$P_L = \beta \frac{\omega_1(\phi - V_b)^2}{S_{max}} \tag{7.1.9}$$

where Q_δ is given by (2.3.18) and is evaluated at the output frequency. S_{max} is the maximum junction elastance, V_b is the breakdown voltage, and ω_1 is the input frequency.

The tables also include the normalized source and load resistances, R_{in} and R_L, and average junction elastances at the input and output frequencies, $S_{0,1}$ and $S_{0,n}$, where n is the output harmonic number; these elastances must be resonated by the source and load networks. Table 7.2 includes $S_{0,2}$, the elastance at the idler frequency in the tripler, which must be resonated by the idler. The tables also give the normalized bias voltage, $V_{dc,n}$:

$$V_{dc,n} = \frac{\phi - V_{dc}}{\phi - V_b} \tag{7.1.10}$$

and V_{dc} is the actual (not normalized) bias voltage. V_{dc} is easily adjusted empirically to optimize the multiplier's efficiency; therefore, it is not a very important design parameter. When R_{in}, R_L, $S_{0,1}$, and $S_{0,n}$ are known, designing the input and output matching circuits requires matching the simple source and load models shown in Figure 7.2. Reference 7.3 has more extensive tables that include data for designing multipliers at higher harmonics, with different drive levels and idler configurations and with various values of γ between 0.0 and 0.5. Many of these configurations, however, are practical only at low frequencies or describe types of diodes that no longer are made; consequently, they are often not practical at microwave frequencies. The design process will be illustrated by the following example.

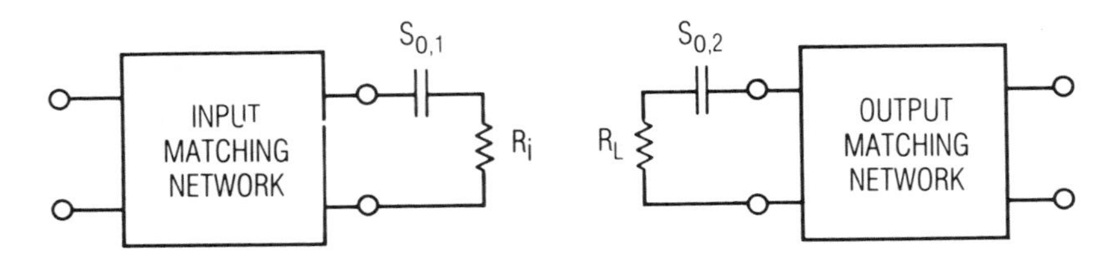

Figure 7.2 Input and output models of the varactor frequency multiplier.

Example 1

We will design a 10–20 GHz frequency doubler using Burkhardt's data, and check the design via a harmonic-balance analysis. Although a p^+n varactor would be the logical choice for this application, the multiplier will be designed to use a Schottky-barrier varactor instead of a p^+n diode. We do this for two reasons: first, to illustrate some of the problems with the use of Schottky-barrier varactors, and second, to compare the design calculations with harmonic-balance data from an existing program (that program is a modified version of one called DIODEMX, described in References 2.1 and 5.2; it does not include a diffusion-capacitance model and, therefore, is limited to use with Schottky-barrier diodes). Because the multiplier is a doubler, no idler is required, but it would still be prudent to short-circuit the diode at high frequencies to prevent power dissipation at the third and higher harmonics.

We begin by selecting a diode that has the parameters $C_{j0} = 0.3$ pF, $\phi = 0.9$ V, $V_b = -9.0$ V, $\gamma = 0.5$, and $R_s = 4.0\ \Omega$. The maximum junction voltage that can be achieved without significant conduction is 0.7 V. We calculate immediately that the minimum junction capacitance is 0.09 pF and the maximum is 0.61 pF, giving $S_{\min}$ and $S_{\max}$ equal to $1.64 \cdot 10^{12}\ F^{-1}$ and $1.11 \cdot 10^{13}\ F^{-1}$, respectively, and a dynamic cutoff frequency of 376 GHz or a dynamic Q of 19 at 20 GHz. We may have to suffer some inaccuracy, because this value of Q_δ is lower than the minimum value of 50 required for the Burkhardt analysis. This situation is unavoidable because a dynamic Q of 50 at 20 GHz would imply a cutoff frequency of 1000 GHz, a value generally beyond the state of the art in varactor diodes of this type. We also find that the drive level $D = 0.98$ from (7.1.7), and that the normalizing parameter for the input and output resistances R_{in} and R_L, $\omega_1/S_{\max}$, is $5.66 \cdot 10^{-3}$.

Table 7.1 Doubler
$\gamma = 0.5$

Drive:	1.0	1.3	1.6
α	9.95	8.3	8.3
β	0.0227	0.0556	0.0835
$R_{in}\omega_1/S_{max}$	0.080	0.098	0.0977
$R_L\omega_1/S_{max}$	0.1355	0.151	0.151
$S_{0,1}/S_{max}$	0.50	0.37	0.28
$S_{0,2}/S_{max}$	0.50	0.40	0.34
$V_{dc,n}$	0.35	0.28	0.24

Source: C.B. Burckhardt, "Analysis of Varactor Frequency Multipliers for Arbitrary Capitance Variations and Drive Level," *Bell Syst. Tech. J.*, 44:675. The tables are reprinted with permission from the *Bell System Technical Journal*, copyright 1965 AT&T.

Table 7.2 Tripler
(idler at $2\omega_1$)
$\gamma = 0.5$

Drive	1.0	1.3	1.6
α	11.6	9.4	9.8
β	0.0241	0.0475	0.0700
$R_{in}\omega_1/S_{max}$	0.137	0.168	0.172
$R_L\omega_1/S_{max}$	0.0613	0.0728	0.0722
$S_{0,1}/S_{max}$	0.50	0.36	0.26
$S_{0,2}/S_{max}$	0.50	0.38	0.31
$S_{0,3}/S_{max}$	0.50	0.38	0.30
$V_{dc,n}$	0.32	0.24	0.18

Source: C.B. Burckhardt, "Analysis of Varactor Frequency Multipliers for Arbitrary Capitance Variations and Drive Level," *Bell Syst. Tech. J.*, 44:675. The tables are reprinted with permission from the *Bell System Technical Journal*, copyright 1965 AT&T.

From Table 7.1, with $D = 1.0$, we find $\alpha = 9.95$ and $\beta = 0.0277$. Substituting these into (7.1.8) and (7.1.9), we find that the conversion efficiency ϵ is 0.589, or -2.3 dB, and $P_L = 14.8$ mW, or 11.7 dBm. It is important to note that these quantities include only the loss in the series resistance and do not include circuit loss, idler loss, or loss at unwanted harmonics in the embedding circuits. Next we find normalized values of R_{in} and R_L from Table 7.1, and quickly determine that $R_{\text{in}} = 14.1\ \Omega$ and $R_L = 23.9\ \Omega$. Similarly, we find both $1/S_{0,1}$ and $1/S_{0,2}$ to be 0.18 pF; the source and load impedances are therefore $14.1 + j88$ and $23.9 + j44$, respectively. Finally, the normalized bias voltage is obtained from the table, and (7.1.10) gives $V_{\text{dc}} = -2.7$ V.

Figure 7.3 shows the input-output power characteristic of the multiplier and the dc diode current calculated via harmonic balance (the embedding impedances at all harmonics are assumed to be zero in this calculation). Both the output power and conversion efficiency increase with input power level, but a minimum loss of 3.8 dB occurs at an input level of 11 dBm, rather than at the design value of 14 dBm. Above this input level, the output power continues to rise, but the conversion loss also increases. The output power saturates above 11 dBm, where the resistive junction begins to rectify the input power, as evidenced by the onset of dc diode current precisely at 11 dBm. At 14 dBm, the diode current is 2.3 mA and the bias is -3.3 V, so the diode is delivering almost 7 mW of dc power to the bias source, more power than in the second-harmonic output! This power must come from the RF input signal, so it represents a waste of available input power and a very expensive way to keep a bias battery well charged. In a p^+n diode, the rectified current would be much lower, and both the efficiency and the output power would be greater at levels above 11 dBm.

The optimum source and load impedances found from the harmonic-balance calculation are $15 + \text{j}104$ and $20 + \text{j}55$, respectively. These agree well with the values of $14 + \text{j}88$ and $24 + \text{j}44$ obtained from the tables. Although it is not a very important parameter, the dc bias voltage of -3.3 V from the harmonic-balance calculation agrees reasonably well with the value of -2.7 V found from the value of $V_{\text{dc},n}$ in Table 7.1.

The rest of the design involves realizing the matching circuits. It is, of course, necessary that the input and output matching circuits do not interact. The output matching circuit and the idlers in Figure 7.1 must present an open circuit to the diode at the input frequency and vice versa. The highly reactive source and load impedances of the frequency doubler in the design example imply that the diode is a high-Q device. It is therefore difficult to achieve wideband operation with a varactor multiplier. Furthermore, in a high-order multiplier, high-Q idler resonators are necessary

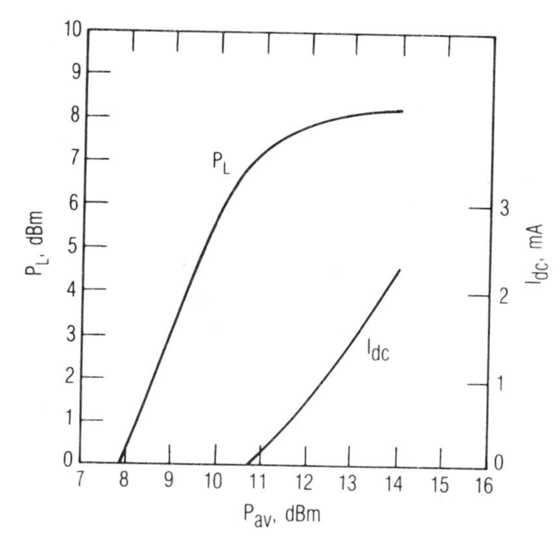

Figure 7.3 Output power and rectified diode current in the multiplier described in Example 1. These data were obtained from harmonic-balance analysis.

to achieve good conversion efficiency. The presence of these idlers further limits the multiplier's bandwidth because efficiency drops significantly as the frequency deviates from the idler's resonant frequency. Accordingly, in systems where broadband operation is necessary, some other type of multiplier (e.g., one using a resistive diode or active FET) may be preferable.

A limitation of this simple design process is that the diode's *C*/*V* characteristic must follow (2.3.5) at all voltages between breakdown voltage and ϕ. Many modern varactor diodes do not have such ideal *C*/*V* characteristics, especially at high reverse voltages. In an epitaxial Schottky-barrier diode, a high reverse voltage may deplete the epilayer before breakdown occurs, and beyond this voltage the variation in capacitance is minimal. In *p^+n punch-through* or *dual-mode* varactors (Section 2.3.4), the varactor's *C*/*V* characteristic is purposely tailored (by adjustment of the doping profile) to minimize the junction-capacitance variation at high reverse voltages; this characteristic minimizes sensitivity to input level and may also enhance stability somewhat. When such devices are used in a multiplier, it is best to use the voltage at which the *C*/*V* curve begins to limit in place of V_b or Q_B in (7.1.7), (7.1.9), and (7.1.10) and to reduce the efficiency given by (7.1.8), which is quite optimistic under even the best circumstances.

An important question that often arises in the design of multiplier systems is whether it is better to realize a high-order multiplier via a single stage or by a cascade of two or more low-order multipliers. We find from the data in Reference 7.3 that, in theory, a cascade of low-order multipliers usually has greater efficiency than a single high-order multiplier. However, before concluding that this is the case in practice, we must consider the additional losses in cascading two multipliers (it is invariably necessary to use an isolator between them) and especially the additional cost of designing, manufacturing, and testing two separate components and their interconnecting hardware. When these practical considerations are included, the answer to this question is not nearly as clear. It must be answered on an ad hoc basis, in view of the requirements of the system in which the multiplier is to be used.

7.1.4 Bias and Stability

The stability of any nonlinear circuit is difficult to assess analytically (stability has been addressed in Chapter 3). However, we have observed that most stability problems encountered in varactor frequency multipliers are the result of practical design deficiencies and are rarely inherent in the nature of the component. Thus, it is best to examine stability from a practical viewpoint.

The best way to ensure stable operation is to control the broadband embedding impedance characteristic very carefully. Specifically, the input source and output load must not vary with input or output level (e.g., one must not drive a mixer's LO port directly from a multiplier; an isolator should be used); the input and output networks must not have any spurious resonances, especially near harmonics or subharmonics of the input or output frequencies; and the idler resonances must be implemented effectively. In general, the simplest effective matching circuits are least likely to introduce instability. It is also important to have a spectrally clean excitation; the excitation signal must not have significant spurious signals, harmonics, or noise.

Even a p^+n varactor has some dc current caused by rectification at high input levels; by introducing a resistor in the diode's dc return path, we can use this current to bias the diode. The resistor also helps to reduce the sensitivity of the output power level to the input power level: As input drive is increased, the resulting increase in dc current further reverse-biases the diode, reducing the multiplier's efficiency. The design of the bias circuit often has a strong effect on stability; the use of a high-impedance bias source usually results in better stability than does a voltage source. Furthermore, low-frequency resonances in the bias circuit are a frequent cause of instability; these should be avoided.

7.2 FREQUENCY MULTIPLIERS USING STEP-RECOVERY DIODES

The SRD is used to achieve efficient high-order frequency multiplication. The key to its operation is its very strong capacitive nonlinearity, which is realized almost exclusively by charge-storage effects. The SRD multiplier operates by generating a very fast voltage pulse once for each cycle of the input voltage; that pulse is then applied to a filter that converts it to a sinusoidal output voltage. Without the need for idlers, SRD multipliers can achieve conversion efficiency on the order of $1/n$, where n is the harmonic number; they are, however, narrowband components and are not capable of operating at output frequencies above approximately 20 GHz.

7.2.1 Multiplier Operation

Because it is consistent with the way SRD multipliers are most often operated, Hamilton and Hall's description of SRD multiplier operation (Reference 7.4) is widely accepted. A description of the operation of the SRD in a slightly different circuit is given by Hedderly (Reference 7.5) in a paper that contains more useful information about the factors that limit efficiency.

We begin by treating the SRD multiplier as an ideal circuit, one that is lossless and has an ideal diode. First, we describe the multiplier circuit as a pulse generator and, then, show how the pulse generator is modified to achieve a sinusoidal output. The ideal SRD has the C/V characteristic shown in Figure 7.4: The reverse-bias capacitance is small and independent of voltage, the forward-bias capacitance is infinite, and other parasitics, such as series resistance, are negligible. We assume that the forward characteristic begins at $V = 0$, although in reality it begins at $V = \phi$; this assumption simplifies the analysis and makes little difference in the results. We also assume that the voltage across the diode never exceeds the reverse breakdown voltage (this requirement limits output power).

In an ideal diode, all forward current creates stored charge at the junction without causing a change in voltage; this stored charge must be removed by reverse current before a reverse voltage is possible. If the carrier recombination time of a practical diode is long compared to the inverse of the input frequency, very little of the stored charge recombines (i.e., very little charge contributes to resistive conduction) before it is removed, and the diode is nearly ideal in this respect. In the following derivation, we assume that all charge is stored and is recoverable and that the diode can switch from forward to reverse conduction instantaneously after the stored charge is removed.

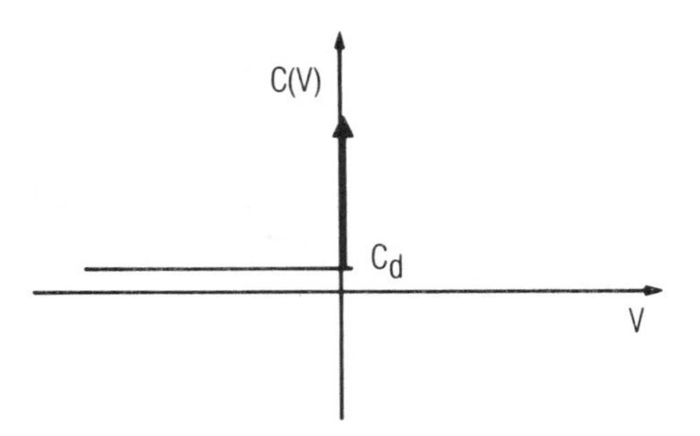

Figure 7.4 *C/V* characteristic of an ideal step-recovery diode.

Figure 7.5 shows the circuit of the pulse generator. The excitation consists of a sinusoid at frequency ω_1 plus dc bias V_{dc}, and the source impedance $Z_s(\omega)$ is assumed to be zero at dc and all harmonics of ω_1; therefore, the input voltage $V_1(t)$ is sinusoidal. The phase of $V_1(t)$ is chosen so that the beginning of the SRD's conduction occurs at $t = 0$. Then,

$$V_1(t) = V_1 \sin(\omega_1 t + \alpha) + V_{dc} \tag{7.2.1}$$

where α is the phase angle of $V_1(t)$ and V_{dc} is the dc component; normally $V_{dc} < 0$. During this interval, the diode is forward-biased, so its capacitance is infinite, and it is effectively a short circuit; the equivalent circuit is shown in Figure 7.6(a). The current is found directly to be

$$I_L(t) = I_L(0) + \frac{V_1}{\omega_1 L}(\cos(\alpha) - \cos(\omega_1 t + \alpha)) + \frac{V_{dc}}{L}t \tag{7.2.2}$$

where $I_L(0)$ is the initial current in the inductor at the beginning of the conduction cycle. The second term in (7.2.2) is the sinusoidal component, where α is a phase angle; and the third term is the linear current ramp generated by the bias source. The voltage waveform $V_1(t)$ and the resulting current $I_L(t)$ are shown in Figure 7.7; when the current is positive, charge is stored in the SRD, and when it is negative, reverse-conduction removes this stored charge. At the end of the conduction interval, the stored charge, Q_s, is zero:

$$Q_s = \int_0^{T-T_t} I_L(t)\,dt = 0 \tag{7.2.3}$$

where T is the excitation period and T_t is the length of the impulse period. At $t = T - T_t$, all the stored charge has been removed and the diode switches to its reverse-bias state. At this point, the impulse interval begins.

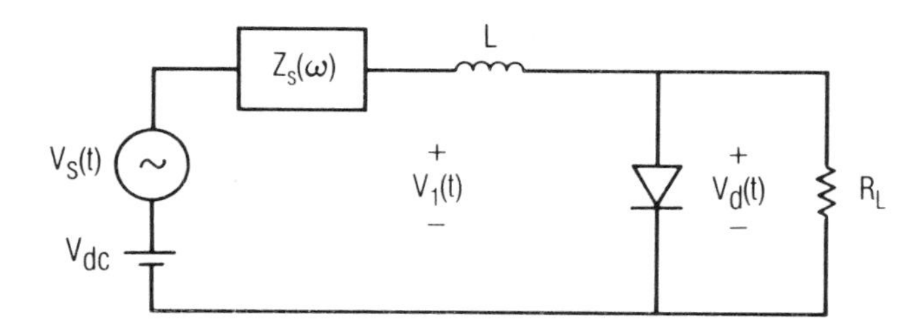

Figure 7.5 Pulse-generator circuit using a step-recovery diode.

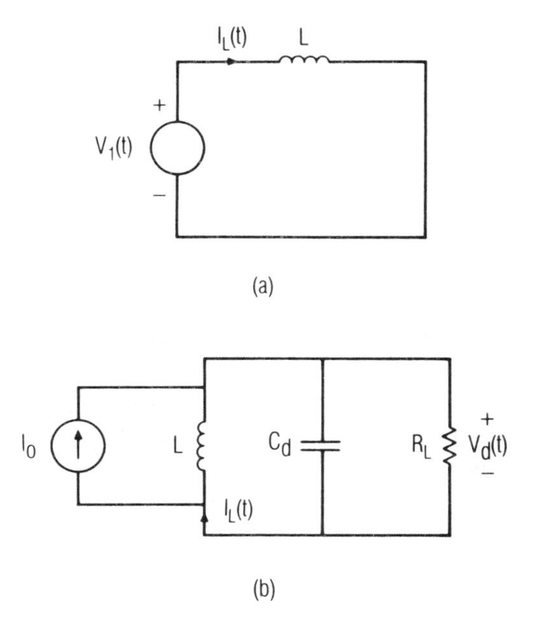

Figure 7.6 Equivalent circuit of the SRD multiplier (a) during the conduction interval; (b) during the impulse interval.

At the instant the diode switches, the current in the inductor is the excitation current for the harmonic-generating impulse. Therefore, we adjust V_{dc} so that the diode switches at the instant when $I_L(t)$ has its maximum negative value. At that instant, $dI_L(t)/dt = 0$, so the voltage across the inductor is zero and the diode voltage $V_d(t)$ is zero; $V_1(t)$ is the sum of these voltages, so it must also be zero. Because the SRD switches when $V_1(t) = 0$, the multiplier has the equivalent circuit of Figure 7.6(b), in which the voltage source has been eliminated and the inductor current

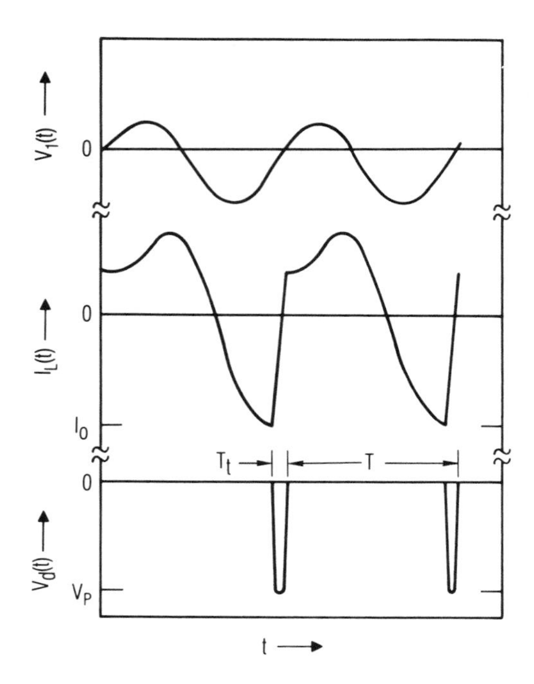

Figure 7.7 Voltage and current waveforms in the SRD impulse generator.

$I_L(T - T_t)$ is the only excitation (we will call this current I_0 for simplicity; I_0 is a negative quantity). The diode capacitance is now C_d, a relatively small depletion capacitance.

The response $V_d(t)$ of the circuit in Figure 7.6(b) is a damped sinusoid at the resonant frequency of L and C_d:

$$V_d(t) = I_0 \left(\frac{L}{C_d(1 - \zeta^2)} \right)^{1/2} \exp\left(\frac{-\zeta\, \omega_n t'}{(1 - \zeta^2)^{1/2}} \right) \sin(\omega_n t') \tag{7.2.4}$$

where $t' = t - T + T_t$; that is, t' is time measured from the beginning of the impulse interval. The loaded resonant frequency, ω_n, of the tuned circuit in Figure 7.6(b) is

$$\omega_n = \left(\frac{1 - \zeta^2}{LC_d} \right)^{1/2} \tag{7.2.5}$$

and the damping factor ζ is

$$\zeta = \frac{1}{2R_L}\left(\frac{L}{C_d}\right)^{1/2} \tag{7.2.6}$$

This sinusoid does not last very long; as soon as $V_d(t)$ reaches zero, at the end of one-half cycle, the diode again switches to its high-capacitance state and $V_d(t)$ is clamped at zero. Thus, the output voltage consists of a very short-lived pulse, a half sinusoid at the loaded resonant frequency of the circuit in Figure 7.6(b). The pulse waveform is shown in Figure 7.7; the peak voltage is

$$V_p = -I_0\left(\frac{L}{C_d}\right)^{1/2} \exp\left(\frac{-\pi\zeta}{2(1-\zeta^2)^{1/2}}\right) \tag{7.2.7}$$

and the pulse width T_t is

$$T_t = \frac{\pi}{\omega_n} \tag{7.2.8}$$

The current in the inductor during this interval is

$$I_L(t) = I_0 + \frac{1}{L}\int_{T-T_t}^{T} V_d(t)\,\mathrm{d}t \tag{7.2.9}$$

so that

$$I_L(t) = I_0 \exp\left[\frac{-\zeta\omega_n t'}{(1-\zeta^2)^{1/2}}\right]\left[\cos(\omega_n t') + \frac{\zeta\sin(\omega_n t')}{(1-\zeta^2)^{1/2}}\right] \tag{7.2.10}$$

The power P_o in the pulse train is

$$P_o = \frac{1}{T}\int_{T-T_t}^{T} \frac{V_d^2(t)}{R_L}\,\mathrm{d}t = \frac{\omega_1 V_p^2}{4\omega_n R_L} \tag{7.2.11}$$

The input quasi-impedance (see Section 3.1.1, Example 1) of the multiplier circuit $Z_{\text{in}}(\omega_1)$, including the inductor L, is the ratio of the fundamental-frequency components of $V_1(t)$ and $I_L(t)$. Because the impulse interval is so short, it is tempting to ignore it in approximating $Z_{\text{in}}(\omega_1)$; however, it is only during this interval that power is removed from the circuit, so ignoring the impulse interval gives the trivial result that $Z_{\text{in}}(\omega_1) = \omega_1 L$. This is, however, a good approximation of the imaginary part of the input impedance.

The real part of the input impedance can be found from power considerations. Because the diode and inductor are assumed lossless, the power dissipated in the real part of the input impedance must equal the output power. We express the fundamental component of $I_L(t)$ as I_1; then,

$$|I_1|^2 = \frac{V_1^2}{(\omega_1 L)^2 + R^2} \tag{7.2.12}$$

where $R \equiv \mathrm{Re}\{Z_i(\omega_1)\}$. The input power is P_{in}, and

$$P_{\mathrm{in}} = \frac{1}{2}|I_1|^2 R = P_o = \frac{V_p^2 \omega_1}{4 R_L \omega_n} \tag{7.2.13}$$

We find from other analyses that in most well designed multipliers $R \approx \omega_1 L$; then substituting (7.2.12) into (7.2.13) gives

$$R = \frac{4 V_p^2 L^2 \omega_1^3}{V_1^2 R_L \omega_n} \tag{7.2.14}$$

Real diodes, of course, have a series resistance R_s that must be added to R. Finally, the estimate of the input impedance is

$$Z_{\mathrm{in}}(\omega_1) \approx j\omega_1 L + R_s + \frac{2 V_p^2 L^2 \omega_1^3}{V_1^2 R_L \omega_n} \tag{7.2.15}$$

Equation (7.2.15) is not very useful for design purposes because it is difficult to estimate V_p/V_1 without calculating the complete current waveform. Hamilton and Hall give expressions for the real and imaginary parts of the input impedance; however, it appears that they have calculated the inverses of the real and imaginary parts of the input admittance instead. Their tabulated results can be approximated as

$$G_i \approx \frac{1}{\omega_1 L(1.2 + \zeta)} \tag{7.2.16}$$

and

$$B_i \approx \frac{-1}{\omega_1 L(0.7 + \zeta)} \tag{7.2.17}$$

We now have a circuit (Figure 7.5) that generates a pulse train, $V_d(t)$. The spectrum of $V_d(t)$ has components at many harmonics of ω_1, and the envelope of that spectrum has its first zero at $3\omega_n$. This circuit can be used effectively as a *comb generator*, a circuit that generates a large number of harmonically related tones.

However, more often we wish to generate a single output frequency as efficiently as possible. In this case, it is not enough just to filter the output; we must use a resonant network that does not dissipate appreciable power at unwanted harmonics and does not upset the pulse waveform too seriously. Note that the value of R_L in the impulse generator does not affect the shape of the pulse; even making $R_L \rightarrow \infty$ has no effect except to increase V_p (because $\zeta \rightarrow 0$ in (7.2.7)). Thus, an open circuit at unwanted harmonics and an appropriate resistance at the desired output frequency are the desired terminations. The resonant network that realizes these terminations is an ideal series LC resonator. The SRD frequency multiplier is shown, in its conceptual form, in Figure 7.8; the box marked f_N is the resonator.

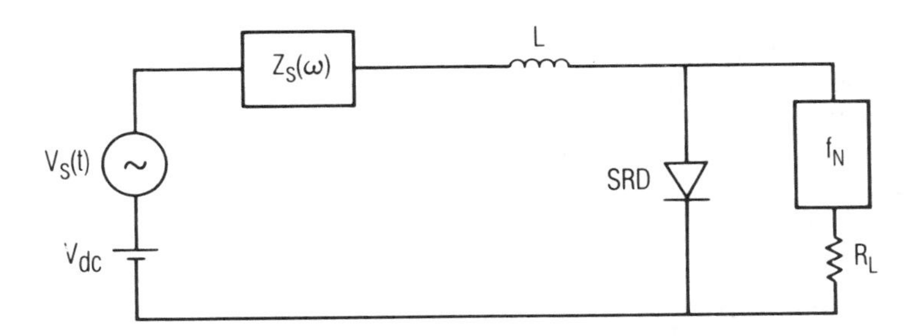

Figure 7.8 Circuit of an SRD frequency multiplier. f_N is an ideal series LC resonator tuned to the Nth harmonic of the input frequency.

We must be careful to recognize that the circuit in Figure 7.8 is not equivalent to that of Figure 7.5 because the resonator changes the diode's termination at unwanted harmonics to an open circuit rather than to a finite resistance. Because the diode is a short circuit over most of the excitation period, this change has less effect than we might imagine. The main practical effect is to make the multiplier operate as if it were an impulse generator having a lower damping factor than the value given by (7.2.6); that is, when terminated in a resonant network, the multiplier is less stable than when terminated in a resistance. Accordingly, it is generally good practice to design an SRD frequency multiplier to have a damping factor of approximately 0.6–0.7, rather than the value of 0.4–0.5 that would provide stable operation in an impulse generator.

If the multiplier uses an ideal diode and is ideally terminated, all the energy of the each impulse is converted to output power at the desired harmonic frequency. Under these conditions the output power is given by (7.2.11). Unfortunately, because of the distinct paucity of ideal conditions in the world of microwave electronics, the efficiency is considerably lower. The most serious reduction in efficiency comes from the series resistance of the diode. Power is dissipated in the diode's forward resistance not only at the input and output frequencies but at all other harmonics as well. Power is also dissipated at these harmonics in the losses in the input matching circuit, the inductor, and the output resonator; in particular, the high loaded Q of the output resonator, necessary in order to reject unwanted harmonics, can make the resonator very lossy.

Another important cause of loss is the recombination current in the diode. Even if the carrier recombination time is long, a fraction of the injected charge recombines and cannot generate output power. This phenomenon has the same effect as adding a resistance in parallel with the diode during the pulse interval. Similarly, the transition time of the diode is always finite and lengthens the pulse interval. The increased pulse length reduces the magnitude of the higher harmonics and thus reduces efficiency. The effects of finite pulse length have not been analyzed adequately in any published reports; it is probable that these most strongly limit the SRD frequency multiplier's efficiency.

7.2.2 Multiplier Design

Designing a step-recovery diode multiplier is a relatively straightforward application of the equations in Section 7.2.1. It is most important to select an appropriate diode and the proper damping factor. The process will be illustrated by the following example.

Example 2

We design an SRD multiplier to generate 20 mW at 4 GHz from a 1 GHz excitation. The circuit of the multiplier is shown in Figure 7.9. The diode's recombination time must be long compared to the period of the input excitation, so $\tau >> 10^{-9}$ s, in fact 10^{-8} s would not be too great. The ideal pulse length is one-half period at the output frequency; thus

$T_t = 1.25 \cdot 10^{-10}$ s. The diode's transition time must be considerably shorter than this, no more than approximately 70 to 100 ps. Estimating the optimum value of reverse capacitance, C_d, is a controversial subject among multiplier designers; this controversy is not unexpected, because the criteria for selecting C_d are mostly empirical. The range of suggested values for the diode's reactance, under reverse bias, varies from 10 or 20 Ω at the output frequency to more than double this value; the best choice is probably an intermediate value that gives a reasonable input impedance without making V_p too great. From (7.2.16) and (7.2.17), we see that the input impedance is proportional to $\omega_1 L$, a reactance that must resonate with C_d; thus, increasing C_d reduces L and thus reduces input impedance. We begin by choosing $C_d = 1.0$ pF and $\zeta = 0.5$, a good compromise between pulse length (low ζ) and stability (high ζ); from (7.2.5) and (7.2.6), we have $L = 1.19$ nH and $R_L = 35\ \Omega$. We find the input admittance from (7.2.16) and (7.2.17) and convert to impedance; the result is $Z_{in}(\omega_1) = 4.2 + j6.0$, a low but reasonable value.

Equation (7.2.11) can be used to find the peak impulse voltage, V_p; V_p must be kept below the diode's reverse breakdown voltage. If the multiplier had 100 percent efficiency, (7.2.11) would be directly applicable and could be solved for V_p. However, we expect loss on the order of at least 6 dB; most of the loss is caused by inefficiencies in converting the pulse energy to output power. Accordingly, it would be more realistic, from a design standpoint, to use input power instead of output power in determining V_p. Therefore, we use 80 mW instead of 20 mW in (7.2.11). This gives $V_p = 6.7$ V, considerably below the breakdown voltage of virtually all practical SRDs.

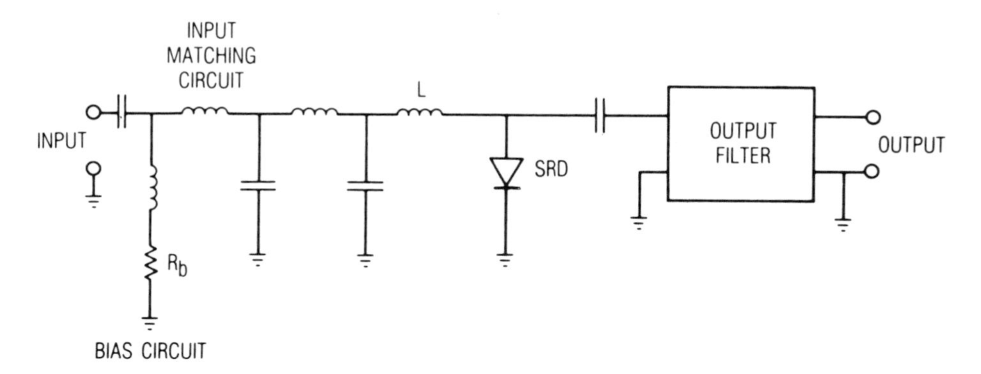

Figure 7.9 The SRD frequency multiplier designed in Example 2.

Designing the input matching circuit may be difficult because of the low input impedance; for this reason a multistage matching network is usually necessary. A low-pass structure consisting of series inductors and parallel capacitors has the required short-circuit output impedance at harmonics of ω_1. In order to prevent instability, the matching and bias circuits must have no spurious resonances; all capacitors must have series resonant frequencies well above the input frequency, and it is best not to use a bypass capacitor in the bias circuit. Because of the matching circuit's low output impedance, the currents in the matching elements are relatively great; these elements must be high-quality parts (i.e., they must have high Qs) or the loss in the matching circuit may be excessive.

There are many ways to design a load network, and in general, the simplest designs are best. A lumped-element series resonator is usually not realizable at 4 GHz, so a distributed equivalent network must be used. One possibility is to connect the diode directly to a filter that has the desired out-of-band characteristics; another is to couple it loosely through a capacitor to a narrowband filter. It is wise to design this circuit to provide the impedance transformation between the standard 50 Ω coaxial load impedance and R_L. Experienced designers of SRD multipliers report that some types of resonant networks give better efficiency and stability than others for reasons that are not always clear. For example, Hamilton and Hall recommend a resonant transmission-line section; this structure, however, can introduce instability if the line impedance is low. Other possibilities are a quarter-wave coupled-line section or a weak capacitive coupling to a quarter-wave resonator.

Simple resonant structures often have inadequate Q to reject the harmonics closest to the output frequency; in this case the multiplier should be followed by a loosely coupled high-Q resonator or filter. If the output circuit has been designed to match R_L to 50 Ω, the multiplier can be tested easily without this filter in place and the filter can be tested without the multiplier; this practice significantly eases the testing of both components.

A disturbing property of SRD multipliers is that they cannot be analyzed as readily as varactor or resistive multipliers. Harmonic-balance methods are the logical choice for analyzing virtually all types of frequency multipliers; however, harmonic-balance analysis of SRD multipliers is troublesome because the large number of harmonics involved, the strong reactive nonlinearity of the diode, and the possibility of instability make convergence precarious. Accordingly, the algorithm used for such analysis should be selected primarily for good convergence; the reflection algorithm is probably a good choice. One might well question the worth of harmonic-balance analysis of SRD multipliers designed for high-order operation (i.e., those having output harmonics greater than the tenth); the difficulty in modeling the diode, accounting for all the losses, and determining the embedding impedances at all significant harmonics might be so difficult in most cases that the results would probably be meaningless. Time-domain analysis of the impulse-generator circuit of Figure 7.5 is often successful (see, e.g., Reference 7.6); however, extending time-domain analysis to the complete harmonic generator would probably be unsuccessful (except, perhaps, at low frequencies) because of the standard set of problems inherent in the time-domain analysis of microwave circuits. These problems are discussed in Chapters 1 and 3.

7.3 Resistive Diode Frequency Multipliers

Resistive diode (i.e., Schottky-barrier diode) frequency multipliers have not been employed widely in microwave systems. The reason for their lack of use is that they are significantly less efficient than varactor multipliers and are limited in output power. Furthermore, their efficiency decreases rapidly as the harmonic number increases, so resistive diode multipliers are rarely practical for generating harmonics greater than the second. However, resistive multipliers have three significant advantages over reactive multipliers. First, they are capable of very wide bandwidths; multipliers having flat responses over full waveguide bandwidths can be realized relatively easily. Second, they are very stable; although the Schottky-barrier diodes used in resistive multipliers have nonlinear junction capacitances, the resistive junction provides enough loss to prevent parametric oscillation. Third, it is nearly impossible to fabricate p^+n varactors that can be used at frequencies above 100 GHz, so most millimeter-wave frequency multipliers, even those using Schottky-barrier varactors, are driven into conduction; thus, they are, in reality, hybrids of resistive and reactive multipliers. Because of these advantages, resistive frequency multiplication may occasionally be an attractive option in the design of a microwave system.

7.3.1 Approximate Analysis and Design of Resistive Doublers

Figure 7.10 shows a canonical representation of a resistive multiplier. The diode symbol represents an ideal diode, one that has no junction capacitance. (We shall see later that the nonlinearity of the junction capacitance is generally insignificant in these multipliers.) The series resistance, R_s, is shown separately from the diode. R_i is the source impedance at f_1, and R_L is the load impedance at $2f_1$. The blocks marked f_1 and $2f_1$ are ideal parallel-resonant filters; that is, they have infinite impedance at

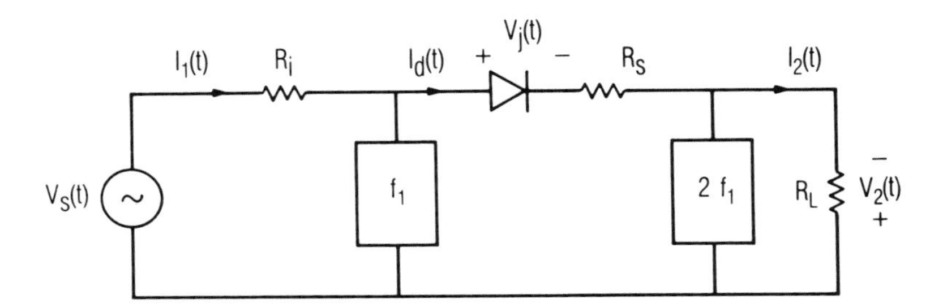

Figure 7.10 Circuit of a resistive frequency doubler: f_1 and $2f_1$ are ideal parallel LC resonators tuned to the fundamental frequency and its second harmonic.

frequencies f_1 and $2f_1$, respectively, and zero impedance at all other frequencies. These resonators represent the frequency-selective parts of the matching circuits in a practical multiplier. Because of the properties of these resonators, voltage components at only these two frequencies exist across the diode-R_s combination, and only fundamental-frequency and second-harmonic currents circulate in the input and output loops, respectively. V_1 is the magnitude (peak value) of the fundamental component of the diode junction voltage, $V_j(t)$, and V_2 is the magnitude of the second-harmonic voltage across R_L. Similarly, I_1 and I_2 are the peak values of the fundamental and second-harmonic components of the diode current, $I_d(t)$. The source voltage $V_s(t)$ is a sinusoid at frequency f_1; dc bias may also exist.

We can understand the operation of the multiplier by first imagining that the diode is short-circuited at all harmonics except the fundamental, a condition that can be established by letting $R_L = 0$, and that the diode is pumped to a high peak current ($\geqslant$25 mA) by $V_s(t)$. Under these conditions, the current waveform, shown in Figure 7.11, is a series of pulses, in phase with the positive excursion of $V_s(t)$ and shaped much like half-cosine pulses. The diode's dc bias is invariably a few tenths of a volt and positive, and the duty cycle of the pulses is close to 50 percent. We assume that the current waveform is adequately approximated as a series of half-cosine pulses and, from Fourier analysis, find that the peak value of the fundamental current component is

$$I_1 = 0.5\, I_{\max} \tag{7.3.1}$$

Similarly, we find that the second-harmonic current component is

$$I_2 = \frac{2}{3\,\pi}\, I_{\max} \approx 0.2\, I_{\max} \tag{7.3.2}$$

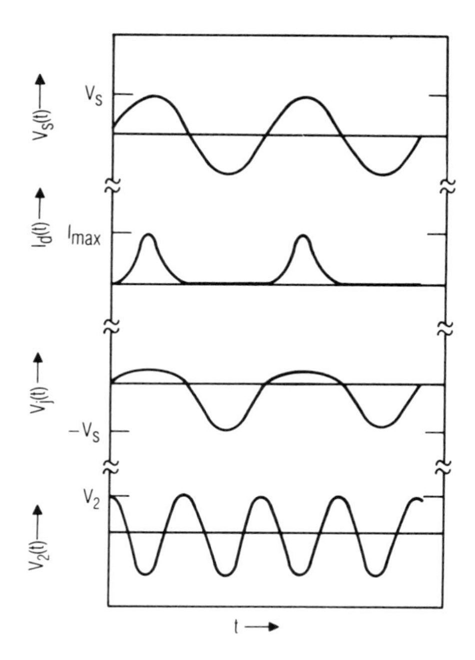

Figure 7.11 Voltage and current waveforms in the resistive doubler.

It is also worth noting that the dc component of $I_j(t)$, I_{dc}, is

$$I_{dc} = \frac{1}{\pi} I_{max} \tag{7.3.3}$$

where I_{max} is the peak value of the current pulses of $I_j(t)$. Because of the source resistance, the junction voltage $V_j(t)$ has more harmonic components than just the first and second. In the time domain, $V_j(t)$ is a clipped sinusoid; if $R_s \ll R_i$, the magnitude of the fundamental component of $V_j(t)$ is

$$V_1 = 0.5(V_s + V_f) \approx 0.5V_s \tag{7.3.4}$$

where V_s is the peak value of $V_s(t)$ and V_f is the forward voltage of the diode, approximately 0.6 V.

Now imagine that R_L is slowly increased from its zero value. I_2 circulates in R_L and generates a voltage $V_2(t)$, the second-harmonic output, shown in Figure 7.11. While R_L is small, I_2 remains approximately constant, so the second-harmonic output power increases with R_L. However, the phase of $V_2(t)$ is such that it reduces the peak positive value of $V_j(t)$ and thus reduces the peak value of $I_d(t)$, I_{max}. This reduction in I_{max} in turn reduces the value of I_2, and eventually a point is reached where the output power levels off and begins to decrease. If R_L is increased further, V_2 also increases, and eventually the second-harmonic component of the junction current becomes evident as a dip in the peak of the current pulse.

The effect of the magnitude of R_L on the shape of the current pulse is shown in Figure 7.12. It appears at first that the current pulse in Figure 7.12(c) (large R_L) has a strong second-harmonic component; however, this second-harmonic component in fact is relatively weak because the peak current I_{max} is much lower when R_L is large than when R_L is optimum. Harmonic-balance studies of resistive multipliers indicate that optimum efficiency is achieved at the value of R_L where this dip just begins to form.

In order to design a multiplier, we need to determine the input resistance at f_1, the optimum output load resistance R_L, and the output power as a function of input power. The input quasi-impedance of the junction is the ratio of the fundamental-frequency voltage to current at the junction:

$$R_j = \frac{V_1}{I_1} = \frac{V_s}{I_{max}} \tag{7.3.5}$$

The input impedance is simply the sum of this impedance and the series resistance:

$$R_{in} = R_j + R_s \tag{7.3.6}$$

The multiplier's input power equals the sum of the real power of the junction plus the power dissipated in R_s, at all the harmonics, minus the output power. If $R_s \ll R_j$ the fundamental-frequency power dominates; then,

$$P_{in} \approx \frac{1}{2} V_1 I_1 + \frac{1}{2} I_1^2 R_s = \frac{1}{8} I_{max}^2 (R_j + R_s) \tag{7.3.7}$$

We shall see that the efficiency of a resistive multiplier is invariably very low because most of the input power is dissipated in the diode junction and in the series resistance at the fundamental frequency and very little is converted to harmonics. When the input is matched, the power available from the source is equal to P_{in}; then,

$$P_{av} = P_{in} \approx \frac{1}{8} I_{max}^2 (R_j + R_s) \tag{7.3.8}$$

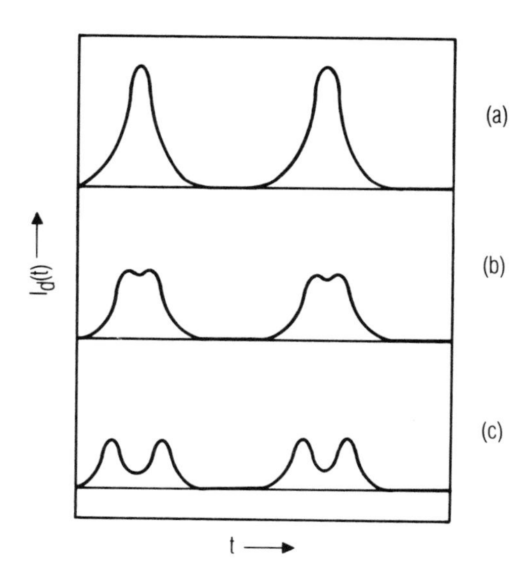

Figure 7.12 Current waveforms in the diode: (a) $R_L = 0$; (b) optimum R_L; (c) R_L greater than optimum. The peak current is greatest in (a), lowest in (c).

We now consider the output. At the peak of the excitation cycle, all the voltage components across the diode must equal the forward voltage. Summing these voltages around the output loop gives

$$V_f = V_1 - V_2 + V_{dc} - I_{max}R_s \tag{7.3.9}$$

or

$$V_2 = V_1 - I_{max}R_s + V_{dc} - V_f \tag{7.3.10}$$

where V_{dc} is the dc bias. In (7.3.9) and (7.3.10), we have assumed that only the first- and second-harmonic components in $V_j(t)$ are significant. The quantity $V_{dc} - V_f$ is no more than a few tenths of a volt; it can be neglected, and we find V_2 to be

$$V_2 = V_1 - I_{max}R_s \tag{7.3.11}$$

From (7.3.1) and (7.3.5), we can express (7.3.11) in the more convenient form:

$$V_2 = 0.5 I_{max}(R_j - 2R_s) \tag{7.3.12}$$

We find from harmonic-balance calculations that the value of V_2 given by (7.3.12) is too great; it results in a high value of R_L and a current waveform similar to that shown in Figure 7.12(c). This results from the diode's exponential I/V characteristic, which causes the current to be very sensitive to junction voltage. The value of V_2 given by (7.3.12) is not precisely correct for two reasons: first, V_1 is itself approximate; second, it is determined only at a single instant, the peak of the excitation cycle, and does not include effects of high V_2 over the entire junction voltage waveform. We find empirically that a better value of V_2 is approximately one-third that given by (7.3.12). Therefore,

$$V_2 \approx 0.167\, I_{max}(R_j - 2R_s) \tag{7.3.13}$$

The load impedance is

$$R_L = \frac{V_2}{I_2} = 0.833\,(R_j - 2R_s) \tag{7.3.14}$$

Equation (7.3.2) was used to express I_2 in (7.3.14).

The output power is

$$P_L = 0.5\, I_2^2\, R_L = 0.0167\, I_{max}^2\,(R_j - 2R_s) \tag{7.3.15}$$

where (7.3.2) has been used to express I_2. The maximum available conversion gain $G_{av,max}$ is found from (7.3.15) and (7.3.8):

$$G_{av,max} = \left.\frac{P_L}{P_{av}}\right|_{\text{input matched}} = \frac{0.0167\, I_{max}^2\,(R_j - 2R_s)}{0.125\, I_{max}^2\,(R_j + R_s)} \tag{7.3.16}$$

or

$$G_{av,max} = 0.133\,\frac{(R_j - 2R_s)}{(R_j + R_s)} \tag{7.3.17}$$

A clear implication of Equation (7.3.17) is that resistive multipliers suffer from low efficiency. Even if the parasitic series resistance R_s is zero, (7.3.17) implies that the maximum conversion gain of a resistive doubler is only 0.133, or -8.8 dB. This high loss is the unavoidable result of power dissipation in the diode junction. Of course, (7.3.17) is approximate, so the −8.8 dB limit must also be considered approximate; however, it is difficult to see any way that the efficiency could be more than 1 dB or so greater than this limiting value. It is possible that the nonlinear junction capacitance may improve the efficiency slightly in some cases; however, harmonic-balance calculations show that such improvement is rarely more than 1 dB. The procedure for designing a doubler is illustrated by the following example.

Example 3

We will design a 20–40 GHz frequency doubler. Schottky-barrier diodes are not produced specifically for multiplier use, but good mixer diodes are acceptable and readily available. A typical 4-micron diode has $R_s = 6.0\ \Omega$ and $C_{j0} = 0.05$ pF. Initially, we will ignore the junction capacitance; later we will include it in the circuit and will design the matching circuit to compensate for it.

We begin by recognizing that, because of the low conversion efficiency, virtually all the input power is dissipated in the diode. A 4-micron Schottky diode has a thermal resistance of approximately 2000°C per watt. We wish to limit the temperature rise in the junction to approximately 50°C, a prudent limit, so the power dissipation can not exceed 0.025 W, or 14 dBm. We therefore choose the nominal available input power to be 10 dBm to allow for the effects of input power variation over the input frequency range, changes in environmental temperature, and dc bias power, as well as to maintain a dB or two of margin.

A second consideration is that the dc junction current must be limited. In order to achieve high output power and efficiency, we wish to have a high value of I_{max}; I_{max}, however, is limited by the fact that I_{dc} should not exceed approximately 10 mA in a 4-micron diode. Using (7.3.3), we select $I_{max} = 30$ mA. In practice, it may be necessary to provide dc bias in order to achieve this value of I_{max} at the prescribed 10 dBm power level. Equations (7.3.8) and (7.3.6) give

$$R_i = R_{in} = R_j + R_s = 89\ \Omega \tag{7.3.18}$$

and with $R_s = 6\ \Omega$, $R_j = 83\ \Omega$. The conversion loss is found from (7.3.17) to be −9.7 dB, and with 10 dBm of input power, the output power, P_L is 0.3 dBm or 1.07 mW. The load resistance, R_L, is found directly from (7.3.14) to be 59 Ω.

We now account for the junction capacitance. In a manner analogous to the design of the LO circuit in a diode mixer, we assumed that the junction capacitance can be approximated as a lumped capacitance equal to C_{j0}, in parallel with the junction. Thus, at the input frequency the diode is equivalent to an 89 Ω resistor in parallel with 0.05 pF, and at the output it is equivalent to a 59 Ω resistor in parallel with the 0.05 pF capacitor.

Four final details should be examined. First, we have considered only a multiplier having short-circuit embedding impedances. The performance of a multiplier having other terminations might be significantly different. However, especially in high-frequency multipliers, the short-circuit case is usually valid because, regardless of the diode's terminating impedances, the junction capacitance short-circuits the resistive junction at the higher harmonics of the input frequency. Second, it may be surprising that we never explicitly considered the diode's I/V characteristic. We did, however, account for it implicitly in our assumptions about the shape of the current pulse and $V_j(t)$. The unstated assumption was that the diode does not have an ideal rectifying characteristic (i.e., it is not a short circuit under forward bias) and neither does it have an unusually "soft" I/V characteristic (this would have rendered invalid the assumptions about $V_j(t)$ and $I_d(t)$). The justification for this assumption is no more or less than the author's years of staring at calculated diode current and voltage waveforms. Third, it may seem cavalier to assume that the desired value of I_{max} is achieved at the desired input level. Of course, P_{av} and I_{max} cannot be selected independently unless dc bias is used; dc bias can be varied to adjust the waveforms to achieve I_{max} and P_{av} simultaneously. Some judgment is necessary here; if we attempt to achieve a value of I_{max} that is unreasonable in view of P_{av}, the $I_d(t)$ and $V_j(t)$ waveforms will not approximate those in Figure 7.11, and the results will be unsatisfactory. Finally, the assumption in (7.3.8) that all the input power is dissipated in the diode and the conclusion that the efficiency is low may seem like a circular argument. It is not, because this assumption was used only to find an expression for the input power; the output power was determined from other considerations.

This design process is extremely simple, but it is also based on some rather drastic approximations. We can check these results, and perhaps learn a little more about multipliers, by performing harmonic-balance analyses. The harmonic-balance analysis of a diode frequency multiplier is essentially identical to the large-signal analysis of a mixer diode; the only difference is that the desired data are not the junction capacitance and conductance waveforms but the harmonic output power. The harmonic-balance analysis of this multiplier was performed by the same means as that used for the varactor multiplier in Example 1: The large-signal part of the computer program DIODEMX (References 2.1 and 5.2) was modified so that its output includes the second-harmonic current I_2, and power, P_L.

In the first analysis, we ignored the junction capacitance and assumed the diode to be purely resistive (we did not eliminate R_s, however). The results of this effort tell us how good the design really is. In a second analysis, we included the junction capacitance; the results of this analysis tell us whether the capacitance can indeed be modeled as we have suggested, and also whether its nonlinearity affects the conversion efficiency.

The results of these analyses are shown in Table 7.3. The first two columns of the table compare the approximate design and the harmonic-balance calculation that used $C_{j0} = 0$. The agreement between the two sets of results is remarkably good; the harmonic-balance calculation shows slightly better conversion loss, which is due primarily to a greater value of I_2. This difference is not surprising, in view of the high sensitivity of I_2 to the shape and duty cycle of the current pulses in $I_j(t)$. The third column shows the results of the harmonic-balance calculation that included C_{j0}. In this calculation, the input was precisely matched and the output impedance was varied to optimize the conversion loss at 10 dBm; the input was found to be modeled accurately as an 84-Ω resistor in parallel with C_{j0}, and the output was a 50-Ω resistor in parallel with $0.8C_{j0}$. The conversion loss in this case is very close to that of the purely resistive case; however, the load resistance is lower and I_2 is greater. This is not an unexpected result because the output power is relatively insensitive to load resistance, whereas I_2 is not. Thus, when the load impedance was varied in order to optimize the output power, it was not unlikely that a value of I_2 widely different from that of the purely resistive case would be obtained. The nonlinear junction capacitance did little either to improve or reduce conversion performance as long as the diode was matched.

Table 7.3 Comparison of Approximate and Harmonic-Balance Designs

	Approximate	*Harmonic Balance* (resistive diode)	*Harmonic Balance* (including C_j0)
Conversion Loss	−9.7 dB	−8.5 dB	−8.5 dB
R_i	89	82[1]	84[2]
R_L	59	59[1]	50[3]
I_{max}	30 mA	28 mA	28 mA
I_2	6.0 mA	6.9 mA	9.1 mA
I_{dc}	9.5 mA	9.7 mA	8.7 mA

Notes:

[1]In this calculation, the source and load impedances were not optimized; they were the same as those of the approximation. R_L is the design value; R_j is calculated.

[2]The input impedance was equivalent to 84 Ω in parallel with C_{j0}.

[3]The output impedance was equivalent to 50 Ω in parallel with $0.8C_{j0}$.

7.4 BALANCED MULTIPLIERS

It is a common practice to realize diode frequency multipliers in balanced structures. Balanced multipliers have significant advantages compared to single-ended multipliers; the most important are increased output power and the inherent rejection of certain unwanted harmonics. The input or load impedance of a balanced multiplier in some cases differs by a factor of two from that of a single-diode multiplier; therefore, a balanced multiplier sometimes provides more satisfactory input or load impedance.

Diode multipliers are sometimes interconnected via hybrids, but for economy they are more often used in the antiparallel or series forms described in Section 5.2.1. The antiparallel connection, shown in Figure 5.17, is probably the simplest form of a balanced multiplier; it rejects even harmonics of the input frequency and, consequently, can be used only as an even-order multiplier. In an antiparallel-diode multiplier, each diode effectively short-circuits the other at the second harmonic, so each diode acts as a type of idler for the other. This circuit does not reject the fundamental frequency, however, so it requires an output filter.

In theory, the antiparallel circuit can be used to realize either resistive or reactive multipliers. However, because the stability of a varactor multiplier is sensitive to slight unbalance between the diodes, varactor multipliers are not often realized as antiparallel circuits. It is thoroughly practical, however, to realize resistive multipliers this way, although the restriction to third-harmonic operation in the resistive multiplier results in low efficiency.

Figure 7.13 shows a multiplier realized in microstrip according to this concept; it consists of an input low-pass filter, a coupled-line output filter, and two diodes. The input filter is separated from the diodes by a section of transmission line that is one-quarter wavelength long at the output frequency; in this way the output impedance of the filter, which is very low at the output frequency, is transformed to a very high impedance at the diode and, therefore, does not affect the output matching. The input impedance of the coupled-line filter is very high at the input frequency. The element values of these filters can be modified so that they provide the necessary matching as well as filtering. Reverse bias can be provided to the diodes by series resistors; these must be bypassed effectively at RF frequencies or imbalance and poor efficiency will result. Forward bias requires a voltage source in place of the resistors.

The design of the multiplier is relatively straightforward. One begins by designing a single-diode prototype circuit that has short-circuit embedding impedances at the even harmonics of the input frequency and, in the

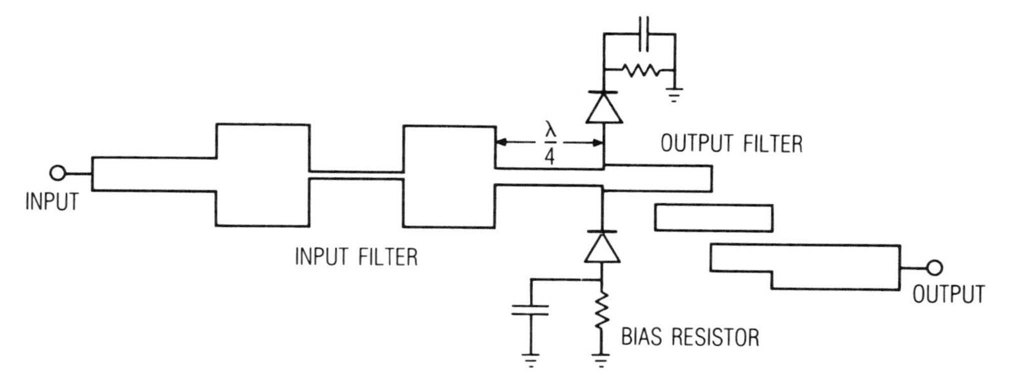

Figure 7.13 A microstrip realization of a resistive frequency multiplier using antiparallel diodes.

process, determines the input and load impedances. Because two devices are connected in parallel, the source and load impedances for the pair are half those necessary for a single device, so the filters and matching circuits are designed around these lower values. The input power for the pair must also be twice that of the single-diode prototype, although the conversion loss will be the same.

In contrast to the antiparallel circuit, the series circuit of Section 5.2.1 and Figure 5.23 rejects odd harmonics of the fundamental frequency (one of which is the fundamental frequency itself); in many multipliers, the third harmonic is usually relatively weak, so the series-diode multiplier often can be used without an output filter. The series interconnection is sometimes difficult to realize in strip-transmission media, but it can be realized very easily in waveguide or fin line.

Figure 7.14 shows a realization in waveguide of a series-diode multiplier; this circuit is popular because it requires neither input nor output filters. The two diodes are connected in series across the waveguide, and the excitation is applied to the node connecting them. The input excitation pumps the diodes out of phase; because their fundamental-frequency currents are not in phase, those currents do not excite the waveguide (furthermore, the width of the waveguide can usually be selected so that the waveguide is cut off at the fundamental frequency). The diodes' second-harmonic currents, however, are in phase, and they excite a TE_{10} mode very efficiently. Current components at higher odd harmonics are also out of phase; although these harmonics are usually weak, some harmonics may excite higher-order modes in the waveguide unless the waveguide is designed to prevent them.

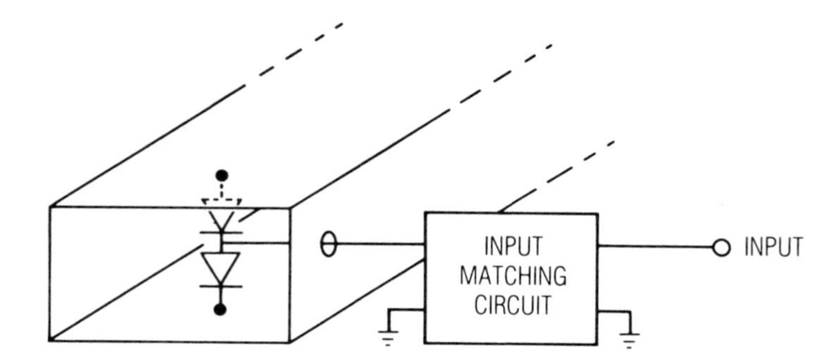

Figure 7.14 A frequency multiplier, with a waveguide output, using antiseries diodes.

The currents in the diodes are equal and have the same direction at the node connecting the diodes; thus, none of the second-harmonic current circulates in the input circuit. Furthermore, the wire connecting that node to the input circuit is perpendicular to the electric field of the TE_{10} mode, so the second-harmonic energy in the waveguide is not coupled to the input. Therefore, an input filter normally is unnecessary. The diodes are in parallel at the input port, but they are in series across the waveguide output port. Thus, the input matching circuit should be designed for half the input impedance that would be used for a single-diode multiplier, and the output load impedance presented to the pair should be twice the load impedance presented to a single-diode circuit. The diode mounting structure can be modeled in a manner identical to that of a mixer diode; that model is described in Sections 6.3.3 and 6.4.

The performance of such multipliers can be impressive. Reference 7.7 describes a realization of a frequency doubler that is identical in concept to that of Figure 7.14. This multiplier uses beam-lead diodes and a fin-line transition to waveguide and achieves an average of 13 dB conversion loss and 7 dBm output power over a 66–94 GHz output band. Although varactor multipliers in this frequency range usually can achieve better conversion efficiency, they cannot achieve such a wide instantaneous bandwidth. The performance of this multiplier is adequate for use in many applications, as, say, a low-power transmitter, a mixer LO source, or in test instruments.

REFERENCES

[7.1] M.T. Faber, J.W. Archer, and R.J. Mattauch, "A Frequency Doubler with 35 Percent Efficiency at W Band," *Microwave J.*, Vol. 28, No. 7, July 1985, p. 145.

[7.2] J.M. Manley and H.E. Rowe, "Some General Properties of Nonlinear Elements," *Proc. IRE,* Vol. 44, 1956, p. 904.

[7.3] C.B. Burckhardt, "Analysis of Varactor Frequency Multipliers for Arbitrary Capacitance Variations and Drive Level," *Bell Syst. Tech. J.,* Vol. 44, 1965, p. 675.

[7.4] S. Hamilton and R. Hall, "Shunt-Mode Harmonic Generation Using Step-Recovery Diodes," *Microwave J.*, Vol. 10, No. 4, April 1967, p. 69.

[7.5] D.L. Hedderly, "An Analysis of a Circuit for the Generation of High-Order Harmonics Using an Ideal Nonlinear Capacitor," *IEEE Trans. Electron Devices,* Vol. 9, 1962, p. 484.

[7.6] S. Goldman, "Computer Aids Design of Impulse Multipliers," *Microwaves and RF,* Vol. 22, No. 10, October 1983, p. 101.

[7.7] C. Nguyen, "A 35 Percent Bandwidth *Q*- to *W*-Band Frequency Doubler," *Microwave J.*, Vol. 30, No. 9, September 1987, p. 232.

CHAPTER 8

MESFET SMALL-SIGNAL AMPLIFIERS

The GaAs MESFET has made possible the practical design of small-signal amplifiers at frequencies well into the millimeter-wave region. The best of these amplifiers have spectacular noise figures, close to those of parametric amplifiers, and have good gain and linearity. Furthermore, the MESFET is adaptable to monolithic integration, allowing the development of compact, low-cost, highly reliable microwave circuits and systems.

This chapter is concerned with small-signal nonlinear distortion phenomena in MESFET amplifiers. Such amplifiers are designed primarily to have low noise figures or specific values of small-signal gain, and their linearity usually must be optimized within gain and noise constraints. However, in many systems, especially broadband receivers, sensitivity may be limited by distortion products generated by strong interfering signals as well as noise; in the design of such systems, we must make a prudent trade-off between noise and linearity. The distortion phenomena of greatest concern are saturation, intermodulation distortion, harmonic distortion, and AM/PM conversion; these terms are defined in Section 1.3. We shall see that Volterra-series analysis is applicable to all these phenomena, although harmonic-balance analysis is preferable for determining single-tone saturation effects.

8.1 REVIEW OF LINEAR AMPLIFIER THEORY

8.1.1 Stability Considerations in Linear Amplifier Design

When used in a linear amplifier, a GaAs MESFET is usually operated in its common-source configuration. In its simplest form, a FET amplifier consists of the MESFET, an input matching network, and an output matching network. Bias circuitry must also be included; however, a well designed bias network does not affect the RF matching of the device, so we will not consider the bias circuit further. The circuit model of the FET amplifier is shown in Figure 8.1(a).

The MESFET is treated in the design process as a two-port circuit and is described by a set of two-port parameters, usually *S* or *Y* parameters (*S* parameters are preferred in most microwave applications). The *S* parameters of a MESFET vary with dc bias and, therefore, must be measured at the bias voltages at which the device will be operated. If the matching networks are lossless (we will assume that they are), they can be represented as lumped impedances or reflection coefficients at the operating frequency, and we can redraw the circuit in Figure 8.1(a) to form the canonical equivalent circuit shown in Figure 8.1(b).

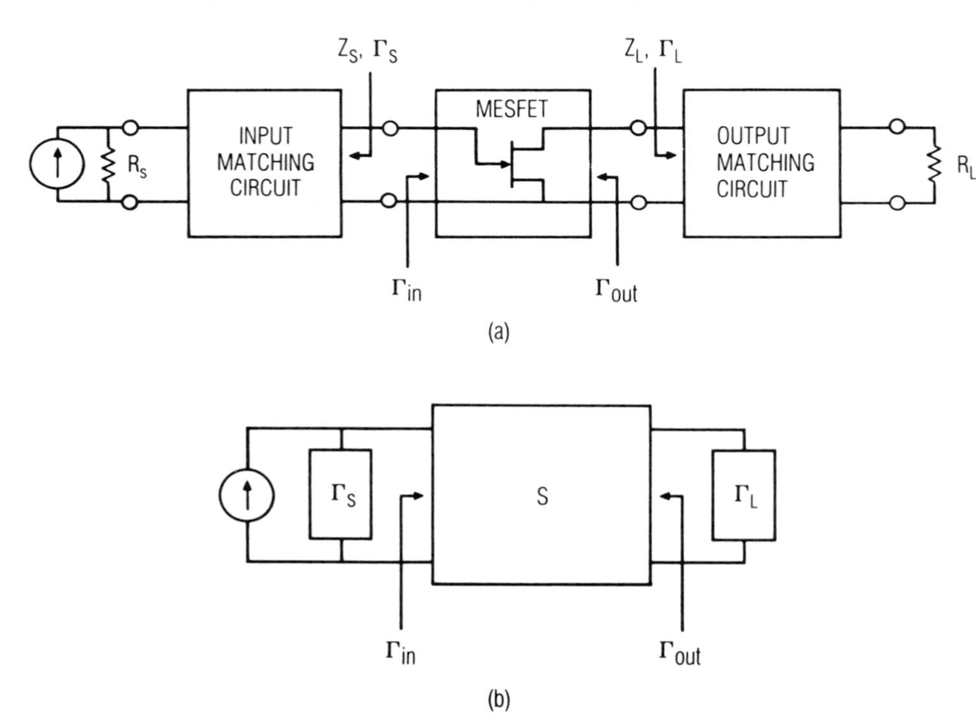

Figure 8.1 (a) Small-signal amplifier consisting of input and output matching circuits and a MESFET; (b) canonical model of the amplifier.

As an alternative to a two-port, we can represent the MESFET by a lumped-element quasistatic equivalent circuit; because the elements of the equivalent circuit can be treated as nonlinear, this representation can be used for nonlinear as well as linear analysis. If the lumped-element model is well conceived, its S parameters can be calculated easily and they should agree well with those measured from the device. A widely used small-signal equivalent circuit of a GaAs MESFET is shown in Figure 8.2.

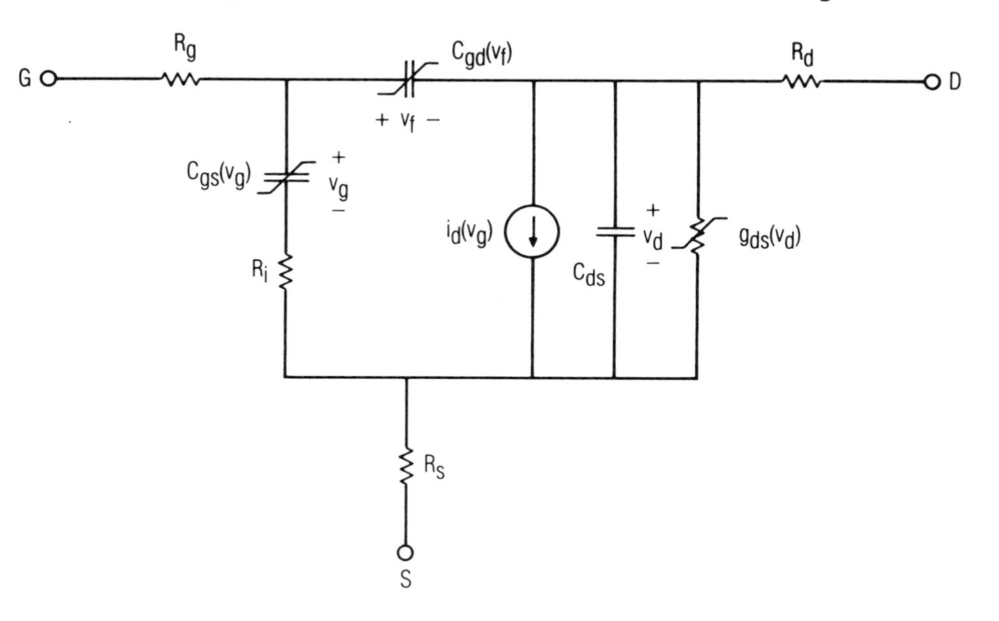

Figure 8.2 Small-signal nonlinear MESFET equivalent circuit. Four elements (C_{gs}, C_{gd}, i_d, and g_{ds}) are nonlinear, although C_{gd} often can be treated as a linear element.

The nonzero value of $S_{1,2}$ implies that the device has feedback; that feedback is a consequence of elements C_{gd} and R_s in Figure 8.2. A high value of R_s tends to stabilize the device (although it reduces gain and increases noise figure), but C_{gd} degrades stability; MESFETs having large values of C_{gd} generally have poor stability. A linear amplifier can be made to oscillate only if the input or output impedance has a negative real part; in any specific device, those port impedances are functions solely of the MESFET's S parameters and its source and load terminations.

At some frequencies, it is possible to find combinations of source and load impedances that result in an output or input impedance having a negative real part and that, therefore, can cause oscillation; at other frequencies, passive terminations that can cause oscillation cannot be found. In the former case, we say that the MESFET is *conditionally stable* or, equivalently, *potentially unstable;* in the latter, the MESFET is *unconditionally stable.* Generally, a MESFET is conditionally stable at low frequencies and unconditionally stable at the high end of its useful frequency range.

Stability in FET amplifiers is important for reasons beyond the natural desire to prevent oscillation: It affects the criteria for which the amplifier can be designed. When the amplifier is unconditionally stable, it is always possible to achieve a conjugate match simultaneously at the input and output; when the amplifier is conditionally stable, it is generally impossible to achieve a simultaneous conjugate match (in theory it is sometimes possible to achieve a simultaneous conjugate match in a conditionally stable device, but in practice the necessary conditions never occur in practical small-signal MESFETs). Furthermore, many of the characteristics of a MESFET that improve performance (the most important being high transconductance, low C_{gs}, and low R_s) also raise the minimum frequency at which the device is unconditionally stable. Thus, most high-quality MESFETs are only conditionally stable at microwave frequencies.

When a device is unconditionally stable, the design process can be very simple: we calculate the source and load impedances that result in a simultaneous conjugate match (so-called *SCM conditions*) and design matching networks that present these impedances to the gate and drain of the device. The gain that results is the *maximum available gain,* or MAG (Section 1.5). SCM conditions may not be desirable, however, for several reasons: (1) the device may be conditionally stable; (2) a value of gain other than the MAG may be desired; (3) an input mismatch must be used if optimum noise figure is desired; or (4) in a broadband amplifier, we may wish to mismatch either the source or load at low frequencies in order to obtain a flat passband. In these cases, a unique set of source and load impedances that result in the desired gain generally does not exist. Consequently, our design procedure must allow us to select source and load terminations that result in a specific value of gain and, in some cases, an acceptable noise figure. The design procedure must also prevent us from inadvertently using source or load impedances that result in instability.

The conditions necessary and sufficient for unconditional stability are that the Linvill stability factor, K, is greater than 1.0 and that the magnitude of the determinant of the S matrix, $|\Delta_S|$, is less than 1.0. If either of these conditions is not met, the device is conditionally stable. SCM conditions can be found if $K > 1$, regardless of $|\Delta_S|$, although this situation rarely occurs in practice and never occurs in practical GaAs MESFETs. The determinant of the S matrix is

$$\Delta_S = S_{1,1}S_{2,2} - S_{2,1}S_{1,2} \tag{8.1.1}$$

and the stability factor, K, is

$$K = \frac{1 - |S_{1,1}|^2 - |S_{2,2}|^2 + |\Delta_S|^2}{2|S_{1,2}S_{2,1}|} \tag{8.1.2}$$

It is interesting to note that an amplifier having lossless input and output matching circuits has the same value of K as the MESFET it uses.

If the device is conditionally stable, we need to know the input and output terminations that can cause oscillation, the source and load reflection coefficients for which

$$|\Gamma_{in}| > 1.0 \tag{8.1.3}$$

and

$$|\Gamma_{out}| > 1.0 \tag{8.1.4}$$

where Γ_{in} and Γ_{out} are the respective input and output reflection coefficients of the device. These are given by the following relations:

$$\Gamma_{in} = S_{1,1} + \frac{S_{2,1}S_{1,2}\Gamma_L}{1 - S_{2,2}\Gamma_L} \tag{8.1.5}$$

$$\Gamma_{out} = S_{2,2} + \frac{S_{2,1}S_{1,2}\Gamma_s}{1 - S_{1,1}\Gamma_s} \tag{8.1.6}$$

The solutions of (8.1.3) and (8.1.4) are regions in the plane of the load and source reflection coefficients, respectively, and can be plotted conveniently on a Smith chart. The borders of the regions are circles; the values of Γ_L that border the stability region defined by (8.1.3) and (8.1.5) is called the *output stability* circle. Its center, C_L, is

$$C_L = \frac{(S_{2,2} - \Delta_S S_{1,1}^*)^*}{|S_{2,2}|^2 - |\Delta_S|^2} \tag{8.1.7}$$

and its radius r_L is

$$r_L = \left| \frac{S_{1,2}S_{2,1}}{|S_{2,2}|^2 - |\Delta_S|^2} \right| \tag{8.1.8}$$

The input stability circle defines the boundaries of the region in which Γ_s satisfies (8.1.4). Its center and radius, C_s and r_s, are

$$C_s = \frac{(S_{1,1} - \Delta_S S_{2,2}^*)^*}{|S_{1,1}|^2 - |\Delta_S|^2} \tag{8.1.9}$$

and

$$r_s = \left| \frac{S_{1,2} S_{2,1}}{|S_{1,1}|^2 - |\Delta_S|^2} \right| \tag{8.1.10}$$

Equations (8.1.7) through (8.1.10) give the boundaries of the stability regions, but they do not indicate whether the region that ensures stability is inside or outside the stability circle. The stable region is determined easily from the following considerations: if $|S_{1,1}| < 1$, then the point $\Gamma_L = 0$, the center of the Smith chart that defines the output stability circles, must be in the stable region; similarly, if $|S_{2,2}| < 1$, the point $\Gamma_s = 0$ in the input plane must be within the stable region. In practical GaAs MESFETs that do not employ external feedback, the outside of the circle is invariably the stable region.

8.1.2 Amplifier Design

Designing a small-signal MESFET amplifier primarily involves selecting the appropriate source and load impedances (or reflection coefficients) and designing the input and output matching circuits to present those impedances to the device. If the device is unconditionally stable and maximum gain is desired, the process of determining source and load reflection coefficients is straightforward. The reflection coefficients that give a simultaneous conjugate match, $\Gamma_{s,m}$ and $\Gamma_{L,m}$, are

$$\Gamma_{s,m} = \frac{B_1 \pm (B_1^2 - 4|C_1|^2)^{1/2}}{2C_1} \tag{8.1.11}$$

and

$$\Gamma_{L,m} = \frac{B_2 \pm (B_2^2 - 4|C_2|^2)^{1/2}}{2C_2} \tag{8.1.12}$$

where

$$B_1 = 1 + |S_{1,1}|^2 - |S_{2,2}|^2 - |\Delta_S|^2 \tag{8.1.13}$$

$$B_2 = 1 + |S_{2,2}|^2 - |S_{1,1}|^2 - |\Delta_S|^2 \tag{8.1.14}$$

$$C_1 = S_{1,1} - \Delta_S S_{2,2}^* \tag{8.1.15}$$

and

$$C_2 = S_{2,2} - \Delta_S S_{1,1}^* \tag{8.1.16}$$

Under SCM conditions, the transducer gain is the maximum available gain; it is

$$G_t = \text{MAG} = \frac{|S_{2,1}|}{|S_{1,2}|}\left[K - (K^2 - 1)^{1/2}\right] \tag{8.1.17}$$

If the device is conditionally stable, or if it is unconditionally stable but the desired gain is less than the maximum available gain, the source and load reflection coefficients that give the desired gain are not unique. If the source reflection coefficient is specified, the locus of load reflection coefficients that give the desired gain is a circle in the reflection-coefficient (Smith chart) plane; similarly, if the load is specified, the source reflection coefficients lie on a circle. Although the desired gain can often be achieved without a conjugate match at either the input or output, it is usually desirable to match at least one port; having one port well matched allows stages to be cascaded easily and with minimal gain variation over a passband.

An amplifier having one conjugate-matched port can be designed by employing the concepts of available gain G_a and power gain G_p (Section 1.5). These quantities are defined as

$$\begin{aligned} G_p &= \frac{\text{power delivered to the load}}{\text{power delivered to the network}} \\ &= \frac{1}{1 - |\Gamma_{\text{in}}|^2} |S_{2,1}|^2 \frac{1 - |\Gamma_L|^2}{|1 - S_{2,2}\Gamma_L|^2} \end{aligned} \tag{8.1.18}$$

and

$$G_a = \frac{\text{power available from the network}}{\text{power available from the source}}$$

$$= \frac{1 - |\Gamma_s|^2}{|1 - S_{1,1}\Gamma_s|^2} |S_{2,1}|^2 \frac{1}{1 - |\Gamma_{out}|^2} \tag{8.1.19}$$

where Γ_{in} and Γ_{out} are given by (8.1.5) and (8.1.6). From (8.1.18), we can see that G_p is independent of Γ_s; therefore, designing an amplifier to have a specific value of power gain requires only selecting Γ_L. However, the quantity that we loosely call gain is in fact transducer gain G_t, which is defined as

$$G_t = \frac{\text{power delivered to the load}}{\text{power available from the source}} \tag{8.1.20}$$

One of the many possible expressions for G_t is

$$G_t = \frac{1 - |\Gamma_s|^2}{|1 - S_{1,1}\Gamma_s|^2} |S_{2,1}|^2 \frac{1 - |\Gamma_L|^2}{|1 - \Gamma_{out}\Gamma_L|^2} \tag{8.1.21}$$

If the input is conjugately matched, then the power delivered to the network equals the power available from the source, and from (8.1.18) and (8.1.20), $G_t = G_p$. Similarly, (8.1.19) indicates that the available gain is independent of Γ_L; achieving the desired value of G_a requires only selecting Γ_s. If the output is matched, the power delivered to the load equals the power available from the network, and $G_t = G_a$. Thus, one can achieve a specified value of G_t by designing the amplifier to have G_p or G_a equal to the desired value of G_t and then conjugate-matching the input or output, respectively. The design procedure is as follows:

1. Select the desired transducer gain G_t.
2. Decide which port is to be matched.
3. If the input is to be matched, select Γ_L to achieve $G_p = G_t$; then find $\Gamma_s = \Gamma_{in}^*$ from (8.1.5).
4. If the output is to be matched, select Γ_s to achieve $G_a = G_t$; then find $\Gamma_L = \Gamma_{out}^*$ from (8.1.6).

The remaining problem is to find the values of Γ_L that provide the specified G_p or the values of Γ_s that provide the specified G_a; these quantities lie on circles in the load or source planes, respectively. The center and radius of the Γ_L circle, called the *power gain circle*, are, respectively,

$$C_p = \frac{g_p(S_{2,2} - \Delta_S S_{1,1}^*)^*}{1 + g_p(|S_{2,2}|^2 - |\Delta_S|^2)} \tag{8.1.22}$$

and

$$r_p = \frac{(1 - 2Kg_p|S_{1,2}S_{2,1}| + g_p^2|S_{1,2}S_{2,1}|^2)^{1/2}}{1 + g_p(|S_{2,2}|^2 - |\Delta_S|^2)} \tag{8.1.23}$$

where K is given by (8.1.2) and $g_p = G_p/|S_{2,1}|^2$. The loci of Γ_s that give constant available gain are also circles, and their centers and radii are given by the similar relations:

$$C_a = \frac{g_a(S_{1,1} - \Delta_S S_{2,2}^*)^*}{1 + g_a(|S_{1,1}|^2 - |\Delta_S|^2)} \tag{8.1.24}$$

and

$$r_a = \frac{(1 - 2Kg_a|S_{1,2}S_{2,1}| + g_a^2|S_{1,2}S_{2,1}|^2)^{1/2}}{1 + g_a(|S_{1,1}|^2 - |\Delta_S|^2)} \tag{8.1.25}$$

Comparing (8.1.24) and (8.1.9), we can see that the centers of the input stability circle and available-gain circle lie on the same line; similarly, the centers of the output stability circle and power-gain circle lie on the same line. Moreover, although it is not obvious from the equations, the gain and stability circles intersect at the edge of the reflection-coefficient plane.

Example 1

A MESFET has the following S parameters at 10 GHz:

$$S_{1,1} = 0.8\ \angle -85^\circ \qquad S_{1,2} = 0.10\ \angle 45^\circ$$

$$S_{2,1} = 1.7\ \angle 125^\circ \qquad S_{2,2} = 0.65\ \angle -70^\circ$$

We wish to find the input and output stability circles and gain circles that represent G_p and G_a values of 10 dB. We first use (8.1.2) to find K, and calculate Δ_S from (8.1.1). We find that $K = 0.2705$ and $\Delta_S = 0.393\ \angle -140^\circ$, so the device is conditionally stable. If the device were unconditionally stable there would be no need to find the stability circles. Equations (8.1.7) and (8.1.8) give the output stability circle; its center and radius are 1.32 $\angle 83°$ and 0.634, respectively. Similarly (8.1.9) and (8.1.10) give the center and radius of the input stability circle; these are, respectively, 1.145 $\angle 92°$ and 0.350.

We now calculate the gain circles. First, we find $g_p = g_a = 3.16/1.7^2 = 1.093$; then using the K and Δ_S found earlier, we use (8.1.22) and (8.1.23) to find the power-gain circle; its center is 0.636 $\angle 83°$ and its radius is 0.526. Similarly, we use (8.1.24) and (8.1.25) to find the available-gain circle's center and radius, 0.718 $\angle 92°$ and 0.378, respectively. These circles are plotted on a Smith chart in Figure 8.3.

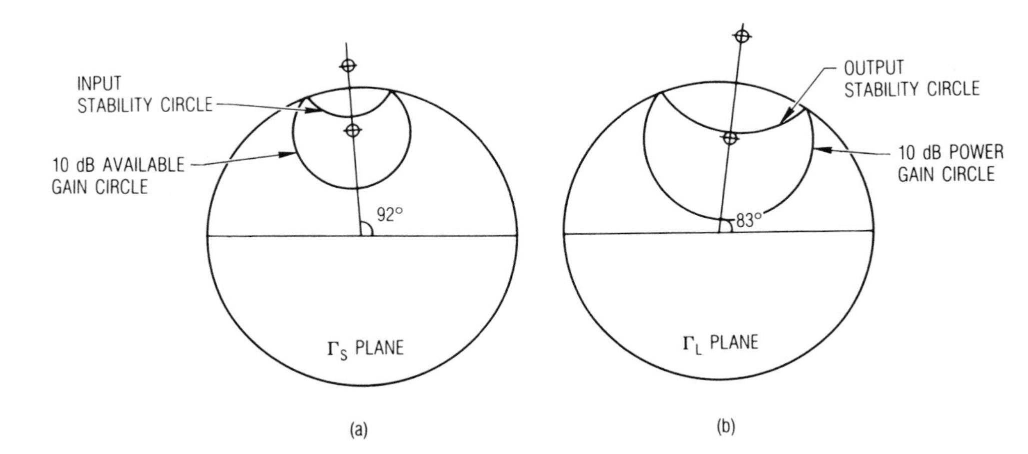

Figure 8.3 Stability and gain circles of the FET in Example 1: (a) input plane; (b) output plane.

We have now identified a range of values of Γ_s or Γ_L that must be used in order to achieve a specified value of transducer gain; however, we still have no clear rationale for selecting one value over another. Clearly, it is wise to pick a value that is not too close to the stability circle, or a small source or load mismatch may cause oscillation. A consideration in the design of a low-noise amplifier is that Γ_s should be as close as possible to the value that optimizes the noise figure; thus, we would pick Γ_s to optimize the noise figure and would choose $\Gamma_L = \Gamma_{\text{out}}^*$ to match the output. A third criterion (the one we all have been waiting for!) is to pick Γ_s or Γ_L to optimize linearity, perhaps within constraints on gain and noise figure. The remainder of this chapter is devoted to an examination of that criterion.

8.2 NONLINEAR ANALYSIS OF SMALL-SIGNAL FET AMPLIFIERS

The nonlinear analysis of the FET amplifier requires the use of the lumped-element equivalent circuit of Figure 8.2, along with appropriate source and load networks. The Volterra series is the logical means to analyze the circuit to evaluate such small-signal nonlinear effects such as harmonics, intermodulation, or AM/PM conversion; because the circuit includes feedback elements and reactive nonlinearities, power-series analysis cannot be used. For Volterra-series analysis to be used, the significant nonlinear elements must be characterized via a power series in terms of their small-signal control voltages.

8.2.1 Nonlinearities in the MESFET Equivalent Circuit

The equivalent circuit in Figure 8.2 shows four nonlinear elements: the gate-drain capacitance, C_{gd}; the gate-source capacitance, C_{gs}; the controlled current source, i_d; and the drain-source conductance, g_{ds}. When used in a small-signal amplifier, a MESFET is always operated well into its saturation region; the description of the GaAs MESFET's large-signal model in Section 2.4.2 showed that in saturation C_{gs} depends only weakly upon v_d, and C_{gd} depends so weakly upon v_g and v_d that it often can be treated as a linear element. Thus, C_{gs} is shown in Figure 8.2 as a function of only v_g, and C_{gd} (if it is treated as a nonlinear element) as a function of only the voltage v_f across it.

The nonlinearities of the capacitances usually do not dominate in the establishment of the small-signal nonlinear performance of the circuit; the dominant element is usually $i_d(v_g)$ or (infrequently) $g_{ds}(v_d)$. Therefore, we can often take some reasonable liberties with the nonlinear characterization of these less significant elements. In particular, it is a common practice to treat C_{gs} as a Schottky-barrier capacitance, even though it sometimes follows the ideal Schottky-barrier C/V curve only moderately well. The Schottky-barrier characteristic invariably represents a nonlinearity that is far too strong to represent C_{gd} accurately. When the C/V or Q/V characteristic of an element has been determined, the incremental power-series representation can be found from (2.2.14) through (2.2.17).

The controlled current source, i_d, and the gate-drain conductance, g_{ds}, represent the MESFET's channel current, a single nonlinearity that has the two control voltages, v_g and v_d. Equation (2.2.8) gives a Taylor-series characterization of a multiply controlled nonlinearity; we can identify

v_1 in (2.2.8) as v_g, v_2 as v_d, and the MESFET's dc I/V characteristic, $I_d(V_g, V_d)$, as f (as in Chapter 2, we let capital letters represent large-signal voltages and currents and lower-case letters represent incremental ones). If we ignore the cross terms in (2.2.8) (those that include the product term v_1v_2; these are at least an order of magnitude smaller than the other terms), we can treat the nonlinearity as two nonlinear elements in parallel, one depending upon v_g and the other on v_d. The equation can then be split into two parts, one representing the dependence on v_1 and the other representing the dependence on v_2. After substituting and rearranging (2.2.8), we obtain

$$i_d = \frac{\partial I_d}{\partial V_g} v_g + \frac{1}{2}\frac{\partial^2 I_d}{\partial V_g^2} v_g^2 + \frac{1}{6}\frac{\partial^3 I_d}{\partial V_g^3} v_g^3 + \ldots$$

$$+ \frac{\partial I_d}{\partial V_d} v_d + \frac{1}{2}\frac{\partial^2 I_d}{\partial V_d^2} v_d^2 + \frac{1}{6}\frac{\partial^3 I_d}{\partial V_d^3} v_d^3 + \ldots \qquad (8.2.1)$$

where the derivatives are evaluated at the dc bias points V_{g0} and V_{d0}. The terms in (8.2.1) that contain v_g represent a nonlinear controlled current source, and the ones that contain v_d represent a nonlinear conductance. The former is, of course, the current source, $i_d(v_g)$, in Figure 8.2, and the latter is $g_{ds}(v_d)$.

An unfortunate complication of this neat situation is that the drain conductance, $g_{ds}(v_d)$, often depends upon frequency, and the value of g_{ds} obtained from dc I/V measurements is usually much lower than that measured at high frequencies. For this reason, it is probably better to determine $g_{ds}(v_g)$ by fitting the equivalent circuit to measured S parameters (Section 2.5). It is important to remember that the Volterra-series analysis requires a series expansion of this element's incremental I/V characteristic, *not* of its G/V characteristic.

8.2.2 Nonlinear Phenomena in FET Amplifiers

The nonlinear phenomena of greatest concern in FET amplifiers are AM/PM conversion, harmonic generation, intermodulation, and saturation. The first three of these phenomena can be analyzed by Volterra techniques; saturation characteristics can be found by either Volterra-series or harmonic-balance analysis. For saturation calculations, harmonic-balance analysis is probably preferable to Volterra-series analysis because the harmonic-balance approach can include the effects of the bias circuit and strong nonlinearities in the MESFET model. These effects are often the dominant ones in establishing saturation characteristics and are generally not modeled by the Volterra series. Nevertheless, in situations where saturation effects are dominated by weak nonlinearities, especially the MESFET's nonlinear transconductance, Volterra-series analysis is an acceptable method for determining saturation characteristics.

We use the macroscopic ("black box") circuit model of the amplifier; this model is shown in Figure 8.4. The excitation is the signal $v_s(t)$, which consists of Q sinusoidal components:

$$v_s(t) = \frac{1}{2} \sum_{\substack{q=-Q \\ q \neq 0}}^{Q} V_{s,q} \exp(j\omega_q t) \tag{8.2.2}$$

The response $i(t)$, the output current, is

$$\begin{aligned} i(t) = \sum_{n=1}^{N} \frac{1}{2^n} \sum_{q1=-Q}^{Q} \sum_{\substack{q2=-Q \\ q1\ ...\ qn \neq 0}}^{Q} \cdots \sum_{qn=-Q}^{Q} & V_{s,q1} V_{s,q2} \cdots \\ & \cdot V_{s,qn} H_n(\omega_{q1}, \omega_{q2}, \ldots, \omega_{qn}) \\ & \cdot \exp[(\omega_{q1} + \omega_{q2} + \ldots \omega_{qn})t] \end{aligned} \tag{8.2.3}$$

The current $i(t)$ is the sum of all the nth-order output currents $i_n(t)$; an nth-order output current is the sum of all current components that arise from mixing between n input frequencies. The function $H_n(\omega_{q1}, \omega_{q2}, \ldots, \omega_{qn})$, called the *nth-order nonlinear transfer function*, relates the output current at the frequency $\omega_{q1} + \omega_{q2} + \ldots \omega_{qn}$ to the individual components of $v_s(t)$ at those frequencies. In this section, we assume that the nonlinear transfer functions of the circuit are known, and we show how they can be used to evaluate a circuit's nonlinear behavior. The next section is concerned with analyzing the MESFET equivalent circuit in order to determine those transfer functions.

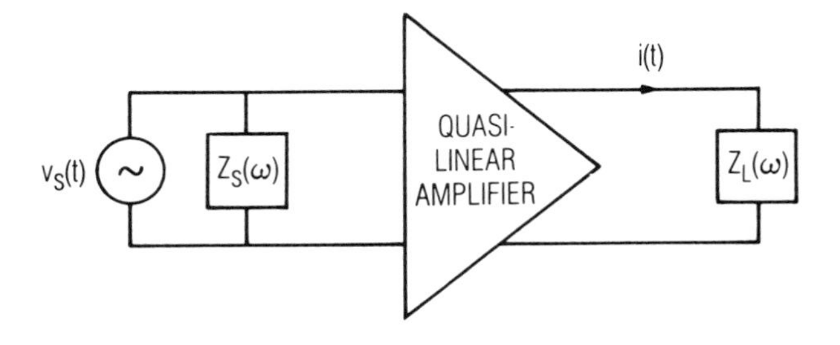

Figure 8.4 Quasilinear amplifier model.

Saturation and AM/PM Conversion

When the amplifier is driven into saturation by a single sinusoidal signal at frequency ω_1, the output current at ω_1 can be found by evaluating (8.2.3) under conditions of a single-tone excitation and by retaining only the terms at ω_1. The result is

$$I(\omega_1) = V_{s,1}H_1(\omega_1) + \frac{3}{4} V_{s,1}V_{s,1}V_{s,1}^{*}H_3(\omega_1, \omega_1, -\omega_1) \tag{8.2.4}$$

where $I(\omega)$ is the component of the output current $i(t)$ at ω. In (8.2.4), we have considered only the positive-frequency part of $I(\omega)$ (i.e., $I(\omega)$ is a phasor) and have limited the summation over n to $N = 3$; components of order greater than 3 are neglected. The coefficient of 3 in the second term of (8.2.4), and similar coefficients in the following equations, may be confusing. They arise because there are multiple identical terms in (8.2.3) at any particular mixing frequency.

Although it may not be obvious, (8.2.4) predicts that as $|V_{s,1}|$ increases, $|I(\omega_1)|$ saturates and begins to decrease. Equation (8.2.4) is valid if $|V_{s,1}|$ remains small enough that $|I(\omega_1)|$ does not decrease with an increase in $|V_{s,1}|$; beyond that point, higher-order terms in the series must be included. The next highest-order component at the same frequency is fifth order; such high orders become significant only as the amplifier is driven strongly into saturation. We define the relative distortion $D(\omega)$ as the ratio of the total output current to the linear (first order) part. $D(\omega)$ represents the fractional deviation from linear operation:

$$D(\omega) = \frac{I(\omega)}{V_{s,1}H_1(\omega_1)} \tag{8.2.5}$$

and substituting (8.2.4) into (8.2.5) gives

$$D(\omega_1) = 1 + \frac{3}{4}|V_{s,1}|^2 \frac{H_3(\omega_1, \omega_1, -\omega_1)}{H_1(\omega_1)} \tag{8.2.6}$$

Equation (8.2.6) indicates that $D(\omega_1)$ can be expressed as the sum of two phasors, as shown in Figure 8.5. If $V_{s,1}$ is very small, $D(\omega_1) = 1$, which indicates linear operation. As $V_{s,1}$ increases, however, $D(\omega_1)$ changes in both magnitude and phase; in FET amplifiers the phase of H_3/H_1 is always such that $|D(\omega_1)|$ decreases, which indicates that the gain decreases (i.e., the output power begins to saturate). The existence of a nonzero phase shift θ shows that, as the device begins to saturate, the

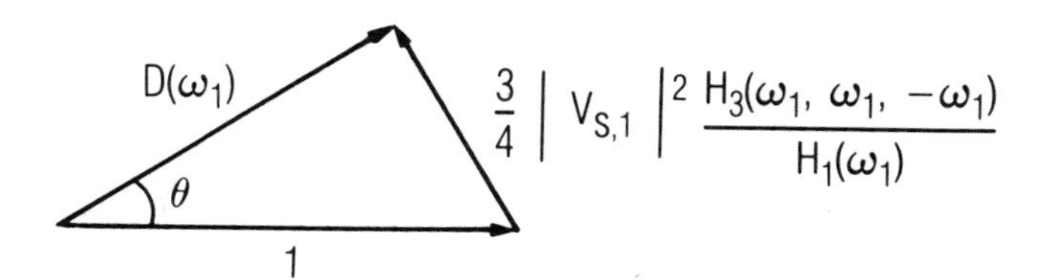

Figure 8.5 Relative distortion vector $D(\omega_1)$ describing saturation and AM/PM conversion, where $|D(\omega_1)|$ is the gain compression and θ is the phase deviation.

phase shift also begins to deviate from its value when $V_{s,1}$ is small; this phenomenon is called *AM/PM conversion.*

Harmonic Generation

Again we consider a single-tone excitation at ω_1. The positive-frequency component of $I(\omega)$ in (8.2.3) at the nth harmonic of ω_1 is:

$$I(n\omega_1) = 2^{-n+1} V_{s,1}^n H_n(\omega_1, \omega_1, \ldots, \omega_1) \tag{8.2.7}$$

for example, the second harmonic output current is

$$I(2\omega_1) = \frac{1}{2} V_{s,1}^2 H_2(\omega_1, \omega_1) \tag{8.2.8}$$

and the third harmonic is

$$I(3\omega_1) = \frac{1}{4} V_{s,1}^3 H_3(\omega_1, \omega_1, \omega_1) \tag{8.2.9}$$

A harmonic can also have a component at a higher order; for example, the second harmonic can include a fourth-order component:

$$I(2\omega_1) = \frac{1}{2} V_{s,1}^2 H_2(\omega_1, \omega_1) + \frac{1}{2} V_{s,1}^3 V_{s,1}^* H_4(\omega_1, \omega_1, \omega_1, -\omega_1) \tag{8.2.10}$$

Note that four identical terms in (8.2.3) contribute to the second term in (8.2.10). We can, of course, pick the phase of $V_{s,1}$ arbitrarily without losing generality, so the conjugate quantity is not significant. In general, an even harmonic can have components at all even orders, and an odd harmonic can have components at all odd orders. The components at orders greater than the lowest, however, are only significant when $V_{s,1}$ approaches saturation.

The relative distortion of the lowest-order component at the nth harmonic is, from (8.2.5),

$$D(n\omega_1) = \frac{2^{-n+1} V_{s,1}^{n-1} H_n(\omega_1, \omega_1, \ldots, \omega_1)}{H_1(\omega_1)} \tag{8.2.11}$$

Because the harmonic output frequency is not equal to the frequency of the linear output, $D(n\omega_1) = 0$ in linear operation.

Intermodulation Distortion

Intermodulation involves the effects of mixing between the fundamental frequencies and harmonics when two or more excitation frequencies exist. If the excitation contains the frequencies $\omega_1, \omega_2, \omega_3, \ldots$, the output may contain the frequencies $m\omega_1 + n\omega_2 + p\omega_3 + \ldots$ where m, n, and p are positive or negative integers. Many of these mixing products are potentially troublesome, but the case that is universally annoying is the one in which two excitation frequencies exist, ω_1 and ω_2, and the intermodulation distortion product is at $2\omega_1 - \omega_2$. Then,

$$I(2\omega_1 - \omega_2) = \frac{3}{4} V_{s,1}^2 V_{s,2}^* H_3(\omega_1, \omega_1, -\omega_2) \tag{8.2.12}$$

Higher-order current components can contribute to a mixing product at $2\omega_1 - \omega_2$. Thus,

$$\begin{aligned} I(2\omega_1 - \omega_2) &= \frac{3}{4} V_{s,1}^2 V_{s,2}^* H_3(\omega_1, \omega_1, -\omega_2) \\ &\quad + 5 V_{s,1}^3 V_{s,1}^* V_{s,2}^* H_5(\omega_1, \omega_1, \omega_1, -\omega_1, -\omega_2) \\ &\quad + \frac{15}{2} V_{s,1}^2 V_{s,2} V_{s,2}^{*2} H_5(\omega_1, \omega_1, \omega_2, -\omega_2, -\omega_2) \end{aligned} \tag{8.2.13}$$

As with the other distortion products, the components of order greater than 3 represent saturation effects and are not significant at very small $V_{s,1}$ and $V_{s,2}$. The relative distortion, when $V_{s,1}$ and $V_{s,2}$ are small, is

$$D(2\omega_1 - \omega_2) = \frac{3}{4} V_{s,1} V_{s,2}^* \frac{H_3(\omega_1, \omega_1, -\omega_2)}{H_1(\omega_1)} \tag{8.2.14}$$

The relative distortion, as it is defined for intermodulation and harmonic generation, is an important quantity. Its magnitude squared is the ratio of the power in the distortion component to the linear power or, more colloquially, the signal-to-distortion ratio. This is an important quantity in specifying a system and, as shown in Chapter 4, can be used to define the intermodulation intercept point of a system or component.

8.2.3 Calculating the Nonlinear Transfer Functions

A method of calculating the nonlinear transfer functions in a multinode circuit is described in Section 4.2.6. In this section, we will apply that method to the MESFET equivalent circuit in Figure 8.2.

We begin by redrawing the circuit in Figure 8.2 as shown in Figure 8.6, as a multiport network having a port in parallel with each nonlinear element as well as the input and output; a port has been included at C_{gd} in case it is to be treated as a nonlinear element. Figure 8.6 is linearized, only the linear parts of the nonlinear elements are included. Nonlinear current sources placed at each port will account for their nonlinearities. Figure 8.6 also includes source and load impedances, $Z_s(\omega)$ and $Z_L(\omega)$, and an excitation source at Port 5, $v_s(t)$. The inductance of the bond wire between the source and ground is often a significant parameter: unlike the inductances of the gate and drain bond wires, it cannot be absorbed into the source or load impedances; for this reason, we have included an inductor, L, in series with the source. As in Chapter 4, we limit nonlinear conductances and capacitances to the third degree and we assume that their incremental I/V and C/V characteristics can be expressed as in (4.2.115) and (4.2.116). The excitation, $v_s(t)$, is expressed, as usual, by (4.2.76) or (4.2.77).

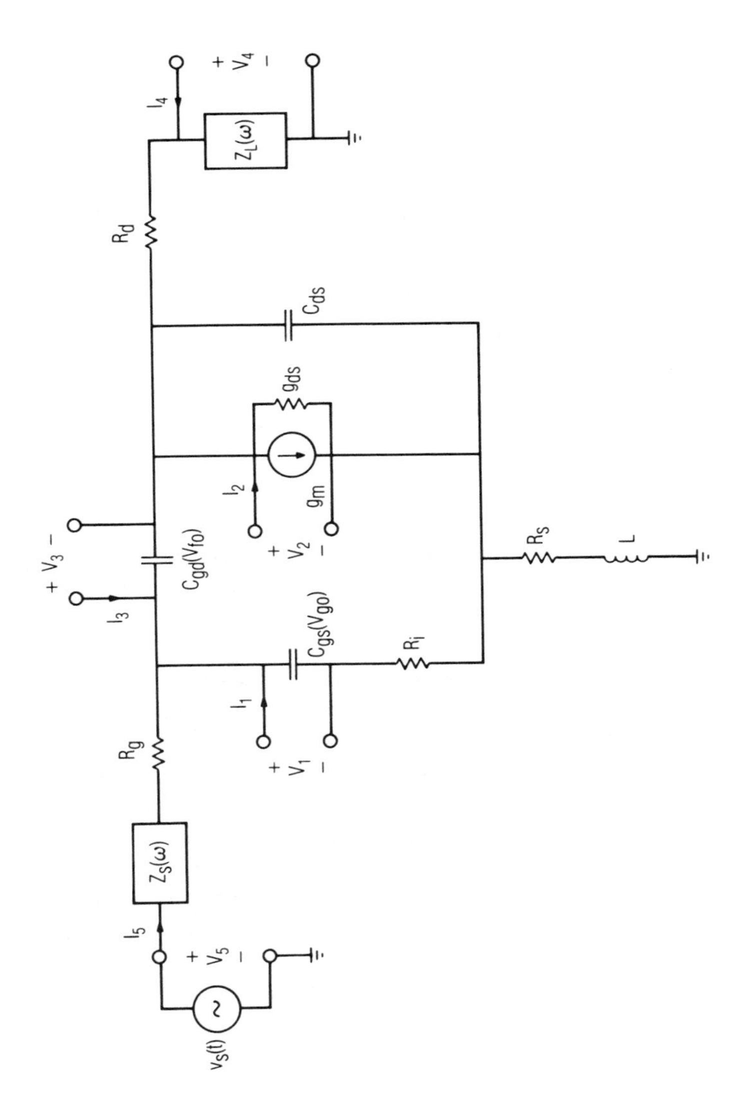

Figure 8.6 Linearized MESFET equivalent circuit, including source and load impedances and the excitation. Ports are defined at each nonlinear element; nonlinear current sources will be placed at these ports.

The Y parameters of the circuit are

$$Y_{1,1}(\omega) = \frac{1}{R_i} + j\omega C_{gs}$$

$$Y_{1,2}(\omega) = Y_{1,3}(\omega) = Y_{3,1}(\omega) = -\frac{1}{R_i}$$

$$Y_{1,4}(\omega) = Y_{4,1}(\omega) = Y_{1,5}(\omega) = Y_{5,1}(\omega) = 0$$

$$Y_{2,1}(\omega) = -\frac{1}{R_i} + g_m$$

$$Y_{2,2}(\omega) = \frac{1}{R_i} + \left[\left(\frac{1}{R_d} + \frac{1}{Z_s(\omega) + R_g}\right)^{-1} + R_s + j\omega L\right]^{-1} + j\omega C_{ds} + g_{ds}$$

$$Y_{4,4}(\omega) = \left[\left(\frac{1}{R_s + j\omega L} + \frac{1}{Z_s(\omega) + R_g}\right)^{-1} + R_d\right]^{-1} + \frac{1}{Z_L(\omega)}$$

$$Y_{5,5}(\omega) = \left[\left(\frac{1}{R_s + j\omega L} + \frac{1}{R_d}\right)^{-1} + Z_s(\omega) + R_g\right]^{-1}$$

$$Y_{2,3}(\omega) = Y_{3,2}(\omega) = \frac{1}{R_i} - Y_{2,5}(\omega)$$

$$Y_{2,4}(\omega) = Y_{4,2}(\omega) = \left(\frac{1}{Z_L(\omega)} - Y_{4,4}(\omega)\right) \cdot \left(\frac{Z_s(\omega) + R_g}{R_s + R_g + j\omega L + Z_s(\omega)}\right)$$

$$Y_{2,5}(\omega) = Y_{5,2}(\omega) = -Y_{5,5}(\omega)\left(\frac{R_d}{R_s + R_d + j\omega L}\right)$$

$$Y_{3,3}(\omega) = Y_{5,5}(\omega) + \frac{1}{R_i} + j\omega C_{gd}$$

$$Y_{3,4}(\omega) = Y_{4,3}(\omega) = \left(Y_{4,4}(\omega) - \frac{1}{Z_L(\omega)}\right) \cdot \left(\frac{R_s + j\omega L}{R_s + R_g + j\omega L + Z_s(\omega)}\right)$$

$$Y_{3,5}(\omega) = Y_{5,3}(\omega) = -Y_{5,5}(\omega)$$

$$Y_{5,4}(\omega) = Y_{4,5}(\omega) = -Y_{3,4}(\omega) \tag{8.2.15}$$

where g_m is the small-signal transconductance of the device; that is, the first coefficient in the series form of $i_d(v_g)$, and g_{ds}, C_{gd}, and C_{gs} are the linear parts of their respective nonlinear elements. First, we must find the first-order voltage components across each nonlinear element; generally, we want to know the output voltage, the voltage at Port 4, as well. These are found in the usual manner by solving the Y equations to obtain the unknown voltages. From (4.2.111) and (4.2.112), the linear (first-order) voltages at each port and at ω_q are

$$\begin{bmatrix} V_{1,1,q} \\ V_{2,1,q} \\ V_{3,1,q} \\ V_{4,1,q} \end{bmatrix} = -\begin{bmatrix} Y_{1,1} & Y_{1,2} & Y_{1,3} & Y_{1,4} \\ Y_{2,1} & Y_{2,2} & Y_{2,3} & Y_{2,4} \\ Y_{3,1} & Y_{3,2} & Y_{3,3} & Y_{3,4} \\ Y_{4,1} & Y_{4,2} & Y_{4,3} & Y_{4,4} \end{bmatrix}^{-1} \begin{bmatrix} Y_{1,5} \\ Y_{2,5} \\ Y_{3,5} \\ Y_{4,5} \end{bmatrix} [V_{5,1,q}] \tag{8.2.16}$$

where $V_{p,1,q}$ is the first-order voltage at Port p at frequency ω_q. At each excitation frequency ω_q, the Y matrix is first found from (8.2.15); $V_{5,1,q}$ is set to $V_{s,q}$, the phasor magnitude of the component of $v_s(t)$ at ω_q; and the vector of first-order port voltages is found from (8.2.16). The result is a table of values of the first-order voltages at each port and at each excitation frequency.

The second-order port voltages are found via the method of nonlinear currents: the term $v_s(t)$ is set to zero ($V_{5,2,k} = 0$; $k = 1 \ldots K$), nonlinear current sources at the second-order frequencies are placed at each port, and the second-order port voltages are found from the network's Y matrix. No current source is placed at Port 4, the output port, because there is no nonlinear element in parallel with it. The resulting circuit is shown in Figure 8.7, in which the FET equivalent circuit of Figure 8.6 is represented by the multiport network labeled Y, and the nth-order nonlinear current at Port p is $I_{p,n}$. The derivation of the second-order source currents is given in (4.2.114) through (4.2.125). At Port p and at the mixing frequency, $\omega_{2,k} = \omega_{q1} + \omega_{q2}$, the second-order current component in a two-terminal nonlinear conductance is

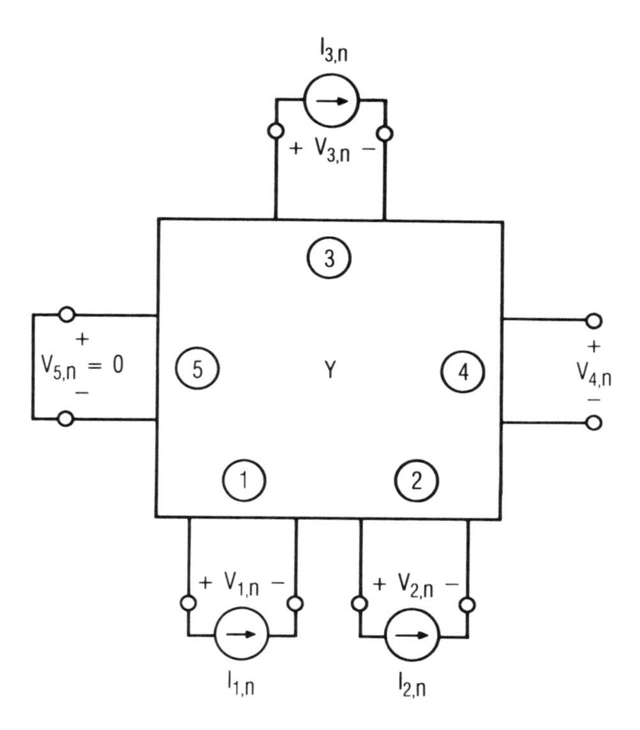

Figure 8.7 Equivalent circuit for determining mixing products of second and higher orders. The multiport represents the linearized MESFET equivalent circuit and the source and load impedances; $I_{p,n}$ is the nth-order nonlinear current source at Port p.

$$I_{p,2,k} = \frac{1}{2} g_{p,2}\, t_{2,k} V_{p,1,q1} V_{p,1,q2} \qquad (8.2.17)$$

and in a capacitance is

$$I_{p,2,k} = \frac{1}{2} \mathrm{j}\omega_{2,k} C_{p,2} t_{2,k} V_{p,1,q1} V_{p,1,q2} \qquad (8.2.18)$$

The coefficient $t_{2,k}$ is 1.0 when $q1 = q2$ (i.e., a harmonic) and is 2.0 otherwise. $C_{p,2}$ and $g_{p,2}$ are the second-degree coefficients in the series expansions of the incremental C/V and I/V characteristics, respectively, of the nonlinearity at Port p. Because there are two nonlinearities at Port 2, g_{ds} and i_d, the nonlinear current at that port has two parts: one represents

g_{ds} and has the form of (8.2.17); the other represents $i_d(v_g)$ and has the form:

$$I_{2,2,k} = \frac{1}{2} g_{2,2} t_{2,k} V_{1,1,q1} V_{1,1,q2} \tag{8.2.19}$$

that is, the general form of (8.2.17) is valid for $i_d(v_g)$, but the voltage components at Port 1 are used to find the second-order source current.

The second-order voltage components are found from the Y-matrix equations, although now the second-order current sources are the excitations. At each second-order mixing frequency, $\omega_{2,k}$, (4.2.113) becomes

$$\begin{bmatrix} V_{1,2,k} \\ V_{2,2,k} \\ V_{3,2,k} \\ V_{4,2,k} \end{bmatrix} = - \begin{bmatrix} Y_{1,1} & Y_{1,2} & Y_{1,3} & Y_{1,4} \\ Y_{2,1} & Y_{2,2} & Y_{2,3} & Y_{2,4} \\ Y_{3,1} & Y_{3,2} & Y_{3,3} & Y_{3,4} \\ Y_{4,1} & Y_{4,2} & Y_{4,3} & Y_{4,4} \end{bmatrix}^{-1} \begin{bmatrix} I_{1,2,k} \\ I_{2,2,k} \\ I_{3,2,k} \\ 0 \end{bmatrix} \tag{8.2.20}$$

The Y matrix must be formed, and (8.2.20) must be evaluated, at each mixing frequency of interest, $\omega_{2,k}$, $k = 1 \ldots K$. Note that if $\omega_{2,k} < 0$, the Y parameters must be evaluated at the negative frequency.

Third-order voltages are found in a similar manner, although the expression for the source currents is a little stickier. The third-order mixing frequencies, $\omega_{3,m}$, arise from direct mixing of excitation frequencies, $\omega_{3,m} = \omega_{q1} + \omega_{q2} + \omega_{q3}$, or from mixing between an excitation frequency and a second-order component, $\omega_m = \omega_q + \omega_{2,k}$. Equations (4.2.135) and (4.2.136) give the third-order source currents; we include the subscript p, designating the port number, which was not used in Chapter 4. Thus, in the case of a conductance:

$$I_{p,3,m} = \sum_{\omega_q + \omega_{2,k} = \omega_{3,m}} g_{p,2} V_{p,1,q} V_{p,2,k} + \frac{g_{p,3} t_{3,m}}{4} \cdot \underset{(\omega_{q1} + \omega_{q2} + \omega_{q3} = \omega_{3,m})}{V_{p,1,q1} V_{p,1,q2} V_{p,1q3}} \tag{8.2.21}$$

and in the case of a capacitor:

$$I_{p,3,m} = j\omega_{3,m} \left[\sum_{\omega q + \omega 2,k = \omega 3,m} C_{p,2} V_{p,1,q} V_{p,2,k} + \frac{C_{p,3} t_{3,m}}{4} \cdot \underset{(\omega_{q1} + \omega_{q2} + \omega_{q3} = \omega_{3,m})}{V_{p,1,q1} V_{p,1,q2} V_{p,1,q3}} \right] \tag{8.2.22}$$

The coefficient $t_{3,m}$, given by (4.1.21), is the number of terms in the triple summation in (4.2.129) that are identical; thus, if all three subscripts are identical ($q_1 = q_2 = q_3$), $t_{3,m} = 1.0$; if any two are identical, $t_{3,m} = 3.0$; otherwise, $t_{3,m} = 6.0$. Again, the port voltages are found via the matrix equations:

$$\begin{bmatrix} V_{1,3,m} \\ V_{2,3,m} \\ V_{3,3,m} \\ V_{4,3,m} \end{bmatrix} = - \begin{bmatrix} Y_{1,1} & Y_{1,2} & Y_{1,3} & Y_{1,4} \\ Y_{2,1} & Y_{2,2} & Y_{2,3} & Y_{2,4} \\ Y_{3,1} & Y_{3,2} & Y_{3,3} & Y_{3,4} \\ Y_{4,1} & Y_{4,2} & Y_{4,3} & Y_{4,4} \end{bmatrix}^{-1} \begin{bmatrix} I_{1,3,m} \\ I_{2,3,m} \\ I_{3,3,m} \\ 0 \end{bmatrix} \tag{8.2.23}$$

where the Y parameters have been found from (8.2.15) at $\omega_{3,m}$.

Nonlinear transfer functions can be expressed in terms of output voltage or load current; the nth-order load current at some mixing frequency, ω, is simply

$$I_{L,n}(\omega) = V_4(\omega)/Z_L(\omega) \tag{8.2.24}$$

We can find second- or third-order nonlinear transfer functions by substituting the excitation voltage components into the equation appropriate for their order; that is,

$$I_{L,1}(\omega_1) = V_{s,1} H_1(\omega_1) \tag{8.2.25}$$

$$I_{L,2}(\omega_1, \omega_2) = \frac{1}{2} V_{s,1} V_{s,2} H_2(\omega_1, \omega_2) \tag{8.2.26}$$

or

$$I_{L,3}(\omega_1, \omega_2, \omega_3) = \frac{1}{4} V_{s,1} V_{s,2} V_{s,3} H_3(\omega_1, \omega_2, \omega_3) \tag{8.2.27}$$

where ω_1, ω_2, and ω_3 are distinct, noncommensurate frequencies. Of course, we can find the output power directly at the mixing frequency of interest as

$$P_{L,n}(\omega) = \frac{1}{2} |I_{L,n}(\omega)|^2 \operatorname{Re}\{Z_L(\omega)\} \quad (8.2.28)$$

8.3 OPTIMIZING THE LINEARITY OF MESFET AMPLIFIERS

We now examine the problem of designing amplifiers to have optimum linearity. This task is not as simple as it might seem at first because there are at least two complications. The first is that there is no universally valid figure of merit for amplifier linearity. The output intercept point for third-order intermodulation (IM) is often accepted as a figure of merit, but it is valid as a figure of merit only when the output power is the fixed quantity. This is often the case in an IF or predetection amplifier, especially when manual or automatic gain control is available to keep the output power constant. However, in many amplifiers (e.g., low-noise preamplifiers used in a microwave receiver), the input power is the fixed quantity; in this case, the input intercept point is a more appropriate figure of merit. The second problem is that linearity is often not the most important quality in an amplifier, and other qualities, usually noise figure and gain, may be more important. Unfortunately, the conditions that optimize linearity may not satisfy constraints on noise figure or gain. When this situation arises, the designer must make a prudent trade-off between the conflicting requirements.

In order to examine the effects of circuit parameters on linearity, we will consider a rather ordinary Ku-band MESFET operating at 10 GHz, and we will be particularly concerned with the $2f_2 - f_1$ third-order IM product. Although it is true that other nonlinear phenomena and IM products of other orders do not necessarily behave in precisely the same manner as this IM product, it is generally true that the techniques that minimize third-order IM also reduce (and in some cases minimize) the deleterious effects of other nonlinear phenomena. Furthermore, the $2f_2 - f_1$ product is by itself one of the most troublesome manifestations of nonlinearity; minimizing it is very important to system designers.

In this section, we will attempt to answer, at least partially, the following questions:

1. How and why does the MESFET's dc bias affect its linearity?
2. How do we select source and load impedances that minimize intermodulation?

3. How do we design an amplifier to have a specific gain or noise figure and optimize linearity within these constraints?
4. What is the effect of the MESFET's source and load terminations at second-order IM frequencies?
5. Which of the MESFET's nonlinearities most strongly affect a FET amplifier's intermodulation performance? Can the device be designed to minimize those nonlinearities that are most significant?

We could pose many more questions regarding the linearity of MESFET amplifiers. Answering these five, however, will do much to clarify the process of designing MESFET amplifiers to have good linearity.

8.3.1 Modeling the MESFET

The MESFET we will use for all the calculations in the rest of this chapter is a low-noise Ku-band device similar to the Avantek AT10650-00. This FET has a recessed channel and a $300\mu \times 0.5\mu$ gate that is rectangular in cross section. We find the element values of the MESFET's equivalent circuit in the manner described in Section 2.5, by adjusting the element values empirically until good agreement is obtained between the S parameters calculated from the circuit and those measured from the device. Because its voltage dependence is weak when the MESFET is in saturation, C_{dg} is treated as a fixed linear element. We find the C/V characteristic of C_{gs} and the G/V characteristic of g_{ds} by repeating the process of fitting the model to S-parameter data at a number of different gate and drain bias voltages; in order to characterize the nonlinear transconductance, we measure the large-signal dc I/V characteristic of the controlled drain-current source, $I_d(V_g)$, at $V_d = 3$ V (it is necessary to find $I_d(V_g)$ by first measuring the terminal I/V characteristic $I_d(V_{gs})$ and then removing the voltage drop across R_s). $I_d(V_g)$ is then used to determine the incremental function $i_d(v_g)$: We derive the power-series expression for $i_d(v_g)$ by performing a least-squares fit of a third-degree polynomial to the $I_d(V_g)$ curve near the bias point. We model C_{gs} simply as a Schottky-barrier capacitance.

The I/V characteristic of the controlled current source and the G/V characteristic of g_{ds} are shown in Figure 8.8; the incremental currents in these elements, i_d and i_{ds}, respectively, are

$$i_d(v_g) = 0.03311\, v_g + 0.00401\, v_g^2 - 0.00426\, v_g^3 \qquad (8.3.1)$$

and

$$i_{ds}(v_d) = 0.00367\ v_d - 0.000370\ v_d^2 + 0.000144\ v_d^3 \tag{8.3.2}$$

These series expressions have been truncated after the third-degree terms and, thus, allow small-signal mixing products up to third-order to be evaluated. The gate and drain bias voltages are -1.5 V and 3.0 V, respectively; these dc bias voltages give $C_{gs} = 0.25$ pF and (although it is not required in the IM calculations) a dc drain current of 23 mA. The first-degree terms in (8.3.1) and (8.3.2) are the linear parts of the nonlinear elements; these would be used instead of the nonlinear form in a quasilinear equivalent circuit of the device. Equation (8.3.1) implies that the MESFET's linear transconductance is 33.1 mS, and that the drain-source resistance is 1/3.67 mS $= 272\ \Omega$. The values of the other circuit elements, including the inductance in series with the source, are shown in Figure 8.9, the MESFET's small-signal equivalent circuit.

The S parameters of the MESFET are modeled by the lumped-element linear circuit (Figure 8.9 with the nonlinear elements replaced by their linear parts), by using a general-purpose computer program that performs linear microwave circuit analysis. At 10 GHz, the S parameters are

$$S_{1,1} = 0.83\ \angle -90^\circ \qquad S_{1,2} = 0.092\ \angle 41^\circ$$
$$S_{2,1} = 1.71\ \angle 109^\circ \qquad S_{2,2} = 0.630\ \angle -51^\circ$$

From these parameters, we calculate the stability factor, K, to be 0.49, so the device is conditionally stable; therefore, simultaneous conjugate match conditions do not exist, MAG is undefined, and we must use the power-gain or available-gain techniques (or some other criteria, if appropriate) to design the amplifier. Although, theoretically, any value of gain can be achieved, practical considerations and other constraints (not the least of which is IM performance) will limit the gain to approximately 10 dB or less.

8.3.2 Bias Effects

It is well known that setting the dc drain current of a GaAs MESFET to approximately $0.5I_{dss}$ maximizes its gain and IM intercept points; the gain and intercept points achieved at this bias level are 1–2 dB greater than those obtained at the bias that optimizes noise figure, approximately $0.1I_{dss}$–$0.2I_{dss}$. Part of the reason for this effect is immediately evident from an inspection of Figure 8.8(a); the I/V curve is clearly more linear near $0.5I_{dss}$ than near $0.2I_{dss}$, and its slope (the MESFET's transconductance) also increases with I_d. Thus, the ratios of the magnitude of the first-degree

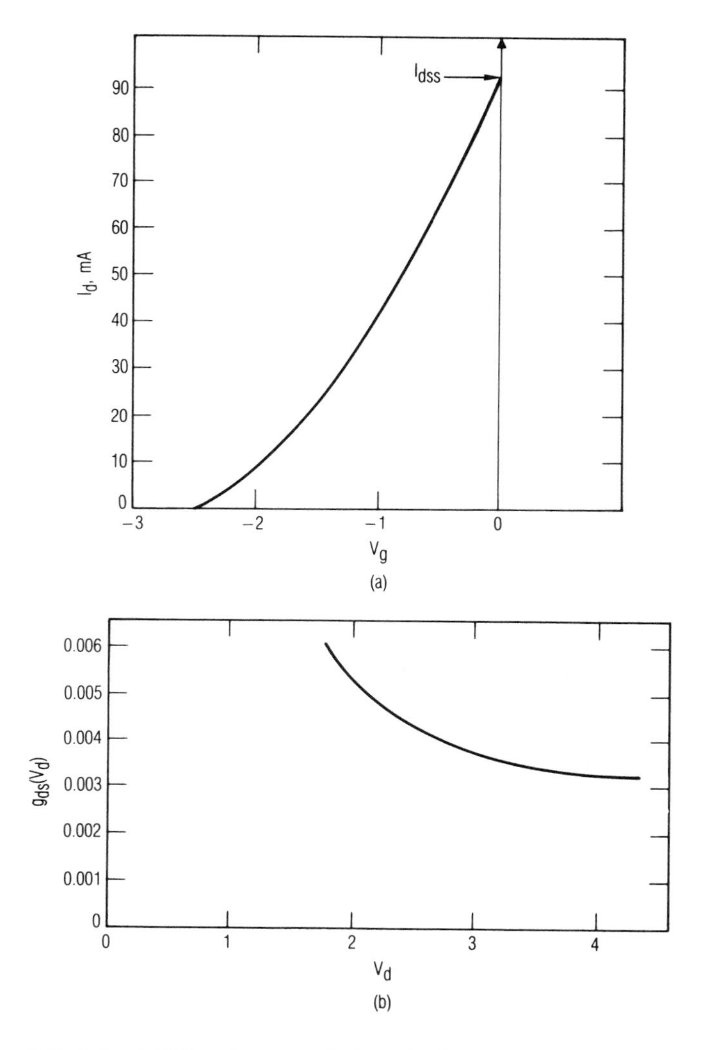

Figure 8.8 (a) $I_d(V_g)$ characteristic of the experimental MESFET at $V_{ds} = 3$ V; (b) $g_{ds}(V_d)$ characteristic of the MESFET.

coefficient in (8.3.1) to the magnitudes of the higher-degree coefficients is greater, which implies a greater signal-to-distortion ratio at the amplifier's output. For this reason, the primary means to improve a FET amplifier's linearity has always been to increase its dc drain current, although this approach inevitably compromises the amplifier's noise figure.

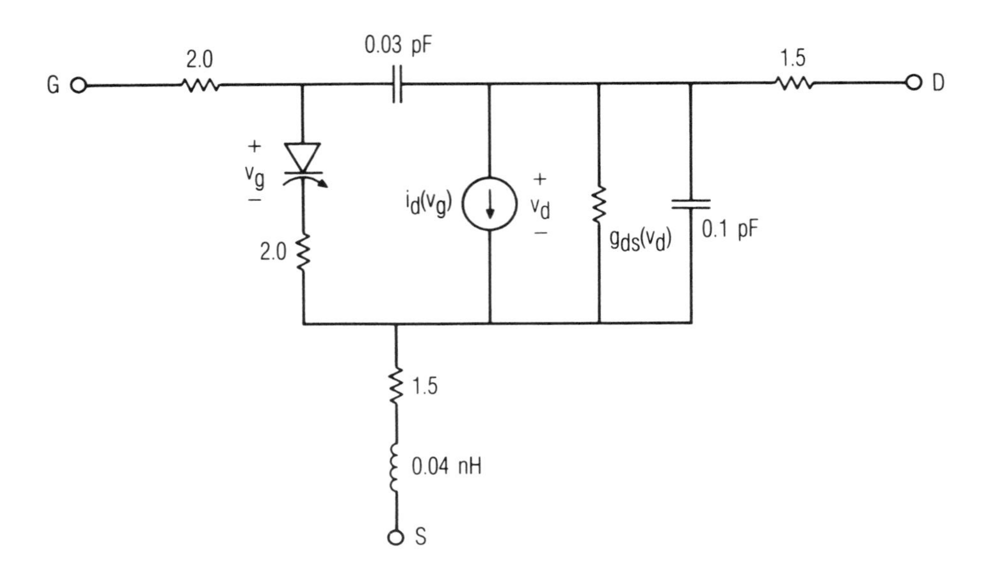

Figure 8.9 Complete incremental nonlinear equivalent circuit of the experimental MESFET.

Dc bias has another efffect on the linearity of the amplifier. Even if increasing the drain current increases the transconductance without changing the linearity of the curve, the change still increases the output IM intercept point. This situation would exist even if increasing the drain current multiplied all the coefficients in (8.3.1) by the same factor. The effect of such a change would be to "scale up" the output power of the device by 1 or 2 dB, so that all linear and IM powers would be increased by the same factor. Because the intercept point is a point on the extrapolated linear and IM output power curves, it would be increased by that same factor of 1 or 2 dB.

A third reason why high drain current improves IM performance is that, for reasons that will be examined in more depth in Chapter 9, increasing drain current increases the output power capability of the MESFET. Intermodulation is generated not only by the small-signal nonlinearity of the device, as manifested by the curvature of the characteristics in Figure 8.8 but also by large-signal nonlinearities. Examples of the latter include the MESFET turning off completely when the large-signal gate voltage drops below the turn-on voltage or inability of the large-signal drain current and voltage to exceed certain maximum and minimum values that depend upon the device and the design of the amplifier. These limits establish the output-power capability of the amplifier; the clipping that results from attempts to exceed these limits can generate large-signal intermodulation, which can be minimized by the use of high bias current.

8.3.3 Effect of Source and Load Impedances

The selection of source and load impedances that optimize linearity has always been a major concern in the design of FET amplifiers. Various researchers (References 8.1 through 8.5) have shown both theoretically and empirically that the appropriate selection of these impedances, particularly the load impedance, strongly affects the output intercept point of a microwave amplifier. The power-series analysis of a simplified FET equivalent circuit in Section 4.1.3 is consistent with this idea; in particular, (4.1.42) and (4.1.43) imply that the IM intercept point of the simplified FET equivalent circuit is a function solely of the load resistance and the power-series coefficients of the controlled current source. In the case of a real MESFET, the situation is more complicated, but the idea that the load impedance and the linearity of the controlled current source primarily establish the FET amplifier's output intercept point is still fundamentally valid.

Figure 8.10(a) shows the effect on the amplifier's output intercept point of varying the source impedance while the load impedance is held constant at a value that gives nearly optimum IM performance, $Z_L = 36 + j68$. The figure shows two curves: in the dashed one, the real part of Z_s is held constant and IP_3 is plotted as a function of the imaginary part; the solid curve shows the opposite case, where the imaginary part of Z_s is constant. The intercept point is remarkably insensitive to Z_s, varying only 0.5 dB over a very wide range of source impedances. In constrast, the amplifier's gain varies between 7.3 and 10.1 dB over the same range of source impedances.

Figure 8.10(b) shows the opposite case: The source impedance is held constant and the load impedance is varied. This figure shows that in this circuit the intercept point does not vary strongly with the imaginary part of Z_L, provided that the real part is close to the optimum value, but it does vary strongly with the real part of Z_L. It is interesting to note that the amplifier's gain varied less than 0.6 dB between $Z_L = 18 + j68$ and $Z_L = 100 + j68$, but the third-order output intercept point varied more than 3 dB. The amplifier's gain is 9.1 dB at the peak of the constant-real curve, and its intercept point is 23.9 dBm.

Some researchers (Reference 8.1) have suggested that selecting $\Gamma_L = S_{2,2}^*$ optimizes (or at least comes close to optimizing) the output intercept point. In this amplifier, that value of Γ_L results in an intercept

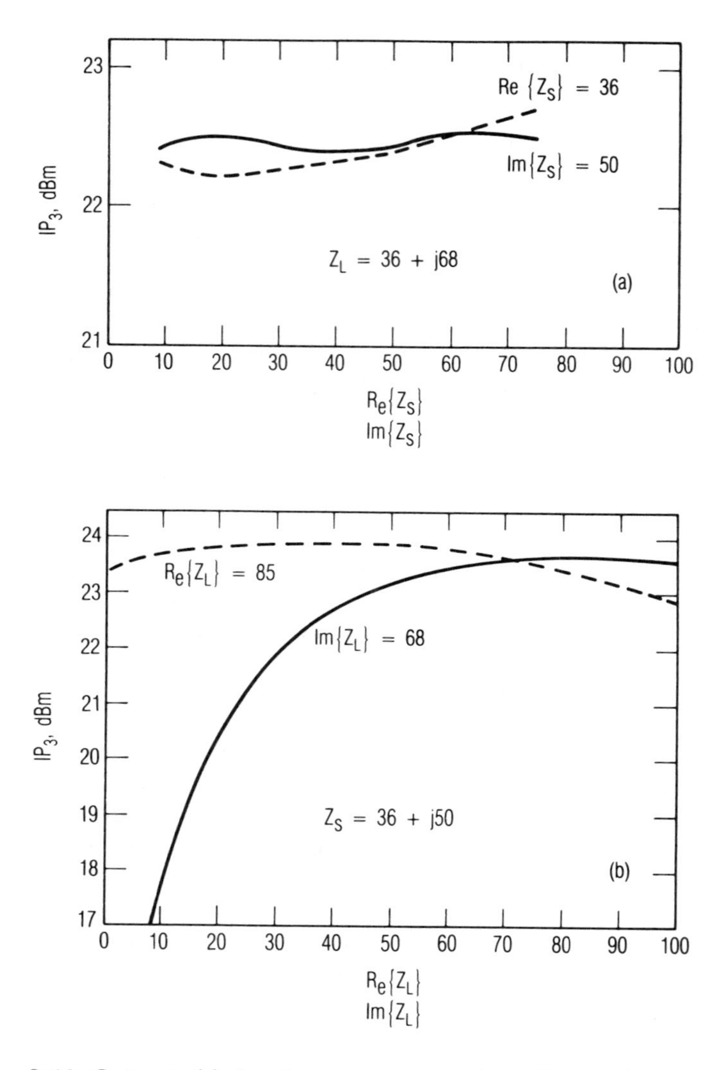

Figure 8.10 Output third-order intercept point (a) as a function of the source impedance; (b) as a function of the load impedance.

point of 22.4 dBm, not a bad result but still approximately 2 dB below the capability of the device. Two troublesome aspects of applying this recommendation to our device are (1) that $S^*_{2,2}$ is uncomfortably close to the output stability circle, and (2) that $S^*_{2,2}$ is still far enough from the optimum value of Γ_L that we probably could not obtain the optimum value by tuning the amplifier empirically. Furthermore, the value of this rule may change with frequency or in different devices. Such general, empirical rules often have many exceptions and thus are valuable only if they are used judiciously.

We can conclude from these results that, in designing an amplifier in which the output intercept point is to be minimized, the selection of the load impedance is critical. After the load impedance is determined, the source impedance can then be selected to achieve the desired gain or noise figure or to provide a trade-off between the gain and noise figure if both requirements cannot be satisfied simultaneously. If the amplifier is designed according to these criteria (the load impedance selected to provide minimum IM and the source selected to provide the required gain or noise figure) then neither the input nor the output is conjugately matched to the source or load; therefore, the amplifier must use some type of isolator or be realized in a hybrid-coupled configuration. We must remember that the dissipation loss of the output coupler or isolator reduces the intercept point dB for dB; thus, if the optimum load is close to the conjugate-match value, we might do just as well to match the output and eliminate the isolator or coupler.

8.3.4 Effect of Constraints on the Gain, Match, and Noise Figure

In Section 8.1, we saw that we can design a FET amplifier to have a specific value of transducer gain by first designing it to have that same value of power gain or available gain and then matching one port. If the output port is to be matched, the source impedance (or, equivalently, the source reflection coefficient) of the amplifier is chosen to achieve the desired available gain; conversely, if a conjugate match at the input port is desired, the load impedance is selected to achieve the desired power gain. The values of source or load impedance that result in a specific value of available or power gain lie on a circle in the Γ_s or Γ_L plane.

Although all the values of Γ_s or Γ_L on one of these circles give the same gain, they do not give the same intermodulation intercept point. This should be clear in the case of a power-gain design, in which the input is matched and Γ_L is selected from the power-gain circle: a wide range of Γ_L values can be used but there is no guarantee that the optimum value lies along the constant-gain circle. However, Γ_L is not fixed in an available-gain design either; because of the requirement that the output port be matched, Γ_L varies as Γ_s is varied. Consequently, neither the available-gain nor the power-gain design processes guarantee that the optimum value of Γ_L can be used.

Nevertheless, intermodulation performance still can be optimized within the constraints of one matched port and a specified value of gain.

Because there is considerable variation in intercept point with values of Γ_s or Γ_L that lie along the gain circles, it is important to select the source or load reflection coefficient optimally. This selection can be made by drawing the gain circle and then calculating the amplifier's intercept point at a range of Γ_s or Γ_L values along the circle.

Figure 8.11(a) shows the power-gain circles of our device representing gains of 7, 10, and 12 dB; Figure 8.11(b) shows the available-gain circles. Both figures include the stability circles as well. Any point on a circle can be expressed as the sum of two vectors:

$$\Gamma_{s,L} = \Gamma_c + \Gamma_0 \tag{8.3.3}$$

where Γ_c is the vector from the center of the plane to the center of the gain circle, and Γ_0 is the vector from the center of the gain circle to the circle itself. The magnitude of Γ_0 is thus equal to the circle's radius, a constant quantity, so the angle of Γ_0 alone, represented by θ_0, defines any position on the circle. This concept is illustrated in Figure 8.11(a).

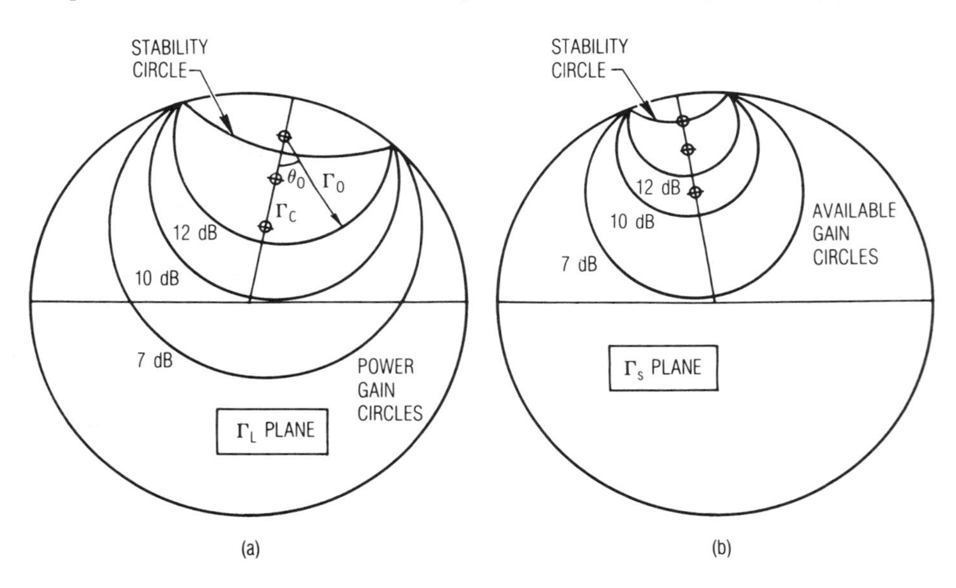

Figure 8.11 Stability circles and gain circles of the MESFET: (a) load plane; (b) source plane.

Figure 8.12(a) shows the third-order output intercept point of the amplifier as a function of gain and Γ_L, using θ_0 to indicate points on the circles; Figure 8.12(b) shows the intercept point as a function of Γ_s. It is

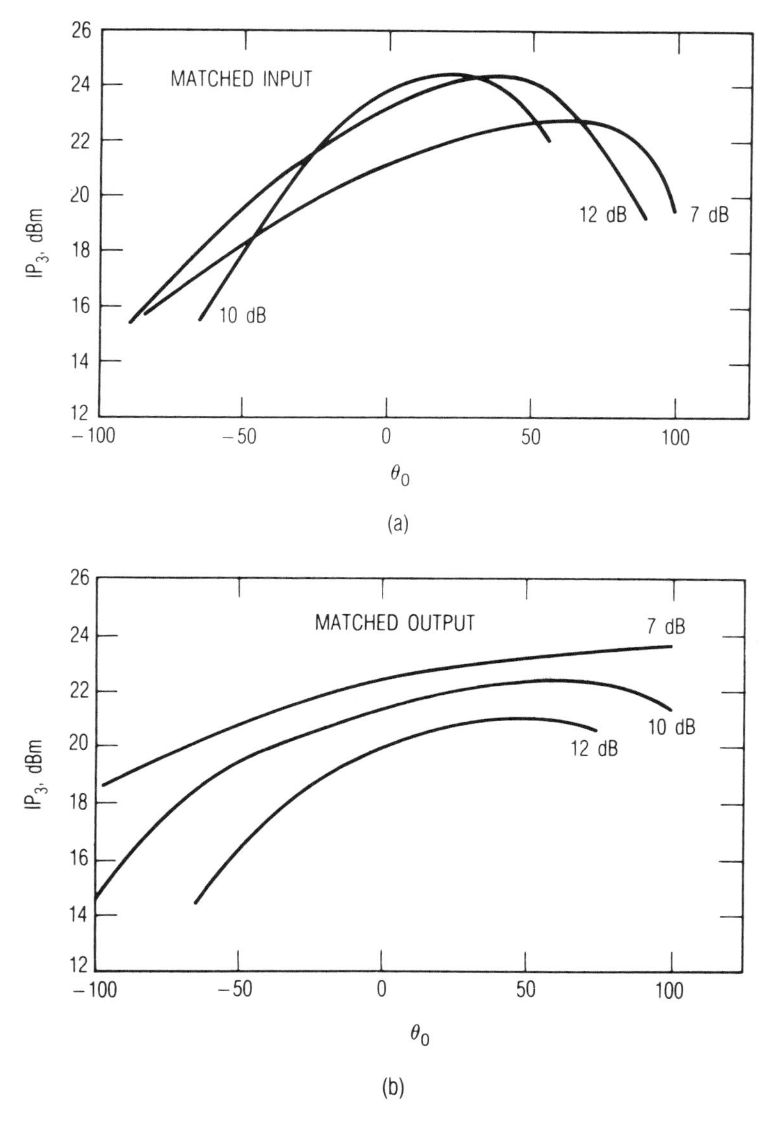

Figure 8.12 Output third-order intercept point as a function of load impedance, (a) when specific values of gain are prescribed and the input is matched; (b) when the output is matched. The variable θ_0 represents points on the stability circles, as shown in Figure 8.11(a).

interesting to note that in this case the best IM performance is achieved in an amplifier having a matched input and a relatively high value of gain, from 10 to 12 dB. When the output is matched, the intercept point is only slightly lower (3 dB lower in the worst case) but it varies less strongly with θ_0. Furthermore, the peaks of all the curves are located in approximately the same place, from $\theta_0 = 20°$ to $\theta_0 = 50°$, and the intercept point decreases, rapidly in some cases, as Γ_s or Γ_L approaches the stability circle. A competent designer who has no knowledge of matching considerations for optimum IM would certainly pick Γ_s and Γ_L far from the stability circle; however, he might select them to have a negative θ_0, which would result in poor IM performance. This situation is particularly insidious because an amplifier designed in this manner would probably work very well in all other respects.

Performing a trade-off between the noise figure, gain, and intermodulation in this design process is staightforward. Matching the input of a MESFET amplifier invariably results in a noise figure that is much greater than the minimum value; in order to minimize the noise figure, we must be free to vary the amplifier's source impedance, so the available-gain design process must be employed. We first draw the gain circles as in Figure 8.11(b) and then draw circles of the constant noise figure on the same chart (noise figures and noise figure circles are not within the scope of this book; further information can be found in Reference 8.6). Finally, we add the values of the third-order intercept point periodically along the gain circles. Having this information, we can immediately determine the gain, noise figure, and intermodulation intercept point of the amplifier that results from any proposed value of Γ_s.

We must be careful to recognize that these results represent only one third-order IM product, the one at $2f_2 - f_1$, and apply strictly to only one device at only one frequency, approximately 10 GHz. The situation may change somewhat at different frequencies or in different MESFETs. Therefore, it is important to perform calculations similar to these in every design in which intermodulation or other small-signal nonlinear effects are of concern.

8.3.5 Effect of Source and Load Terminations at Low-Order Mixing Frequencies

In Chapter 4 (e.g., (4.2.52)), we saw that the second-order nonlinear transfer function $H_2(\omega_1, \omega_2)$ often is a part of the third-order transfer function $H_3(\omega_1, \omega_2, \omega_3)$. This situation is manifested in the case we are examining by the contribution of second-order voltages in the circuit at $f_2 - f_1$ and $2f_2$ to the nonlinear source currents at $2f_2 - f_1$. Therefore, it seems possible that the termination of the MESFET's input and output at these second-order frequencies might affect the intermodualtion performance at the third-order IM frequency. FET amplifiers are normally not designed to have some particular termination at these second-order frequencies; any sensitivity of third-order IM levels to those terminations could partially explain any variability in intercept point in different amplifiers using the same device.

The calculations in the previous subsection were made with the source and load impedances at $f_2 - f_1$ and $2f_2$ set to zero. We now set the source and load impedances at these second-order mixing frequencies to other values and calculate the IM levels in a well conceived design. We select $Z_s = 10.3 + j45.6$ and $Z_L = 100 + j6$ at the design frequency of 10 GHz; these impedances give 9.7 dB gain and an intercept point that is nearly optimum, 24.4 dBm. Varying the second-order terminations has no effect on the small-signal gain.

Table 8.1 shows the result of changing the terminations at the second-order frequencies from short circuits to open circuits. The first and second columns of the table show the source and load impedances at the second-order mixing frequencies; the impedances at $f_2 - f_1$ and at $2f_2$ were set to identical values. The third column shows the change in third-order IM output power (ΔP_{IM}) from the values obtained with short-circuit terminations. The greatest change in third-order IM output power is less than 1 dB; a change of this magnitude is clearly insignificant (note that ΔP_{IM} is independent of excitation level). Thus, we can conclude that in this case the termination of these mixing products is of no concern.

Table 8.1 Effect of Low-Frequency Terminations on IM Output Power

Z_s	Z_L	ΔP_{IM} (dB)
0 + j0	0 + j0	0.00
0 + j0	50 + j0	+0.20
0 + j0	∞	+0.60
50 + j0	0 + j0	+0.30
∞	0 + j0	+0.80
∞	∞	+0.70

The sensitivity of third-order IM output power to second-order impedances is so slight here that it is difficult to conceive of an amplifier in which these terminations would have a significant effect. This minimal effect is a consequence of the fact that the circuit's dominant nonlinearity is the controlled current source, so the effects of second-order voltages must be related to the magnitudes of those voltages at the control-voltage nodes, the terminals of C_{gs}. However, the controlled current source, which generates most of the second-order energy, is only weakly coupled to C_{gs} via C_{gd}; thus, the second-order voltage across C_{gs} is relatively small, regardless of source or load impedances. In other types of circuits, it is possible that these terminations might be significant.

8.3.6 Effects of Individual Nonlinear Elements

The significance of the individual nonlinear elements in the MESFET's equivalent circuit can be found by replacing each of the nonlinear elements with a linear one and by recalculating the IM level. These changes affect only the IM performance; they have no effect on such linear parameters as the small-signal gain. Again, we use the case where $Z_s = 10.3 + j45.6$ and $Z_L = 100 + j6$, and we examine all possible combinations of linear and nonlinear elements. The seven possible combinations, and the changes in IM output level engendered by linearizing the nonlinear circuit elements, are shown in Table 8.2.

Table 8.2 Change in IM Output Power Due to Linearizing of Circuit Elements

Case No.	C_{gs}	g_{ds}	i_d	ΔP_{IM} (dB)
1	NL	NL	NL	0.00
2	lin	NL	NL	−0.29
3	NL	lin	NL	−1.34
4	NL	NL	lin	−8.66
5	NL	lin	lin	−7.60
6	lin	NL	lin	−12.22
7	lin	lin	NL	−2.52

In Table 8.2 lin means that only the linear part of the element's C/V or I/V expansion is used in the calculation; NL means that the first three terms of its incremental nonlinear expansion were used. Some worthwhile conclusions can be drawn from Table 8.2. First, the nonlinearity of $i_d(v_g)$ is clearly the dominant element in this MESFET: in the cases where i_d is linear (Cases 4, 5, and 6), the amplifier has significantly lower IM levels than in those in which i_d is nonlinear. Most studies of intermodulation in MESFETs have drawn the same conclusion, although in at least one study (Reference 8.5) g_{ds} was found to be dominant. Linearizing g_{ds} or C_{gs} has minimal effect if i_d is nonlinear. However, if i_d is linear, linearizing

C_{gs} has a very significant effect (compare Cases 4 and 6), because then the output IM current resulting from the nonlinearity of C_{gs} is a much greater fraction of the total output IM current.

The results in Table 8.2 are generally consistent with the logical expectation that IM levels increase with the number of nonlinear elements. The only exceptions are Cases 4 and 5, in which the circuit having a linear g_{ds} has a slightly greater IM output level than the circuit having a nonlinear g_{ds}. The reason for this result is probably that the amplifier's IM output currents generated by g_{ds} are out of phase with those generated by C_{gs}; thus, cancellation occurs. This phenomenon is not unusual; in some types of circuits, such as multistage amplifiers, IM cancellation can cause significant reductions in IM output over narrow ranges of input level or frequency. Unfortunately, such cancellation invariably occurs over a very small range of frequencies and other operating conditions, so it is rarely a practical means of improving IM performance.

Because it is clear (from these results and many other studies) that the dominant nonlinearity in a MESFET is that of the controlled current source, any changes in the device's structure that increase the linearity of $i_d(v_g)$ must also improve IM performance. This I/V characteristic can be linearized significantly by the use of a nonuniform doping profile in the MESFET's channel; a peripheral benefit of a nonuniform profile is that the $C_{gs}(v_g)$ characteristic is also linearized somewhat. Adjusting the doping profile is difficult if conventional epitaxial growth techniques are used, but such techniques as ion implantation and molecular-beam epitaxy can control the profile adequately. Improvements in the third-order intercept point on the order of 4 dB have been reported (Reference 8.7); this is consistent with the 8 dB improvement in IM output level that can be obtained, as shown in Table 8.2, by linearizing the $i_d(v_g)$ characteristic.

8.3.7 Conclusions

The most significant conclusions one can draw from these calculations are that there is a clear optimum value of Γ_L that optimizes the output intercept point, and that, for a given device, properly selecting Γ_L is the most important factor in optimizing the linearity of an amplifier. That the linearity of the $I_d(V_g)$ characteristic usually dominates the amplifier's IM performance is also very important, because this characteristic can be measured very easily at dc. This allows us to select MESFETs that have good IM performance on the basis of dc characteristics; dc screening can be particularly valuable in low-noise devices because the linearity of $I_d(V_g)$ at low values of I_d varies considerably among device types.

The Volterra-series theory is perhaps a little abstruse, and writing a computer program that implements it is no small task (the author's program, still at an early stage of development, has approximately 2400 lines of source code). However, once the analytical tools are developed, it is not difficult to identify the characteristics that are important in optimizing an amplifier's linearity. Thus, these results illustrate another point: that Volterra-series analysis is entirely practical for calculating IM performance and for optimizing the linearity of a MESFET amplifier.

REFERENCES

[8.1] C.Y. Ho and D. Burgess, "Practical Design of 2–4 GHz Low Intermodulation Distortion GaAs FET Amplifiers with Flat Gain Response and Low Noise Figure," *Microwave J.,* Vol. 26, February 1983, p. 91.

[8.2] R.A. Minasian, "Intermodulation Distortion Analysis of MESFET Amplifiers Using the Volterra Series Representation," *IEEE Trans. Microwave Theory Tech.,* Vol. MTT-28, 1980, p. 1.

[8.3] R.S. Tucker, "Third-Order Intermodulation Distortion and Gain Compression in GaAs FETs," *IEEE Trans. Microwave Theory Tech.,* Vol. MTT-27, 1979, p. 400.

[8.4] F.N. Sechi, "Design Procedure for High-Efficiency Linear Microwave Power Amplifiers," *IEEE Trans. Microwave Theory Tech.,* Vol. MTT-28, 1980, p. 1157.

[8.5] G.M. Lambrianou and C.S. Aitchison, "Optimization of Third-Order Intermodulation Product and Output Power from an X-Band MESFET Amplifier Using Volterra Series Analysis," *IEEE Trans. Microwave Theory Tech.,* Vol. MTT-33, 1985, p. 1395.

[8.6] G. Gonzales, *Microwave Transistor Amplifiers,* Prentice-Hall, Englewood Cliffs, NJ, 1984.

[8.7] J.A. Higgins and R.L. Kuvas, "Analysis and Improvement of Intermodulation Distortion in GaAs Power FETs" *IEEE Trans. Microwave Theory Tech.,* Vol. MTT-28, 1980, p. 9.

Chapter 9

MESFET Power Amplifiers

GaAs MESFETs are not limited to small-signal operation; MESFETs can be used as power amplifiers at frequencies well into the millimeter-wave region. Below 10 GHz, single MESFET chips can provide output power on the order of several watts, and they can be combined to achieve even higher power levels. Their gain, power, noise figure, and bandwidth are adequate to replace traveling-wave tubes (TWTs) in many applications.

As with small-signal amplifiers, the single-tone properties of power amplifiers (gain, output power, and impedances) are of fundamental concern. Although linear theory has some use in the design of power amplifiers, linear theory by itself is usually inadequate for determining all the properties of a power amplifier that we need to know. It is therefore necessary to take into account the MESFET's nonlinearities as well. For this reason, harmonic-balance techniques are the logical method for analyzing MESFET power amplifiers, and this chapter is primarily concerned with applying harmonic-balance analysis to the FET power amplifier.

9.1 POWER MESFETS

9.1.1 Structure of Power MESFETs

Power MESFETs must be designed to survive much greater electrical stresses than small-signal FETs. A power device must support high channel current, survive high drain-gate voltages, endure high temperatures, and dissipate a large amount of heat. Furthermore, like a small-signal device, it must have good gain, linearity, and efficiency, and often must be useful at high frequencies. Because of these severe and often conflicting requirements, designing power MESFETs is at least as difficult as designing any other soild-state device.

A MESFET's output power capability is established primarily by three factors: (1) its drain-gate breakdown voltage, (2) its maximum channel current, and (3) its thermal properties. Obtaining high power from a FET involves maximizing breakdown voltage and channel current, as well as maintaining good heat-dissipation properties, while avoiding the introduction of excessive resistive or capacitive parasitics. We can increase channel current arbitrarily by simply increasing the width of the MESFET's gate; however, increasing gate width increases many device parasitics, especially the gate-source capacitance and, unless measures are employed to keep it low, the gate resistance. For a MESFET's gain to remain constant with changes in gate width, the gate resistance (R_g in Figure 2.16) must decrease in proportion to the change in gate width. Although it is possible to decrease R_g by changing the geometry of the FET, the gate resistance of a MESFET usually cannot be reduced in proportion to the increase in gate width, so gain decreases as gate width increases; consequently, power MESFETs usually have relatively low gain. The usual practice of operating FET power amplifiers at least 1 dB into saturation further reduces gain. Thus, a power MESFET's gain is often marginal at high frequencies, and its maximum operating frequency decreases with gate width.

In order to keep gate resistance low in wide-gate devices and to obtain good thermal properties, the MESFET is designed as a number of *cells,* individual small FETS, connected in parallel. The gates of the individual cells may have multiple feed points or they may be arranged as a number of small sections in order to minimize gate resistance. This cell structure has a price, however: interconnecting the cells requires extra wiring, and this wiring can introduce additional inductance and capacitance. In the past, the cells had to be interconnected by wires when the device was installed in the amplifier or in a package; installing the large number of such wires was a delicate process and involved a risk that the device might be damaged by the assembler. Modern MESFETs often have *air-bridge* interconnections, metal jumpers that interconnect different cells, that are fabricated as part of the device.

The use of multiple cells and multiple short gate segments places difficult requirements upon the manufacturing process. Since even one flaw in one gate segment can ruin the entire device, each FET must have a perfect gate, sometimes several millimeters wide. Because the difficulty of fabricating flawless gates decreases with increasing gate length, the gates of power MESFETs usually are relatively long, longer than those of small-signal devices. And, as gate length is a significant factor in establishing a MESFET's input capacitance and transconductance, its long gate contributes to the power FET's relatively low gain.

The gate-drain breakdown voltage of a power device establishes a fundamental limit to its power capability; therefore, it must be much greater than that of a small-signal device. We can maximize the breakdown voltage by choosing the ohmic contact technology carefully, and by using a recessed-gate structure and fairly wide gate-drain spacing. The gate-drain spacing cannot be increased arbitrarily, however, because it increases the drain series resistance (R_d in Figure 2.16). The full drain current passes through R_d; if R_d is too great, it can significantly reduce the gain, efficiency, and output power.

The presence of inductance in series with the MESFET's source can reduce the gain of a FET amplifier. The source inductance L_s adds a resistive component R_{Ls} having an approximate value of $R_{Ls} = g_m L_s / C_{gs}$, as well as an inductive component of value L_s, in series with the gate. These additional elements reduce the MESFET's maximum available gain and make impedance matching more difficult. If L_s is fixed, R_{Ls} remains approximately constant with changes in gate width; most of the other resistive parasitics in the input decrease with gate width, however, so source inductance becomes more significant as gate width increases. Furthermore, mutual inductance prevents the source inductance from decreasing in proportion to the number of wires used to ground the source. Therefore, source inductance has a particularly strong effect on the gain in power devices, so a low-inductance ground connection is critical to the performance of a power FET. One highly effective way to reduce source inductance is to include *via holes,* metalized holes connecting the source metalization to the underside of the chip, in the design of the FET. Simpler to manufacture, although somewhat less effective, is a *wraparound metalization structure,* in which a source metalization covers the periphery of the chip and grounds the source regions at the chip's edges. The use of either of these structures can do much to maximize gain at high frequencies.

The third factor that limits a MESFET's output power is the chip's ability to dissipate heat. Thermal properties of GaAs devices are especially worrisome because GaAs has poor thermal conductivity, significantly lower than that of silicon. Furthermore, a power MESFET must dissipate quite a lot of heat; the dc-RF efficiency of a power FET is rarely above 40 percent, and some types of amplifiers dissipate more power in the absence of RF output than when in operation. Consequently, a power device must dissipate, in the form of heat, 1.5 to 4 times its RF output power, and often must do so in the presence of one or more other chips dissipating equal amounts of heat. Because the heat dissipation can be so great, the chip must be designed carefully to minimize its thermal resistance: the cells must not be placed too close together, the chip must be made quite thin (some large chips are thinned to from 0.001 to 0.002 inches), and often a thick gold layer must be plated onto the chip's underside. The resulting thermal resistance between the channel and the mounting surface may be from 1 to 2 W/°C (in the case of a large chip) to 50 W/°C or more (for single-cell, medium-power devices). The resulting increase in channel temperature may be several tens of degrees C at full power.

9.1.2 Modeling Power MESFETs

Because the FET power amplifier is a large-signal component, we use the large-signal model of the MESFET described in Section 2.4.2. The lumped quasistatic equivalent circuit is shown in Figure 2.16; this circuit includes three nonlinear elements (C_g, C_d, and I_d), all of which are functions of the gate and drain voltages, V_g and V_d, respectively. It is important to note that MESFETs are often driven into their linear regions (to drain voltages below the knee of the I/V curve) when used in power amplifiers; therefore, when we calculate amplifier performance, it is often important to include the voltage dependence of C_d, which varies most strongly when the device enters its linear region. The resistances and the drain-source capacitance, C_{ds}, are treated as linear elements. Gate and drain inductances and package parasitics, not shown in Figure 2.16, can sometimes be included in the source and load networks. If not, they must be included in the complete model.

It is a common practice to use the empirical form of $I_d(V_g, V_d)$ of (2.4.5) in power-amplifier analyses. It is often helpful to modify (2.4.5) to include the dependence of V_t, the MESFET's turn-on voltage, on V_d via an appropriate empirical expression. $C_g(V_g, V_d)$ is often modeled as an ideal Schottky-barrier capacitance (i.e., independent of V_d), although it is sometimes worthwhile to modify the ideal characteristic of (2.3.5) to account for the drop in capacitance when $V_g < V_t$. Modeling the dependence of C_d on V_g and V_d is somewhat more difficult; usually an empirical expression must be used. Reference 9.1 is a good example of an intelligent approach to the modeling of power MESFETs.

Although Figure 2.16 describes a single-cell device, the same circuit can be used to describe a multicell device. If the multicell device has N identical cells and there are no additional parasitics, we need only divide the resistances of a single cell by N and multiply the capacitances and I_d by N to generate the multicell model. It is usually valid to assume that the cells of a multicell device are identical. However, in practice, we can rarely ignore the parasitic capacitances and inductances associated with air bridges or other overlay metalizations and not part of the individual cells. Often, some of these parasitics can be absorbed into the source and load impedances or into other circuit elements. Nonetheless, because of the large capacitances and small resistances of multicell power devices, directly measuring the S parameters of large devices is difficult; consequently, large devices are usually modeled by scaling measurements of their individual cells.

Because power MESFETs are often operated at elevated channel temperatures, their I/V and C/V characteristics and S parameters measured at room temperature may not be valid. Therefore, it is worthwhile to attempt to measure the nonlinearities as close to the operating temperature as possible.

9.2 FUNDAMENTAL CONSIDERATIONS IN POWER-AMPLIFIER DESIGN

Figure 9.1 shows a simplified circuit of a FET power amplifier. We will derive some of the fundamental properties and limitations of power amplifiers via this circuit. The circuit consists of a MESFET, excitation and gate-bias sources, a tuned circuit, and load, R_L. The drain-bias voltage is V_{dd}, and the gate bias is adjusted so that, in the absence of excitation, the dc drain current is I_{dd}. Initially, we will assume that the MESFET is an ideal transconductance amplifier, that it has no resistive or reactive parasitics (so $V_{gs} = V_g$ and $V_{ds} = V_d$) and that the tuned circuit is resonant at the excitation frequency.

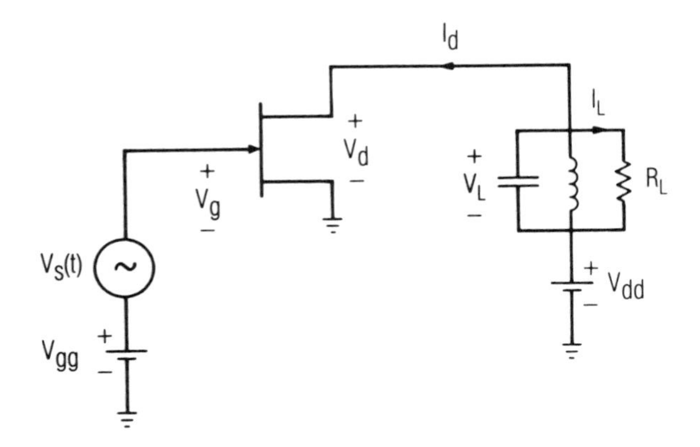

Figure 9.1 Equivalent circuit of an ideal FET power amplifier.

The application of a sinusoidal excitation $V_s(t)$ to the gate generates an RF component of drain current, $\Delta I_d(t)$. If the tuned circuit is resonant at the RF frequency, that current must pass entirely through R_L. The RF component of the drain voltage, $\Delta V_d(t)$, is equal to the voltage drop across R_L; that is:

$$V_L(t) = \Delta V_d(t) = -\Delta I_d(t) R_L \tag{9.2.1}$$

Each curve in the MESFET's drain I/V characteristic, shown in Figure 9.2, represents a range of values of V_d and I_d that can exist when the gate voltage, V_g, has a specified value; (9.2.1) expresses an additional constraint on V_d and I_d. Thus, the drain voltage and current must simultaneously satisfy both (9.2.1) and the I/V curve for V_g; these values of V_d and I_d are found at the point where the I/V curve and (9.2.1) intersect. Figure 9.2 shows (9.2.1) plotted on top of the MESFET's drain I/V curves; when the FET is excited by $V_s(t)$, $V_d(t)$ and $I_d(t)$ must always lie along the straight line, called a *load line*.

In a power amplifier, we wish to maximize the power delivered to R_L. This power is clearly maximum when both $V_L(t) = \Delta V_d(t)$ and $I_L(t) = -\Delta I_d(t)$ have their maximum excursions. If we recognize that V_d and I_d can not be less than zero, these maximum excursions occur when $|V_L(t)| = V_{dd}$ and $|I_L(t)| = I_{dd}$; the geometry of the load line dictates that these conditions are met when $R_L = V_{max,A}/I_{max} = V_{dd}/I_{dd}$. Then, if $V_s(t)$ and V_{gg} are chosen appropriately, the drain voltage varies from zero to $V_{max,A} = 2V_{dd}$, and the drain current varies from zero to $I_{max} = 2I_{dd}$. The $V_d(t)$ and $I_d(t)$ waveforms in this case are shown in Figure 9.3.

The output power, P_L, under these conditions is

$$P_L = 0.5\,|V_L(t)|\,|I_L(t)| = 0.5\,V_{dd}\,I_{dd} \tag{9.2.2}$$

Usually, we wish to maximize the output power of a specified transistor. In this case, $V_{max,A}$ and I_{max} are the devices's maximum drain voltage and current, and the maximum output power is

$$\begin{aligned} P_L &= \frac{1}{2}\left(\frac{1}{2}V_{max,A}\right)\left(\frac{1}{2}I_{max}\right) \\ &= \frac{1}{8}V_{max,A}\,I_{max} \end{aligned} \tag{9.2.3}$$

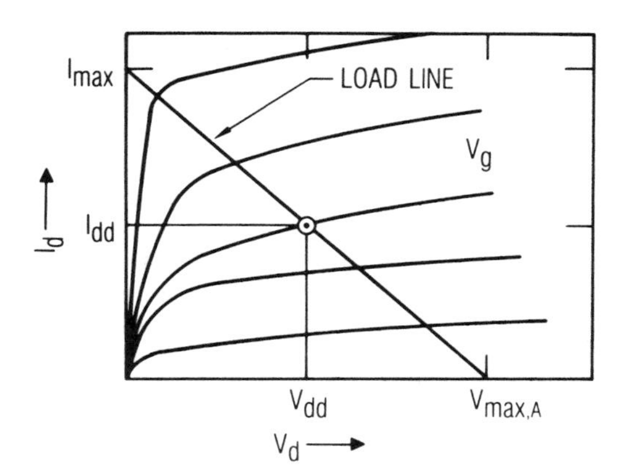

Figure 9.2 Drain *I/V* characteristics and the load line of the MESFET in Figure 9.1.

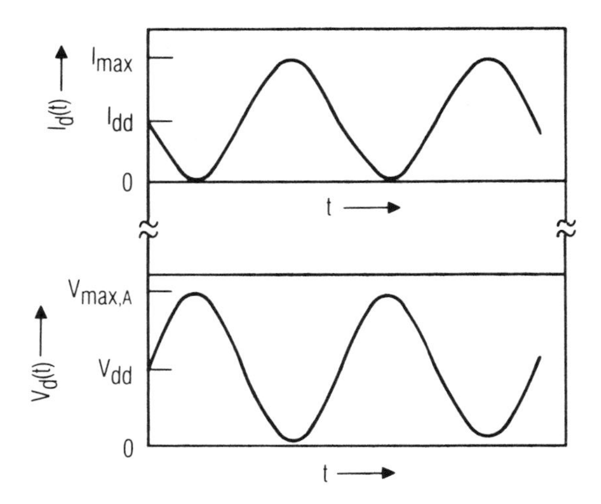

Figure 9.3 Drain voltage and current waveforms in the ideal Class-A FET power amplifier; the bias voltages, excitation, and load resistance are chosen optimally, causing both $I_d(t)$ and $V_d(t)$ to vary between zero and their maximum values.

The dc current remains constant at I_{dd} at all excitation levels; therefore, the dc power $P_{dc} = V_{dd} I_{dd}$ an the dc-RF conversion efficiency is

$$\eta_{dc} = \frac{P_L}{P_{dc}} = \frac{0.5 V_{dd} I_{dd}}{V_{dd} I_{dd}} = 50 \text{ percent} \tag{9.2.4}$$

An amplifier operated in this manner is called a *Class-A amplifier* (although this arcane terminology was originally used to describe vacuum-tube amplifiers, it has been transferred with little modification from vacuum tubes to bipolar transistors and finally to MESFETs). In theory the maximum efficiency of such an amplifier is 50 percent, so the transistor in a Class-A amplifier dissipates at least as much power in the form of heat as it delivers to the RF load.

Two factors complicate this simple reasoning. The first is that it is not possible in practice to vary the drain voltage and current all the way to the peak of the load line, where $I_d = I_{max}$ and $V_d = 0$, because of the knee in the uppermost I/V curve in Figure 9.2. The result of this limitation is that $|V_L(t)|$ cannot quite equal V_{dd}, and $|I_L(t)|$ must be less than I_{dd}, so both the output power and efficiency are somewhat lower than the values given by (9.2.3) and (9.2.4). The second is that the MESFET is nonlinear, so the $I_d(t)$ waveform is generally not sinusoidal. The tuned circuit constrains $I_L(t)$ to be sinusoidal, however, so the assumption that $I_L(t) = -\Delta I_d(t)$ is not precisely correct and in fact $|I_L(t)| < |\Delta I_d(t)|$, which further limits output power and efficiency. Nevertheless, because the purpose of this derivation is to illustrate fundamental properties of power amplifiers, we will continue to assume that $I_d(t)$ can reach I_{max} and that the FET is linear. We will modify these assumptions when we face the problem of accurately designing practical power amplifiers.

Two undesirable characteristics of the Class-A amplifier are its relatively low efficiency and its dissipation of a great amount of power even when it is not excited; in fact, Class-A amplifiers dissipate more power under quiescent (unexcited) conditions than when they are operating. Thus, a Class-A amplifier must be disigned either to safely dissipate its quiescent power or to be turned off when not in use. Both alternatives are unacceptable in many applications.

Many of the disadvantages of Class-A operation are circumvented by Class-B operation. The gate-bias voltage of an ideal Class-B amplifier is set at the turn-on voltage V_t; therefore, the FET's quiescent drain current is zero, so the FET dissipates no power in the absence of excitation. The bias point is thus V_{dd} on the voltage axis of the FET's I/V curves. It is not possible to draw a true load line describing the single-device amplifier in

Figure 9.1 when the amplifier is biased to achieve Class-B operation because the harmonic components of I_d, which are substantial in a Class-B amplifier, do not circulate in R_L; therefore, (9.2.1) is not valid here.

During the half cycle, when $V_s(t)$ is positive, $V_g(t) > V(t)$ and the drain conducts; during the other half cycle, $V_g(t) < V_t$ and the drain current is zero. The drain current $I_d(t)$ is therefore a pulse train, and each pulse has the half-cosine shape shown in Figure 9.4. The dc drain current is the average value of the half-cosine waveform. From Fourier analysis, we find that, under full excitation, $I_{dc} = I_{max}/\pi$, and the amplifier's dc power is

$$P_{dc} = V_{dd} \frac{I_{max}}{\pi} \tag{9.2.5}$$

Because the tuned circuit allows only the fundamental and dc components of drain voltage to exist, the ac part of $V_d(t)$, which is equal to $V_L(t)$, is a continuous sinusoid. The tuned circuit also allows only the fundamental component of $I_d(t)$ to pass through R_L. The power delivered to the load is:

$$P_L = 0.5 I_1 \, |V_L(t)| \tag{9.2.6}$$

where $I_1 = |I_L(t)|$ is the magnitude of the fundamental component of $I_d(t)$. From Figure 9.4, $|V_L(t)| = |\Delta V_d(t)| = V_{dd}$, and from Fourier analysis, $I_1 = 0.5 I_{max}$. Then,

$$P_L = \frac{1}{2}\left(\frac{1}{2} I_{max}\right) V_{dd} = \frac{1}{4} I_{max} V_{dd} \tag{9.2.7}$$

and the dc-RF efficiency is

$$\eta_{dc} = \frac{P_L}{P_{dc}} = \frac{\pi}{4} = 0.78 \tag{9.2.8}$$

Theoretically, the Class-B amplifier has a maximum efficiency of 78 percent, much better than the 50 percent limit of the Class-A amplifier. It has achieved this improvement by allowing the channel to conduct during only half the period of the excitation; during the time that the FET is turned off, the FET dissipates no power. However, the peak value of the Class-B amplifier's drain current is twice the value of $\Delta I_d(t)$ in the Class-A amplifier, so the fundamental-frequency component of the output current is the same in both types of amplifiers.

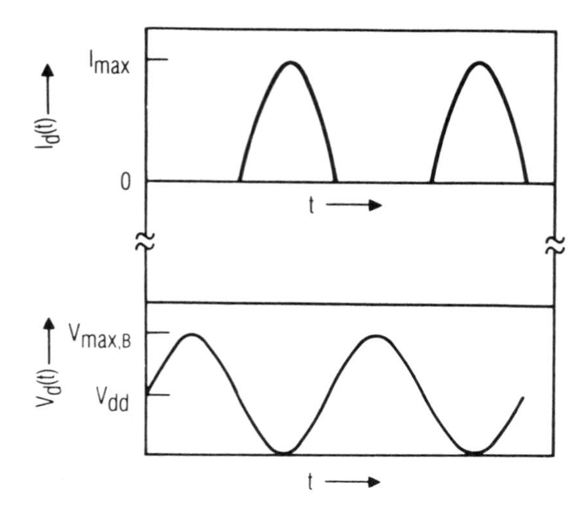

Figure 9.4 Drain voltage and current waveforms in the ideal Class-B amplifier. The drain conducts in sinusoidal pulses because the gate is biased at V_t.

To find the maximum output power in terms of the device's limitations, we let the maximum drain voltage be $V_{\max,B}$ and note that $V_{\max,B} = 2V_{dd} = 2\,|V_L(t)|$. Then,

$$P_L = \frac{1}{2}\left(\frac{1}{2}\,V_{\max,B}\right)\left(\frac{1}{2}\,I_{\max}\right) = \frac{1}{8}\,V_{\max,B}I_{\max} \tag{9.2.9}$$

which is the same as that of the Class-A amplifier if $V_{\max,A} = V_{\max,B}$.

In order to achieve the maximum output power, the load resistance R_L must be such that

$$I_1R_L = 0.5I_{\max}R_L = |V_L(t)| = V_{dd} \tag{9.2.10}$$

so

$$R_L = \frac{2V_{dd}}{I_{\max}} = \frac{V_{\max,B}}{I_{\max}} \tag{9.2.11}$$

and we see that the load resistance of the Class-B amplifier is the same as that of the Class-A. Furthermore, because in both amplifiers the load resistance and the fundamental component of the load current are the same, the output power must also be the same.

Because the maximum drain voltage is limited by gate-drain avalanche breakdown, $V_{max,A}$ is always greater than $V_{max,B}$. In the Class-A amplifier, the maximum drain-gate voltage occurs when $V_d = V_{max,A}$ and $V_g = V_t$. Thus, if V_a is the drain-gate avalanche breakdown voltage:

$$V_{max,A} = V_a - |V_t| \tag{9.2.12}$$

The Class-B amplifier is biased at $V_{gg} = V_t$, so the maximum negative excursion of V_g is $2V_t$. Then,

$$V_{max,B} = V_a - 2|V_t| \tag{9.2.13}$$

so $V_{max,B}$ is less than $V_{max,A}$ by an amount equal to $|V_t|$. Accordingly, the maximum output power of a Class-B amplifier is slightly lower than that of a Class-A amplifier using the same device.

The difference in maximum output power between Class-A and Class-B amplifiers is not the most significant one; there is a much greater difference in their gains. The gate voltage of a Class-A amplifier varies between zero and V_t; in a Class-B amplifier the gate voltage varies between zero and $2V_t$. More input power is required to achieve the Class-B amplifier's wider gate-voltage variation but the output power is nearly the same; thus, Class-B amplifiers have inherently lower gain than Class-A.

Another disadvantage of the Class-B amplifier is that is generates a high level of harmonics in the drain current by switching the FET on and off during each excitation cycle. If the device is terminated in the same impedance at the fundamental and second-harmonic frequencies, the second-harmonic output of an ideal Class-B amplifier is only 7.5 dB below the fundamental output (for reasons that will be examined in Section 9.3, the second-harmonic output of a practical amplifier is usually considerably lower). One solution to the problem of harmonics is to use a "push-pull" configuration, in which the excitation is applied out of phase to the inputs of two Class-B amplifiers and the outputs are combined out of phase. The phase shift of the output combiner must be 180 degrees at the harmonic frequencies as well as the fundamental frequency. This configuration, in conjunction with an appropriate design of the output matching network, can reduce significantly the levels of even harmonics.

In order to avoid the Class-B amplifier's inherently low gain and because the turn-off characteristic of power FETs are often very "soft," we rarely operate power FETs in a true Class-B mode. So-called Class-B microwave amplifiers are usually biased near $0.1 I_{dss}$ and are actually operated in a mode somewhere between Class B and Class A. Conversely, Class-A amplifiers are often not operated in a classical Class-A mode: they are sometimes biased to a minimal current level and driven well into saturation. Both types of operation are called *Class AB* and both represent a compromise between the extremes of either class: Class-AB amplifiers usually have better efficiency than Class-A amplifiers and better gain than Class-B amplifiers.

The requirements placed on microwave amplifiers tend to favor Class-A amplifiers over Class-B ones for most applications. Microwave amplifiers are usually used to amplify continuous signals of constant amplitude (e.g., frequency-modulated or phase-shift-keyed digital signals) in which the high quiescent power dissipation of the Class-A amplifier is not a significant problem. However, the gains of power MESFETs are often very low, especially compared to those of small-signal devices, so the gain advantage of the Class-A amplifier is significant. Although the theoretical dc-RF efficiency of the Class-B amplifier is attractive, the *power-added* efficiency, not the dc-RF efficiency, is usually more important.

Power-added efficiency is the ratio of the additional power provided by the amplifier to dc power; it is defined as

$$\eta_a = \frac{P_L - P_{\text{in}}}{P_{\text{dc}}} \tag{9.2.14}$$

where P_{in} is the RF input power. We can show easily that

$$\eta_a = \eta_{\text{dc}}\left(1 - \frac{1}{G_p}\right) \tag{9.2.15}$$

where G_p is the power gain; $G_p = P_L/P_{\text{in}}$. Equation (9.2.15) implies that the low gain of the Class-B amplifier somewhat offsets the advantage of high dc-RF efficiency: practical Class-B amplifiers usually have, at best, only slightly better power-added efficiency than Class-A amplifiers. Class-B amplifiers are most valuable for amplifying pulsed signals having low duty cycles, where their low average current requirements are a distinct advantage.

9.3 DESIGN OF MESFET POWER AMPLIFIER

In designing FET power amplifiers, we follow the general procedure used in the preceding three chapters: We employ the usual components of approximation and engineering judgement to generate an initial design, then optimize that design via numerical techniques. The numerical process we use to optimize the FET power amplifier is harmonic balance. Because the Class-A amplifier is ideally a linear component, its initial design can employ linear circuit theory, usually very successfully. This is not the case with the Class-B amplifier, however, so we must be more careful with its design.

9.3.1 Approximate Design of Class-A FET Amplifiers

The first step in the design of a power amplifier is to select an appropriate MESFET. Most manufacturers of such devices know the output-power capabilities of their FETs, and this information is listed prominently on the specification sheets along with other traditionally optimistic claims. As a general rule, at frequencies below 12 GHz, most high-quality devices can produce output powers, as a function of gate width, of 0.3 W/mm at the 1-dB compression point and 0.5 W/mm when saturated. However, many experimental devices have produced power densities approaching 1 W/mm and many prosaic devices cannot achieve even 0.25 W/mm.

In designing the amplifier, we recognize that an ideal Class-A amplifier is, after all, a linear component. Therefore, although a practical Class-A amplifier is to some degree nonlinear, we should be able to rely fairly heavily on linear-amplifier theory in the initial approximate design. The fundamental problem in designing a Class-A amplifier, as in designing a small-signal linear amplifier, is to pick the appropriate source and load impedances: in a power amplifier, the load impedance must be selected to achieve the desired output power and the source impedance must provide a conjugate input match. Additionally, we must select a bias point that results in both adequate power and good efficiency.

We use the load-line approach described in Section 9.2 to select the real part of the load admittance. However, in order to select the load conductance properly, we must take into account the limits on the drain voltage and current as explained in Section 9.2. Figure 9.5 shows the terminal *I/V* characteristics of a power MESFET (i.e., with I_d expressed as a function of the terminal voltages V_{gs} and V_{ds} rather than as a function of the internal voltages V_g and V_d). We would prefer to have a plot of the internal *I/V* characteristics, the function $I_d(V_g, V_d)$, that does not include the voltage drops across the drain and source resistances. However, such curves are difficult to generate, and recognizing that this initial design is, after all, approximate, we shall accept a plot of the MESFET's terminal *I/V* characteristics as an approximation of the internal ones.

V_{min}, the minimum drain-source voltage, is limited to approximately 1.5 V by the knee of the *I/V* curve at $V_g = 0.6$ V; I_{max} is similarly limited. Because of a combination of effects, primarily the variation in V_t with V_d

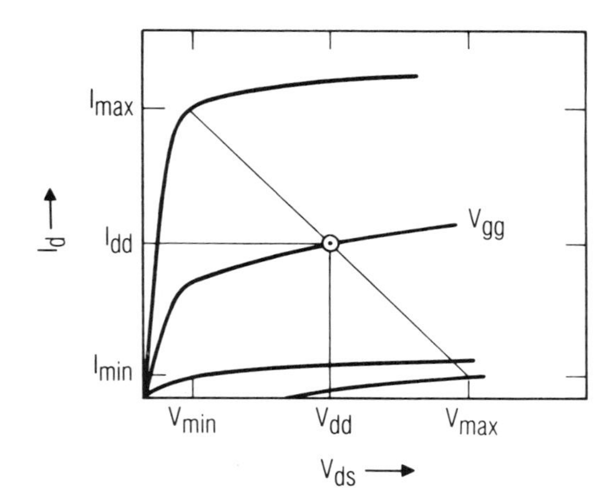

Figure 9.5 Drain *I*/*V* characteristics of a MESFET and the amplifier's load line. Because of the knee of the uppermost *I*/*V* characteristic, the minimum voltage is greater than zero. The optimum bias points are halfway between the maximum and minimum values of both voltage and current.

and the gate-drain avalanche limitation, V_d usually cannot be driven to the point where $I_d = 0$. Thus, there is a finite drain current, I_{min}, at V_{max}, the maximum value of V_d. V_{dd}, the dc drain-source voltage, is selected precisely halfway between V_{max} and V_{min}; I_{dd}, the quiescent dc drain current, is halfway between I_{max} and I_{min}. The gate-bias voltage that establishes this bias point is read directly from the *I*/*V* curves. We draw the load line superimposed on the *I*/*V* curves so that it connects these points; the load conductance is equal to the slope of the load line:

$$G_L = \frac{V_{max} - V_{min}}{I_{max} - I_{min}} \tag{9.3.1}$$

When an unpackaged MESFET is biased in its saturation region, the dominant component of its output admittance is the drain-source capacitance, C_{ds}. Because we wish to present a real load of conductance G_L to the terminals of the controlled current source I_d, the susceptance of the load must resonate with C_{ds}. Thus, the initial estimate of the load admittance is

$$Y_L = G_L - j\omega C_{ds} \tag{9.3.2}$$

If a packaged FET is used, determining the load impedance is complicated somewhat by the presence of the package parasitics but the underlying principle, presenting a real conductance of value G_L to the terminals of the current source, is still the same.

Because the load impedance at the terminals of the current source is real, the ac part of the drain voltage, $V_d(t)$, (which equals the load voltage, $V_L(t)$) and the load current, $I_L(t) = -I_d(t)$ are in phase (again, we write $I_d[V_g(t), V_d(t)]$ as $I_d(t)$ for simplicity). The output power is their product:

$$P_L = \frac{1}{2}\left[\frac{1}{2}(V_{\max} - V_{\min})\right]\left[\frac{1}{2}(I_{\max} - I_{\min})\right] \tag{9.3.3}$$

Some of this power is dissipated in the drain and source resistances, so for this reason as well as the others discussed in Section 9.2, (9.3.3) represents a slightly optimistic estimate. In a well designed MESFET, the power loss in these resistances is less than 1 dB.

The drain current of a Class-A amplifier should remain constant at the dc value under all excitation levels up to approximately the 1-dB gain compression point. As the amplifier is driven further into saturation, the $I_d(t)$ waveform becomes distorted and its average current changes. Below the compression point, the dc power equals the product of V_{dd} and I_{dd}; above the compression point, the dc power is usually greater but much of it is converted to RF output power. Therefore, the quiescent dc power can be considered an upper limit to the power dissipated by the device. If the amplifier has high gain and is to be operated only under excitation, the power dissipated by the device is approximately the difference between the output power and dc power. Designating the power dissipation P_d and the thermal resistance of the device from the channel to the mounting surface θ_{jc}, we find the temperature of the channel, T_{ch} to be

$$T_{ch} = T_a + P_d\theta_{jc} \tag{9.3.4}$$

where T_a is the temperature of the mounting surface. Equation (9.3.4) presupposes that the junction between the device and the mounting surface is thermally perfect; flaws in that junction, such as solder voids, can change the thermal resistance significantly or can cause "hot spots" on the surface of a large chip.

The input of the power FET amplifier is designed to be conjugately matched; therefore, we need to know the input impedance of the terminated MESFET. We can determine this impedance by using small-signal

S parameters and (8.1.5) and by repeating the calculation at appropriate intervals across the frequency range of interest. Finally, the small-signal gain can be found from (8.1.21) and stability factors and circles can be found from the appropriate equations, (8.1.2) and (8.1.7) through (8.1.10); the load impedance that optimizes output power is usually well within the stable region. Harmonic-balance analyses show that the input impedance varies only slightly with power level up to the point where the MESFET's gate begins to rectify the input signal significantly. Furthermore, in a well-designed amplifier, a good margin of small-signal stability is usually adequate to guarantee large-signal stability.

After the source and load impedances are determined, the matching networks can be designed. Designing the matching network is complicated by the fact that the source and load impedances are usually very low and it is best to short-circuit the drain at the harmonics of the excitation frequency. The latter requirement is not very severe in the case of Class-A amplifiers because the second and higher harmonic currents are not very great, but we shall see that it is very important in Class-B amplifiers. However, the combination of low impedances and high current densities requires careful consideration. The gate and drain currents in a power amplifier can be on the order of a few amperes, so even very small resistances can cause significant power dissipation. Capacitors, even those used for such prosaic purposes as dc blocking, must have high Qs, and inductors should not be made from narrow microstrips or fine wire (gold ribbon is a good material for inductors that must carry high currents). The topology of the matching circuit can often be selected to minimize the currents in components with relatively high loss.

We can estimate the source and load impedances at harmonics of the excitation frequency by analyzing the matching circuit via a general-purpose computer program for microwave circuit analysis. At high frequencies, the accuracy of such calculations deteriorates because of transmission-line discontinuities and the presence of high-order modes; however, at the same time, the large capacitances in the MESFET tend to short-circuit the external harmonic terminations and it becomes less important to terminate these harmonics accurately. With care, the matching circuits can usually be analyzed to obtain reasonable estimates of embedding impedances at frequencies as high as 18–20 GHz.

The final step in the design process is to check the circuit via a harmonic-balance analysis. The primary purposes of this step are to ascertain that the required output power will be achieved, to determine the harmonic output power and the degree of saturation, and to time-tune the bias voltages and load and source impedances to optimize the performance.

If the approximate design has been performed carefully, the harmonic-balance analysis will probably indicate that the approximate design is not far from the optimum.

9.3.2 Approximate Design of Class-B FET Amplifiers

The design of the Class-B amplifier parallels that of the Class-A amplifier. The load impedance of an ideal Class-B amplifier is the same as that of an ideal Class-A amplifier having the same output power, and we find from harmonic-balance analyses that a real MESFET's optimum Class-B and Class-A load impedances are often identical. In general, however, it is not possible to estimate the linear gain or input impedance of a Class-B amplifier from small-signal S parameters; instead, we use the MESFET's circuit model to estimate the input impedance and the harmonic-balance analysis to determine gain.

The maximum value of V_d allowable in a Class-B amplifier is somewhat lower than that of a Class-A amplifier. In FET amplifiers that are limited by gate-drain avalanching, the output power in Class-B operation is lower than that in Class-A operation. However, if the amplifiers are not limited by avalanche breakdown, the output powers of both classes are nearly identical. Thus, we can use the same procedure to select the load impedance of a Class-B amplifier as is used for a Class-A amplifier, as long as V_{max} is chosen to have its Class-B value.

The dc drain current of a Class-A amplifier under full excitation can be estimated as I_{max}/π. The dc power dissipation is

$$P_d = V_{\text{dd}} \frac{I_{\text{max}}}{\pi} \tag{9.3.5}$$

This estimate of the dc drain current is reasonable at the 1-dB saturation point; however, because the drain current varies with drive level, it is not valid at other levels. Furthermore, because of the inherently low gain of the Class-B amplifier, the RF input power may be relatively high and therefore may contribute significantly to power dissipation. Equation (9.3.4) is a valid expression for the channel temperature of a Class-B amplifier as well as a Class-A amplifier.

In an ideal Class-A amplifier, the gate-source voltage V_g varies between V_t and the threshold of gate conduction, approximately 0.5 V. In a Class-B amplifier, V_g varies between $\approx 2V_t$ and the same maximum voltage. Therefore, in order to deliver the same output power, the Class-B amplifier requires approximately twice the voltage across C_{gs} as the Class-A; accordingly, we might conclude that the Class-B input power must be 6 dB greater, so the gain must be 6 dB lower. This conclusion is troubling because many power MESFETS exhibit little more than 6 dB gain in Class-A operation. Fortunately, the situation is not quite that bad for several reasons: First, even in the ideal case, the difference in voltage is usually slightly less than a factor of two; second, the Class-B amplifier is often biased slightly above V_t, so it has a small quiescent drain current, which reduces the difference in the variation of V_g even further; and third, because the gate-bias voltage is more negative, C_{gs} is lower in Class B than in Class A. As a result, the difference in gain between Class-B and Class-A amplifiers using the same FET is usually from 3 to 5 dB, still significant but not as disastrous as 6 dB.

An adequate initial estimate of the input impedance of the Class-B amplifier is:

$$Z_{\text{in}} = R_g + R_i + R_s + \frac{\langle g_m \rangle L_s}{C_{gs}(V_t)} + \mathrm{j}\left[\omega\, L_s - \frac{1}{\omega\, C_{gs}\,(V_t)}\right] \tag{9.3.6}$$

that is, the sum of the impedances in the input loop of the MESFET. The term $\langle g_m \rangle$ is the transconductance averaged over the excitation cycle; $\langle g_m \rangle \approx 20$ percent of the peak transconductance. This impedance can be calculated more precisely by the harmonic-balance analysis; like that of the Class-A amplifier, the input impedance of the Class-B amplifier is not unduly sensitive to RF input level as long as the FET is not driven to the point where the gate junction rectifies the input signal.

9.3.3 Design Examples and Performance Study

To further investigate the design and performance of FET power amplifiers, we shall design Class-A and Class-B amplifiers using a common MESFET and optimize the design via harmonic-balance analysis. Although we shall see that the approximate design procedure is remarkably good when applied to Class-A amplifiers, it is not quite as reliable when applied to Class-B amplifiers, and in neither case does it produce all the information we would like to have with adequate accuracy.

The device we shall use is an experimental X-band power MESFET that has a gate width of 2.4 mm and an output power capability of slightly less than 1 W. The amplifier will be operated at 10 GHz. The equivalent circuit of the MESFET is shown in Figure 9.6(a), and its *I/V* characteristics are shown in Figure 9.6(b). We model the gate-source capacitance as an ideal, uniformly doped Schottky barrier having a zero-voltage capacitance

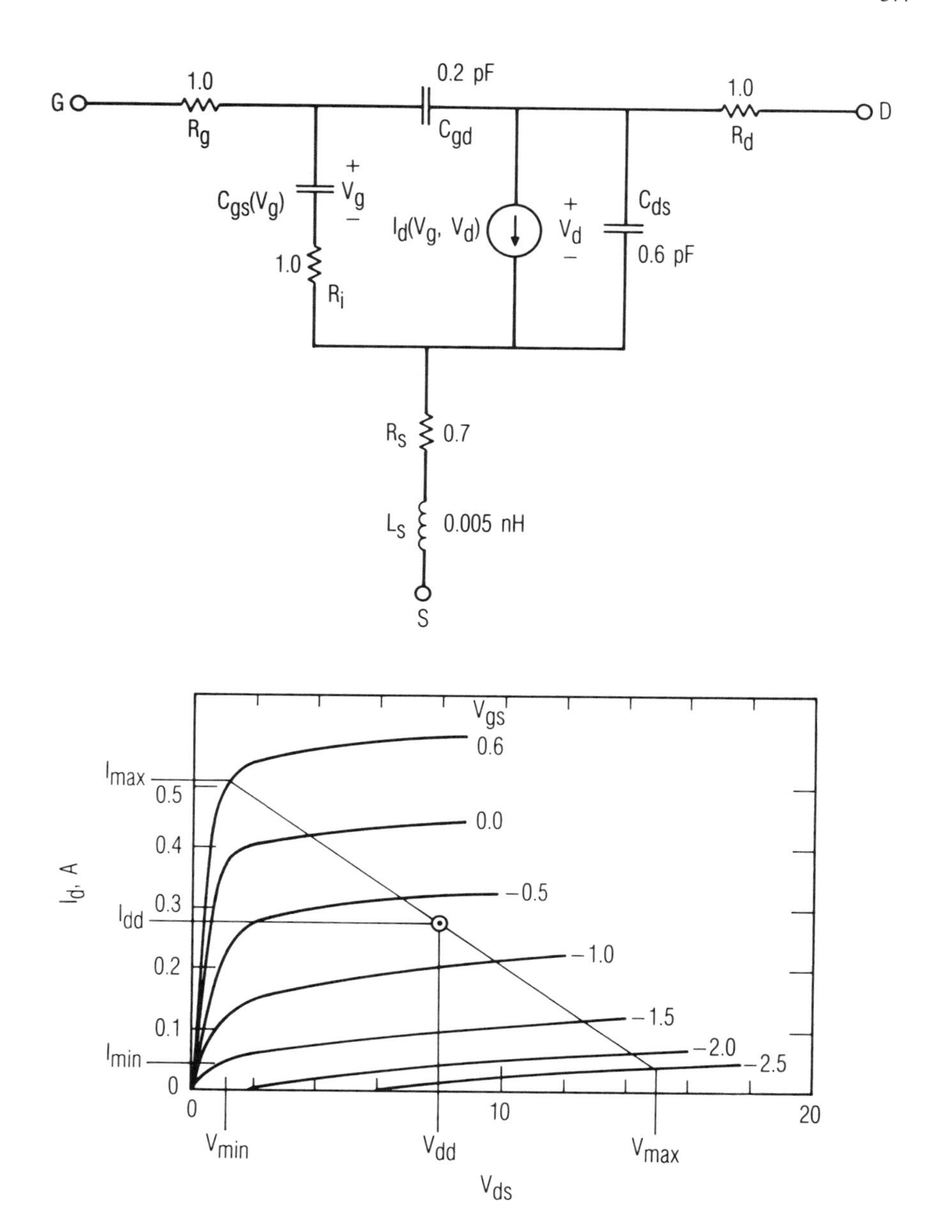

Figure 9.6 (a) Equivalent circuit of the 2.4-mm power MESFET used in the design examples; (b) the MESFET's *I*/*V* characteristics, including the load line.

of 4.0 pF, and dependent upon only one control voltage, V_g; lacking a good measurement of C_{gd} in the MESFET's linear range, we treat C_{gd} as a linear component. These simplifications will affect somewhat the accuracy of the calculations when the MESFET is strongly driven, but they will not affect the accuracy under moderate excitation. The I/V characteristic, $I_d(V_g, V_d)$ is expressed via (2.4.5), and the coefficients A_0 through A_3 are found via a least-squares fit to the measured I/V data at $V_{ds} = 8$ V, after the voltage drop across R_s is removed. The final parameter, α, is found easily by trial and error. We obtain

$$\begin{aligned} A_0 &= 0.5304 \\ A_1 &= 0.2595 \\ A_2 &= -0.0542 \\ A_3 &= -0.0305 \end{aligned}$$

and

$$\alpha = 1.0$$

Other parameters are not shown in the figures are $V_t = -2.5$ V, $\phi = 0.7$ V, $C_{gs}(0) = 4.0$ pF.

Example 1

We begin designing the Class-A amplifier by determining the load impedance. Figure 9.6(b) shows the load-line superimposed on the drain I/V characteristics. This load line was constructed under the constraint that $V_{dd} = 8.0$ V; such constraints on dc power are common in practical designs and they place an additional limitation on the power that can be obtained from a device. Because of this constraint on V_{dd}, we will not be able to obtain the maximum possible output power from the device, but because 8 V is close to the maximum value of V_{dd}, the power we will obtain will also be close to the maximum. From the I/V curves, we obtain $V_{max} = 14.7$ V, $V_{min} = 1.3$ V, $I_{max} = 500$ mA, $I_{min} = 40$ mA, and $I_{dd} = 270$ mA. Thus, we have

$$G_L = \frac{0.500 - 0.040}{14.7 - 1.3} \tag{9.3.7}$$

or $G_L = 0.034$ S. The output power, from (9.3.3), is:

$$P_L = \frac{1}{2}\left[\frac{1}{2}\,(14.7 - 1.3)\right]\left[\frac{1}{2}\,(0.500 - 0.040)\right] \tag{9.3.8}$$

Thus, $P_L = 0.771$ W or 28.9 dBm. For the reasons given in Section 9.2, we expect this estimate to be optimistic by approximately 1 dB, so we anticipate a more realistic 1-dB compression point of approximately 28 dBm. Finally, from (9.3.2), the load admittance is

$$Y_L = G_L - \mathrm{j}\omega C_{\mathrm{ds}} = 0.034 - \mathrm{j}0.05 \tag{9.3.9}$$

and inverting (9.3.9) gives the load impedance, $Z_L = 9.3 + \mathrm{j}13.5$. We use a general-purpose computer program that performs linear analysis to find the input impedance and small-signal gain. First, we must find C_{gs} at the bias point; from Figure 9.6(b), $V_g = -0.7$ V and

$$C_{\mathrm{gs}} = \frac{C_{\mathrm{gs}}(0)}{\left(1 - \dfrac{V}{\phi}\right)^{1/2}} = \frac{4.0\ \mathrm{pF}}{\left(1 - \dfrac{-0.7}{0.7}\right)^{1/2}} = 2.8\ \mathrm{pF} \tag{9.3.10}$$

The transconductance is found from the I/V curves or by differentiating $I_d(V_g, V_d)$; it is 315 mS. Using these values in a linear analysis of the circuit in Figure 9.6(a) gives $Z_{\mathrm{in}} = Z_s^* = 2.3 - \mathrm{j}4.3$ and $G_t = 10.0$ dB. It is important to note that Z_s and Z_L are the impedances presented to the gate and drain terminals of the FET, so the matching circuits must include the effects of bond-wire inductances, package parasitics, and any other parasitics not included in Figure 9.6(a). These impedances also do not include the effects of loss; loss in the matching circuits not only reduces the gain and power levels but also affects the source and load impedances presented to the MESFET.

We now have enough information to perform a harmonic-balance analysis. Because we have not designed matching circuits, we will assume that those circuits present the optimum termination, a short-circuit, to the gate and drain at all harmonics; the effect of nonzero embedding impedances will be examined later. The harmonic-balance analysis shows that the performance of the initial design is indeed very good: the power at the 1-dB compression point is 27.6 dBm and the power-added efficiency is 22 percent. The saturated output power (defined as the power level where η_a is maximum) is 29.2 dBm; the gain at this output level is 6.2 dB and the efficiency is 28 percent.

Optimizing the design involves adjusting the load impedance empirically to maximize output power, adjusting the gate bias to optimize power and efficiency, determining the input impedance at full output, and resetting the source impedance to the conjugate of that value. We find that the optimum design is very close to the initial one; the optimum load impedance is 9.0 + j12.0, the source impedance at the 1-dB compression point is 2.5 + j4.5, and the optimum gate bias is −0.8 V. The output power is 27.8 dBm at the 1-dB compression point (1.1 dB below the initial estimate) and the saturated output power is 29.5 dBm; the gain at saturation is 6.4 dB and the power-added efficiency is 30 percent. The dc current at saturation is 290 mA, and the dc power is 2.3 W. The output power and power-added efficiency, η_a, are shown in Figure 9.7 as functions of input power level.

The peak drain-gate voltage predicted by the harmonic-balance analysis is 15.4 V at the 1-dB compression point. This voltage is slightly less than $V_{\text{max}} + |V_t| = 17.2$ V because the peak values of $V_d(t)$ and $V_g(t)$ are only 13.3 V and −2.1 V, respectively. The amount 17.2 V is well below the breakdown voltage of the device, which implies that greater output power may be attainable if V_{dd} is increased and that the same output power can be obtained in Class-B operation by using the same value of V_{dd}.

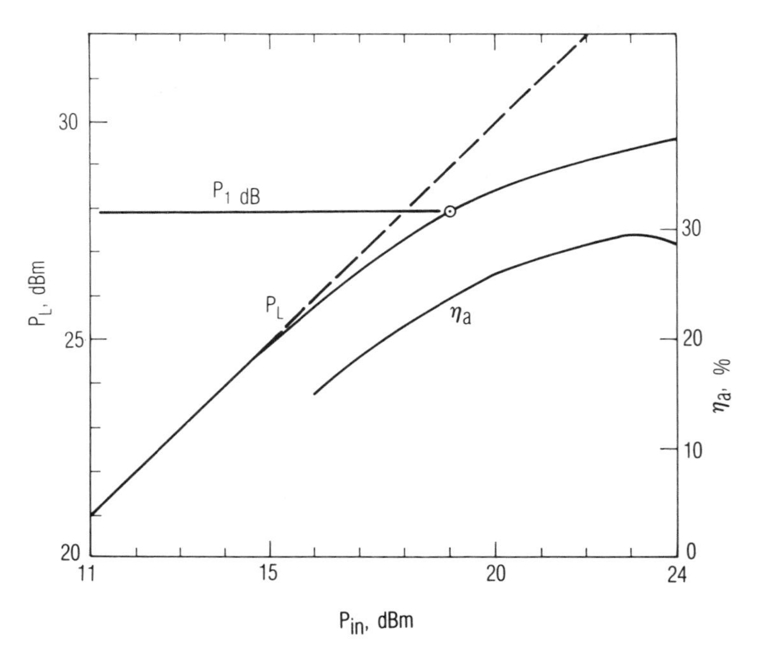

Figure 9.7 Output power, P_L, and power-added efficiency, η_a, of the optimized Class-A amplifier, calculated via harmonic balance.

Example 2

We now design a Class-B amplifier by modifying the previous design. The first modification is to reexamine the bias voltages. We retain the drain bias V_{dd} of 8 V and adjust the gate bias V_{gg} to -2.0 V so that there is a slight quiescent current, approximately $0.1 I_{dss}$, or 40 mA.

Again, we assume that the harmonic source and load impedances are short circuits, and we are therefore concerned with the fundamental-frequency terminations only. Because the Class-A and Class-B load impedances should be the same, we choose the same load impedance for the Class-B stage as the Class-A. The gate-source capacitance of the Class-B amplifier is lower than that of the Class-A because the voltage across C_{gs} is more negative. Consequently, the imaginary part of the input impedance should be greater. From (9.3.6), we obtain an initial estimate of $Z_s = Z_{in}^* = 2.7 + j7.6$. These modifications are all that are necessary to convert the Class-A amplifier to a Class-B amplifier.

Once more, we optimize the design via harmonic-balance analysis. Although the drain voltage and current ranges are slightly different (e.g., $I_{min} = 0$ in Class B), we find that the optimum Class-B load impedance is equal to the optimum Class-A load and the input impedance at the 1-dB compression point is $2.7 - j5.2$. P_L and η_a, as a function of P_{in}, are shown in Figure 9.8. We see that the saturated output power is 29.3 dBm, nearly identical to that of the Class-A amplifier, but the output at the 1-dB gain compression point is 28.9 dBm, approximately 1 dB greater than that of the Class-A amplifier. The gain at saturation is 3.0 dB, and although the dc efficiency of the Class-B amplifier is much better than that of the Class-A, the power-added efficiency has the same maximum value, 30 percent. However, because the dc current varies with input power, the Class-B amplifier achieves its maximum efficiency over a much broader range than does the Class-A.

The small-signal gain of an ideal Class-B amplifier is undefined because the FET is turned off when quiescent. When the FET has a small quiescent drain current, the small-signal gain can be defined; however, that gain may be a meaningless parameter because the transconductance at the low drain current may be very low. Here we define the "small-signal" gain of the Class-B amplifier as the gain derived from the straight part of the P_{in}/P_L characteristic, which, from Figure 9.8, is 5.2 dB.

The peak drain-gate voltage of the Class-B amplifier at the 1-dB compression point is 18.8 V, more than 3 V greater than that of the Class-A but still below the breakdown voltage of most power devices. The range of $V_d(t)$ in the Class-B amplifier is 1.0–14.1 V, somewhat greater than that of the Class-A, and the peak value of $V_g(t)$ is -4.1 V, predictably greater

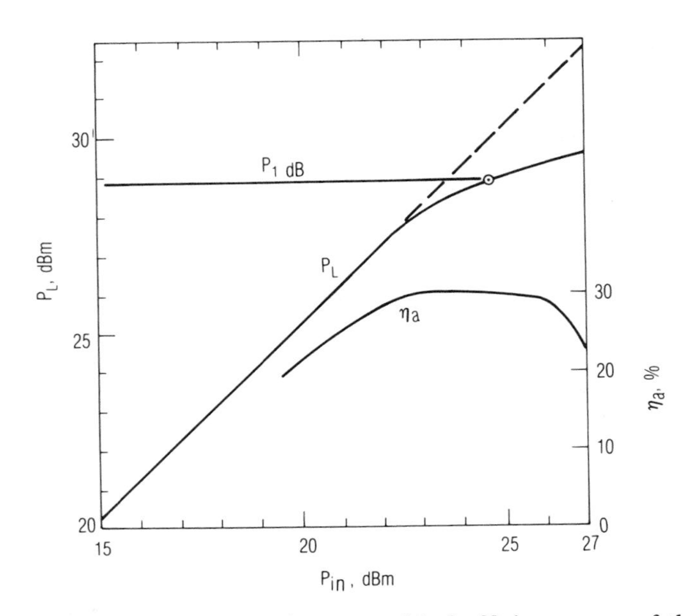

Figure 9.8 Output power, P_L, and power-added efficiency, η_a, of the optimized Class-B amplifier, calculated via harmonic balance.

than that of the Class-A amplifier and very close to the theoretical value of twice the bias voltage. The drain-current pulse has a peak value of 490 mA, and it has a slight dip at its peak because at that instant $V_d(t)$ drops slightly below the knee of the *I/V* curve.

The results of this design exercise have validated our earlier conclusions about the properties of Class-A and Class-B amplifiers, that Class-B amplifiers generally have greater dc efficiency and poorer gain than Class-A and, when they are not limited by drain-gate breakdown, approximately equal power. It is interesting to note that experimental studies of Class-B amplifiers (Reference 9.2) draw identical conclusions. We also saw that power FET amplifiers can be designed quite effectively via a process that uses both linear and nonlinear circuit theory. Table 9.1 summarizes the results of the design study.

Table 9.1 Amplifier Performance Comparison (frequency = 10 GHz)

Parameter	*Approximate Init. Design*	*Harmonic Balance Calculations* Init. Design	*Opt. Class-A*	*Opt. Class-B*
V_{dd}	8.0	8.0	8.0	8.0
V_{gg}	−0.7	−0.7	−0.8	−2.0
I_{dc}	270	297	280	199
$V_d(t)$ range	1.3–14.7	1.6–13.3	1.6–13.3	1.0–14.1
$I_d(t)$ range	40–500	30–500	30–500	0–490
$V_{f,max}$	17.2	15.4	15.4	18.8
G_t (small-signal, dB)	10.0	10.0	9.9	5.2
Z_{in}	2.3 − j4.3	2.3 − j4.3	2.5 − j4.5	2.7 − j5.6
Z_L	9.3 + j13.5	9.3 + j13.5	9.0 + j12.0	9.0 + j12.0
P_L	28.9	27.6	27.8	28.8
η_a	30%	22%	24%	30%
$\eta_{a,max}$	NE	28%	30%	30%
$P_L(\eta_{a,max})$	NE	29.2	29.5	28–29.3[1]
$G_t(\eta_{a,max})$	NE	6.2	6.4	3–5[1]

Notes: All entries are in dB, dBm, mA, V, and ohms, as appropriate, and unless noted, all entries refer to the 1-dB compression point. NE means the parameter was not estimated in the approximate design.
[1] η_a is constant over a wide range of power levels in a Class-B amplifier.

9.3.4 Effect of Nonzero Harmonic Terminations

We now examine the effect of deviations from the ideal conditions of the previous two examples. In particular, we wish to examine the effect of load impedances other than a short circuit at harmonics of the excitation frequency. We examine the effects on both the Class-A and Class-B amplifiers; however, because the harmonic components of the drain current are greater in the Class-B amplifier than in the Class-A, we expect the performance of the Class-B amplifier to be more sensitive to harmonic terminations.

It is possible to understand intuitively the effect of nonzero harmonic terminations. The terminating impedance, $Z_L(\omega)$, is simply the amplifier's load impedance transformed to the drain of the FET through the output matching circuit. All the harmonic components of the drain current circulate in $Z_L(\omega)$, so if $Z_L(\omega)$ has a nonzero real part at a harmonic frequency, there can be output power at that harmonic. This output power is a manifestation of harmonic distortion and is undesirable for many rather obvious reasons. However, even if $Z_L(\omega)$ is purely reactive at some harmonic, and thus there is no harmonic output power, V_d has a voltage component at that harmonic; this component can cause the fundamental output power to be reduced. The power is reduced because, in general, output power is limited by the need to keep $0 < V_d(t)\ V_{max}$. The addition of a harmonic component to $V_d(t)$ may force the fundamental component of $V_d(t)$ to be reduced, so that $V_d(t)$ remains within the prescribed limits. Harmonic-balance analysis is the only accurate means to assess the effects of harmonic terminations in a power FET amplifier.

In order to assess the effects of harmonic terminations on the Class-B amplifer, the harmonic-balance analysis was performed repeatedly with a wide variety of terminations. Most terminations had surprisingly little effect on the amplifier. For example, terminating the drain in an open circuit at all harmonics did not change the fundamental output power at all, and an open circuit at only the second harmonic reduced the output level by only 0.2 dB. The worst-case reduction in fundamental output power caused by a purely resistive second-harmonic load was only 0.25 dB.

The reason for the low sensitivity to harmonic terminations appears to be that the relatively large value of C_{ds}, 0.6 pF, effectively short-circuits the channel at harmonics of the excitation frequency. Therefore, unless the second-harmonic termination resonates with C_{ds}, it has a minimal effect on fundamental output. However, if the load does resonate with C_{ds}, the results can have a serious effect on the performance of the amplifier. Figure 9.9(a) shows the effect of a purely reactive second-harmonic termination on the output power of the optimized Class-B amplifier. The power decreases by more than 1.5 dB (approximately 30 percent) at a specific value of reactance, 10 Ω; additionally, when this termination is used, the drain current and voltage waveforms are significantly distorted.

If the second-harmonic termination has a real part, output power at that frequency is possible; Figure 9.9(b) shows the effect of adding a conductance in parallel with the susceptance (-0.1 S) that resonates C_{ds}.

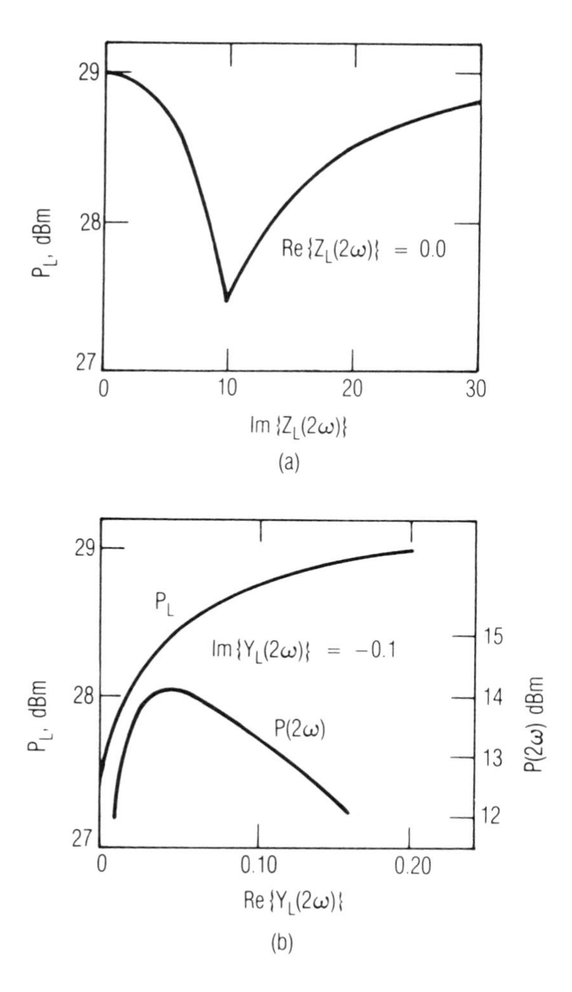

Figure 9.9 The effect of the second-harmonic load impedance on fundamental output power and harmonic distortion in the Class-B amplifier: (a) effect of a purely reactive termination; (b) fundamental and second-harmonic output powers as a function of Re{$Y_L(2\omega)$} when Im{$Y_L(2\omega)$} has its worst-case value. The input level is 25 dBm.

Although the fundamental output power rises monotonically with $\mathrm{Re}\{Y_L(2\omega)\}$, the harmonic output peaks near $\mathrm{Re}\{Y_L(2\omega)\} = 0.05$. The worst-case harmonic level is 14 dB below the fundamental output; earlier we noted that this harmonic is approximately 7 dB below the output in an ideal Class-B amplifier. However, the actual second-harmonic output is much lower than 7 dB because the FET is driven into its linear range at the peak of the excitation cycle. As a result, the drain current pulse is not simply flattened but actually has a conspicuous dip at the point that would otherwise be its peak. This dip decreases its second-harmonic component significantly, and even though C_{ds} is resonated by the imaginary part of the load susceptance, the second-harmonic output power is reduced considerably.

The effect of nonzero harmonic terminations in the Class-A amplifier is even less spectacular than in the Class-B, for two reasons: First, as in the Class-B case, C_{ds} short-circuits the drain; and second, the harmonic components of the drain current are much smaller because in Class-A operation the drain current is not switched on and off. Thus, harmonics are generated primarily by the nonlinearities in $I_d(V_g, V_d)$, and these are relatively weak compared to a switching function. Consequently, the harmonic terminations have virtually no effect on the fundamental output power; in the worst case a harmonic termination decreases the fundamental output power by less than 0.2 dB. However, the harmonic distortion, although relatively low, varies significantly with Y_L. Figure 9.10 shows the effect on the second-harmonic output power of varying the load conductance $\mathrm{Re}\{Y_L(2\omega)\}$ while maintaining the susceptance constant at -0.1. Although the output power remains constant within 0.1 dB over this range of load impedances, the second-harmonic output power varies by several dB.

This result, that the fundamental output power is relatively independent of changes in harmonic terminations, has several important implications. One of the properties of a quasilinear component was precisely this: that we could determine its fundamental-frequency performance without considering nonlinearities or harmonic source and load impedances. The results of this exercise indicate that the Class-A amplifier could almost be considered a quasilinear component; although it was necessary to account for the nonlinearities in obtaining the last dB or so of accuracy in calculating its gain, efficiency, and output power, it was not necessary to consider embedding impedances at harmonic frequencies. The second implication is that, in spite of the nasty things we said about

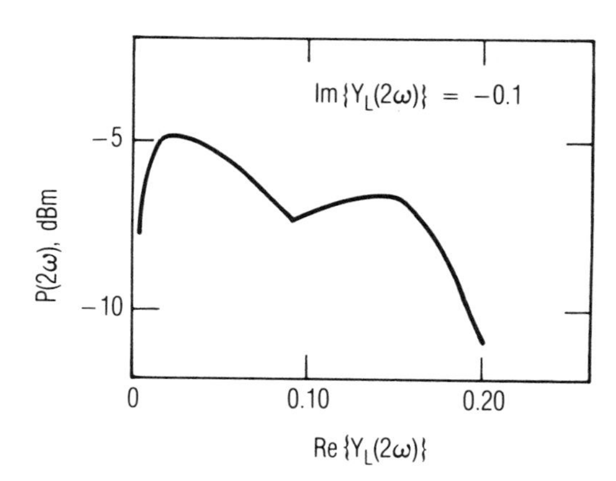

Figure 9.10 Harmonic distortion levels in the optimized Class-A amplifier, as a function of the second-harmonic load conductance $\text{Re}\{Y_L(2\omega)\}$, at an input level of 19 dBm. The fundamental output power is constant at 28.9 dBm over this range of $\text{Re}\{Y_L(2\omega)\}$.

load-pull characterization in Chapter 1, we would characterize this device quite successfully via load-pull measurements. The major disadvantage of load-pull characterization is that it does not account for harmonic terminations; this disadvantage evaporates if the device is insensitive to such terminations.

We must remember, however, that the insensitivity to harmonic terminations resulted from the large value of C_{ds} in this particular device. Other MESFETs may have much lower values of C_{ds}, and in these, the harmonic terminations may have much more significant effects. For example, it is a common practice to use small-signal MESFETs as "power amplifiers," in the sense that they are biased and their loads are selected to optimize output power rather than gain, noise, or other parameters. In these devices, C_{ds} is very small and the harmonic terminations may be much more significant.

Finally, we note that we have not considered the harmonic terminations at the input. The reason for this neglect is that the source impedances, at the fundamental and harmonic frequencies, have very little effect on the output power of the amplifier; in addition, the source impedances at harmonics do not have much effect, even on gain. It is conceivable that the harmonic source impedances might have some effect on the intermodulation or harmonic-distortion characteristics of a power amplifier but the results of the studies, described in the previous chapter, of the effects of source impedance on intermodulation make such phenomena seem unlikely.

9.4 HARMONIC-BALANCE ANALYSIS OF FET POWER AMPLIFIERS

Before leaving the subject of FET power-amplifier analysis, we should examine the details involved in applying harmonic-balance techniques to power-amplifier circuits. The harmonic-balance calculations used to generate the data in Section 9.3 were performed on an IBM PC-AT computer using a special-purpose program, developed by the author, that analyzes MESFET circuits. Although C_{gd} was treated as a linear element in performing these calculations, the program allows C_{gd} to be treated as a nonlinear element if the user can model it appropriately; the other nonlinear elements are I_d and C_{gs}. I_d and C_{gs} are described by analytical expressions; I_d is given by (2.4.5) and C_{gs} is modeled as an ideal Schottky-barrier capacitance having one control voltage, V_g. The program uses Newton's method as a convergence algorithm and includes eight harmonics of the excitation frequency plus dc. It requires approximately 10 seconds to perform each Newton iteration; most of the time is used in solving (3.1.37) to obtain the updated voltage vector. The convergence properties are generally very good, although slow convergence has been encountered occasionally in frequency-multiplier circuits.

This program follows very closely the methods described in Chapter 3. Figure 9.11(a) shows the equivalent circuit of the MESFET, including source and load terminations, bias sources, and the excitation source; in order to retain generality, we have treated C_{gd} as a nonlinear element. Figure 9.11(b) shows the circuit divided into nonlinear and linear subcircuits. Ports 1 through 3 are those to which the nonlinear elements are connected; Ports 4 and 5 are the ports to which the excitation and bias voltages are applied. We note that $V_1 \equiv V_g$ and $V_2 \equiv V_d$.

The Y parameters of the linear subcircuit are found readily in the usual manner. They are

$$Y_{1,1}(\omega) = -Y_{1,2}(\omega) = -Y_{1,3}(\omega) = \frac{1}{R_i}$$

$$Y_{1,4}(\omega) = Y_{1,5}(\omega) = 0$$

$$Y_{2,2}(\omega) = \frac{1}{R_i} + \left[\left(\frac{1}{Z_d} + \frac{1}{Z_g}\right)^{-1} + R_s + j\omega L\right]^{-1} + j\omega C_{ds}$$

$$Y_{4,4}(\omega) = \left[\left(\frac{1}{Z_d} + \frac{1}{R_s + L_s}\right)^{-1} + Z_g\right]^{-1}$$

$$Y_{5,5}(\omega) = \left[\left(\frac{1}{Z_g} + \frac{1}{R_s + L_s}\right)^{-1} + Z_d\right]^{-1}$$

$$Y_{2,4}(\omega) = -Y_{4,4}(\omega)\frac{Z_d}{R_s + j\omega L_s + Z_d}$$

$$Y_{2,5}(\omega) = -Y_{5,5}(\omega)\frac{Z_g}{R_s + j\omega L_s + Z_g}$$

$$Y_{2,3}(\omega) = -Y_{2,4}(\omega) + Y_{1,1}(\omega)$$

$$Y_{3,3}(\omega) = Y_{4,4}(\omega) + Y_{1,1}(\omega)$$

$$Y_{3,4}(\omega) = -Y_{4,4}(\omega)$$

$$Y_{3,5}(\omega) = Y_{5,5}(\omega)\frac{R_s + j\omega L_s}{R_s + j\omega L_s + Z_g}$$

$$Y_{4,5}(\omega) = -Y_{3,5}(\omega) \qquad (9.4.1)$$

where

$$Z_d = Z_L(\omega) + R_d$$

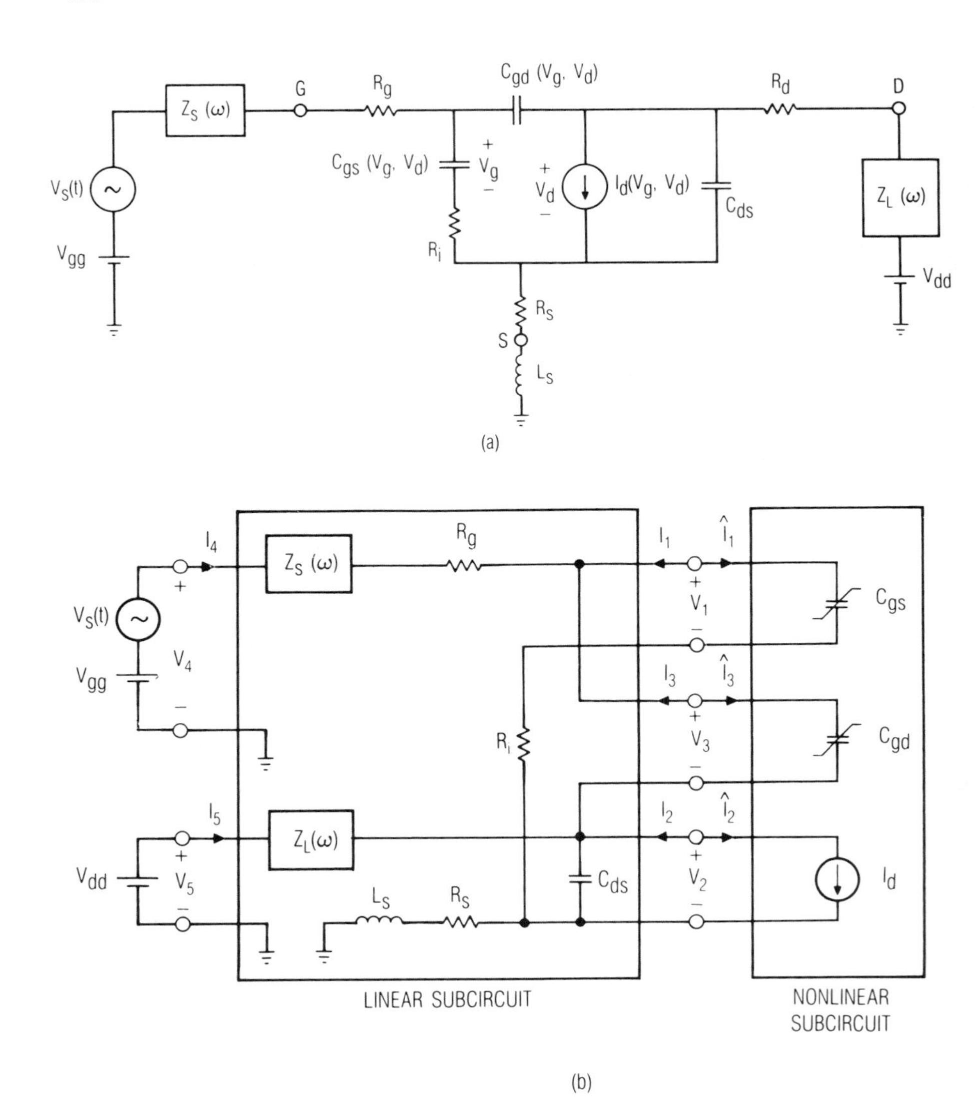

Figure 9.11 (a) Equivalent circuit of the FET power amplifier; (b) the same circuit separated into linear and nonlinear subcircuits.

and

$$Z_g = Z_s(\omega) + R_g$$

The remaining Y parameters are found from the reciprocal property of the linear subcircuit, which mandates that $Y_{m,n} = Y_{n,m}$.

If we wish to use Newton's method, we need to formulate the Jacobian. From (3.1.38):

$$\mathbf{J}_F = \mathbf{Y}_{3\times3} + \frac{\partial \mathbf{I}}{\partial \mathbf{V}} + \mathrm{j}\boldsymbol{\Omega}\frac{\partial \mathbf{Q}}{\partial \mathbf{V}} \tag{9.4.2}$$

where $\mathbf{Y}_{3\times3}$ is defined by (3.1.8) and (3.1.9), $\boldsymbol{\Omega}$ by (3.1.16), and the derivative terms by (3.1.39) through (3.1.41). The Jacobian has the form:

$$\mathbf{J}_F = \begin{bmatrix} \mathbf{J}_{1,1} & \mathbf{J}_{1,2} & \mathbf{J}_{1,3} \\ \mathbf{J}_{2,1} & \mathbf{J}_{2,2} & \mathbf{J}_{2,3} \\ \mathbf{J}_{3,1} & \mathbf{J}_{3,2} & \mathbf{J}_{3,3} \end{bmatrix}$$

that is, it is composed of a 3×3 set of submatrices, where

$$\mathbf{J}_{m,n} = \mathbf{Y}_{m,n} + \mathbf{G}_{m,n} + \mathrm{j}\boldsymbol{\omega}\mathbf{C}_{m,n} \tag{9.4.3}$$

$\mathbf{Y}_{m,n}$ has the form shown in (3.1.6), and $\boldsymbol{\omega}$ has the same form as $\boldsymbol{\Omega}$ in (3.1.16) except that it includes only one cycle of $0 \ldots K\omega_p$ along the main diagonal. The terms $\mathbf{G}_{m,n}$ and $\mathbf{C}_{m,n}$ ar Fourier coefficients of the derivative waveforms as shown in (3.1.39) through (3.1.41). These matrices have the same form as the conversion matrices used in small-signal analysis, shown in (3.2.24) and (3.2.32). The fact that the small-signal conversion matrices and the submatrices of the Jacobian have the same form may seem surprising; they are the same because they represent the same thing, a linear perturbation of the large-signal I/V or Q/V characteristics.

As in most other applications of harmonic balance, the preferred convergence algorithms are Newton's method and the reflection algorithm. The most important considerations in using Newton's method are generating an initial estimate of the solution, constructing the Jacobian matrix efficiently, and solving the linear equations to obtain the updated voltage vector. The author's program generates the initial estimate by decomposing the MESFET into a unilateral two-port: C_{gd} is removed and added to C_{gs}, R_s and L_s are included in the input circuit, and C_{gs} is linearized. The drain-current waveform $I_d(t)$ is then calculated under the assumption that $V_d(t) = V_{dd}$ (i.e., the FET remains in saturation throughout the excitation cycle) by first finding the fundamental-frequency component of $V_g(t)$ and then evaluating $I_d(V_g(t), V_{dd})$. Harmonic components of $V_g(t)$ are not estimated. Harmonic components of $V_d(t)$ are found by Fourier-transforming $I_d(t)$ and by multiplying the embedding impedances, in parallel with C_{ds}, by the harmonic components of I_d. Finally, other necessary components, such as the harmonic components of V_f (the voltage across C_{gd}), are found from these voltages. This initial estimate is satisfactory as long as the FET is not driven strongly into power saturation; if the FET is too strongly driven to allow convergence, the circuit is analyzed first at a lower power level and then the power is increased stepwise until the desired input level is reached. Computing the initial estimate takes less than 1 second.

In order to use Newton's method efficiently, it is essential that the voltage dependences of the nonlinear elements be described by functions that can be differentiated analytically. If a numerical MESFET model is used, constructing the Jacobian requires evaluating the derivative waveforms numerically; this process requires an large amount of computation, which may be prohibitive if accurate results are to be obtained on a small computer. When derivative waveforms must be determined numerically, constructing the Jacobian is likely to be the most time-consuming part of a very slow analysis. The requirement of analytical expressions places some practical limitations on the models that can be used successfully with Newton's method.

If analytical expressions are used in the MESFET model, the most time-consuming part of the analysis is solving the linear equations to obtain the new estimate of the harmonic voltage vectors. This step is time-consuming because the Jacobian matrix is complex and has dimensions $N \times H$, where N is the number of nonlinear elements and H is the number of harmonics including dc. Thus, if we have three nonlinear elements (C_{gs}, C_{gd}, and I_d) and wish to consider harmonics up to the eighth, we must solve a 27×27 system of linear equations. The system of equations is invariably well conditioned (Section 3.3.3), so solving the equations rarely involves any unanticipated difficulties; any of the well-known techniques for solving linear equations can be employed.

Curtice (Reference 9.3) describes the use of Newton's method in a generalized harmonic-balance technique that analyzes multitone circuits.

He employs a partial update, in which the inverted Jacobian of (3.1.37) is multiplied by a constant, β, which he calls an *update factor*, in determining the updated voltage vector; that is, (3.1.37) becomes

$$\mathbf{V}^{p+1} = \mathbf{V}^{p} - \beta \left. \frac{\partial \mathbf{F}(\mathbf{V})}{\partial \mathbf{V}} \right|_{\mathbf{V}=\mathbf{V}^{p}}^{-1} \mathbf{F}(\mathbf{V}^{p}) \tag{9.4.4}$$

Thus, a value of $\beta < 1$ reduces uniformly the change in the elements of $\mathbf{V}$ in each iteration. Curtice claims that a value of β near 0.5 minimizes the number of iterations necessary to achieve convergence; a smaller value causes the change in $\mathbf{V}$ in each iteration to be very small, slowing the solution process, but a larger β causes $\mathbf{V}$ to oscillate, also slowing convergence. The use of the update factor $\beta < 1$ is most helpful in analyzing circuits that may have convergence problems, such as amplifiers driven into saturation or frequency multipliers.

Convergence problems encountered in Newton's method usually result from an inadequate initial estimate of the port voltage vector; however, when the circuit is driven strongly into gain saturation, the amplifier's voltage and current waveforms are often so complex that an adequate estimate may not be possible. In this case, we can sometimes obtain convergence by first analyzing the circuit at a reduced excitation level and using the resulting voltage vector as the initial estimate in a second analysis at a higher excitation level. The excitation power is raised stepwise (perhaps 1 dB per step) in this manner until the desired excitation level is reached, and at each step, the initial estimate is the port voltage vector obtained from the previous iteration.

The reflection algorithm is another good choice for use in FET analysis. An important advantage of the reflection algorithm is that it does not require the calculation of derivative waveforms, so it can be used efficiently with numerical models of the MESFET. Furthermore, it generally requires a relatively small amount of computation per iteration and has modest memory requirements; these are significant advantages when small computers are used to analyze large circuits.

Hwang and Itoh (Reference 9.4) have described how a method originally introduced by the author (References 2.1 and 2.7) can be used to analyze FET power amplifiers. They have included a modification based on the splitting method of Hicks and Khan (Reference 3.9). When applied to the analysis of FET amplifiers, the method reduces the number of iterations necessary to achieve convergence by a factor of approximately 3. This method is analogous to Curtice's partial update of the voltage vector used in Newton's method.

Convergence problems encountered in the reflection algorithm usually arise in the solution of the differential equations that describe the nonlinear subcircuit not the iterative method itself. This phenomenon is especially evident when the entire FET is used as the nonlinear subcircuit. Accordingly, reducing the excitation power level and "creeping up" on the solution rarely works as well with the reflection algorithm as it does with Newton's method; the cure for convergence problems in this case is to improve the solution of the nonlinear equations. This solution process involves integrating a set of time-domain nonlinear differential equations; improving that process may be as simple as reducing the time increment in the integration. The use of a partial update such as the one suggested by Hwang and Itoh may also improve convergence by reducing the range of excitation voltage waveforms applied to the nonlinear subcircuit in each iteration.

Throughout this chapter, we have assumed that the MESFET's gate-channel junction is not driven to the point of rectification. Gate rectification, indicated by the presence of dc gate current, is often unacceptable in high-reliability amplifiers. In large, strongly driven power devices, however, some gate rectification is inevitable, but it should never be great enough to affect the amplifier's RF performance. In harmonic-balance analysis, we can account for gate rectification by including diodes in the equivalent circuit, in parallel with C_{gs} an C_{gd}. If diodes are not included, it is important to check the results of the harmonic-balance analysis to ascertain that the gate voltage is not great enough to cause substantial rectification; if it is, the results of the analysis may be misleading.

9.5 PRACTICAL CONSIDERATIONS

Although we have already discussed some of the practical aspects of designing and realizing FET power amplifiers, a few matters that deserve attention still remain. These include some aspects of the design of matching circuits and thermal considerations.

The very large dimensions of power MESFET chips introduce several practical difficulties. Because good output power and efficiency requires that all the cells operate at full power, it is critically important that all cells in the device have equal excitation. If the MESFET chip is very wide (large chips can be several millimeters in width), the bond wires from the cells near the center of the chip to the microstrip line are often shorter than those from the cells that are close to the chip's ends. Thus, the outer cells and inner cells are matched differently. Even if the chip is no wider than the microstrip to which it is connected and the bond wires have equal lengths, the connections to the outer edges of the microstrip have source impedances different from those close to the center, and these unequal impedances may cause the cells' drive levels to be nonuniform. A symptom of unequal drive is the existence of dc gate current at power levels well below saturation.

A simple way to avoid the problem of unequal drive is to use a tree structure in the input microstrip. The input microstrip is split into two branches, those are then split in two, and the process continues until there are enough branches to provide a separate source microstrip to each cell. The branches are made narrow enough so that the source impedance does not vary over the width of the strip.

Another solution to the problem of unequal drive in large chips is to realize a power amplifier as a combination of smaller, single-device amplifiers. The process of connecting two or more amplifiers together to increase their output power capability is called *power combining*; two amplifiers may be power-combined by means of hybrids, as discussed in Chapter 5, or several amplifiers may be combined via more complex circuits. This approach is relatively expensive because it requires the design of power-combining circuitry as well as the individual amplifiers. It is often necessary at high frequencies, however, because the input impedance of a wide, high-power device may be too low to allow adequate matching.

The use of chip devices requires special assembly equipment that is often not available to the amplifier designer, who therefore may prefer to use packaged devices. Packaged devices are usually practical at low frequencies (below 8–12 GHz, depending upon power level), but at higher frequencies the package parasitics may complicate an already difficult matching problem. One solution to the problem of package parasitics is for the device manufacturer to place some of the matching components inside the package. These raise the input and optimum load impedances to a level that can be matched easily by the external circuit. Such devices are said to be *internally matched*; some internally matched power FETs are so carefully designed that it is not necessary to use any external matching circuits. A disadvantage of internal matching is that the internal circuit has a specific, limited frequency range that cannot be adjusted by the user.

The thermal design of the FET power amplifier must be performed carefully so that the MESFET's channel temperature is minimized. Temperature affects both the performance and the reliability of an amplifier: The MESFET's transconductance varies approximately in inverse proportion with temperature, and its mean time to failure increases exponentially with temperature. Although little can be done to change the thermal design of the MESFET itself, the circuit designer can do much to minimize the temperature increase caused by factors under his or her control. If an unpackaged chip is used, the chip must be soldered effectively to the mounting surface; a packaged device must be screwed or soldered in place, according to its design. The housing in which the device is mounted must be designed to provide good heat transfer to its outer surface, and in many cases, a separate heat dissipator, sometimes including forced-air cooling, must be provided.

Hot areas on the surface of the device can be caused by solder voids under the chip or, more commonly, flaws in fabrication that were not evident during conventional testing. These "hot spots" can lead to early failure of the device, however, and it is imperative that devices used in high-reliability applications be free of them. The most commonly used method of identifying such problems is by performing an infrared scan across the surface of the device. Another popular method is by using a liquid crystal material that can be deposited directly on the chip and observed under polarized light.

REFERENCES

[9.1] W.R. Curtice and M. Ettenberg, "A Nonlinear GaAs FET Model for Use in the Design of Output Circuits for Power Amplifiers," *IEEE Trans. Microwave Theory Tech.*, Vol. MTT-33, 1985, p. 1383.

[9.2] J.R. Lane et al., "High-Efficiency 1-, 2-, and 4-W Class-B FET Power Amplifiers," *IEEE Trans. Microwave Theory Tech.*, Vol. MTT-34, 1986, p. 1318.

[9.3] W.R. Curtice, "Nonlinear Analysis of GaAs MESFET Amplifiers, Mixers, and Distributed Amplifiers Using the Harmonic Balance Technique," *IEEE Trans. Microwave Theory Tech.*, Vol. MTT-35, 1987, p. 441.

[9.4] V.D. Hwang and T. Itoh, "An Efficient Approach for Large-Signal Modeling and Analysis of the GaAs MESFET," *IEEE Trans. Microwave Theory Tech.*, Vol. MTT-35, 1987, p. 396.

CHAPTER 10

FET FREQUENCY MULTIPLIERS

GaAs MESFET frequency multipliers are probably at the top of the list of infrequently used microwave electronic components. However, were their use completely unwarranted, this chapter would not have been written. Like many other infrequently used components, FET frequency multipliers have significant advantages over competing technologies (in this case, diode multipliers): FET multipliers can achieve broad bandwidths and conversion gain that is greater than unity. Furthermore, because small-signal MESFETs can be used to realize efficient multipliers, a high-frequency FET multiplier chain usually consumes little dc power and dissipates little heat; this is an important advantage in space-based systems. For example, receiver LO chains that use diode multipliers often require high-power, high-gain driver amplifiers; such amplifiers often consume more dc power than the rest of the receiver.

This chapter is primarily concerned with low-power "Class-B" multipliers; these operate in a manner analogous to that of a Class-B power amplifier. Such multipliers are very stable and have good gain, efficiency, and output power. Other modes of operation can provide higher gain than the Class-B multiplier. However, the high gain is often the result of feedback effects, which may make the multiplier unstable. Consequently, the Class-B multiplier is usually the more practical form of FET frequency multiplier for most applications.

10.1 DESIGN PHILOSOPHY

In the past, frequency multipliers were often used to generate high levels of microwave RF power. High-power multipliers were important components because microwave solid-state power amplifiers did not yet exist; power amplification at microwave frequencies could be provided only by tubes, which were expensive and unreliable and required high dc power. Accordingly, a "high-power" multiplier chain (which rarely had an output power greater than a fraction of 1 W) consisted of a power amplifier (often a UHF bipolar amplifier) that delivered several watts to a cascade of varactor or SRD multiplier stages.

Today, solid-state power amplification at microwave frequencies is possible, so high-power multiplier chains are used less frequently. Instead, the functions of power amplification and signal generation are usually separated: Signals having the required frequencies are generated at relatively low powers and, if greater power is needed, those signals are amplified. Keeping these functions separate has two important advantages: First, it minimizes the consumption of dc power and the generation of heat and allows separation of the components that dissipate the most heat from those that may be temperature-sensitive. Second, operating the multipliers at low power reduces the levels of spurious signals and harmonics. Furthermore, many systems do not require high-power signals: most frequency-multiplier chains are used in low-power systems, like mixer LOs, in test instruments or frequency synthesizers, or as low-power drivers for transmitters. The output power of such chains is often between 0 and 10 dBm.

When used as frequency multipliers, small-signal MESFETs can usually achieve unity (or greater) conversion gain over broad bandwidths while maintaining good dc-RF efficiency at these low power levels. In contrast, diode multipliers always exhibit loss: Varactor multipliers are lossy narrowband components that operate best at moderate to high power levels; resistive (Schottky-diode) multipliers are more broadband but have even greater loss and a limited power-handling ability. Thus, the medium- to high-power driver amplifiers required by such multipliers generate RF power that is eventually wasted in the diodes and matching circuits; it is not unusual for a driver amplifier and diode multiplier chain to require several watts of dc power to generate a few milliwatts of RF power.

The low-power Class-B multipliers we examine in this chapter generate low-level RF output power (normally below 10 dBm) at low harmonics, have at least unity gain, and may have high output frequencies, sometimes in the millimeter-wave region. The design approach we shall develop is, of course, applicable to FET multipliers operating at higher powers and lower frequencies; designing a high-power multiplier requires only using a larger MESFET (i.e., one having greater gate width) and providing greater input power. Like the Class-B power amplifier discussed in the previous chapter, the gate of a frequency-multiplier FET is biased near V_t, the channel conducts in pulses having a duty cycle near 50 percent, and the gate and drain are short-circuited at all unwanted harmonics of the excitation frequency. The primary difference between the Class-B multiplier and the Class-B amplifier is that the multiplier's output is taken at a harmonic rather than at the fundamental frequency.

10.2 APPROXIMATE DESIGN OF FET FREQUENCY MULTIPLIERS

Following the pattern of the previous chapters, we begin with an approximate design procedure, use it to generate an initial design, and then optimize that design via harmonic balance. We use the same approach as in Chapter 9; we begin by examining the properties of a large-signal multiplier circuit that uses an ideal FET, then modify the circuit to include a MESFET having a minimal set of parasitic elements. This latter circuit forms the basis of the approximate design procedure.

Figure 10.1 shows the circuit of a FET frequency multiplier that uses an ideal MESFET. This circuit appears to be identical to the one in Figure 9.1, which represented a power amplifier; the only difference is that in Figure 10.1 the output resonator is tuned to the nth harmonic of the excitation frequency, not the fundamental. The resonators shown in the figure short-circuit the MESFET's gate at all frequencies except the excitation frequency, ω_p, and short-circuit the drain at the fundamental frequency and all harmonics except the output frequency, $n\omega_p$. Unless we state otherwise, we will assume throughout this section that a short-circuit termination is optimum; in Section 10.4.1 we will examine this assumption further.

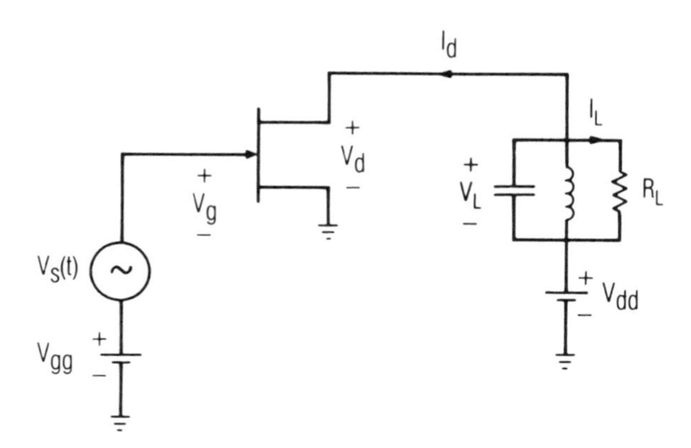

Figure 10.1 Circuit of an ideal FET frequency multiplier.

For reasons that will be clear shortly, the gate-bias voltage in an efficient FET mixer must be equal to or less than (more negative than) the turn-on voltage, V_t. Thus, the MESFET's channel conducts only during the positive half of the excitation cycle, and the drain conducts in pulses; the shape of the pulses is approximately a rectified cosine or half cosine. In this derivation, we assume that the drain-current waveform can be modeled as a train of half-cosine pulses (this assumption is justified by the results of harmonic-balance analyses). The duty cycle of the pulses varies with V_{gg}; if $V_{gg} = V_t$, the duty cycle is 50 percent, but if $V_{gg} < V_t$ (the usual situation), the MESFET is turned off over most of the excitation cycle, so the duty cycle is less than 50 percent.

Figure 10.2 shows the voltage and current waveforms of the ideal MESFET used as a frequency doubler. Because the output resonator eliminates all voltage components except the one at the nth harmonic, the drain voltage, $V_d(t)$, is a sinusoid having frequency $n\omega_p$. For best efficiency and output power, the drain voltage must vary between V_{max} and V_{min}; V_{min} is the value of drain voltage at the knee of the drain I/V curve when the gate voltage has its maximum value $V_{g,max}$. V_{max} and V_{min} are established by the same considerations as those used in power amplifiers; V_{dd}, the dc drain voltage, is halfway between V_{max} and V_{min}. The gate voltage varies between $V_{g,max}$, the peak gate voltage (limited to approximately 0.5 V by rectification in the gate-channel Schottky junction), and $2V_{gg} - V_{g,max}$, a relatively high reverse voltage. The drain current peaks at the value I_{max}, and the current pulses have the time duration t_0; $t_0 \leq T/2$, where T is the period of the excitation. If we make $t = 0$ equal to the point where the current is maximum, the Fourier-series representation of the current has only cosine components:

$$I_d(t) = I_0 + I_1 \cos(\omega_p t) + I_2 \cos(2\omega_p t) + I_3 \cos(3\omega_p t) + \dots \qquad (10.2.1)$$

When $n \geq 1$, the coefficients are

$$I_n = I_{max} \frac{4t_0}{\pi T} \left| \frac{\cos(n\pi t_0/T)}{1 - (2nt_0/T)^2} \right| \qquad (10.2.2)$$

and when $n = 0$:

$$I_0 = I_{max} \frac{2t_0}{\pi T} \qquad (10.2.3)$$

When $t_0/T = 0.5/n$, $n \neq 0$, (10.2.2) is indeterminate. Then, I_n is

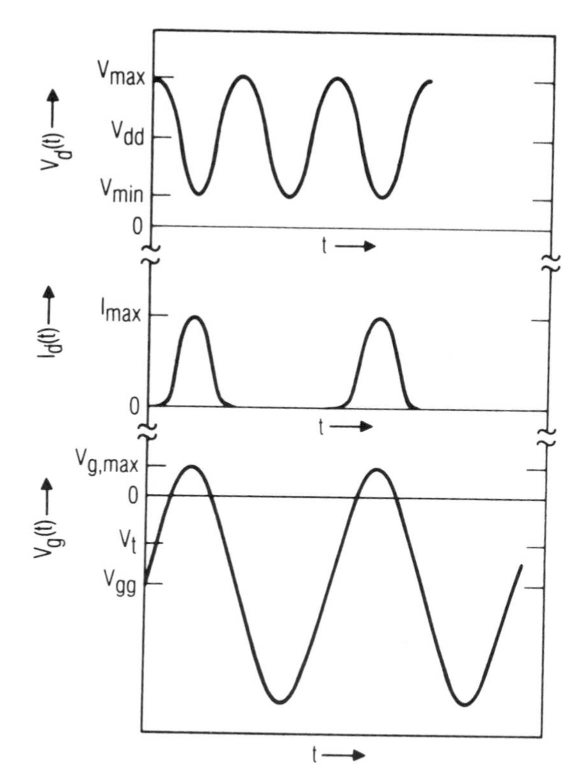

Figure 10.2 Voltage and current waveforms in the ideal FET frequency multiplier.

$$I_n = I_{\max} \frac{t_0}{T} \tag{10.2.4}$$

Because the tuned circuit in Figure 10.1 is an open circuit at the output frequency ω_n, all of the nth-harmonic current, I_n, circulates in R_L and contributes to output power. Accordingly, in order for the FET multiplier to achieve maximum output power and efficiency, we must maximize I_n. Equation (10.2.2) shows that we have only one degree of freedom for doing so: varying t_0/T. Figure 10.3 shows a plot of $I_n/I_{\max}$ as a function of t_0/T when $n = 2$ through $n = 4$; each of these curves has a clear maximum below $t_0/T = 0.5$. It would appear that, in order to achieve the optimum value of I_n, we need only to adjust V_{gg} so that $I_d(t)$ has the desired period of conduction t_0.

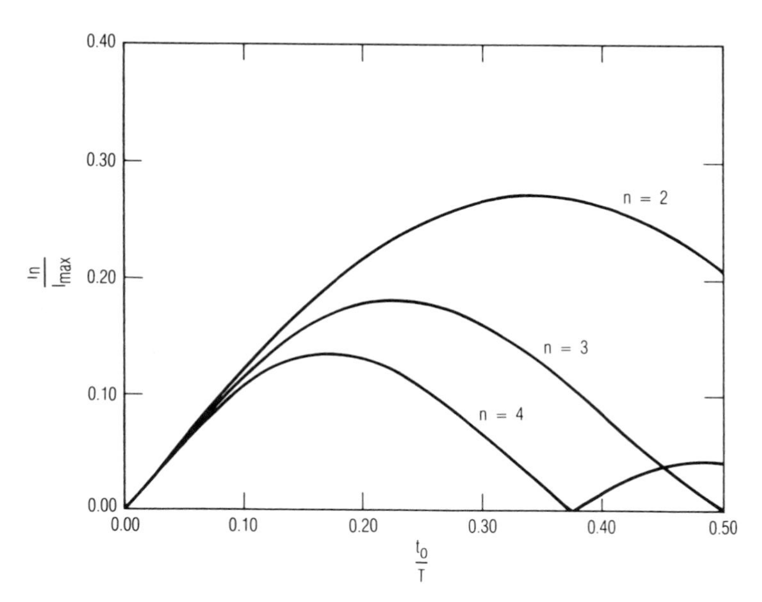

Figure 10.3 Harmonic drain-current components as a function of t_0/T when the drain-current waveform is a half-sinusoidal pulse train.

Unfortunately, two problems arise from this attempt to achieve a short conduction period. The first is that we would have to make $V_{gg} \ll V_t$, and this large bias voltage would make the magnitude of the peak reverse voltage, which is approximately $2V_{gg}$, very great. Ideally, the peak reverse gate voltage occurs at the minimum drain voltage, but because of phase shifts in practical multipliers and the more rapid variation of $V_d(t)$ than $V_g(t)$, the peak drain-gate voltage can be nearly $V_{max} - 2V_{gg}$. If V_{gg} is adjusted to make t_0/T very small, the peak drain-gate voltage may be much greater than the avalanche voltage of the MESFET. The second problem is that, even if the device could survive this high voltage, the input power required to achieve such a wide gate-voltage variation would be so great that the multiplier's conversion gain would be poor. Thus, it is necessary in most cases (especially in a multiplier having an output harmonic greater than the second) to use a value of t_0/T that is greater than the optimum. Selecting t_0/T to achieve an acceptable trade-off between gain and output power is an important part of the design process.

The maximum reverse gate voltage that the MESFET can tolerate establishes one limit on t_0/T. If the gate voltage varies between $V_{g,\max}$ and the peak reverse voltage $V_{g,\min}$, then the phase angle θ_t over which

$V_g(t) > V_t$ is

$$\theta_t = 2\cos^{-1}\left(\frac{2V_t - V_{g,\max} - V_{g,\min}}{V_{g,\max} - V_{g,\min}}\right) \tag{10.2.5}$$

The bias voltage that achieves this value of θ_t is

$$V_{gg} = \frac{V_{g,\min} + V_{g,\max}}{2} \tag{10.2.6}$$

Equation (10.2.5) shows that a large negative value of $V_{g,\min}$ reduces the conduction angle of the FET. It also shows that reducing $V_{g,\max}$ has the same effect and, by decreasing the range of $V_g(t)$, reduces input power. However, decreasing $V_{g,\max}$ is not a good way to achieve a low value of t_0/T; decreasing $V_{g,\max}$ decreases $I_{\max}$, and thus reduces output power. Furthermore, when $V_{g,\max}$ is not as great as possible, the multiplier may not be operated in gain saturation and the output power may vary appreciably with input power (in most practical applications, multipliers are operated in gain saturation in order to stabilize their gains).

The difficulty of achieving a low value of t_0/T can be illustrated by an example. Suppose that a FET has the parameters $V_t = -1.5$ V, $V_{g,\min} = -7.0$ V, and $V_{g,\max} = 0.5$. Equation (10.2.5) indicates that $\theta_t = 2.183$ (125°) and, therefore, $t_0/T = 0.35$. This is the minimum t_0/T that can be achieved with this device if $V_{g,\max}$ is not reduced. Figure 10.3 shows that this value of t_0/T is nearly optimum for a doubler and is not disastrously far from the optimum value for a tripler (although $I_3 < I_2/2$, so a tripler's output power would be more than 6 dB below that of a doubler). However, $t_0/T = 0.35$ is near the zero of I_4, so a fourth-harmonic multiplier having this value of t_0/T would have very low output power and efficiency. If a fourth-harmonic multiplier were desired, it would be better to increase t_0/T to 0.5, although even then the output power would be at least 16 dB below that of the doubler. It is easy to see from this example why the published research shows that successful FET frequency multipliers have most frequently been doublers.

The current in the load resistance, R_L,is I_n. For the voltage V_L across the load to vary between $V_{\max}$ and $V_{\min}$:

$$|V_L(t)| = I_n R_L = \frac{(V_{\max} - V_{\min})}{2} \tag{10.2.7}$$

The optimum load resistance is

$$R_L = \frac{V_{\max} - V_{\min}}{2I_n} \tag{10.2.8}$$

Because I_n is relatively small compared to I_1 in a Class-B amplifier, R_L in a multiplier is usually much greater. The output power at the nth harmonic, $P_{L,n}$, is

$$P_{L,n} = \frac{1}{2} I_n^2 R_L = \frac{1}{2} I_n \frac{V_{\max} - V_{\min}}{2} \tag{10.2.9}$$

As with a power amplifier, the dc drain bias voltage is halfway between $V_{\max}$ and $V_{\min}$; that is,

$$V_{dd} = \frac{V_{\max} + V_{\min}}{2} \tag{10.2.10}$$

The dc power is

$$P_{dc} = V_{dd}\, I_{dc} = V_{dd}\, I_0 \tag{10.2.11}$$

Substituting I_0 from (10.2.3) into (10.2.11) gives

$$P_{dc} = \frac{2t_0}{\pi T} I_{\max} V_{dd} \tag{10.2.12}$$

The dc-RF efficiency is

$$\eta_{dc} = \frac{P_{L,n}}{P_{dc}} \tag{10.2.13}$$

Because the harmonic output current in a multiplier is usually much less than the fundamental current in an amplifier, η_{dc} is usually much lower in a FET multiplier than in a FET amplifier.

We can approximate the RF input power by employing the same set of assumptions that is used to approximate the LO power in a FET mixer (Section 11.1.2): because the drain is short-circuited at the fundamental frequency, the input of the MESFET can be modeled as a series connection of $R_s + R_i + R_g$ and C_{gs} (V_{gg}). The excitation source must generate an RF voltage having the peak value $V_{g,max} - V_{gg}$ across C_{gs}; if the source is matched, the power available from the source must equal P_{in}:

$$P_{av} = P_{in} = \frac{1}{2} (V_{g,max} - V_{gg})^2 \omega_p^2 C_{gs}^2 (R_s + R_i + R_g) \qquad (10.2.14)$$

and the conversion gain is simply $P_{L,n}/P_{av}$. The power-added efficiency of a FET multiplier is

$$\eta_a = \frac{P_{L,n} - P_{in}}{P_{dc}} \qquad (10.2.15)$$

or

$$\eta_a = \eta_{dc} \left(1 - \frac{1}{G_p}\right) \qquad (10.2.16)$$

where G_p is the power gain ($P_{L,n}/P_{in}$) of the multiplier.

A final consideration is the trade-off between V_{max} and $V_{g,min}$. Neither of these parameters can be established independently in any MESFET; V_{max} and $V_{g,min}$ must be chosen so that the drain-gate avalanche voltage is not exceeded. The maximum drain-gate voltage is approximately $V_{max} - V_{g,min}$, so we have the limitation:

$$V_{max} - V_{g,min} < V_a \qquad (10.2.17)$$

where V_a is the drain-gate avalanche voltage. Thus, we can increase $|V_{g,min}|$ by decreasing V_{max}. Decreasing V_{max} decreases the optimum value of R_L, not an undesirable result in view of the fact that R_L is often too great to be realized in practice. It is usually not possible to decrease V_{min} when V_{max} is reduced, however, so from (10.2.9) we see that decreasing V_{max} reduces $P_{L,n}$. The design process is illustrated by the following example.

Example 1

We wish to design a FET frequency doubler that has an input frequency of 10 GHz. Figure 10.4 shows the circuit of the multiplier. The MESFET has the following parameters:

V_a = 12.0 V
V_t = −2.0 V
R_s = 2.0Ω
R_i = 2.0Ω
R_g = 1.0Ω
R_d = 2.0Ω
I_{dss} = 80 mA (at V_{ds} = 3.0V)
L_s = 0.005 nH
C_{gs} = 0.25 pF (at V_{gs} = V_{gg})
C_{gd} = 0.08 pF
C_{ds} = 0.10 pF

We use the expresssion for $I_d(V_g, V_d)$ in (2.4.5), in which the drain current depends on gate voltage as a third-degree polynomial and on drain voltage as a hyperbolic tangent function. The parameters of this expression are found by fitting it to the measured I/V characteristic via a least-squares technique; the parameters are:

A_0 = 0.09670
A_1 = 0.11334
A_2 = 0.04853
A_3 = 0.00801
α = 1.5

From the I/V curves or (2.4.5), we estimate I_{max} = 80 mA and V_{min} = 1.0 V. $V_{g,min}$ and V_{max} must obey the constraint expressed by (10.2.17), so we choose $V_{g,min} = -7.0$ V and $V_{max} = 5.0$ V; we also choose $V_{g,max}$ = 0.2 V, slightly below the lowest value that allows rectification.

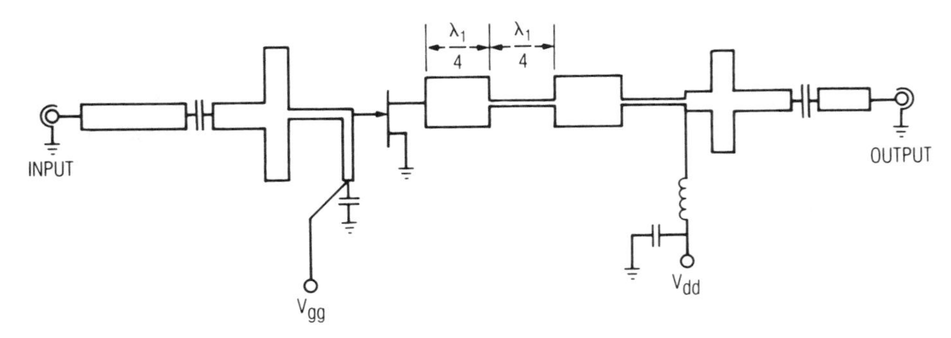

Figure 10.4 Circuit of the FET frequency doubler designed in Example 1.

Equations (10.2.10) and (10.2.6), respectively, give $V_{dd} = 3.0$ V (a convenient value) and $V_{gg} = 3.4$ V; substituting these values into (10.2.5) gives $\theta_t = 2.36$ (135°) or $t_0/T = 0.37$. Figure 10.3 shows that this value of t_0/T is close to the optimum value for a doubler and that $I_2 = 0.27I_{max}$ or 21.6 mA.

Equation (10.2.14) can now be used to find the input power; (10.2.14) implies that $P_{in} = 8.0$ mW or 9.0 dBm. If the input is conjugately matched, the input power is equal to the power available from the excitation source. The output power, $P_{L,2}$, is given by (10.2.9), and we find it to be 21.6 mW or 13.3 dBm; thus, the conversion gain is 4.3 dB. The dc drain current from (10.2.12) is 19.9 mA, which gives 59.7 mW dc power and 36 percent dc-RF efficiency. Finally, R_L is found from (10.2.8) to be 92.6 Ω, and in order to resonate the output capacitance, C_{ds}, there must be a susceptance in parallel with R_L of $-2\omega_p C_{ds}$ or -12.5 mS. Converting this load to an impedance gives $Z_L(2\omega_p) = 39.4 + j45.8\ \Omega$. The estimated input impedance is simply $R_s + R_i + R_g + 1/j\omega_p C_{gs}$ or $5 - j63\ \Omega$.

The rest of the design involves realizing the input and output matching networks. The output matching network is relatively easy to design; it consists of a filter, to short-circuit the drain at the fundamental frequency and unwanted harmonics, followed by matching elements. A half-wave filter is ideal for the output; it consists of a cascade of alternating high- and low-impedance transmission-line sections, each $\lambda/4$ long at ω_p; these sections are $\lambda/2$ long at $2\omega_p$ and $3\lambda/4$ long at $3\omega_p$. Thus, the frequencies of maximum rejection occur at ω_p and $3\omega_p$, but the filter has no rejection at the output frequency $2\omega_p$. From Figure 10.3, we see that $I_4 \approx 0$ at $t_0/T = 0.37$, so the fourth-harmonic output should be very low. The gate's short circuit at the second-harmonic frequency is less critical; a shorted stub $\lambda/4$ long at ω_p is adequate to provide the termination. This stub has no effect on the excitation but is $\lambda/2$ long at $2\omega_p$ and thus short-circuits the gate at this frequency.

The validity of this design was tested by means of a harmonic-balance calculation. In order to have a valid comparison, we compare the performance when the approximate design and the harmonic-balance calculation have the same gate voltage variation. By normalizing the gate voltage rather than the available input power, we can more easily separate the effects of the input- and output-circuit designs. Accordingly, the input power and bias were adjusted in the harmonic-balance calculation until the estimated peak-to-peak voltage of 7.2 V across C_{gs} was achieved.

The multiplier's operating parameters found via the harmonic-balance analysis are compared in Table 10.1 to those determined from the approximate design. The two sets of data agree reasonably well, although the output power calculated via harmonic balance is 1.6 dB lower than the estimated output power. The main reason for the difference is that the current pulse is not precisely one-half sinusoid; the pulse is somewhat distorted, so that its shape appears to be something between a cosine and a triangle. This distortion reduces the magnitude of I_2 and thus decreases the output power at $2\omega_p$. The second-harmonic peak-to-peak voltage across R_L of 3.5 V, instead of 4.0 V, is evidence that I_2 is lower than intended; this difference in voltage alone accounts for 1.2 dB of the difference in output power. The calculated value of t_0/T, 0.44, is slightly greater than the estimated value, 0.37; this difference further reduces I_2.

Table 10.1 FET Frequency Doubler Designs (input frequency = 10 GHz)

	Designs	
Parameter	*Approximate*	*Harmonic Balance*
$V_g(t)$ range	−7.0–0.2 V	−6.9–0.1 V[(1)]
$V_d(t)$ range	1.0–5.0 V	1.1–4.6 V
I_{max}	80.0 mA	91.0 mA
I_{dc}	19.9 mA	22.7 mA
V_{gg}	−3.4 V	−3.3 V
V_{dd}	3.0 V	3.0 V
t_0/T	0.37	0.44
Z_L	39 + j46	39 + j46[(2)]
Z_{in}	5 − j63	3.4 − j56
P_{av}	9.0 dBm	11.0 dBm[(1)]
$P_{L,2}$	13.3 dBm	11.7 dBm
G_t	4.3 dB	0.7 dB
P_{dc}	59.7 mW	68.1 mW
Γ_{in}	0.0	0.65
Input reflection loss	0.0 dB	2.4 dB

Notes:
[(1)] P_{av} is adjusted to achieve the same $V_g(t)$ range in both cases.
[(2)] The load impedance was not optimized; it was set equal to the approximate value.

It is also possible that the load impedance is not precisely optimum; certainly, R_L could be increased to achieve the full peak-to-peak output voltage of 4.0 V; this change would increase the output power approximately 0.6 dB. It is clear that C_{ds} is effectively resonated because the peak of the drain current pulse, $I_d(t)$, occurs almost exactly at the minimum of the drain voltage, $V_d(t)$; this condition implies that the impedance presented to the terminals of the controlled source I_d is entirely real.

The largest difference between the two sets of data is in the input power. The input power needed to achieve the required range of $V_g(t)$ is 11.0 dBm, 2.0 dB higher than the 9.0 dBm estimated. However, because of the rather coarse estimate of the input impedance, the input VSWR of the multiplier using the estimated source impedance is 4.7. This apparently large error is not as disappointing as it may at first appear; it is caused by only a very small error in the real part of the estimated input impedance, 5.0 Ω *versus* 3.4 Ω. Errors of this magnitude are nearly impossible to avoid but they have a remarkably large effect upon input match. Correcting errors of this magnitude is the normal reason for tuning a practical amplifier. Tuning the input to obtain a conjugate match would reclaim 2.4 dB, making the required input power 8.6 dBm, in good agreement with the estimate of 9.0 dBm.

It is a worthwhile exercise to compare this design process to those of the varactor and resistive multipliers in Chapter 7. Although the latter are no more difficult to implement, the approximate design of the FET frequency multiplier is much "cleaner" than those of the diode multipliers: the design is more intuitive, the approximations are not as severe, less empiricism is required, and there is a better initial agreement between the approximate and harmonic-balance analyses. Indeed, after performing harmonic-balance analyses of a FET multiplier and a varactor multiplier, we can see immediately that the performance of the FET multiplier is far less sensitive to virtually every circuit parameter than the varactor. This property, that the FET multiplier is very "designable," is difficult to quantify, but is nevertheless one of the multiplier's most important characteristics.

10.3 HARMONIC-BALANCE ANALYSIS OF FET FREQUENCY MULTIPLIERS

The harmonic-balance analysis of FET frequency multipliers is virtually identical to the analysis of FET power amplifiers in Section 9.4. The same MESFET equivalent circuit is used, so the Y parameters in Section 9.4 can also be used. The same considerations apply to C_{gd}: In the previous example, we assumed that C_{gd} could be treated as a linear element; however, in cases where the drain-gate voltage drops below a few volts during the excitation cycle, it may be necessary to include the nonlinearities of C_{gd}. The two convergence algorithms preferred for power-amplifier analysis, the reflection algorithm and Newton's method, are also those preferred for multiplier analysis (optimization also has been used successfully). The main differences between the power-amplifier and multiplier analyses are that FET frequency multipliers usually employ small-signal or medium-power devices not high-power MESFETs and the output is filtered from the drain current at a harmonic of the excitation frequency not at the fundamental frequency.

In spite of the similarities between the analyses, it is important to recognize that the multiplier analysis has some unique aspects that require special care. One is that the channel of a MESFET used in a FET frequency multiplier conducts in pulses having a low duty cycle and significant harmonic content. This situation makes generating an initial estimate of the solution (if Newton's method is used) a little tricky because the initial estimate must include good estimates of several harmonic components of the drain current. The initial estimate was not very critical in the Class-A amplifier, because the harmonics were small, nor in a well designed mixer, because the drain-source voltage could be approximated successfully as a dc quantity. However, in the FET multiplier, several frequency components of both the drain voltage and current are significant. The quality of the estimate of these components has a strong effect on the speed and success of convergence.

If Newton's method is used, the procedure for making the initial estimate of the harmonic voltages in a power amplifier, outlined in Section 9.4, works well in most cases for FET multipliers. When this procedure is used, the current-error function sometimes oscillates in the first few iterations of the analysis, but as long as the FET is not driven strongly into gain saturation, the oscillation eventually ceases and the error function then decreases monotonically. In Section 9.4, we saw that difficult convergence in analyzing a FET power amplifier could be prevented by reducing the excitation power and repeating the analysis. The results at the reduced excitation level are than used as an initial estimate for the next analysis, which is performed at a higher excitation level. The excitation is increased stepwise in this fashion until the desired power level is reached. However, this process is not directly applicable to a FET multiplier, because the multiplier's gate bias is below V_t and, therefore, the FET does not have drain current until the excitation is relatively strong. A better procedure is to begin the analysis with increased (more positive) gate bias, so $V_{gg} \approx V_t$, and the excitation level reduced. The excitation is increased and the bias is decreased at each step until the analysis can be performed at the desired power and bias levels.

The oscillation in the error function that is often encountered at the outset of the analysis implies that the change in the port-voltage vector at each iteration may be too great; it may therefore be helpful to begin the analysis with an update factor less than 1 (9.4.4) and to increase the update factor as the error function decreases. The magnitude of the factor depends strongly upon the nature of the problem, and optimizing it may require some experimentation; however, as a general rule, the factor should be as large as possible while preventing the oscillation of the error function.

When the update factor is adjusted in this manner, it is usually best to begin the analysis with a relatively small factor of approximately 0.5 and, if convergence progresses well, to increase it gradually to 1.0 over the first five to ten iterations.

The reflection algorithm can also be used to analyze FET frequency multipliers. In most published descriptions of the use of this algorithm with FET circuits (References 2.1, 2.7, 9.4), the circuit was partitioned in such a way that the source and load networks were the linear subcircuits and the entire FET equivalent circuit, including its linear elements, was the nonlinear subcircuit. This approach is usually adequate and has the advantage of minimizing the number of reflection circuits; however, it allows numerical instability to occur during the integration of the FET's nonlinear differential equations. The MESFET's rather strong nonlinearities and its very wide gate-source voltage range make numerical instability considerably more likely in a Class-B frequency multiplier than in other large-signal circuits using MESFETs. If instability is encountered, the solution may be to use a different circuit partition, such as putting only the three nonlinear elements, C_{gs}, C_{gd}, and I_d, in the nonlinear subcircuit.

10.4 OTHER ASPECTS OF FET FREQUENCY-MULTIPLIER DESIGN

The process outlined in the previous section is not the only conceivable approach to the design of FET frequency multipliers. Frequency multipliers have been devised wherein the MESFET's gate and drain terminations are not short circuits and, as with mixers and amplifiers, balanced circuits have been proposed that have significant advantages over single-device structures. These two ideas are examined in this section.

10.4.1 Effect of Gate and Drain Terminations at Unwanted Harmonics

The previous sections were based on the hypothesis that the MESFET's optimum gate and drain terminations at unwanted harmonics are both short circuits. Certainly, the use of short-circuit terminations at these frequencies results in good performance, but we have accepted on faith, not proven, that short-circuit terminations are in some sense the optimum ones. In fact, there have been reports that the use of other terminations, especially an open-circuit drain termination at the fundamental frequency, has advantages over a short circuit. The primary advantage of using other terminations is that greater gain can be achieved, although the increase in gain usually is the result of feedback. Achieving an increase in gain via feedback is rarely desirable because it reduces stability margins and bandwidth.

Dow and Rosenheck (Reference 10.1) have compared FET frequency doublers that use short- and open-circuit drain terminations at the fundamental excitation frequency. Their short-circuit doubler is biased at $V_{gg} \approx V_t$ and designed in a manner similar to that described in this chapter; the open-circuit doubler is biased near $V_{gg} = 0$. They describe one short-circuit frequency multiplier having a 20 GHz input and a 40 GHz output and two multipliers having open-circuit terminations, one a 10–20 GHz doubler and the other a 20–40 GHz doubler. All three multipliers use power MESFETs to achieve output powers between 12 dBm and 18 dBm, depending upon the frequency and the multiplier's mode of operation. The performance of the short-circuit multiplier (which they, like us, call a *Class-B multiplier*) is approximately what we would expect: the 40 GHz multiplier has relatively low gain, −11 dB, but good output power, over 18 dBm. The gain would probably be better using smaller devices, optimized for such high frequencies.

The performance of the multipliers having an open-circuit fundamental-frequency termination (which these authors call a *Class-A multiplier*) are more interesting. Because the MESFETs' drains are open-circuited, the drain-source voltage varies considerably, probably as much as in a Class-A amplifier. This multiplier represents the dual case of the Class-B multiplier, its fundamental-frequency drain current is zero but its fundamental-frequency drain voltage is not. The performance of these multipliers is quite good; the 40 GHz Class-A doubler's output power is only 11 dBm, but it exhibits −2 dB gain, much higher than that of the Class-B doubler. The authors conclude that the properties of Class-A and Class-B multipliers roughly parallel those of amplifiers: Class A provides high gain and moderate efficiency and power, whereas Class B provides lower gain but better efficiency and, at least in this case, better output power.

The complete story leads to a somewhat different conclusion. On closer inspection, the operating characteristics of the Class-A multiplier become troubling, especially to anyone who wants to create a stable electronic system. These difficulties arise from a fundamental conflict with Kirchhoff's current law in the open-circuit drain termination of the Class-A multiplier. The conflict is that the controlled current source, I_d, responding to the fundamental-frequency gate voltage, imposes a constraint that the drain current must have a large fundamental-frequency current component; however, the open-circuit termination makes a fundamental component of the output current impossible. If the MESFET's drain terminals are open-circuited at the fundamental frequency, one of three things must happen: (1) The fundamental current must exist in the other elements of the MESFET equivalent circuit, especially C_{gd} and C_{ds}; (2) there must be enough second-harmonic feedback from the drain to the gate to suppress the fundamental component of V_g so that I_d has a much greater second-harmonic component; or (3) the internal drain voltage, V_d, must collapse over part of the excitation cycle. This collapse allows the MESFET to enter its linear region (and provides the fundamental current via g_{ds}). The first two occurances are inherently feedback effects, and feedback is likely to have a deleterious effect upon the multiplier's stability.

In fact, all three phenomena occur to some degree, depending on the terminating impedance at all the harmonics but especially the fundamental. We can understand the combined effects of these phenomena and the conditions under which they may be serious by examining the results of a study by Rauscher (Reference 10.2). He describes the performance of a small-signal MESFET, normally used as a Ku-band amplifier, operating as a frequency doubler between 15 GHz and 30 GHz. Because of the multiplier's high output frequency and the relatively large value of C_{gd} in this device, feedback effects are very significant and the effect of the drain-terminating impedance on gain and stability is pronounced. The conversion gain is approximately 2 dB when the drain is shorted but rises monotonically with the reactance of an inductive fundamental-frequency termination. When the terminating reactance is only 45 Ω, the multiplier oscillates, and in contrast to the conclusions in Reference 10.1, this multiplier has very low conversion efficiency (-4 dB gain) when the termination is an open circuit. An interesting result is that the multiplier's output power is highly independent of the fundamental-frequency drain termination. We are forced to conclude that using an open-circuit drain termination at the fundamental frequency may have unpredictable results and its effect on stability in particular is likely to be deleterious.

10.4.2 Balanced Frequency Multipliers

Relatively few balanced frequency-multiplier circuits have been proposed. By far the most practical and most commonly used one is the antiseries or "push-push" multiplier, a circuit that has been in use since the days of vacuum tubes. The characteristics of an antiseries FET doubler are particularly good because the structure eliminates the problem of achieving a broadband fundamental-frequency short circuit at the drain.

The general properties of the push-push circuit are described in detail in Section 5.2.1. Although that section describes a circuit consisting of two-terminal nonlinear elements, the circuit using MESFETs, shown in

Figure 10.5, is conceptually identical. The two MESFETs are connected, via individual matching circuits, to two mutually isolated ports of a 180-degree hybrid. The delta port of the hybrid is used as the input and the sigma port is terminated. The gates of the two MESFETs are driven by signals having a 180-degree phase difference; therefore, the fundamental components of the drain currents are out of phase, so each MESFET effectively short-circuits the other at the fundamental frequency and creates a virtual ground at the drain. This inherent short circuit is usually adequate to ensure that the drain voltage has at most a very small fundamental RF component and to implement Class-B operation effectively. The second harmonics of the drain currents in the two MESFETs have no phase difference, however, and therefore the drain-current components at this frequency combine in phase at the output.

This configuration has several advantages over a single-device circuit. One is that the matching circuit can be located close to the drains of the FETs; it need not be separated from them by the intervening filter as in the single-device multiplier. Eliminating the parasitic effects of this filter allows the balanced multiplier to have greater bandwidth than would a single-device multiplier. A second advantage is that, like other balanced circuits, the FET balanced multiplier has 3 dB greater output power than an equivalent single-device circuit. This can be a significant advantage when used at high frequencies with very small devices that have low output power. A third is that it is often easier to realize the load impedance of a balanced multiplier than that of a single-device multiplier. As we saw in Section 5.2.1, the effective load impedance presented to each device in an antiseries circuit is twice the actual load impedance of the balanced circuit (Z_L in Figure 10.5). The load impedance required by a single-device FET multiplier often is relatively high; however, the load impedance of the balanced multiplier need only be half that of the single-device circuit. This property significantly eases the task of matching the output at the second harmonic.

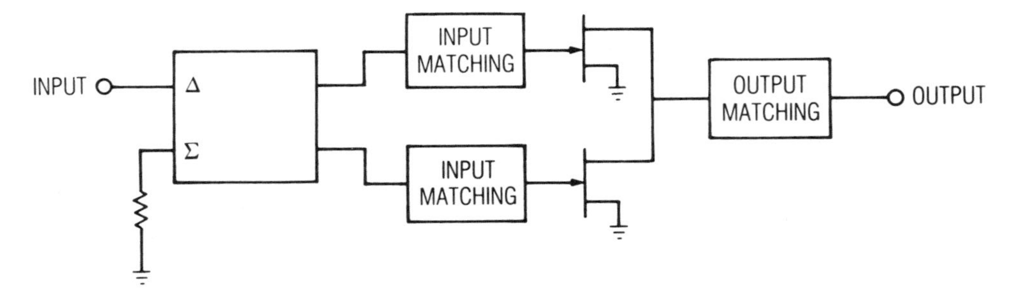

Figure 10.5 Balanced push-push or antiseries FET frequency multiplier.

10.5 SOME PRACTICAL CONCERNS

We saw in Chapter 7 that their very low noise levels was one of the most attractive properties of varactor frequency multipliers, especially in such applications as receiver LO sources. Since their gain, bandwidth, and efficiency make FET multipliers attractive for generating LO signals in communications receivers and phase noise and AM noise are important properties of the receiver's LO, it would seem wise to examine the noise properties of FET frequency multipliers. This is particularly true of receivers used in phase-modulated communications systems because the phase noise of the receiver LO is transferred degree-for-degree to the received signal.

The GaAs MESFET is known to have relatively high levels of $1/f$ noise, and this noise can modulate the phase of a signal applied to the FET (the elements primarily responsible for this modulation are the nonlinear capacitances). This phenomenon occurs in FET oscillators and has been analyzed (Reference 10.3). Unfortunately, a similar analysis of phase noise in FET frequency multipliers has not been performed, and there are few reliable experimental studies of such noise. It is fair to expect, however, that FET frequency multipliers might in some cases increase phase noise significantly, beyond the inevitable $20 \log(n)$, which is the minimum CNR degradation in an ideal frequency multiplier (Section 7.1.1). Other noise sources, such as noise from the multiplier's bias circuits, can introduce phase noise. Accordingly, it would be wise to be as concerned about phase noise in a FET frequency multiplier as in a FET oscillator.

Another important concern is the rejection of the fundamental-frequency output and unwanted harmonics. Rejecting the fundamental output is particularly important because the fundamental component of the drain current in a FET multiplier is much greater than the harmonic components, so the fundamental output power is often not much lower than that of the desired harmonic. If the balance of the hybrid and the individual MESFET multipliers is very good, approximately 20 dB rejection of the fundamental output can be achieved by a balanced circuit alone. However, because most circuits require even more rejection, some degree of high-pass filtering at the output is often needed. A simple filter is usually adequate because the rejection band is invariably far below the output passband. The third harmonic output of a FET doubler is usually very low, so minimal filtering is needed at this frequency. Many types of microwave filters (e.g., a half-wave filter) have frequencies of maximum rejection near 0.5 and 1.5 times the passband frequency; this property makes such filters ideal for use in FET multipliers. The fourth harmonic of a well designed FET frequency doubler is often nearly nonexistent, which is evident from the zero of I_4/I_{max} near the peak of I_2/I_{max} in Figure 10.3. Thus, the fourth and higher harmonics are rarely of concern.

REFERENCES

[10.1] G.S. Dow and L.S. Rosenheck, "A New Approach for mm-Wave Generation," *Microwave J.*, Vol. 26, September 1983, p. 147.

[10.2] C. Rauscher, "High-Frequency Doubler Operation of GaAs Field-Effect Transistors," *IEEE Trans. Microwave Theory Tech.*, Vol. MTT-31, 1983, p. 462.

[10.3] H.J. Siweris and B. Schiek, "Analysis of Noise Upconversion in Microwave FET Oscillators," *IEEE Trans. Microwave Theory Tech.*, Vol. MTT-33, 1985, p. 233.

Chapter 11

FET Mixers

Although the Schottky-barrier diode is used more frequently than the MESFET for realizing mixers, it is possible to design FET mixers that are in many respects superior to diode mixers. X-band mixers having 4–5 dB *single-sideband* (SSB) noise figures, 6–10 dB gain, and 20 dBm third-order intermodulation intercept points are regularly produced, and this performance can be achieved at lower LO power levels than would be required for diode mixers. MESFETs and FET variants (e.g., the *high electron mobility transistor,* or HEMT) can exhibit conversion gain well into the millimeter-wave region. Dual-gate MESFETs reduce the problem of obtaining adequate LO-RF isolation in single-device FET mixers: the RF and LO are applied to separate gates, and the low capacitance between the gates provides approximately 20 dB of LO-RF isolation without the use of filters or hybrids. Balanced FET mixers also can be produced, and these reject spurious responses and LO noise in a manner similar to that of balanced diode mixers.

This chapter describes the analysis and design of single-gate and dual-gate FET mixers. FET mixers can be analyzed via the same large-signal–small-signal approach described in Chapter 3 and applied to diode mixers in Chapter 6. As in the previous chapters, our approach is to begin with an approximate initial design, and then to optimize that design by using numerical methods.

11.1 APPROXIMATE DESIGN OF SINGLE-GATE FET MIXERS

11.1.1 Design Philosophy

In the design of diode mixers, we wished to minimize conversion loss because low conversion loss generally guarantees low-noise operation. In microwave FET mixers, high gain usually is relatively easy to obtain, but this does not automatically ensure that other aspects of performance will be good. Indeed, high mixer gain is often undesirable in receivers because the intermodulation performance of a high-gain mixer, which is usually worse than that of other components, tends to dominate in establishing the intermodulation performance of the receiver. Therefore, in most receiver applications the FET mixer is designed not to achieve the maximum possible conversion gain but to achieve a low noise figure at a specific value of gain.

Figure 11.1 shows a block diagram of a FET mixer. The mixer consists of a MESFET and RF, LO, and IF matching circuits (bias circuits, not shown in the figure, are also required). The matching circuits provide filtering as well as matching. They present the proper terminations to the FET's gate and drain at unwanted mixing frequencies and LO harmonics and provide port-to-port isolation. Although other types of mixers have been proposed, the FET mixers that are optimum in most respects have the LO and RF signals applied to the gate, the IF filtered from the drain, and the time-varying transconductance is the dominant contributor to frequency conversion (these are sometimes called *transconductance mixers* or *transconductance downconverters.* In such mixers the effects of harmonically varying gate-drain capacitance, gate-source capacitance, and drain-source resistance are often deleterious and must be minimized.

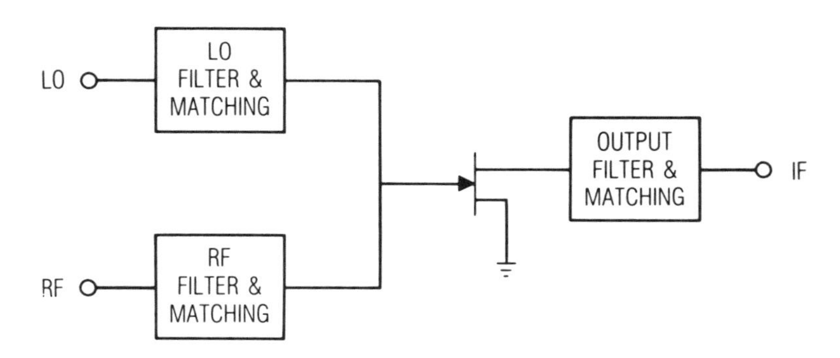

Figure 11.1 Single-gate FET mixer.

Because the time-varying transconductance is the primary contributor to mixing, it is important to maximize the range of the MESFET's transconductance variation and, in particular, the magnitude of the fundamental-frequency component of the transconductance. To maximize the transconductance variation, the MESFET must be biased close to its turn-on voltage, V_t, and must remain in its current-saturation region throughout the LO cycle. Full saturation can be achieved by ensuring that the variation of the large-signal drain voltage $V_d(t)$ under LO pumping is minimal; it is best if it remains constant at its dc value V_{dd}. This condition is achieved by short-circuiting the drain at the fundamental LO frequency and all LO harmonics. If the drain is effectively shorted, the drain LO current, which may have a fairly high peak value, cannot cause any drain-source voltage variation; then, the LO voltage across the gate-drain capacitance is minimal, so feedback is minimal and the mixer is very stable. In this case, the drain current has the same half-sinusoidal pulse waveform as the Class-B power amplifier, and the transconductance waveform is similar.

It is usually best to bias the MESFET at the same drain voltage it would require if it were used in an amplifier, near 3 V for most small-signal devices. Although the optimum gate bias is usually near V_t, fine adjustment of the gate voltage must be made empirically as part of the circuit tuning. A well designed mixer is usually relatively insensitive to small changes in dc drain voltage.

FET mixers are often conditionally stable, so it is impossible to find source and load impedances that simultaneously match the RF input and IF output ports. Even when the mixer is unconditionally stable, the output impedance of a FET mixer having an IF frequency below X-band is very high, on the order of 1000 Ω, and has only a small reactive component; this impedance is much greater than the drain-source resistance of an unpumped, dc-biased FET. Except at low frequencies and over very narrow bandwidths, in practice, it is nearly impossible to obtain a conjugate match to such a high impedance; therefore, it is usually impossible to match the output of a FET mixer. A better choice is to use a resistive load at the IF, its value selected to obtain the desired conversion gain. In this case the mixer's output VSWR will, of course, be relatively high; however, theoretical and practical limitations of impedance matching dictate that the high output VSWR is unavoidable, regardless of the philosophy employed in designing the IF circuit. Nevertheless, a resistive load, if properly implemented, will provide stable operation, flat frequency response, and the desired gain.

Ordinary small-signal MESFETs are generally used to realize single-gate FET mixers. A MESFET designed to be used in low-noise amplifiers within some specific frequency range usually works well as a mixer within the same range. Special situations often affect the choice of a device; for example, it is generally easier to obtain high conversion gain (if high gain is indeed what is desired) and a high output intermodulation intercept point from a device having a relatively wide gate. There is some experimental evidence that good noise figures are more readily obtained by using narrow devices. At very high (i.e., millimeter-wave) frequencies, the number of available devices is likely to be relatively small; most devices are optimized for amplifier use and therefore have very narrow gates. It may be difficult to obtain conversion gain at high frequencies from such devices.

Even though the FET is pumped strongly by the LO, its average transconductance is high and therefore it can amplify as well as mix; this amplification must be minimized in order to achieve good stability and to prevent spurious effects. In particular, the mixer must not have appreciable linear gain at the IF frequency, or spurious inputs at the IF frequency (e.g., noise from the gate-bias circuit) will be amplified and will appear in the output. Similarly, RF and LO amplification can result in instability and spurious responses. The only way to minimize unwanted amplification is to mismatch the device at either the gate or drain at these frequencies; therefore, we should design the mixer to have a short circuit at the gate and drain at all unwanted mixing frequencies and LO harmonics. This precaution helps prevent large-signal instability that might be caused by the pumped nonlinear gate-source capacitance and by ordinary feedback effects.

Achieving adequate LO/IF isolation can be difficult in FET mixers. The LO current in the FET's drain is very great; it may have a peak value of 100 mA even in small-signal devices, so the LO-frequency output power is potentially very high. Unfortunately, it is difficult to design an IF matching circuit that provides very high LO isolation and still meets all the other requirements placed upon it; therefore, in practical mixers the level of the LO leakage from the IF port often is high, sometimes even higher than the applied LO power. This LO leakage may then saturate the IF amplifier or generate spurious signals. Accordingly, it is important that the IF output circuit include sufficient filtering to provide adequate LO-IF isolation. The required rejection depends upon the FET's output power capability and the level of LO leakage that the IF amplifier can tolerate. For example, most small-signal FETs have saturated output levels of at most 10 to 16 dBm. If the leakage is to be kept to -30 dBm or lower, at least 40 to 46 dB of rejection must be provided by the IF circuit. This large amount of rejection may dictate that a separate LO-rejection filter be used.

11.1.2 Design Procedure

We now consider an approximate procedure for designing a FET mixer. As with the components described in the previous chapters, we use this procedure to generate an approximate design, and then optimize the design numerically. The design procedure described in this section is strictly valid only for gate-driven transconductance downconverters, but with a little careful thought, it can be modified to include upconverters or other types of mixers. The following section (Section 11.2) on large-signal–small-signal analysis is more general and does not require these limitations.

The design of the FET mixer must optimize the large-signal LO pumping (it must vary the transconductance over the widest range possible while using as little power as possible) as well as the small-signal operation. We begin with the LO design—recall that the FET is to be short-circuited at the drain at all LO harmonics and at the gate at all harmonics except the fundamental frequency. If the gate and drain are well shorted at unwanted frequencies, it is possible to simplify the FET equivalent circuit to obtain the approximate unilateral equivalent circuit shown in Figure 11.2(a). In generating this circuit we assumed that L_s is negligible; we included the source resistance, R_s, in the input loop; and we recognized that when the drain is shorted, C_{gd} is effectively in parallel with C_{gs}. Usually, $C_{gd} \ll C_{gs}$, so C_{gd} can be neglected. The parallel-tuned circuit at the output is tuned to the IF frequency, and the input-tuned circuit is assumed to be sufficiently broadband to include both the RF and LO frequencies. These resonators short-circuit the drain and gate at all other frequencies.

The input impedance is found from Figure 11.2(a) by inspection to be

$$Z_{\text{in}}(\omega) = R_g + R_s + R_i + \frac{1}{j\omega C_{gs}} \tag{11.1.1}$$

where C_{gs} is the gate-source capacitance at the bias voltage ($V_{gg} = V_t$); $\omega = \omega_p$, the LO frequency; and R_g, R_s, and R_i are respectively the gate, source, and intrinsic resistances of the MESFET.

Ideally, the input matching circuit should match the input impedance of the FET at both the RF and the LO frequencies. However, in many cases the LO and RF frequencies are significantly different, and it is impossible to match the device successfuly at both frequencies. When this conflict exists, it is better to match the device at the RF frequency and to accept a mismatch at the LO frequency: a poor RF match degrades conversion performance, but the only consequence of a poor LO match is to waste a little LO power.

The minimum required LO power can be estimated from Figure 11.2(a), under the assumption that the input is matched at the LO frequency. We assume that the gate is biased at V_t and that the LO voltage at the gate varies between $V_{g,\max}$ (the maximum forward gate voltage that does not allow rectification, approximately 0.5 V to 0.6 V in most MESFETs) and the maximum reverse voltage, $2V_t - V_{g,\max}$. The LO power is

$$P_{\mathrm{LO,min}} = \frac{1}{2}\,(V_{g,\max} - V_t)^2\,\omega_p^2 C_{\mathrm{gs}}^2\,(R_g + R_s + R_i) \tag{11.1.2}$$

If the gate is not matched at the LO frequency, reflection losses must be included. These losses can be substantial and can raise the required LO power by several dB.

If we make the reasonable assumption that the transconductance waveform can be approximated by the pulse train of half-sinusoids shown in Figure 11.2(b), the circuit in Figure 11.2(a) can be analyzed relatively easily to determine its conversion gain. Because the input impedance of a FET is not highly sensitive to signal level (as long as the gate is not driven to the point of rectification), the input impedance of a FET mixer at the RF frequency is virtually the same as the LO input impedance. Therefore, (11.1.1) is a valid expression for RF input impedance as long as $\omega = \omega_1$, the RF frequency. The MESFET's RF input is usually conjugately matched; although it is conceivable that the noise figure could be improved by mismatching, as is done with FET amplifiers, there have been no experimental results indicating that similar techniques improve the noise figure of a FET mixer.

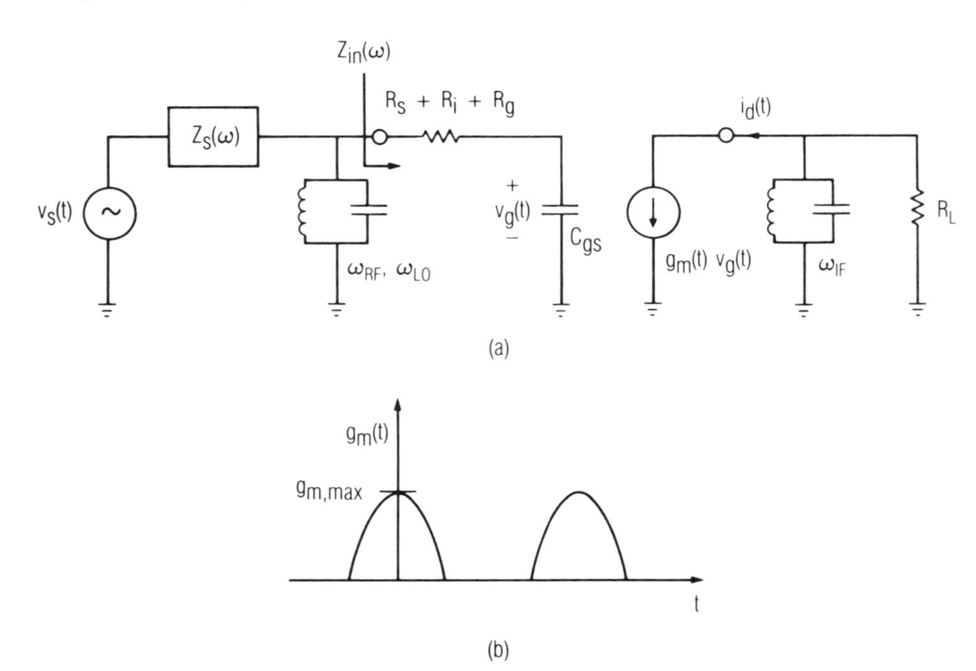

Figure 11.2 (a) Simplified equivalent circuit of the single-gate FET mixer; (b) the MESFET's transconductance waveform when $V_{gg} \approx V_t$.

The RF excitation, $v_s(t)$ in Figure 11.2(a), is

$$v_s(t) = V_s \cos(\omega_1 t) \tag{11.1.3}$$

If the source is matched, $Z_s(\omega_1) = Z_{\mathrm{in}}^*(\omega_1)$ and the small-signal gate voltage is

$$v_g(t) = \frac{V_s \cos(\omega_1 t + \phi)}{2\,\omega_1 C_{gs}(R_s + R_g + R_i)} \tag{11.1.4}$$

The phase shift, ϕ, will not be evaluated because it does not affect the conversion gain. The fundamental-frequency component of $g_m(t)$ (Figure 11.2(b)) is

$$g_{m,1}(t) = \frac{1}{2}\, g_{m,\max} \cos(\omega_p t) \tag{11.1.5}$$

where $g_{m,\max}$ is the peak value of $g_m(t)$. The small-signal drain current $i_d(t)$ is

$$i_d(t) = g_m(t) v_g(t) \tag{11.1.6}$$

The current $i_d(t)$ includes components at the RF and IF frequencies and at all other mixing frequencies shown in Figure 3.15. Substituting (11.1.4) and (11.1.5) into (11.1.6), employing the usual trigonometric identities, and retaining only the terms at the IF frequency gives the IF component of $i_d(t)$, $i_{\mathrm{IF}}(t)$ (note that only the fundamental component, $g_{m,1}(t)$, of $g_m(t)$ contributes to frequency conversion):

$$i_{\mathrm{IF}}(t) = \frac{g_{m,\max} V_s \cos(\omega_0 t)}{8\omega_1 C_{gs}(R_s + R_g + R_i)} \tag{11.1.7}$$

where ω_0 is the IF frequency corresponding to the notation in (3.2.18). The output power is

$$\begin{aligned} P_L(\omega_0) &= \frac{1}{2}\, |i_{\mathrm{IF}}(t)|^2 R_L \\ &= \frac{g_{m,\max}^2 V_s^2 R_L}{128\,\omega_1^2 C_{gs}^2 (R_s + R_g + R_i)^2} \end{aligned} \tag{11.1.8}$$

The available power from the conjugate-matched source is

$$P_{av}(\omega_1) = \frac{V_s^2}{8\ \mathrm{Re}\{Z_s(\omega_1)\}} = \frac{V_s^2}{8(R_s + R_g + R_i)} \tag{11.1.9}$$

and the transducer conversion gain, G_t, is the ratio of (11.1.8) and (11.1.9); that is,

$$G_t = \frac{P_L(\omega_0)}{P_{av}(\omega_1)} = \frac{g_{m,\max}^2 R_L}{16\ \omega_1^2 C_{gs}^2 (R_s + R_g + R_i)} \tag{11.1.10}$$

Equation (11.1.10) is remarkably accurate in theory and in practice, as long as the optimum short-circuit embedding impedances are achieved and the gate is optimally biased; that is, near the FET's turn-on voltage, V_t.

Equation (11.1.10) seems to imply that we can make conversion gain arbitrarily high by increasing the IF load impedance, R_L, or by increasing the device's width (thus, reducing R_s and R_i). These implications are generally valid; however, practical difficulties limit the conversion gain. Problems involving stability and realizability limit R_L from 100 Ω to at most 200 Ω, and the reduction in input impedance limits our ability to increase the gate's width. Furthermore, in many cases, it may not be desirable to have high gain in a mixer, since intermodulation problems may result. It is possible, however, to achieve remarkably high gain (above 10 dB) at X-band in mixers using medium-power devices (having gate widths around 0.6 mm) and high load impedances. Because of the mixer's high output impedance, in some cases it is even possible for a MESFET to achieve greater gain as a mixer than as an amplifier.

The design process is relatively simple. The first task is to estimate the important parameters of the MESFET, $g_{m,\max}$, $R_s + R_i + R_g$, and $C_{gs}(V_t)$. The peak transconductance $g_{m,\max}$ can be found from dc measurements, as can the resistances (a simple method is described in Reference 2.1); C_{gs} can be estimated with adequate accuracy from the FET's S parameters. We should then select a value of R_L that can be achieved in practice and satisfies the gain requirements, as indicated by (11.1.10), and estimate the input impedance via (11.1.1). If the input Q of the device is so high that it cannot be matched over the required bandwidth, reflection losses must also be included in the estimate of the gain. The final step is to design the input and output networks to match the input, to present R_L to the drain at the IF frequency, and to short-circuit the gate and drain at all other significant frequencies.

11.1.3 Matching Circuits in FET Mixers

The input and output matching circuits used in FET mixers have unique requirements, so designing them may require special care. The input matching circuit must not only match the RF source to the MESFET's gate, it must also short-circuit the FET over a wide range of frequencies. It is particularly important that the gate be shorted at the IF frequency. If the IF frequency is much lower than the RF, this short can be realized via the bias-circuit elements. As long as the IF short is realized effectively, the only other critical function of the input matching network is impedance matching at the RF frequency and, if possible, at the LO frequency.

Because of its limited Q, a quarter-wave stub is almost never adequate to short-circuit the drain at the RF and LO frequencies; it is better to realize the IF matching circuit as a low-pass filter connected directly to the MESFET's drain, and to include additional elements to provide the desired IF terminating impedance. The IF matching network is a critical part of a FET mixer, and applying a little creative thought to its design can do much to ensure that the mixer's performance will be very good. A standard, textbook filter design is often not a good choice for the IF filter because a filter having even very high rejection may present a reactive termination rather than a short circuit to the drain. In many cases it is possible for the IF circuit to provide both impedance transformation and filtering functions via a single structure; this approach minimizes circuit loss and complexity. It is usually possible to design a low-pass IF circuit whose performance is relatively insensitive to dimensional tolerances.

11.2 LARGE-SIGNAL–SMALL-SIGNAL ANALYSIS OF FET MIXERS

Like diode mixers, the FET mixer can be analyzed via the large-signal–small-signal approach described in Section 3.2. Example 8 in Section 3.2.2 describes the application of the small-signal conversion-matrix analysis to a simplified equivalent circuit of a FET mixer, Figure 3.19; indeed, this equivalent circuit is probably adequate for most simple downconverters if the mixer's matching circuits are well designed and the RF frequency is below approximately 10 GHz. Other examples of FET mixer designs can be found in References 2.1 and 2.7.

In many other cases, this simplified analysis is not adequate. Two such cases are mixers operating at high frequencies (where the simplified equivalent circuit is not accurate enough) and monolithic circuits (which cannot be adjusted and therefore must be designed very accurately). In these cases, the mixers can be analyzed via a Y-parameter formulation, which is very general and can be applied to a wide variety of types of mixers.

11.2.1 Large-Signal Analysis

The large-signal nonlinear MESFET equivalent circuit shown in Figure 2.16 is applicable to FET mixers as well as to other large-signal components. If the mixer is well designed, the drain voltage is nearly constant over the LO period and therefore the FET remains in its saturation region throughout the LO cycle. Accordingly, the nonlinearities in I_d and C_{gs} can be simplifed significantly, and usually C_{gd} can be treated as a linear element. The equivalent circuit of the FET mixer, under large-signal LO excitation, is shown in Figure 11.3. As in the previous chapters, the source and load matching networks are described by the embedding impedances, $Z_s(n\omega_p)$ and $Z_L(n\omega_p)$, where ω_p is the LO radian frequency.

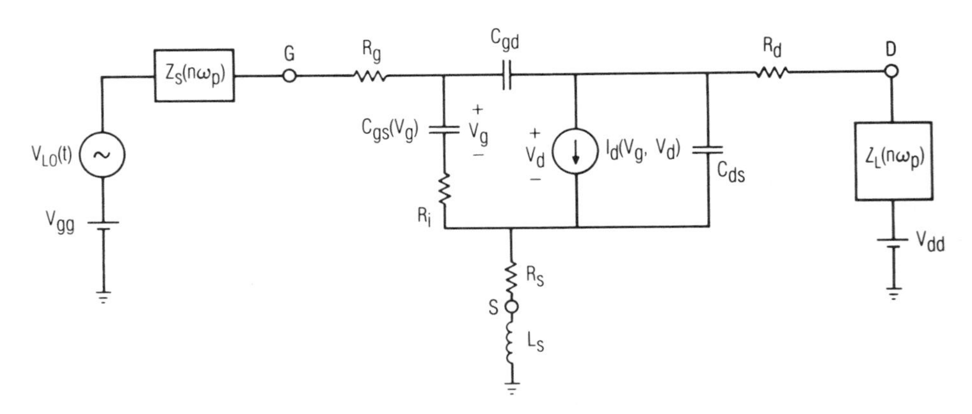

Figure 11.3 Large-signal equivalent circuit of the FET mixer.

The large-signal analysis is best performed via the harmonic-balance method, not unlike the analysis of FET power amplifiers and frequency multipliers. Because the drain in a well designed FET mixer is terminated in a low impedance, the drain voltage contains only a dc component and a small ac component; therefore, the iterative process converges relatively quickly. The LO analysis of FET mixers is one of the less troublesome applications of harmonic-balance analysis.

Although virtually all the convergence algorithms described in Chapter 3 are applicable to FET mixers, the preferred algorithms are Newton's method and the reflection algorithm. References 2.1, 2.7, and 11.1 describe identical applications of the reflection algorithm to FET mixers, and Reference 9.4 describes a way to improve the method's rate of convergence. Reference 11.2 describes the use of Newton's method, as well as some good ideas that expedite the calculation of the Jacobian. In contrast to the power amplifier and FET multiplier analyses, in which the goal was to determine gain and output power, the goal of the mixer LO analysis is to find the waveforms of the control voltages. These voltage waveforms are then used to generate the conductance and capacitance waveforms and finally the conversion matrices.

If Newton's method is used to perform the LO analysis, the equivalent circuit of Figure 11.3 must be put in the form shown in Figure 11.4. In this figure, the FET mixer equivalent circuit has been divided into linear and nonlinear subcircuits and the vectors representing the harmonic voltages at the connecting ports, $\mathbf{V}_1$ and $\mathbf{V}_2$, have been identified. Note that $\mathbf{V}_1$ is equivalent to the control voltage, $\mathbf{V}_g$ and $\mathbf{V}_2 \equiv \mathbf{V}_d$. The linear subcircuit contains all the linear elements in Figure 11.3, including the source and load impedances; the two nonlinear elements C_{gs} and I_d comprise the entire nonlinear subcircuit.

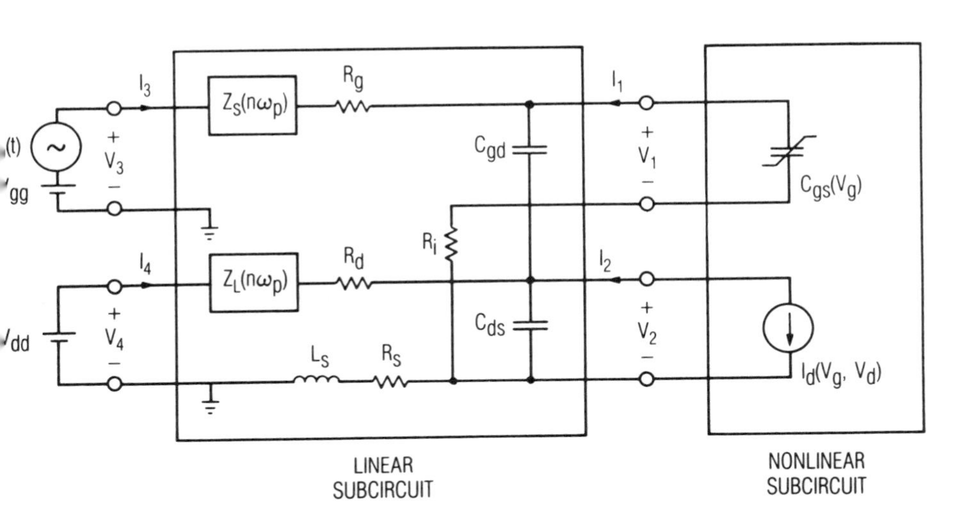

Figure 11.4 Large-signal equivalent circuit of the FET mixer with the circuit divided into linear and nonlinear subcircuits.

The current-error function $\mathbf{F}(\mathbf{V})$ is given by (3.1.20), in which $N = 2$, $\mathbf{I}_G \equiv \mathbf{I}_d$, and $\mathbf{Q}$ is the charge vector, given by (3.1.13) through (3.1.15), representing C_{gs}. In this case, $N = 2$, so there are only two subvectors, $\mathbf{Q}_n$, $\mathbf{I}_n$, and $\mathbf{V}_n$, in each $\mathbf{Q}$, $\mathbf{I}$, or $\mathbf{V}$ vector in (3.1.20). $\mathbf{Y}_{2\times2}$ is formed by eliminating the third and fourth rows and columns of submatrices in $\mathbf{Y}$, the Y matrix of the linear subcircuit, and $\mathbf{I}_s$ is formed from the eliminated columns of $\mathbf{Y}$ as indicated by (3.1.7) through (3.1.9). The Jacobian follows directly from (3.1.38) through (3.1.41).

The initial estimate of the voltage components, required to begin the Newton iterations, can be formed in the same manner as in the FET power amplifier analysis (Section 9.4). The approximations used in the power-amplifier analysis are particularly applicable to FET mixers, so the initial estimate should, in most cases, be very close to the solution.

11.2.2 Small-Signal Analysis

Reference 2.1 includes a set of small-signal equations that can be used to find conversion gain and port impedances of a FET mixer. These equations are lengthy but can be evaluated very rapidly; they allow very efficient analysis and do not require a lot of computer memory. Unless sparse-matrix techniques are employed, the approach to small-signal analysis described in this subsection requires more memory and may not be as efficient. Its advantage is its generality; the same approach can be applied to a wide variety of different circuits and used in a general-purpose program.

Figure 11.5 shows the linear incremental time-varying equivalent circuit of the mixer, devoid of the approximations used to form Figure 11.2(a). The mixing frequencies, ω_q and ω_r, follow (3.2.18); $\omega_q = \omega_0 + q\omega_p$ and $\omega_r = \omega_0 + r\omega_p$. The source admittance at the input frequency, ω_q, has been separated from $Y_s(\omega)$; the remaining part of the source admittance is designated $Y_s'(\omega)$, where $Y_s'(\omega_q) = 0$. Likewise, $Y_L(\omega_r)$, the load admittance at the output frequency, has been separated from $Y_L(\omega)$, leaving $Y_L'(\omega)$ as the reduced load admittance; both $Y_s(\omega_q)$ and $Y_L(\omega_r)$ are zero at frequencies other than ω_q and ω_r, respectively. These admittances have been removed from the embedding networks so they can be treated as variables and selected independently to achieve the desired conversion gain and matching conditions. The nonlinear capacitor, C_{gs}, has been converted, according to (3.2.12), to a time-varying linear capacitor $C_{gs}(t)$. According to (3.2.9) and (3.2.10), the multiply controlled source, I_d, is converted to two time-varying elements: the time-varying conductance $g_{ds}(t)$ and the transconductance $g_m(t)$.

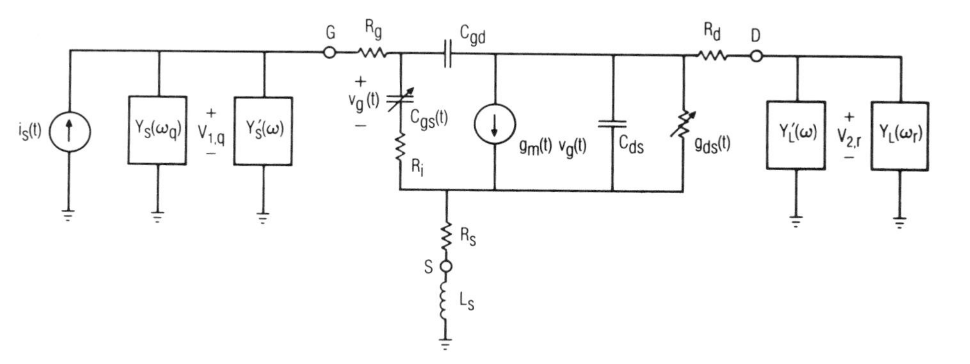

Figure 11.5 Small-signal equivalent circuit of the FET mixer with the source and load admittances separated from the embedding networks.

Figure 11.6 shows the same circuit redrawn so that the time-invariant part of the FET equivalant circuit is represented by a four-port, characterized by its Y matrix. The four ports are the gate and drain terminals and the terminals of $C_{gs}(t)$ and $g_m(t)$; because $g_{ds}(t)$ is in parallel with g_m, it does not require a separate port. The Y matrix of the time-invariant part of the circuit is different from a conventional four-port Y matrix. It relates the voltages and currents not only at the ports but at the mixing frequencies, $\omega_k = \omega_0 + k\omega_p$, $k = 0 \ldots K$, as well. Because the circuit has four ports and $2K + 1$ mixing frequencies, each vector of voltage or current has $4(2K + 1)$ elements. The elements of $\mathbf{Y}$ are submatrices having the form of conversion matrices; however, because they represent a time-invariant circuit, they have the same diagonal form as (3.2.35). The form of the four-port equations is

$$\begin{bmatrix} \mathbf{I}_1 \\ \mathbf{I}_2 \\ \mathbf{I}_3 \\ \mathbf{I}_4 \end{bmatrix} = \begin{bmatrix} \mathbf{Y}_{1,1} & \mathbf{Y}_{1,2} & \mathbf{Y}_{1,3} & \mathbf{Y}_{1,4} \\ \mathbf{Y}_{2,1} & \mathbf{Y}_{2,2} & \mathbf{Y}_{2,3} & \mathbf{Y}_{2,4} \\ \mathbf{Y}_{3,1} & \mathbf{Y}_{3,2} & \mathbf{Y}_{3,3} & \mathbf{Y}_{3,4} \\ \mathbf{Y}_{4,1} & \mathbf{Y}_{4,2} & \mathbf{Y}_{4,3} & \mathbf{Y}_{4,4} \end{bmatrix} \begin{bmatrix} \mathbf{V}_1 \\ \mathbf{V}_2 \\ \mathbf{V}_3 \\ \mathbf{V}_4 \end{bmatrix} \tag{11.2.1}$$

where

$$\mathbf{Y}_{m,n} = \text{diag}[Y_{m,n}(\omega_k)],\ k = -K \ldots -2, -1, 0, 1, 2 \ldots K$$

that is,

$$\mathbf{Y}_{m,n} = \begin{bmatrix} Y^*_{m,n}(-\omega_{-K}) & 0 \ldots & 0 \ldots & & & \ldots 0 \\ \cdot & \cdot & \cdot & & & \cdot \\ 0 & \cdot & \cdot & & & \cdot \\ \cdot & \cdot & \cdot & & & \cdot \\ \cdot & Y^*_{m,n}(-\omega_{-1}) & 0 & & & \ldots 0 \\ 0 & 0 & Y_{m,n}(\omega_0) & & & \ldots 0 \\ \cdot & \cdot & & & & \cdot \\ \cdot & \cdot & 0 & Y_{m,n}(\omega_1) & & \cdot \\ \cdot & \cdot & \cdot & \cdot & & \cdot \\ \cdot & \cdot & \cdot & & \cdot & \cdot \\ \cdot & \cdot & \cdot & & \cdot & \cdot \\ 0 \ldots & 0 \ldots & 0 \ldots\ldots & & & Y_{m,n}(\omega_K) \end{bmatrix} \tag{11.2.2}$$

Each submatrix is a diagonal matrix having $Y_{m,n}(\omega_k)$ in the position corresponding to ω_k. Similarly, the voltage and current subvectors, $\mathbf{V}_n$ and $\mathbf{I}_n$, represent the mixing voltages or currents at the nth port, and thus have the form shown in (3.2.24), (3.2.25), and (3.2.32); that is,

$$\mathbf{V}_n = \begin{bmatrix} V^*_{n,-K} \\ \cdot \\ \cdot \\ \cdot \\ V^*_{n,-1} \\ V_{n,0} \\ V_{n,1} \\ \cdot \\ \cdot \\ \cdot \\ V_{n,K} \end{bmatrix}, \quad \mathbf{I}_n = \begin{bmatrix} I^*_{n,-K} \\ \cdot \\ \cdot \\ \cdot \\ I^*_{n,-1} \\ I_{n,0} \\ I_{n,1} \\ \cdot \\ \cdot \\ \cdot \\ I_{n,K} \end{bmatrix} \tag{11.2.3}$$

The time-varying elements are represented by their conversion matrices; they have the form shown in (3.2.24) and (3.2.32). We account for the time-varying elements by augmenting $\mathbf{Y}$ with their conversion matrices, by adding their admittance-form conversion matrices to the appropriate submatrices of $\mathbf{Y}$. Because $j\mathbf{\Omega}\mathbf{C}_{gs}$, the conversion matrix representing $C_{gs}(t)$, relates $\mathbf{I}_3$ and $\mathbf{V}_3$, it is added to $\mathbf{Y}_{3,3}$; also, because $\mathbf{G}_{ds}$, the matrix representing $g_{ds}(t)$, relates $\mathbf{I}_4$ and $\mathbf{V}_4$, it is added to $\mathbf{Y}_{4,4}$. $\mathbf{G}_m$, representing $g_m(t)$, relates $\mathbf{I}_4$ to $\mathbf{V}_3$, so it is added to $\mathbf{Y}_{4,3}$. Finally, the reduced embedding admittances $Y'_s(\omega)$ and $Y'_L(\omega)$ are converted to the diagonal form of (3.2.35) and are added to $\mathbf{Y}_{1,1}$ and $\mathbf{Y}_{2,2}$, respectively.

The process of augmenting **Y** absorbs the time-varying elements and embedding impedances into the four-port, converting **Y** into a four-port conversion matrix of the entire FET and embedding network (except, of course, $Y_s(\omega_q)$ and $Y_L(\omega_r)$, which were removed). We call this augmented matrix $\mathbf{Y}^a$; $\mathbf{Y}^a$ represents the entire circuit within the dashed lines in Figure 11.6. The augmented matrix $\mathbf{Y}^a$ relates all the voltages and current components at all four ports of the complete network and at all mixing frequencies. Clearly, many of the current components are zero: Ports 3 and 4 are open-circuited at all mixing frequencies, and Ports 1 and 2 are open-circuited at all frequencies except ω_q at Port 1 and ω_r at Port 2; only the RF input current, $I_{1,q}$, and the IF output current, $I_{2,r}$, are not zero. Consequently, these are the only currents of interest. Although nonzero voltages appear at each port and at each mixing frequency, none of these is of interest except $V_{1,q}$ and $V_{2,r}$, the voltage components at the same ports and mixing frequencies as $I_{1,q}$ and $I_{2,r}$.

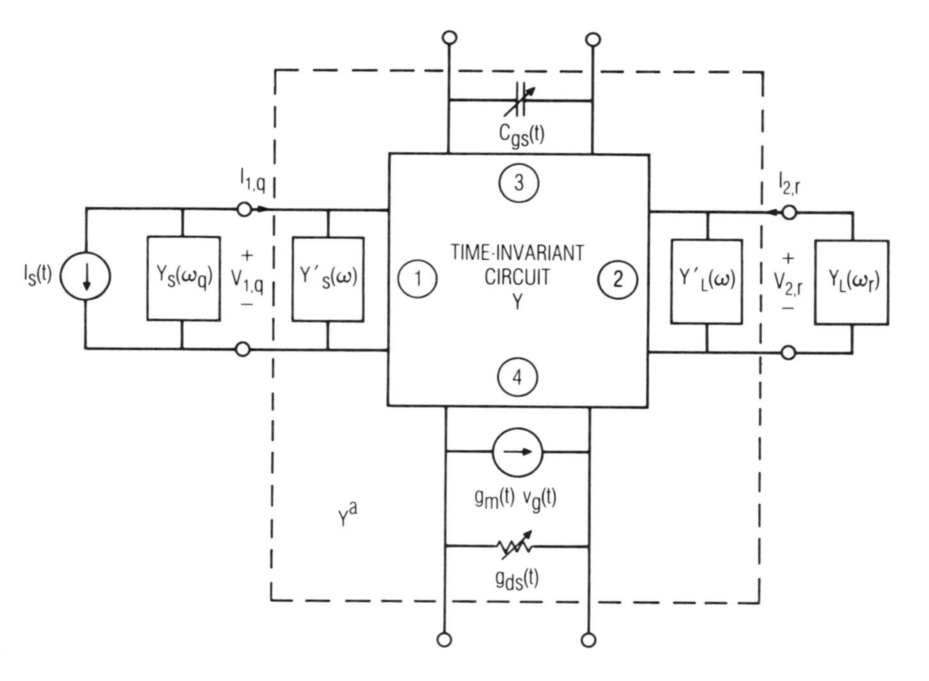

Figure 11.6 Small-signal equivalent circuit of the FET mixer, separated into time-varying and time-invariant parts.

We would like to reduce this system of equations to its simplest form; because we are concerned with only two voltage and current components, the simplest form is a 2×2 matrix. We can reduce this system of equations to a 2×2 set by first inverting $\mathbf{Y}^a$ to obtain the impedance-form conversion matrix, $\mathbf{Z}^a$, and then setting all the currents except $I_{1,q}$ and $I_{2,r}$ equal to zero. All the columns of $\mathbf{Z}^a$ except those corresponding to $I_{1,q}$ and $I_{2,r}$ are multiplied by current components that have zero values, so those columns can be eliminated; all the rows except those corresponding to $V_{1,q}$ and $V_{2,r}$ are then of no interest, so they can also be eliminated. This process reduces $\mathbf{Z}^a$ to a 2×2 Z matrix, which can be inverted again to form a 2×2 Y matrix. This matrix has the form:

$$\begin{bmatrix} I_{1,q} \\ I_{2,r} \end{bmatrix} = \begin{bmatrix} y_{1,1} & y_{1,2} \\ y_{2,1} & y_{2,2} \end{bmatrix} \begin{bmatrix} V_{1,q} \\ V_{2,r} \end{bmatrix} \qquad (11.2.4)$$

The frequency subscripts, q and r in (11.2.4), could be eliminated because they are now extraneous; ω_q is the only frequency of interest at Port 1 and ω_r is the only frequency of interest at Port 2. The admittance equations in (11.2.4) can be treated in a manner identical to the admittance equations of any linear time-invariant two-port. The conversion gain and port impedances can be found via (3.2.60) through (3.2.67).

In the previous derivation, we treated C_{gd} as a linear component. In some cases, for good reasons, the drain may not be terminated in a short circuit at the LO frequency and then the FET may be driven into its linear region over part of the LO cycle. In this case, the variation in C_{gd} may be significant and C_{gd} must be treated as a nonlinear element in both the large- and small-signal analyses. Doing so requires no major changes to either analysis; in the large-signal analysis, it is necessary only to move C_{gd} from the linear to the nonlinear subcircuit, and in the small-signal analysis, C_{gd} must be located at another port in $\mathbf{Y}$. The only effect of these changes is to increase the size of the Y matrices in both analyses.

11.3 DESIGN EXAMPLES

This section will illustrate the design of single-gate FET mixers via two examples; unlike some of the examples in previous chapters, the mixers described here have been built and tested, thus we can present measured performance data. The first example is a rather prosaic mixer that represents one of the author's early designs; although its gain is too high for it to be practical, it is included because it was carefully characterized and tested, so the calculated and measured performance data can be compared. The second is a more exotic design, a millimeter wave HEMT mixer.

Example 1

The design of an 8.0 GHz narrowband mixer will be described. The IF frequency is 1.0 GHz, and the LO frequency is fixed at 7.0 GHz; the desired gain is relatively high, 10 to 11 dB. The MESFET is an X-band device that has a 0.5 μ by 600 μ gate; its circuit parameters are $R_s = 1\ \Omega$, $R_i = 1.5\ \Omega$, $R_g = 1.5\ \Omega$, $C_{gs}(V_t) = 0.5$ pF, and $g_{m,\max} = 0.090$. We begin by estimating $R_L = 7\ \Omega$ and, from (11.1.10), the conversion gain is 11.5 dB. This value does not include circuit losses, so the conversion gain of the completed mixer should be approximately 11 dB. The input impedance, from (11.1.1), is $4.0 + j39.8\ \Omega$.

We now design the matching circuits. We would like to include the filtering and impedance transformation functions in a single structure, and we find that it is possible to do so. First, we design a conventional four-section Chebyschev low-pass filter, using series transmission-line sections to approximate inductors and capacitors. A four-section filter provides a reasonably good short circuit to the drain at the RF and fundamental LO frequencies. Because the filter is asymmetrical, the IF-frequency impedance at the drain end is higher than the 50-Ω IF termination and is approximately 65 Ω at the IF frequency; thus, the filter provides the required impedance transformation as well as the RF-LO frequency short circuit. We use a computer and a general-purpose circuit-optimization program to fine-tune the filter design, primarily to minimize the RF- and LO-frequency impedances at the drain end and to achieve the required 70-Ω IF load. The optimization process also optimizes the terminations at the LO harmonics and compensates for the discontinuities in the microstrip elements.

Because the mixer has a narrow bandwidth, a simple stub-matching circuit suffices for matching the input over a narrow RF band; the circuit is designed with the aid of a Smith chart. The IF short is realized via the gate bias circuit, which consists of a 100-Ω stub one-quarter wavelength long at the RF frequency; an 18-pF capacitor provides the RF short circuit at the end of the stub. At the IF frequency, the capacitor series-resonates the inductive stub, thus shorting the gate to ground. Gate bias is applied through this stub; drain bias is applied through a wirewound inductor and bypass capacitor at the IF output. Blocking capacitors are also required at the input and output.

Figure 11.7 shows the circuit of the mixer. This circuit still requires a simple filter-diplexer to combine the LO and RF signals or two identical circuits could be combined via hybrids to realize a balanced structure.

Figure 11.8 shows the measured conversion loss as a function of LO level and compares it to the loss calculated numerically. The calculated

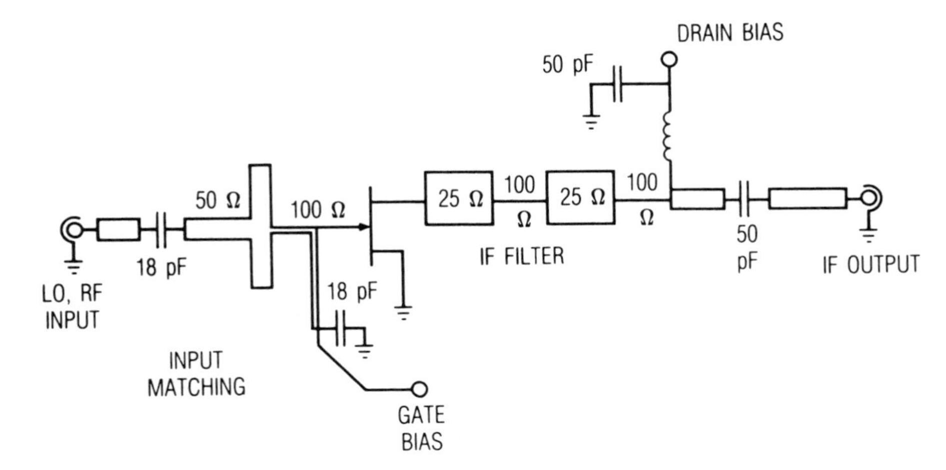

Figure 11.7 Microstrip circuit of the FET mixer in Example 1.

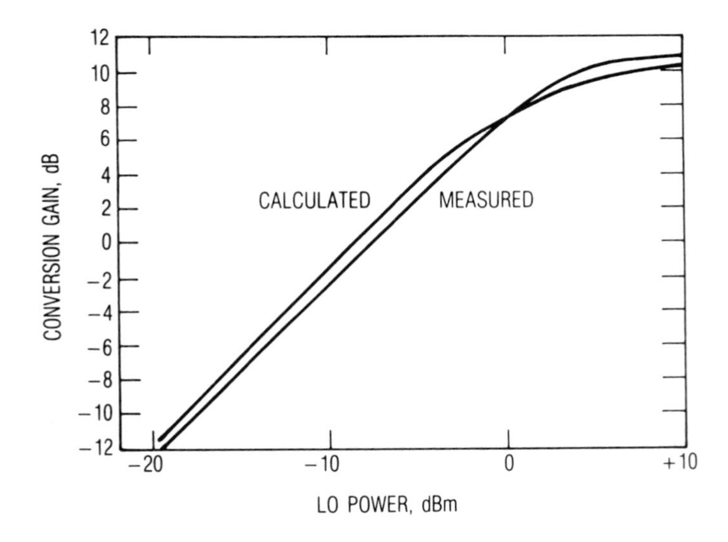

Figure 11.8 Conversion gain of the FET mixer in Example 1.

and measured conversion gains agree within approximately 1 dB over a 30 dB range of LO power, and the peak gain of 11 dB agrees well with the simplified calculations. The input was matched at the RF frequency but no attempt was made to match the LO. Accordingly, the LO power is considerably greater than (11.1.2) predicts, 10 dBm instead of approximately 7.5 dBm. The input return loss at the LO frequency is approximately 3 dB, however, which implies that the reflection loss is 3 dB. Accordingly, the conjugate-match LO power would be approximately 7 dB, in good agreement with the simple theory.

Example 2

Active mixers are not limited to operation at microwave frequencies; FET mixers having conversion gain can be designed at frequencies well into the millimeter-wave region. We now describe the design of a millimeter-wave mixer that uses a FET variant, a high electron mobility transistor or HEMT. The mixer is to operate at 45 GHz and its IF frequency is 3.0 GHz; the mixer must have a 2 GHz RF bandwidth but the IF bandwidth is only 100 MHz. Our goal is to achieve a conversion gain of 1–2 dB, probably the highest value that can be achieved in practice with available devices. The design of this mixer is described in somewhat less detail in Reference 11.3.

The HEMT has been optimized to achieve low-noise amplification at frequencies above 30 GHz. It has gate dimensions of 0.3 $\mu \times 70$ μ, and its maximum transconductance, $g_{m,\max}$ is 0.028 S. The other important parameters are $C_{gs}(V_t) = 0.074$ pF and $R_g + R_s + R_i = 7.8\ \Omega$. Because of the small gate dimensions, C_{gs} is remarkably low; for the same reason, the transconductance is also relatively low, which forces us to use a high value of R_L: According to (11.1.10), R_L must be at least 70 Ω to achieve conversion gain. Because the required IF bandwidth is very low, we can probably design the matching circuit to present a load resistance of more than 100 Ω to the drain; we pick $R_L = 110\ \Omega$, giving a 2 dB gain exclusive of circuit loss. This will be the goal of the IF circuit design. The input impedance at the center of the RF band, from (11.1.1), is 7.8 + j48.

The design of the IF filter is somewhat different from the previous example. We first design an IF filter that effectively shorts the drain over the LO and RF bands and select a filter consisting of a series of alternating low- and high-impedance transmission-line sections. In the initial design, each section is $\lambda/4$ long at f_0, the midpoint of the LO and RF frequency ranges; the characteristic impedances, $Z_{c,L}$ and $Z_{c,H}$, of the low- and high-impedance sections are chosen so that $(Z_{c,L}Z_{c,H})^{1/2} = 50\ \Omega$. Under these conditions, the filter's maximum rejection occurs at f_0 and, if more than three or four sections are used, its input impedance is essentially zero at f_0. At frequencies much lower than f_0 (e.g., the IF), the structure operates approximately as a 50-Ω transmission line. Finally, a short-circuit stub is located at the output side of the filter so that a real impedance of 110 Ω is presented to the HEMT's drain at the IF frequency. The entire circuit is then optimized by computer over the RF, LO, and IF frequency ranges.

The IF circuit is realized in microstrip on a 0.010 inch polished alumina substrate. Although the high dielectric constant of this material causes the filter to be quite small, the use of a material having a lower dielectric constant (e.g., fused silica) would have caused the IF matching stub to be inconveniently long. The IF stub is shorted via a chip capacitor and dc drain bias is applied through the stub.

The design of the input matching circuit is essentially the same as that of the mixer in Example 1. It is a stub matching circuit and the gate-bias circuit provides the IF short. Also, like the 8 GHz mixer, the RF and LO are applied to the gate via a common transmission line, so some type of input diplexer must be used. The input circuit is realized in microstrip on a fused-silica substrate to take advantage of that material's low loss and low dielectric constant; a waveguide-microstrip transition is included on the same substrate. Figure 11.9 shows the mixer's circuit.

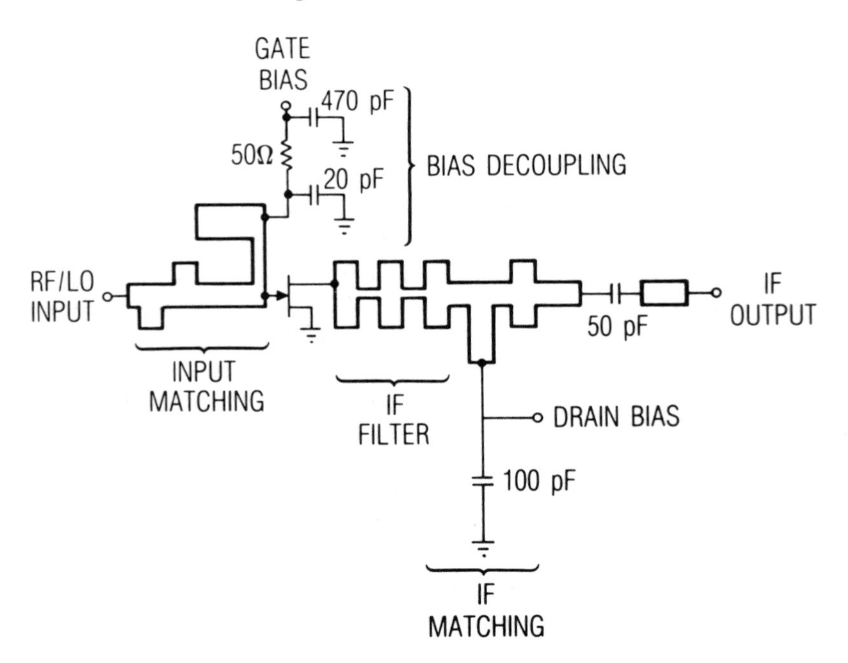

Figure 11.9 Microstrip circuit of the 45 GHz mixer (Example 2). The input waveguide-microstrip transition is not shown.

Figure 11.10 shows the conversion gain of this mixer as a function of LO level at the center of the RF band. The maximum conversion gain is 1.5 dB, in good agreement with (11.1.10), and the conversion gain at the slightly lower LO level (which provides minimum noise figure) is above unity. Figure 11.11 shows the conversion gain at the center of the IF frequency as the LO and RF are swept in tandem across the 2 GHz passband. Because of matching limitations and circuit losses, the average conversion gain is approximately unity, less than the 2 dB predicted by (11.1.10).

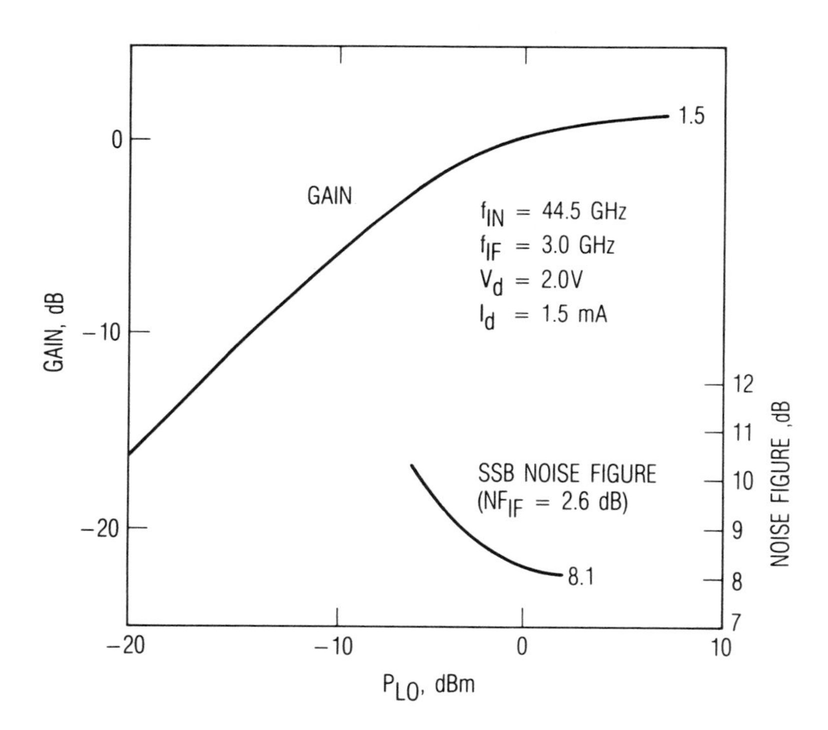

Figure 11.10 Noise figure and gain, as a function of LO power, of the 45 GHz mixer at the center of its RF band. The noise figure includes the effect of an IF amplifier having a 2.6 dB noise figure.

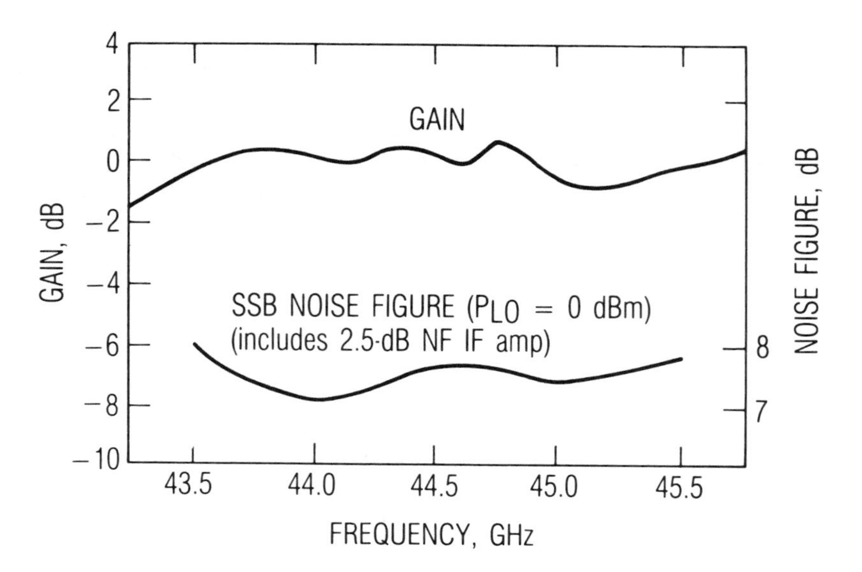

Figure 11.11 Noise figure and conversion gain of the 45 GHz mixer. The IF frequency was fixed and the LO and RF were swept in tandem across their respective bands.

11.4 DUAL-GATE FET MIXERS

Dual-gate FETs have a major advantage over single-gate FETs when used as mixers: The LO and RF can be applied to separate gates and, because the capacitance between the gates is very low, the mixer has good LO-RF isolation. Because of its high LO-RF isolation, a single-device dual-gate FET mixer often can be used in applications where a balanced mixer would otherwise be needed. Balanced FET mixers are also possible and can be used in applications where spurious response and LO noise rejection, as well as good LO-RF isolation, are desired. It is usually simpler to use single-gate FETs in realizing balanced mixers; a balanced dual-gate FET mixer requires separate hybrids for the LO and RF inputs and may therefore be rather complicated. Dual-gate FETs are also used in GaAs integrated circuits, where filters and distributed hybrids are impractical, and good LO-RF isolation may otherwise be difficult to achieve.

Although dual-gate mixers are often perceived to be superior to single-gate mixers, on the whole, the reported performance of dual-gate FET mixers has not been very good. Although dual-gate mixers usually exhibit reasonably good gain, their noise figures have been disappointing, considerably worse than those of single-gate mixers. One reason for this situation is that dual-gate FET mixers have inherent disadvantages compared to single-gate mixers; another is that the dual-gate mixer is a much more complex component than the single-gate mixer, and the subtleties of its operation are not always appreciated by designers (a good explanation of these subtleties can be found in Reference 11.4).

The large-signal–small-signal analysis of single-gate mixers outlined in Section 11.2 has never been applied to dual-gate devices, primarily because of the difficulty of generating equations that adequately describe the dual-gate MESFET's complicated equivalent circuit. In principle, we could circumvent this problem by calculating the Y-matrix, performing the harmonic-balance analysis, and forming a conversion matrix for the entire circuit entirely by numerical means. We could also analyze the dual-gate mixer via a general-purpose program that performs a multitone harmonic-balance analysis; such programs are available commercially but they are relatively slow. A problem in performing such analyses is the lack of good models of the dual-gate MESFET, valid to high frequencies, and it is questionable whether a highly accurate analysis is worth doing unless a comparably accurate dual-gate MESFET model is available. Most of the principles for designing single-gate FET mixers can be applied to dual-gate FET mixers, so the process of designing dual-gate FET mixers is essentially the same as that of designing single-gate mixers. The dual-gate mixer's operation, however, is significantly different from that of the single-gate device.

Figure 11.12 shows a simplified circuit of a dual-gate FET mixer; the dual-gate FET is modeled as two single-gate FETs in series. The LO is applied to Gate 2, the gate of FET 2 (the gate connected to the external drain terminal) and varies V_{gs2}. The RF signal is applied to Gate 1, the gate of FET 1, and mixing therefore must occur via variation of the transconductance between V_{gs1} and I_d. RF and LO sources are connected to Gate 1 and Gate 2 through matching circuits, represented by the embedding impedances $Z_{s,\mathrm{RF}}(\omega)$ and $Z_{s,\mathrm{LO}}(\omega)$, respectively; a series-resonant filter is used to ground Gate 2 at the IF frequency. As with the single-gate FET mixer, the load impedance $Z_L(\omega)$ is a short-circuit at all LO harmonics and mixing frequencies except the IF; this termination guarantees that the LO power is not dissipated in the IF load and that the drain voltage V_{ds} remains constant over the LO cycle.

Figure 11.13 shows the dc drain I/V characteristic of FET 1 in Figure 11.12 as a function of the gate voltages V_{gs1} and V_{gs2} when V_{ds} is fixed at 5.0 V. V_{ds} must be divided between the channels of the two single-gate MESFETs because $V_{ds1} + V_{ds2} = V_{ds}$. When two FETs are connected in series, it is impossible to have a stable operating point if both devices are in current saturation because, in this case, the FETs' channels are equivalent to two current sources in series. Inevitably, one device must be

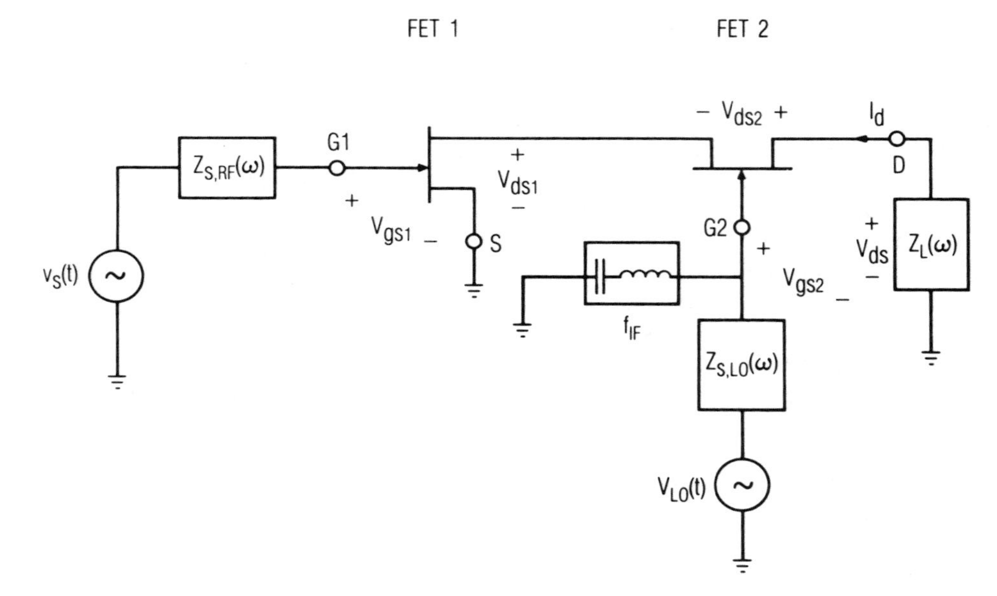

Figure 11.12 Circuit of the dual-gate FET mixer; the dual-gate device is modeled as two single-gate MESFETs in series.

saturated and the other must operate in its linear region; most of V_{ds} is dropped across the saturated FET.

If FET 2 is linear and FET 1 is saturated (the operating point is close to the right side of the set of curves in Figure 11.13), varying V_{gs2} via the LO voltage, while V_{gs1} is constant, does not vary the transconductance between V_{gs1} and the drain current I_d; therefore, no mixing can occur. Significant transconductance variation occurs only when the gate voltages lie within the shaded region of Figure 11.13, the region in which FET 2 is saturated and FET 1 is linear. In this case, the I_d/V_{gs1} transconductance variation occurs primarily because the drain voltage of FET 1 is varied from nearly zero, a value that forces the FET to be in its linear region and its channel to be a low-value resistance, almost to the point of current saturation.

In a dual-gate mixer, mixing occurs primarily in FET 1 because its transconductance and drain-source resistance vary with time while the device is in its linear region. In this mode of operation, the transconductance of FET 1 is relatively low and its low drain-source resistance shunts the IF output, further reducing conversion gain. In contrast, a single-gate device is in current saturation throughout the LO cycle, so its transconductance is greater and its drain-source resistance is very high. For this

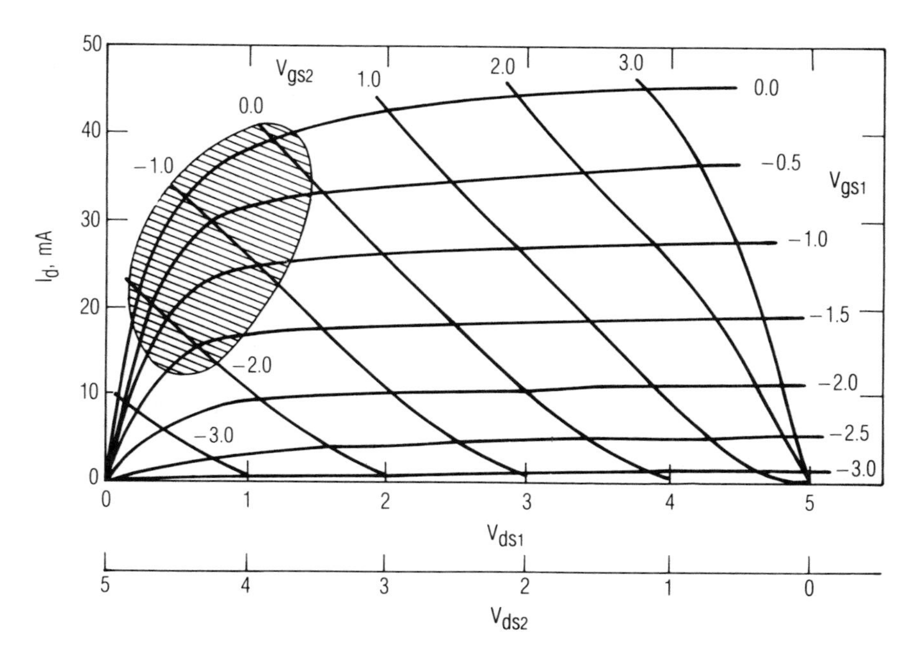

Figure 11.13 *I/V* characteristics of the dual-gate FET when V_{ds} is fixed at 5.0 V.

reason, the single-gate FET is a much more efficient mixer than a dual-gate FET.

In the dual-gate mixer, FET 2 remains in its saturation region throughout the LO cycle and its high transconductance varies only moderately. Consequently, this FET provides some mixing between the RF drain current of FET 1 and the LO, but its primary effect is to amplify FET 1's IF output; the series resonator grounds the gate of FET 2 so FET 2 operates as a common-gate amplifier at the IF frequency. The input impedance of this amplifier is approximately $1/\langle g_m(t)\rangle$, where $\langle g_m(t)\rangle$ is the average transconductance of FET 2. This impedance is usually considerably greater than the IF output impedance of the mixing FET, so the amplifier's input coupling is not optimum and, as a result, its gain is not great. The combination of the mixing FET's poor conversion transconductance and the poor current coupling to the input of the amplifying FET cause the dual-gate mixer's gain and noise performance to be poorer than that of the single-gate FET.

The procedure for designing a dual-gate mixer is much the same as that for designing a single-gate mixer. The dual-gate mixer requires both a carefully designed RF-LO filter at its drain and a resistive IF load. As with the single-gate mixer, the IF output impedance of the dual-gate FET mixer is relatively high, although for a different reason: The high output impedance is a property of a common-base FET amplifier. Thus, good gain can be achieved, in spite of the inherent limitations of the device, by using a relatively high value of IF load resistance. The IF resonator connected to Gate 2 has a critical effect upon the mixer's stability and LO efficiency. If the resonator's reactance at the LO frequency is too low, the LO matching may be poor; however, at some frequency, the combination of the resonator's reactance and the impedance of $Z_{s,\mathrm{LO}}(\omega)$ may cause the mixer to oscillate. We can avoid such problems by making certain that $Z_{s,\mathrm{LO}}(\omega)$ and the resonator do not present a high inductive reactance to Gate 2 outside the LO frequency range. As with the single-gate mixer, source and load impedances, $Z_{s,\mathrm{RF}}(\omega)$ and $Z_L(\omega)$, should be short circuits at unwanted mixing frequencies.

11.5 BALANCED FET MIXERS

11.5.1 Singly Balanced Mixers

A pair of single-gate FET mixers can be combined via quadrature or 180-degree hybrids to create a singly balanced mixer. The properties of balanced FET mixers (LO isolation, spurious-response rejection, and LO noise rejection) are essentially the same as in balanced diode mixers. However, FETs cannot be "reversed," as are diodes, so the structures of singly balanced FET mixers are not entirely analogous to those of singly balanced diode mixers. The balanced FET mixer employs the same type of hybrid and input structure as a diode mixer but, because the IF currents in the individual devices are usually out of phase, the FET mixer requires an IF hybrid to subtract them. Because the output hybrid complicates both the circuit and its layout, the need for an output hybrid is a disadvantage of single-gate FET balanced mixers.

Both the 180-degree and 90-degree (quadrature) FET mixers shown in Figure 11.14(a) and 11.14(b), respectively, require 180-degree output hybrids and, in both mixers, the IF output is derived from the delta port. In Figure 11.14(a), the RF and LO are applied to the sum (Σ) and difference (Δ) ports, respectively; if the ports are reversed, the conversion gain and noise figure are the same but the spurious-response characteristics are not. Because the IF currents are subtracted instead of added, the spurious-response characteristic of a singly balanced FET mixer are precisely the opposite of those of a singly balanced diode mixer, described in

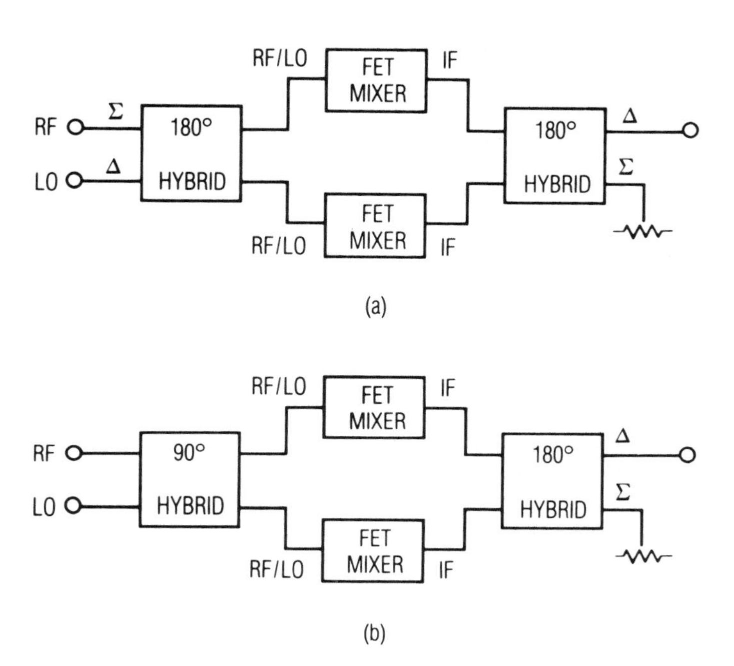

Figure 11.14 (a) 180-degree and (b) quadrature singly balanced FET mixers.

Section 6.4.1. Pumping the devices out of phase, the case shown in Figure 11.14(a), rejects spurious responses arising from odd harmonics of the RF mixing with even harmonics of the LO. If the LO were applied to the sigma port of the input hybrid, the devices would be pumped in phase and the opposite would occur: The mixer would reject mixing products between odd harmonics of the LO and even harmonics of the RF. In both cases, however, the mixer would reject all responses that arise from even harmonics of both the RF and LO.

There are other valid reasons for applying the LO or the RF to a particular port. If the LO is applied to the sigma port, it may be possible to achieve LO rejection via the output hybrid. This property is particularly valuable when the LO frequency is close to the IF frequency, as might occur in an upconverter, and it may not be possible to separate the frequencies by filters. If the LO and IF are both within the output hybrid's bandwidth, the hybrid will combine the IF but will reject the LO. The rejection level depends upon the amplitude and phase balances of both the mixer and the hybrid, but well designed hybrids and mixers should have LO rejection near 20 dB. Conversely, in conventional downconverters, where the LO and RF frequencies are high compared to the IF, there may be an advantage to using the delta port for the LO. The drains of the two FETs can then be connected by a small-value capacitor, which connects the FETs' drains together at the LO frequency but leaves them separate at the IF. Because the LO currents in the FET drains are 180 degrees out of phase at the fundamental frequency and all odd harmonics, each FET effectively short-circuits the other, reducing LO leakage significantly.

The singly balanced quadrature mixer, shown in Figure 11.14(b), has a 90-degree hybrid at the input and the RF and LO are applied to one pair of mutually isolated ports; the other pair of isolated ports is connected to the inputs of two single-device mixers. This configuration has the same properties as a quadrature diode mixer, specifically that the isolation between the RF and LO ports is good only if the inputs of the FETs are well matched at both the LO and RF frequencies. This mixer has the same spurious response properties as the quadrature diode mixer. It rejects only spurious responses associated with even harmonics of both the RF and LO frequencies.

Figure 11.15 shows a circuit of a singly balanced dual-gate FET mixer. The circuit consists of two dual-gate FETs connected in parallel and having a common IF output at the drain; however, the gates of the two dual-gate devices are driven out of phase. By using the techniques described in Chapter 5, we can show that the IF drain currents in both devices are in phase; the LO and RF drain currents, however, are out of phase. Furthermore, the spurious drain currents that result from mixing between even LO harmonics and odd RF harmonics or from odd LO harmonics and even RF harmonics are out of phase, so they do not appear in the output. Spurious responses that arise as mixing products between even harmonics of both the LO and RF are not rejected, however, and AM noise accompanying the LO signal is also not rejected. The out-of-phase signals required at the gates need not be generated by distributed hybrids; active "balun" circuits, which are analogs of electromagnetic baluns, can be used to generate signals having the required phases. Because it is simple and does not require distributed hybrids, this circuit is widely used in monolithic mixers.

11.5.2 Doubly Balanced FET Mixers

Doubly balanced FET mixers have essentially the same characteristics as doubly balanced diode mixers: good port isolation, broad bandwidth, rejection of LO AM noise, and rejection of all spurious responses

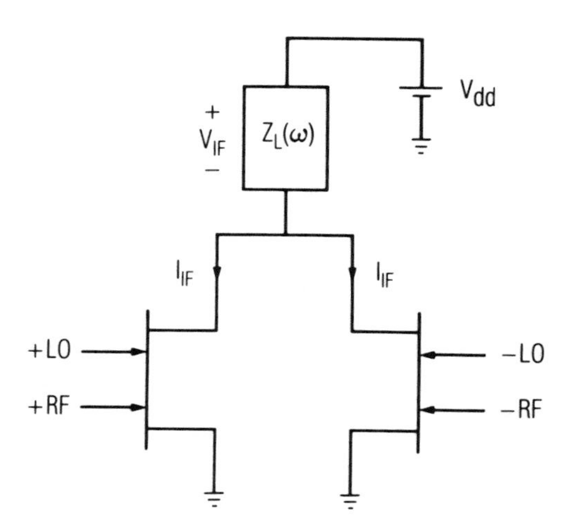

Figure 11.15 Singly balanced dual-gate FET mixers; the gates of the two devices must be excited by RF and LO waveforms that have 180-degree phase differences.

that include an even harmonic of either or both the RF or LO frequencies. Because they do not require distributed hybrids and often achieve good performance as directly coupled circuits (without the need for matching circuits), doubly balanced mixers, often using dual-gate MESFETs, are preferred for use in monolithic integrated circuits.

Figure 11.16(a) shows one realization of a doubly balanced mixer using dual-gate FETs and Figure 11.16(b) shows a simplified equivalent circuit. The doubly balanced mixer consists of two singly balanced mixers of the type shown in Figure 11.15; however, the phases of the RF and LO voltages applied to the pair of FETs on the left, F_1/F_2, are different from the phases at the rightmost pair, F_3/F_4. The result of these differences are that the IF current in the pair on the right is the negative of the IF current in the other pair of FETs, so the IF ports must be combined via a subtracting circuit or balun; again, on-chip active circuits are often used. Similarly, active RF and LO baluns are used to provide a pair of balanced outputs for the gates.

It is important to recognize that this mixer, as well as the singly balanced mixer in Figure 11.15, consists of nothing more than conventional dual-gate mixers in parallel; thus, the procedure for designing such mixers is essentially the same as that for designing a single-device dual-gate mixer. The only significant difference is that the parallel devices short-circuit each other at the LO and RF frequencies, so RF and LO filtering at the drain

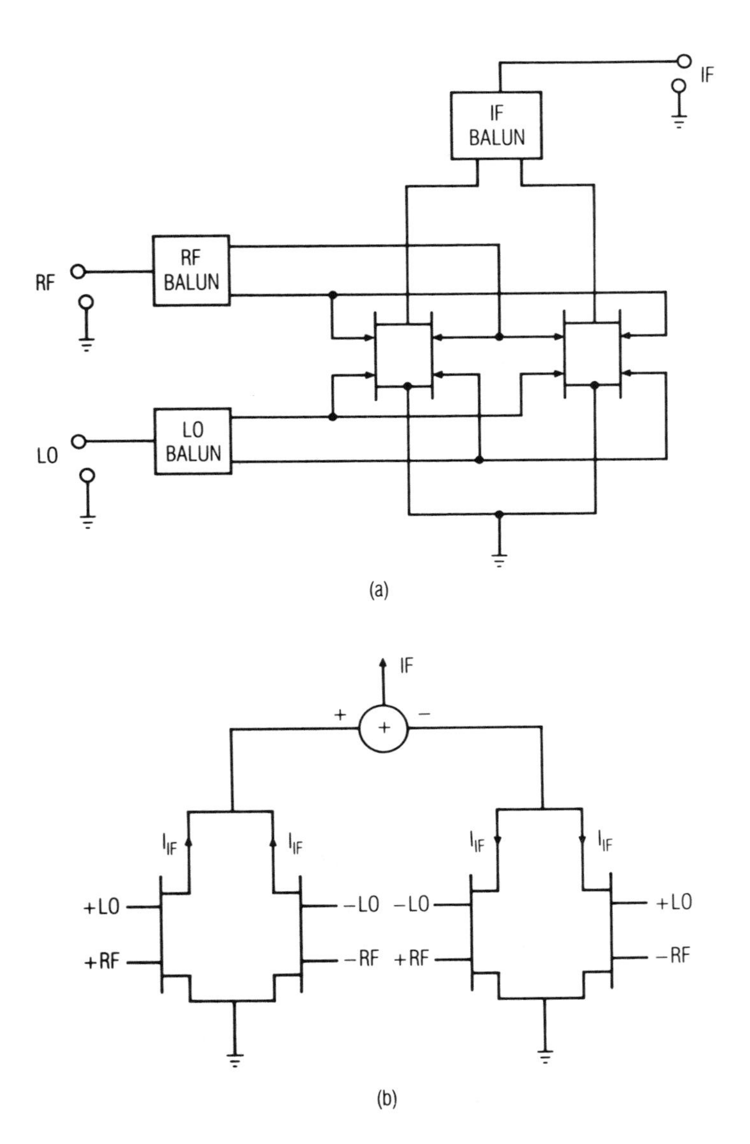

Figure 11.16 (a) Doubly balanced, dual-gate FET mixer; (b) simplified equivalent circuit.

is generally not required, and the load resistance presented to each MESFET is one-fourth the actual load resistance.

REFERENCES

[11.1] J. Dreifuss, A. Madjar, and A. Bar-Lev, "A Novel Method for the Analysis of Microwave Two-Port Active Mixers," *IEEE Trans. Microwave Theory Tech.*, Vol. MTT-33, 1985, p. 1241.

[11.2] C. Camacho-Peñalosa and C.S. Aitchison, "Analysis and Design of MESFET Gate Mixers," *IEEE Trans. Microwave Theory Tech.*, Vol. MTT-35, 1987, p. 643.

[11.3] S.A. Maas, "Design and Performance of a 45-GHz HEMT Mixer," *IEEE Trans. Microwave Theory Tech.*, Vol. MTT-34, 1986, p. 799.

[11.4] C. Tsironis, R. Meierer, and R. Stahlmann, "Dual-Gate MESFET Mixers," *IEEE Trans. Microwave Theory Tech.*, Vol. MTT-32, 1984, p. 248.

CHAPTER 12

OSCILLATORS

Microwave oscillators can be realized in many different forms. Because they have very low phase noise and high dc-RF efficiency, bipolar-transistor oscillators are usually preferred whenever they are practical, that is, at frequencies below approximately 10 GHz. Although their phase-noise levels are significantly greater than those of bipolar oscillators, oscillators using MESFETs and MESFET variants (e.g., the HEMT) can operate at much higher frequencies, well into the millimeter-wave region (Reference 12.1). High-Q dielectric resonators (for fixed-frequency oscillators) and varactors (for *voltage-controlled oscillators,* or VCOs) are often used to reduce the phase noise of FET oscillators to an acceptable level. IMPATT and Gunn oscillators are capable of operation at frequencies well above 100 GHz, but these oscillators often have high noise levels and poor dc-RF efficiency.

This chapter is primarily concerned with the design and nonlinear analysis of high-frequency FET oscillators; however, the techniques we examine are generally applicable to bipolar-transistor oscillators, as well. We begin with the classical approach to negative-resistance oscillators and then reexamine the classical concepts in view of our more sophisticated understanding of nonlinear circuits. Finally, we examine some practical circuits and design techniques.

12.1 CLASSICAL OSCILLATOR THEORY

12.1.1 Negative-Resistance Oscillators and Oscillation Conditions

A general understanding of the operation of electronic oscillators has existed almost as long as active devices. However, more recent work by Kurokawa (Reference 12.2) is usually cited as the basis for the design of modern microwave oscillators. In this work a microwave oscillator is modeled as a one-port in which the real part of the port impedance is negative. The one-port can represent a two-terminal solid-state device that exhibits "negative resistance" (e.g., a tunnel diode) or a two-port that has feedback (e.g., a conditionally stable transistor or amplifier).

An oscillator modeled in this manner is shown in Figure 12.1. The load impedance, $Z_L(\omega)$, is linear but the source impedance, $Z_s(I_0, \omega)$ (the output impedance of the oscillator), is modeled in an unusual fashion: it is a linear impedance that is a function of I_0, the magnitude of the fundamental-frequency component of the output current. The real part of Z_s is negative and decreases with an increase in I_0. Although no linear impedance behaves in this manner, a nonlinear impedance can behave this way if the current and voltage harmonics are ignored. An example of such an impedance is the LO input quasi-impedance of a pumped mixer diode (6.2.1). The small-signal source $v(t)$ in Figure 12.1 represents a perturbation in the voltage across the combined impedances; in practical circuits, it represents noise, an injection-locking signal, or the turn-on transient of the circuit.

If the total resistance in the loop is negative (i.e., if $\text{Re}\{Z_s\} + \text{Re}\{Z_L\} < 0$), the circuit is unstable: if $v(t)$ is perturbed, the response $i(t)$ increases exponentially with time and will become sinusoidal at some frequency, ω_p, where $\text{Im}\{Z_s\} = -\text{Im}\{Z_L\}$. This response is a consequence of the fact that the poles of the total admittance in parallel with $v(t)$ are in the right half of the s plane.

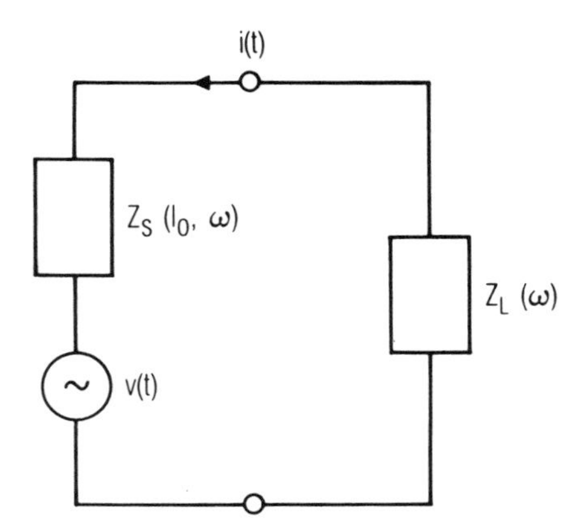

Figure 12.1 The classical model of a negative-resistance oscillator.

As the amplitude of the oscillation increases, $|\mathrm{Re}\{Z_s\}|$ decreases and eventually reaches a point where:

$$Z_s(I_0, \omega_p) = -Z_L(\omega_p) \tag{12.1.1}$$

If I_0 were to increase beyond the point at which (12.1.1) is satisfied, I_0 would decrease and eventually $|\mathrm{Re}\{Z_s\}|$ would rise to the point where (12.1.1) would again be valid. Thus, the value of I_0 that satisfies (12.1.1) is stable, so I_0 remains at that level and oscillation continues at a constant amplitude. The decrease in $|\mathrm{Re}\{Z_s\}|$ with increasing I_0 is an inevitable consequence of the inability of the amplitude of $i(t)$, in practice, to become infinite.

The source could also be described by a nonlinear conductance $Y_s(V_0, \omega)$; then, the oscillation condition is:

$$Y_s(V_0, \omega_p) = -Y_L(\omega_p) \tag{12.1.2}$$

This is the case of a parallel resonance having a total negative conductance, in which the transient excitation comes from a shunt small-signal current source and V_0 is the magnitude of the shunt voltage. The oscillation begins when the real part of the shunt conductance is negative, and $|\mathrm{Re}\{Y_s\}|$ decreases as the oscillation increases until (12.1.2) is satisfied.

In practice, Y_s or Z_s is realized by a solid-state device, which invariably includes nonlinear capacitances, and the average values of those capacitances, and hence $\mathrm{Im}\{Y_s\}$ or $\mathrm{Im}\{Z_s\}$, vary at least slightly with V_0 or I_0. Thus, the frequency at which oscillations begin (when V_0, I_0 are small) is not necessarily the same as that for which (12.1.1) or (12.1.2) is satisfied (and V_0, I_0 are large). However, if a transistor oscillator circuit includes a high-Q resonator, that resonator, not the reactances of the solid-state device, will dominate in establishing the imaginary part of Y_s and the frequency shift will be minimal. Similarly, if the load itself includes a resonator, $\mathrm{Im}\{Y_L\}$ varies rapidly close to resonance, so changes in $\mathrm{Im}\{Y_s\}$ do not cause much frequency deviation.

The oscillation is stable if the sinusoidal voltage or current returns to its steady-state value after it is perturbed. Kurokawa gives another condition for stable oscillation: in terms of impedance, the condition is

$$\frac{\partial R_s}{\partial I}\frac{\partial X_L}{\partial \omega} - \frac{\partial X_s}{\partial I}\frac{\partial R_L}{\partial \omega} > 0 \tag{12.1.3}$$

where $R_s = \mathrm{Re}\{Z_s\}$, $X_s = \mathrm{Im}\{Z_s\}$, $R_L = \mathrm{Re}\{Z_L\}$, $X_L = \mathrm{Im}\{Z_L\}$, and the derivatives are evaluated at $I = I_0$ and $\omega = \omega_p$.

12.1.2 Negative Resistance in Transistors

We noted in Chapter 8 that, in some cases, a two-port could become unstable (it could oscillate) if the load and source impedances were chosen appropriately. In order to allow such oscillation, it must be possible to obtain an input or output impedance having a negative real part (or, equivalently, an input- or output-reflection coefficient greater than unity). This condition can occur only if both $S_{2,1}$ and $S_{1,2}$ are nonzero, if the two-port has forward gain and feedback. In designing small-signal amplifiers, we wish to minimize the effects of feedback; however, in designing oscillators, we introduce additional feedback in order to cause the device to oscillate.

We now heuristically examine negative-resistance or negative-conductance phenomena in transistor circuits by means of a very simple model and apply our understanding of large-signal and small-signal properties of nonlinear circuits to show how $\mathrm{Re}\{Z_s\}$ or $\mathrm{Re}\{Y_s\}$ changes with I_0 or V_0. Figure 12.2(a) shows an ideal MESFET and a feedback network, F; we assume that the magnitude of the voltage gain of F is A_F, its phase shift is 180°, and the port impedances of F are infinite. We use V_d and I_d to designate the static or instantaneous voltage and current at the output terminals; the time-waveforms are $V_d(t)$ and $I_d(t)$. The MESFET has no capacitive or resistive parasitics (it does have an ideal Schottky-barrier gate-source junction); its transconductance is g_m and it is biased at $V_d = V_{\mathrm{dc}}$ and $I_d = I_{\mathrm{dc}}$. Because we are using this example to illustrate only some of the properties of negative resistance in transistor circuits, we have not included a resonator or any other reactive elements; these would be necessary in practice to establish a sinusoidal oscillation.

Via a simple small-signal linear analysis, we can show that the port conductance of the circuit G_s is

$$G_s = -g_m A_F \tag{12.1.4}$$

and G_s is negative for all positive values of g_m and A_F. G_s is the circuit's incremental port conductance in the vicinity of the bias point (V_{dc}, I_{dc}). The large-signal terminal I/V characteristic, $I_d(V_d)$, is graphed on top of the FET's I/V curves. The slope of $I_d(V_d)$ is negative at the bias point, indicating negative conductance.

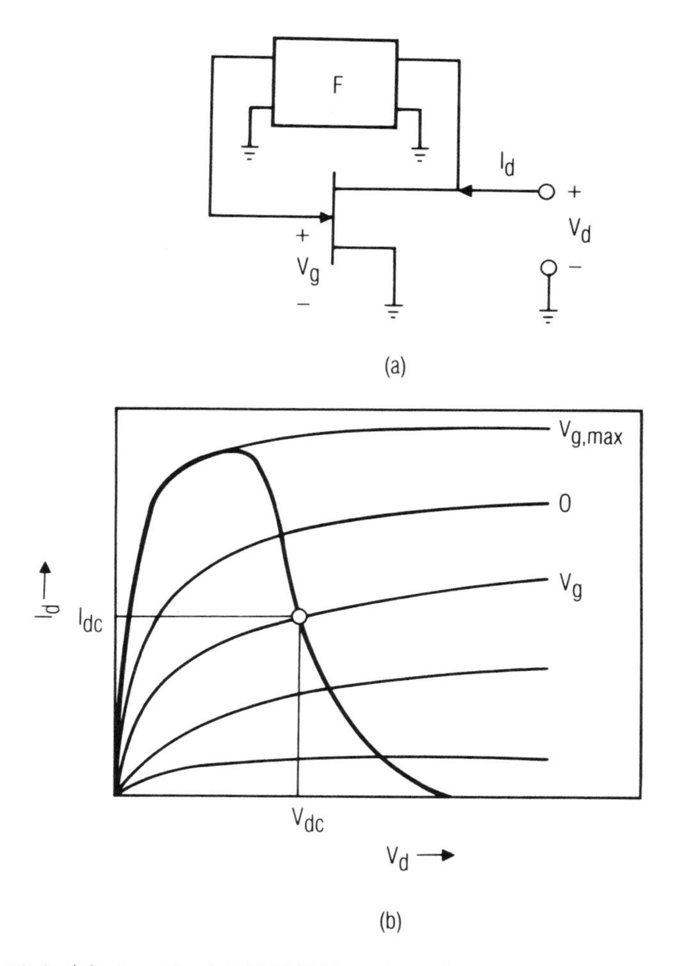

Figure 12.2 (a) An ideal MESFET and a phase-reversing feedback network; (b) the *I*/*V* characteristic at the terminals.

The negative conductance exists only over a limited range of V_d; because of the clamping action of the gate-channel Schottky junction, the gate voltage, V_g, cannot increase beyond $V_{g,\max}$ (approximately 0.6 V); accordingly, $I_d(V_d)$ follows the curve of constant $V_{g,\max}$ at low values of V_d. Similarly, beyond the point where $-A_F V_d = V_t$, I_d is zero. If V_d is increased further, I_d can increase only through avalanching or other second-order effects (e.g., changes in V_t with V_d). Thus, the device has an incremental conductance that is positive at low V_d, is negative over a region of V_d, and then becomes positive (or zero) again at even greater voltages.

Figure 12.3 shows how the ac part of $I_d(t)$, ΔI_d, behaves as the amplitude of the ac part of $V_d(t)$, ΔV_d, increases. When ΔV_d is very small, ΔI_d is also small, is nearly sinusoidal, and is 180° out of phase with ΔV_d. The current waveform is shown in Figure 12.3(a): as V_d rises, I_d decreases, and the circuit exhibits negative conductance over the entire range of $V_d(t)$. The magnitude of the large-signal negative conductance is defined in a manner identical to that of the large-signal "quasi-impedance" that we encountered in the large-signal analysis of diode circuits (e.g., Equation (6.2.1) in Section 6.2.1):

$$G_s = \frac{I_{d,1}}{V_{d,1}} \tag{12.1.5}$$

where $V_{d,1}$ and $I_{d,1}$ are the fundamental-frequency components of $V_d(t)$ and $I_d(t)$, respectively (note that these are not precisely the same as ΔI_d and ΔV_d, which can contain harmonic components). The slope of the terminal *I/V* characteristic, Figure 12.2(b), is relatively constant in the vicinity of the bias point, so as long as V_d is small, G_s does not vary much. If V_d is increased, however, it encounters the lower part of the *I/V* curve, which has a more positive slope. Then, $I_{d,1}$ does not increase as fast as $V_{d,1}$, and $|G_s|$ decreases.

If the amplitude of ΔV_d is increased further, the peaks of $V_d(t)$ eventually exceed the range of the negative-resistance region, and $I_d(t)$ has the waveform shown in Figure 12.3(b). The waveform shows two "dips" at its peaks; these occur when V_d enters the positive-resistance range. At this point, the increase in $I_{d,1}$ with $V_{d,1}$ virtually ceases and, accordingly, $|G_s|$ decreases rapidly, although G_s still remains negative. If V_d increases further, these dips become deeper and, eventually, a point is reached where $I_{d,1} = 0$ and therefore $G_s = 0$.

If ΔV_d is increased even further, the waveform is as shown in Figure 12.3(c). In this case, ΔV_d is so great that $V_d(t)$ remains within the positive-resistance range over most of its period, remaining in the negative-resistance region only briefly while $V_d(t) = V_{dc}$. The only part of the $I_d(t)$ waveform that implies negative resistance is the rising part of the peak that occurs when $V_d \approx V_{dc}$; over the rest of the period, the variations in $I_d(t)$ are in phase with $V_d(t)$. Thus, throughout most of the period, $I_{d,1}$ is in phase with $V_{d,1}$, and consequently, the resulting positive resistance dominates, making $G_s > 0$. We see that, as the amplitude of the oscillation increases, the large-signal conductance G_s increases from its incremental value, which is negative, to zero and finally becomes positive.

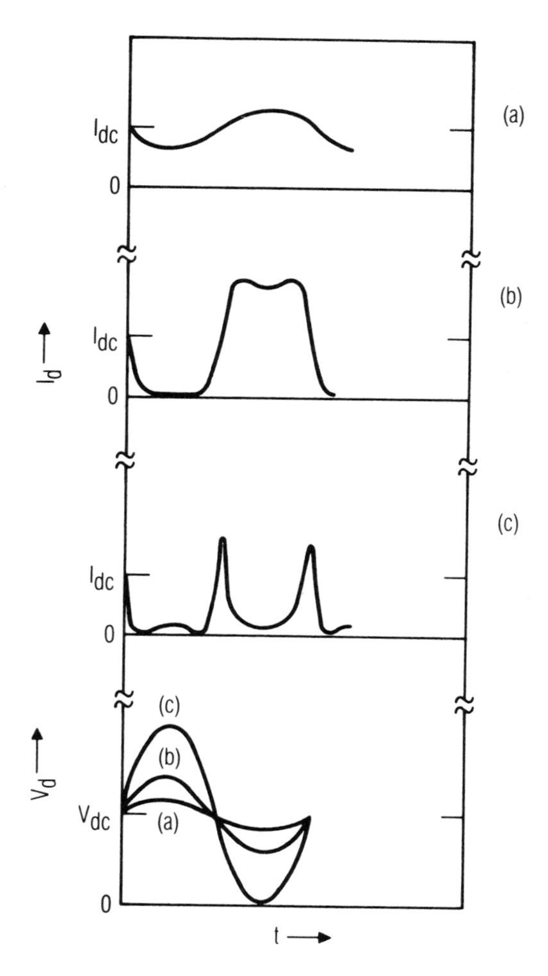

Figure 12.3 The voltage and current waveforms in the circuit of Figure 12.2(a): (a) $V_d(t)$ is entirely within the negative-resistance region; (b) $V_d(t)$ peaks at the edge of the region; (c) $V_d(t)$ peaks well outside the region.

In this example, we have "assumed away" all the reactive parts of the circuit. In a real oscillator, reactive parasitics would exist and some type of resonator would be used to set the oscillation frequency; these elements would add an imaginary part to Y_s with which the load would have to resonate. If the resonator were to have a high Q, its reactance would dominate the imaginary part of Y_s making $\text{Im}\{Y_s\}$ very frequency-sensitive and keeping the oscillation frequency close to the resonant frequency. Thus, the temperature stability of the oscillator would be essentially that of the resonator, and the resonator's narrow bandwidth would act as a filter to minimize phase and *amplitude* (AM) noise.

12.1.3 Oscillator Design via the Classical Approach

Although an oscillator is in reality a large-signal nonlinear component, small-signal linear considerations are usually sufficient to ensure that oscillation conditions are met and to establish very closely the operating frequency. A design based on linear theory is valid because the oscillator, at the onset of oscillation (before the output level rises to the steady-state value), is in fact a linear small-signal component. If the frequency does not change appreciably as the amplitude of the oscillation increases and if precise knowledge of the output power is not needed, small-signal design may be adequate by itself. By using nonlinear analysis, however, we can predict output power precisely and determine the voltage waveforms across critical components in the circuit, such as a tuning varactor. The latter may be very valuable in maximizing power and efficiency or minimizing noise in VCOs.

The classical approach to the design of oscillators involves four steps:

1. selecting the circuit topology;
2. choosing bias conditions that provide adequate output power;
3. adding feedback to maximize the reflection coefficient at at least one port; and
4. selecting source and load impedances that satisfy the oscillation conditions and provide adequate output power.

These steps guarantee only that oscillation will begin at the desired frequency. They do not precisely establish the amplitude or frequency of the large-signal steady-state oscillation. Without the use of nonlinear analysis, accurately estimating the output power can be especially difficult because the factors that limit the output-voltage range are not easy to identify. However, if the oscillator's small-signal output-reflection coefficient can be made very high, a wide range of values of the load-reflection coefficient will statisfy the oscillation conditions, and the output power usually can be maximized empirically within this range.

Most transistor oscillators consist of a positive-feedback amplifier that has a resonator as an input termination. Selecting the oscillator circuit's topology primarily involves selecting the type of amplifier; the choice of amplifier depends strongly on the application of the oscillator. For example, a common-gate circuit is usually preferred for VCOs but, for fixed-frequency oscillators using dielectric resonators, a common-source configuration is often preferred. We shall examine this matter further in Section 12.3.1.

Bias conditions are chosen in a manner similar to that used for the Class-A power amplifier: V_{dc} and I_{dc} are chosen to allow a wide enough variation of the RF voltage and current to provide acceptable output power. If output power is not an imporant consideration, any bias point in the MESFET's saturation region that provides good transconductance is probably acceptable; V_{dc} is often made equal to the drain voltage that the MESFET would have when used as an amplifier and I_{dc} is often set to approximately $0.5 I_{dss}$.

Adding inductance in series with the MESFET's common terminal (in series with the gate in a common-gate configuration or with the source in a common-source configuration) usually increases the magnitude of the input- and output-reflection coefficients to values well above 1.0. If these reflection coefficients are high, the designer has a large degree of freedom in selecting the load impedance and usually obtains well behaved operation. It is also important to adjust the feedback and to design the load network so that oscillation can occur at only a single frequency, so that there are no additional resonances between $Y_s(\omega)$ and $Y_L(\omega)$ within the frequency range for which $\mathrm{Re}\{Y_s\} < 0$.

As with a small-signal amplifier, satisfying the oscillation conditions at either the input or the output is enough to ensure oscillation. Therefore, the load is usually chosen to provide adequate output power and the input termination is chosen to satisfy the oscillation conditions. The design process is illustated by the following example.

Example 1.

We design a 10 GHz VCO having approximately 10 dBm of output power. Because this output level is well within the capability of a small-signal MESFET, we choose a K-band GaAs MSEFET having a 0.4 μm × 200 μm gate. Its small-signal S parameters at 10 GHz in common-source configuration are

$$S_{1,1} = 0.69 \angle -110^\circ, \quad S_{1,2} = 0.07 \angle 62^\circ$$
$$S_{2,1} = 2.57 \angle 104^\circ, \quad S_{2,2} = 0.37 \angle -58^\circ$$

We also choose bias values of $V_{dc} = 3.0$ V and $I_{dc} = 30$ mA, the conditions under which the S parameters were measured. This value of drain current is approximately $0.5 I_{dss}$ and is the value that has been found empirically to give maximum gain in amplifier operation. The oscillator circuit is shown (minus the dc bias and blocking circuitry) in Figure 12.4.

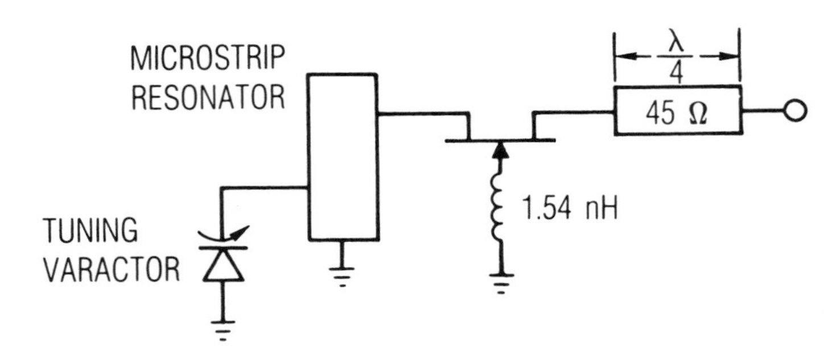

Figure 12.4 The circuit of the oscillator designed in Example 1. The dc bias circuitry is not included.

The MESFET is used in a common-gate configuration; an inductor in series with the gate provides feedback. Because it is relatively easy to obtain broadband negative resistance in a common-gate circuit, this is the preferred configuration for realizing wideband VCOs. The input (the MESFET's source terminal) is terminated by a varactor-tuned resonator, and the output is terminated by a load that ensures good output power. We choose the load impedance on the basis of the output-power requirement; we then design the resonator to satisfy the oscillation conditions.

We first estimate the load conductance. The load conductance is chosen to maximize output power; we find it by making a very rough estimate of the fundamental-frequency RF components of the drain voltage and current. If $I_{dc} = 30$ mA, the fundamental RF current must be less than 30 mA; we estimate it to be 25 mA. In Section 12.1.2, we saw that the fundamental RF drain voltage must be considerably less than V_{dc}. Although we derived this result by considering a common-source circuit, the same is approximately true for the common-gate circuit. Thus, we assume that the peak RF drain voltage is approximately 1 V and the output power, P_{out} is

$$P_{out} = \frac{1}{2} V_{d,1} I_{d,1} = (0.5) \cdot (1.0) \cdot (0.025) = 12.5 \text{ mW} \tag{12.1.6}$$

The real part of the load conductance is

$$G_L = \frac{I_{d,1}}{V_{d,1}} = \frac{0.025}{1.0} = \frac{1}{40}\ \text{S} \tag{12.1.7}$$

or a shunt resistance of 40 Ω. Because the output-reflection coefficient, Γ_{out}, in a common-gate circuit is usually capacitive, the output load that gives maximum power is usually inductive. However, the susceptance found from small-signal S parameters may not be optimum for large-signal operation. Furthermore, $S_{1,2}$ in the common-gate FET has a high value, approximately 1.8, indicating that the output admittance is sensitive to the input termination. Thus, it is probably futile to predict the large-signal load susceptance from small-signal considerations. Moreover, because of the complications introduced by the feedback inductance and the limited range of ΔV_d, the approach to output-load design used for the FET power amplifier is also invalid. Accordingly, in this first-order design, we use a purely resistive load. The lack of a load susceptance may cost a dB or two of output power, but this power loss can be reclaimed by empirical tuning or by optimizing the design via nonlinear analysis.

We now must make certain that oscillation conditions will be satisfied when this value of load resistance is used. To do so, we maximize the input-reflection coefficient when the output termination is a 40-Ω resistor. We find that a feedback (gate) inductance of 1.5 nH maximizes Γ_{in}, giving $\Gamma_{\text{in}} = 3.5 \angle 95°$. The oscillation condition defined by (12.1.1) or (12.1.2), in terms of reflection coefficients, is

$$\Gamma_t \Gamma_{\text{in}} = 1.0 \tag{12.1.8}$$

where Γ_t is the reflection coefficient of the termination at the source terminal of the FET. Thus, Γ_t must have a phase angle of $-95°$ and a magnitude equal to 1/3.5. $|\Gamma_t|$ can, of course, be greater than 1/3.5, because $|\Gamma_{\text{in}}|$ will decrease to satisfy (12.1.8) as the oscillation develops. This is the same phenomenon as the decrease in $\text{Re}\{Z_s\}$ or increase in $\text{Re}\{Y_s\}$ that occurs to satisfy (12.1.1) or (12.1.2). Γ_t is realized by a resonator with frequency adjusted via a tuning varactor, so $|\Gamma_t|$ is close to 1.0 and $\angle\Gamma_t = -95°$ at a frequency close to resonance.

One way to realize the oscillator would be simply to connect a tuning varactor having $\angle\Gamma_t = -95°$ at 10 GHz to the FET's source terminal. The varactor would have a junction capacitance of approximately 0.3 pF, a practical value. This approach maximizes the tuning range of the oscillator;

however, the temperature stability of the oscillator depends strongly upon the FET's S parameters and the varactor's capacitance, both of which are highly sensitive to temperature. Better stability can be achieved by realizing Γ_t via a resonator that has at least a moderately high Q and good stability. Because the phase of a resonator's reflection coefficient varies rapidly with frequency near resonance, the oscillator's frequency (the frequency at which $\angle\Gamma_t = -95°$) will remain close to the resonator's resonant frequency, even if the FET's S parameters drift enough to change $\angle\Gamma_{in}$. A varactor can then be coupled to the resonator in order to vary its resonant frequency and thus to vary the oscillator's frequency.

For this reason, we choose to terminate the FET in a microstrip resonator, and to have the tuning varactor loosely coupled to the resonator. By adjusting the points on the microstrip at which the FET and varactor are connected, we can achieve the range of $\angle\Gamma_t$, required to tune the oscillator over its frequency range. Finally, we design a quarter-wave transformer to transform the standard 50-Ω load to the desired value of 40 Ω.

This circuit configuration is very popular for realizing wideband VCOs. Even simple oscillators of this configuration can achieve remarkably wide tuning bandwidths, often over an octave. We will have more to say about this and other circuits in Section 12.3.1.

In designing the resonator and its tuning circuit, we find a direct trade-off between tuning range and frequency stability, more generally, a trade-off in the design of any VCO between tuning range and phase noise. To achieve high stability (or low noise), we must use a stable high-Q resonator (e.g., a dielectric resonator or a waveguide cavity) and couple the FET and varactor to it very weakly. This weak coupling limits the ability of the tuning varactor to vary the resonator's frequency, however, so the tuning range is narrow. Coupling the varactor more strongly to the circuit increases its effect on the circuit and provides a wider tuning range; unfortunately, such strong coupling also increases the effect of its poor thermal stability, and it may also reduce the Q of the resonator.

Another phenomenon worth noting is that the phase angle of Γ_{in} is not constant with changes in the output load; indeed, if the load VSWR is 2.0, the phase of Γ_{in} varies between 65° and 125° as the phase of the load-reflection coefficient is varied over 360°. As $\angle\Gamma_{in}$ varies, the frequency at which the oscillation conditions are satisfied must also vary, so the oscillator frequency must change. This phenomenon, called *pulling*, is also more serious when the Q of the resonator is low and the varactor is tightly coupled.

Although relatively crude estimates of the output conductance and power were necessary for the initial design, after the oscillator is fabricated, we can use empirical techniques to obtain an improved estimate of the load impedance. One method that has been widely accepted is called a *device-line measurement* (Reference 12.3). In this process, RF power is applied to the output terminals of the unmatched oscillator and the output-reflection coefficient Γ_{out} is measured under large-signal nonoscillating conditions. Because the magnitude of the large-signal reflection coefficient is greater than unity, power is delivered by the oscillator to the measurement system during this test. That power, P_d, is

$$P_d = P_{av}\left(|\Gamma_{out}|^2 - 1\right) \tag{12.1.9}$$

When the oscillator is delivering P_d to the measuring system, it has the large-signal output-reflection coefficient Γ_{out}. This is the same Γ_{out} it would have in normal operation, delivering P_d to a load, Γ_L, in which $\Gamma_{out}\Gamma_L = 1$. Therefore, the load $\Gamma_L = 1/\Gamma_{out}$ must result in an output power P_d. The underpinnings of this argument are essentially the same as those of load-pull theory, which is applied to power amplifiers; the greatest limitation is that harmonic effects are neglected.

12.2 NONLINEAR ANALYSIS OF TRANSISTOR OSCILLATORS

The oscillator design illustrated in Example 1 was adequate only to guarantee that the circuit would oscillate. Because it was based on small-signal linear S parameters, it was not possible to estimate the output power or to find a load impedance that optimized the output power. Thus we were forced, with much embarrassment, to use rather crude estimates in the design of the output network. We would prefer to use our knowledge of nonlinear analysis to define the load impedance and predict the output power more precisely.

The oscillator in Example 1 is a relatively simple circuit using Class-A bias. We saw in Chapter 9 that harmonic effects in lightly saturated Class-A power amplifiers usually have a negligible effect on the fundamental-frequency output power; although an oscillator drives itself more strongly into saturation, it is reasonable to expect that the harmonic effects in simple oscillators are also relatively mild. However, in more complex circuits, this may not be the case; for example, in a frequency-doubling oscillator, the second harmonic output is more important than the fundamental and, in many types of oscillators (e.g., low-noise VCOs), knowledge of the precise voltage waveforms at each circuit node might be valuable in the design process.

Two problems arise when we attempt to analyze oscillators by harmonic-balance techniques: first, the saturation phenomena that limits the amplitude of the oscillation must be modeled very carefully. In particular, because the gate-channel junction of a FET oscillator is driven into conduction over part of the period of oscillation, we should include diodes in the FET equivalent circuit (Figure 2.16) in parallel with C_g and C_d, and the channel current $I_d(V_g, V_d)$ as well as the capacitances must be modeled accurately throughout both the linear and saturation regions. Because the values of the capacitances, C_g and C_d, change dramatically as the FET enters its linear region, it is possible that in some cases the capacitances may have a significant limiting effect; therefore, modeling these capacitances accurately also may be important.

The second problem is more serious. Harmonic-balance analysis is used primarily when a nonlinear circuit is driven from an external source. Thus, it is assumed at the outset of a harmonic-balance analysis that the excitation frequency and its harmonics are known exactly. In an oscillator, the frequency can change as the oscillation develops and, therefore, can be known only approximately. If the frequency used in the harmonic-balance analysis is not the same as the large-signal oscillation frequency, the results may be erroneous or, even worse, the analysis may indicate that the circuit does not oscillate at all!

One solution to the latter problem is to use time-domain analysis. In this case, we would write nonlinear differential equations and integrate them numerically until a steady-state response was obtained. This process suffers from the standard limitations of time-domain analysis: Transmission lines and lumped impedances often cannot be modeled adequately (a serious problem in view of the reliance of the resonators in microwave oscillators on transmission lines), and long time constants may cause the settling time of the transient response to be long. Furthermore, the presence of a high-Q resonator in the oscillator's circuit may introduce numerical instability.

Because of this extra complexity, very few technical papers have addressed the harmonic-balance analysis of transistor oscillators. However, an important exception to this observation is by Rizzoli, Lipparini, and Marazzi (Reference 12.4). Their approach is to use one or more circuit parameters as variables, along with the harmonic voltages across the nonlinear elements. The current-error function (3.1.20) becomes

$$\mathbf{F}(\mathbf{V}, \mathbf{P}) = \mathbf{I}_s(\mathbf{P}) + \mathbf{Y}_{N \times N}(\mathbf{P})\,\mathbf{V} + \mathrm{j}\boldsymbol{\Omega}\mathbf{Q} + \mathbf{I}_G \qquad (12.2.1)$$

where **P** is a set of circuit parameters. The components of **P** as well as the voltage variables are adjusted by the appropriate harmonic-balance algorithm, and convergence is indicated by a minimal value of the components of **F**. The authors emphasize that the variables **V** and **P** can be adjusted simultaneously in the solution algorithm; it is not necessary to solve the harmonic-balance equations to obtain **V** and then to optimize the circuit variables **P**. An advantage of the authors' approach is that, by defining the error function to include output power or other performance parameters, we can include performance optimization in conjunction with the standard harmonic-balance analysis.

Another possibility is to use harmonic-balance analysis to perform a numerical device-line "measurement." The advantages of this approach are that the frequency of oscillation can be specified as part of the large-signal design, a graph of the output power as a function of load impedance can be obtained very easily, and the source impedance of the measurement system can be adjusted to any convenient value. Furthermore, in contrast to conventional device-line measurements, the effects of load impedances at harmonics can be included conveniently.

12.3 PRACTICAL ASPECTS OF OSCILLATOR DESIGN

12.3.1 Circuit Topology

We saw in Section 12.1 that the common-gate configuration is a popular choice for realizing wideband VCOs. The reason for its importance is that we can achieve negative conductance over a wide frequency range by using a very simple feedback circuit, a single inductor. For the same reason, the common-base circuit is favored for realizing bipolar-transistor VCOs. Other circuit topologies are frequently used for realizing other types of oscillators.

Many microwave systems require free-running (not phase-locked) sources that are highly stable. For these applications, fixed-frequency oscillators using dielectric resonators are ideal. Dielectric resonators are made from modern ceramic materials, which usually include titanium oxides. These resonators have high dielectric constants, so they are usually much smaller than waveguide or coaxial resonators; using a dielectric resonator instead of a metal cavity allows for a considerable reduction in the size and weight of a stable microwave oscillator. Because these materials have very low loss, the Q_s of dielectric resonators are nearly as high as those of metal cavities. Furthermore, the temperature coefficients of the dielectrics are very low and can be controlled by adjustment of the composition of the ceramic. These resonators can even be made to have prescribed temperature coefficients that compensate for drift in the oscillator circuit.

When used in a microstrip circuit, the dielectric resonator is coupled magnetically to the microstrip line. We can adjust the coupling coefficient by varying the distance from the resonator to the edge of the microstrip or by placing the resonator on top of a dielectric spacer (e.g., fused silica). When coupled to a single line, the resonator open-circuits the line at its resonant frequency; when coupled to two microstrips, a dielectric resonator can also be operated in a transmission mode. The resonator must be shielded or radiation loss will reduce its Q significantly; the usual method for shielding the resonator is to mount it (along with the circuit) in a metal cavity. The resonant frequency can be adjusted somewhat by a tuning screw located above the resonator. The theory of dielectric resonators is a separate discipline; that theory and further information on the use of dielectric resonators in filters as well as in oscillators can be found in Reference 12.5.

Figure 12.5 shows a dielectric-resonator oscillator that has a common-source toplogy. The feedback element is the admittance, Y, in series with the MESFET's source; in contrast with the common-gate topology, Y must be capacitive in order to achieve $|\Gamma_{\text{in}}| > 1$. The resonator is coupled to a microstrip line that is connected to the MESFET's gate; the line is terminated in a resistor equal to the characteristic impedance. This transmission structure has a bandstop characteristic, so the reflection coefficient, Γ_t, is unity at resonance and nearly zero at all other frequencies. We can adjust the phase of Γ_t by varying the placement of the resonator along the transmission line. We denote the electrical distance from the FET to the open-circuit plane of the resonator as θ; then from Chapter 10 of Reference 12.5:

$$\Gamma_t = \frac{\gamma}{\gamma + 1 + 2\,\mathrm{j}Q_u\delta}\exp(-2\mathrm{j}\theta) \tag{12.3.1}$$

where γ is the coupling coefficient and Q_u is the unloaded Q of the resonator. The parameter $\delta = (\omega - \omega_0)/\omega_0$ is the fractional bandwidth, where ω_0 is the resonant frequency.

This oscillator can be designed in a manner nearly identical to that of Example 1. The feedback element, Y (often an open-circuit microstrip stub), is designed to maximize Γ_{in}; then γ and θ are chosen so that $|\Gamma_t| > 1/|\Gamma_{\text{in}}|$ and $\angle\,\Gamma_t = -\angle\,\Gamma_{\text{in}}$ at resonance. Although the feedback may cause $|\Gamma_{\text{in}}|$ to be greater than unity over a very broad bandwidth, the resistive termination causes the FET to be stable, thus preventing oscillation at unwanted frequencies.

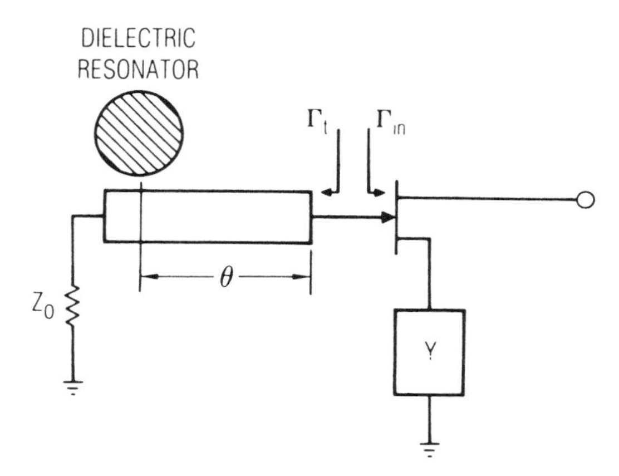

Figure 12.5 A dielectric-resonator oscillator using a FET in a common-source configuration.

Dielectric resonators can be used in circuits having other topologies. When the tunable microstrip resonator in the circuit of Figure 12.4 is replaced by a dielectric resonator, the circuit can be used as a fixed-frequency oscillator. Figure 12.6 shows two other circuits. In Figure 12.6(a), the resonator has been placed in the source circuit and an open-circuit stub has been connected to the MESFET's gate. The properties of this circuit are similar to those of the circuit in Figure 12.5; the main difference is that the resonator allows feedback to exist only at the resonant frequency. The terminating resistor stabilizes the FET at frequencies even slightly removed from ω_0. In the circuit of Figure 12.6(b), the resonator is used in a transmission mode to provide feedback between the drain and gate. In order to minimize pulling, output power is derived from the source; the drain could also be used for the output.

Figure 12.7 shows a FET doubling oscillator. The circuit consists of two fundamental-frequency oscillators of the type shown in Figure 12.5; both share a common resonator. The MESFETs are coupled to the resonator in such a way that the fundamental-frequency components of the voltages at their gates have a phase difference of 180°; the currents in the FETs' channels are therefore out of phase as well. Because the drains are connected in parallel, they effectively short-circuit each other at the fundamental frequency, so the output contains only even harmonics; in practice, the second harmonic is the only significant one. This circuit is similar to the antiseries or "push-push" FET frequency doubler described in Section 10.4.2. The general properties of antiseries circuits are described in Section 5.2.1 and a description of the use of this circuit as a millimeter-wave source can be found in Reference 12.6.

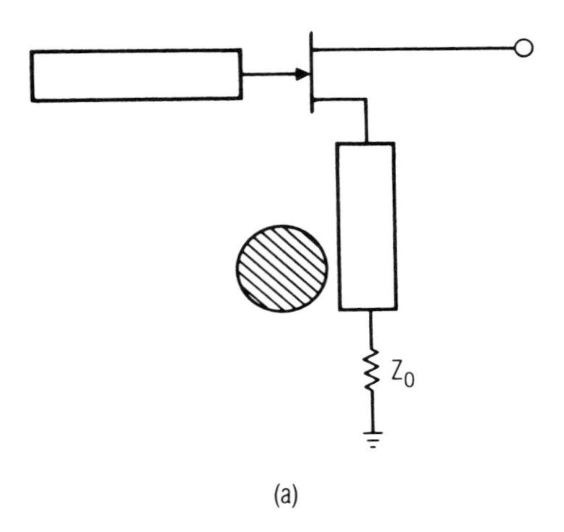

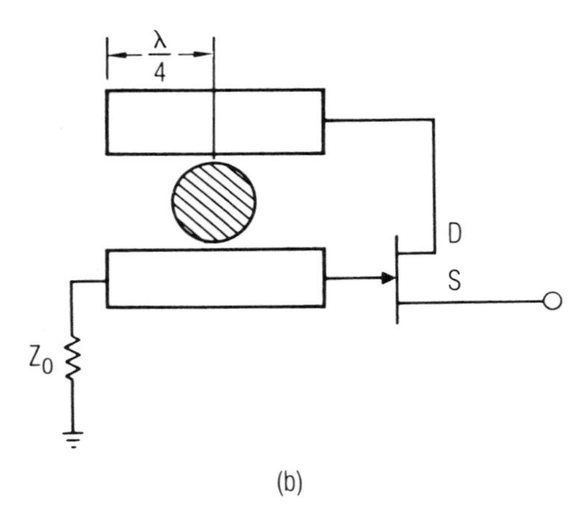

Figure 12.6 Two circuit topologies for a dielectric-resonator oscillator: (a) a common-source circuit having the resonator in the feedback circuit; (b) an oscillator using a transmission resonator to provide drain-to-gate feedback.

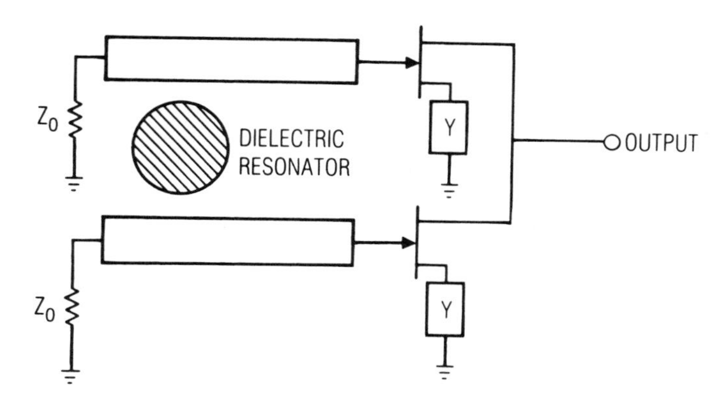

Figure 12.7 A doubling oscillator using a dielectric resonator. The resonator is resonant at half the oscillator's output frequency.

12.3.2 Performance Parameters

Two fundamental characteristics of an oscillator are its frequency and output power. However, other, not as obvious, performance parameters also have a strong effect on the performance of an RF or microwave system and therefore are extremely important. A few of these are described in this section.

Phase Noise

Noise processes in semiconductor devices can modulate the phase of an oscillator and create noise sidebands in its output spectrum. This phase deviation is a serious problem in systems where phase information is crucial, as in radar and coherent communication systems. Because high-frequency components of that noise are attenuated by the resonator, noise processes that generate low-frequency components are of most concern: $1/f$ noise is a particularly significant contributor to phase noise; accordingly, devices having low $1/f$ noise levels are usually preferred for use in oscillators. Bipolar transistors have significantly lower $1/f$ noise levels than do GaAs MESFETs, so they are often preferred for use at frequencies within their operating range.

We can minimize phase noise not only by using a low-noise device but also by using a resonator having a high loaded Q; in VCOs, we can use a high-Q varactor. We can also achieve low phase noise by phase-locking the oscillator to a low-noise source, such as a crystal oscillator. Because phase noise can also be caused by noise originating in the power supply or coupled to the dc bias circuits, power-supply filtering should not be overlooked as a means to minimize phase noise.

Pushing and Pulling

In Section 12.1, we saw that changes in the load impedance could effect the oscillation frequency by changing the phase of Γ_{in}. This phenomenon is called *pulling*. To some degree, pulling is inevitable; it occurs because the feedback necessary to make oscillation possible increases $S_{1,2}$ and thus increases the sensitivity of Γ_{in} to the load impedance. Nevertheless, pulling can be minimized. Beyond the obvious solutions of using an output isolator or buffer amplifier, a high-Q resonator is highly effective in reducing pulling.

Similarly, changes in dc bias voltage can change the transistor's S parameters and Γ_{in}, thus changing the oscillation frequency. This phenomenon is called *pushing*; the straightforward way to minimize pushing is to maintain adequate regulation in the oscillator's bias circuits. As with pulling, pushing is minimized by a high-Q resonator. Pushing is not always undesirable; it is sometimes used to obtain voltage-tuning capability in a narrowband VCO.

Thermal Stability

Variations in the oscillator's temperature can simultaneously affect both the resonator's resonant frequency and the transistor's Γ_{in}, thereby changing the frequency of oscillation. Temperature changes can also reduce output power and, in extreme cases, can even cause oscillation to cease entirely. Some degree of temperature compensation can often be included in the bias circuits, and the resonator can also be designed to compensate for changes in the oscillator. Temperature stability is related more strongly to the stability of the resonator's resonant frequency than to its Q.

Post-Tuning Drift

When the frequency of a VCO is changed, the RF current and voltage waveforms throughout the oscillator also change and even the dc bias current can change slightly. As a result, the heat dissipated in the transistor and the tuning varactor, the dc voltages on blocking capacitors, and the dc currents in coupling inductors all change as well; a small time interval is required before the circuit returns to a steady-state condition. During this time, the frequency may drift; this phenomenon is called *post-tuning drift*.

In a well designed oscillator, the primary cause of post-tuning drift is heat dissipation in the varactor. Thus, careful thermal design of the varactor can reduce post-tuning drift significantly. If the varactor is mounted in a package, it should be mounted on a large metal surface; a beam-lead or chip device mounted on a substrate should be bonded to the substrate metalization over as large an area as possible.

Harmonics and Spurious Outputs

A well designed transistor oscillator is usually free of spurious outputs that are not harmonically related to the frequency of oscillation. However, because the transistor is driven into saturation, most oscillators have significant harmonic outputs. These can be minimized somewhat by Class-A bias (Section 9.2) but, in most cases, the designer has little control of the harmonic levels unless an output filter is used. The doubling oscillator in Figure 12.7 suffers also from subharmonics; even if the circuit is well balanced, these are often only a few dB below the desired output.

REFERENCES

[12.1] M. Sholley et al., "HEMT mm-Wave Amplifiers, Mixers, and Oscillators," *Microwave J.*, Vol. 26, August 1985, p. 121.

[12.2] K. Kurokawa, "Some Basic Characteristics of Broadband Negative Resistance Oscillator Circuits," *Bell Sys. Tech. J.*, Vol. 48, 1969, p. 1937.

[12.3] W. Wagner, "Oscillator Design by Device Line Measurement," *Microwave J.*, Vol. 22, February 1979, p. 43.

[12.4] V. Rizzoli, A. Lipparini, and E. Marazzi, "A General-Purpose Program for Nonlinear Microwave Circuit Design," *IEEE Trans. Microwave Theory Tech.*, Vol. MTT-31, 1983, p. 762.

[12.5] D. Kajfez and P. Guillon (eds.), *Dielectric Resonators*, Artech House, Norwood, MA, 1986.

[12.6] A. M. Pavio and M.A. Smith, "A 20–40 GHz Push-Push Dielectric Resonator Oscillator," *IEEE Trans. Microwave Theory Tech.*, Vol. MTT-33, 1985, p. 1346.

APPENDIX 1

RELATIONSHIPS BETWEEN FOURIER SERIES AND PHASORS

In sinusoidal steady-state analysis, we consider circuits that have voltages and currents of the form:

$$v(t) = V_a \cos(\omega t + \phi) \tag{A1.1}$$

or

$$i(t) = I_a \cos(\omega t + \phi) \tag{A1.2}$$

We consider only the voltage $v(t)$ in (A1.1); the following derivation can be applied analogously to $i(t)$. The voltage waveform, $v(t)$, can be expressed as

$$\begin{aligned} v(t) &= \mathrm{Re}\{V_a \exp(\mathrm{j}\omega t + \phi)\} \\ &= \mathrm{Re}\{V \exp(\mathrm{j}\omega t)\} \end{aligned} \tag{A1.3}$$

where $V = V_a \exp(\mathrm{j}\phi)$. We call V a *phasor* representing the sinusoidal voltage, $v(t)$. V can be written in Cartesian form:

$$V = V_r + \mathrm{j}V_i \tag{A1.4}$$

where

$$|V| = V_a = (V_r^2 + V_i^2)^{1/2} \tag{A1.5}$$

and:

$$\angle V = \phi = \tan^{-1}(V_i/V_r) \tag{A1.6}$$

Note that the magnitude of the phasor equals the peak value of the sinusoidal component it represents. Also, the $\tan^{-1}$ function in (A1.6) must be a four-quadrant function; that is, it takes into account the quadrant implied by the signs of V_i and V_r. Then, in general, $\tan^{-1}(-y/x) \neq -\tan^{-1}(y/-x)$. Substituting (A1.4) into (A1.3) gives

$$v(t) = \mathrm{Re}\{(V_r + \mathrm{j}V_i)\exp(\mathrm{j}\omega t)\} \tag{A1.7}$$

and using the expression:

$$\exp(\mathrm{j}\omega t) = \cos(\omega t) + \mathrm{j}\sin(\omega t) \tag{A1.8}$$

(A1.7) becomes

$$v(t) = V_r\cos(\omega t) - V_i\sin(\omega t) \tag{A1.9}$$

In the case of a nonsinusoidal waveform, each harmonic can be represented by a phasor. Thus,

$$v(t) = V_{0,a} + V_{1,a}\cos(\omega t + \phi_1) + V_{2,a}\cos(2\omega t + \phi_2) + V_{3,a}\cos(3\omega t + \phi_3) + \ldots \tag{A1.10}$$

can be represented by a set of phasors:

$$\begin{aligned} V_0 &= V_{0,a} \\ V_1 &= V_{1,a}\exp(\mathrm{j}\phi_1) = V_{1,r} + \mathrm{j}V_{1,i} \\ V_2 &= V_{2,a}\exp(\mathrm{j}\phi_2) = V_{2,r} + \mathrm{j}V_{2,i} \\ V_3 &= V_{3,a}\exp(\mathrm{j}\phi_3) = V_{3,r} + \mathrm{j}V_{3,i} \end{aligned} \tag{A1.11}$$

and, from (A1.7) and (A1.8), (A1.10) can be written

$$v(t) = V_0 + \sum_{n=1}^{\infty} V_{n,r}\cos(n\omega t) - V_{n,i}\sin(n\omega t) \tag{A1.12}$$

One common form of the Fourier series is

$$v(t) = \frac{1}{2} A_0 + \sum_{n=1}^{\infty} A_n \cos(n\omega t) + B_n \sin(n\omega t) \tag{A1.13}$$

and, by comparing (A1.13) to (A1.12), we see that

$$\begin{aligned} V_0 &= 1/2 A_0 \\ V_{n,r} &= A_n \\ V_{n,i} &= -B_n \end{aligned} \tag{A1.14}$$

and, from (A1.5) and (A1.6):

$$\begin{aligned} |V_n| &= (A_n^2 + B_n^2)^{1/2} \\ \angle V_n &= \tan^{-1}(-B_n/A_n) \end{aligned} \tag{A1.15}$$

A Fourier-series form that is used throughout Chapter 3 is

$$v(t) = \sum_{n=-\infty}^{\infty} C_n \exp(jn\omega t) \tag{A1.16}$$

Noting that $C_{-n} = C_n^*$, we can write (A1.16) in the form:

$$\begin{aligned} v(t) = {} & C_0 + C_1 \exp(j\omega t) + C_1^* \exp(-j\omega t) \\ & + C_2 \exp(j2\omega t) + C_2^* \exp(-j2\omega t) + \ldots \\ = {} & C_0 + 2|C_1| \cos(\omega t + \phi_1) + 2|C_2| \cos(2\omega t + \phi_2) + \ldots \end{aligned} \tag{A1.17}$$

and, from (A1.3), $C_n = |C_n| \exp(j\phi_n)$.

Comparing (A1.17) to (A1.11) gives

$$\begin{aligned} V_0 &= C_0 \\ V_n &= 2C_n \end{aligned} \tag{A1.18}$$

and, from (A1.14):

$$C_n = 1/2\,(A_n - jB_n) \tag{A1.19}$$

We also note that

$$C_n = \frac{1}{T} \int_{-T/2}^{T/2} v(t) \exp(-jn\omega t)/dt \tag{A1.20}$$

which is used in forming the Jacobian matrix in Chapter 3 (e.g., (3.1.40), (3.1.41), and (3.1.44)).

INDEX